Lecture Notes in Physics

New Series m: Monographs

The Editorial Policy for Monographs

The series Lecture Notes in Physics reports new developments in physical research and teaching - quickly, informally, and at a high level. The type of material considered for publication in the New Series m includes monographs and multiauthored topical volumes presenting original research or new angles in a classical field. The timeliness of a manuscript is more important than its form, which may be preliminary or tentative. Manuscripts should be reasonably self-contained. They will often present not only results of the author(s) but also related work by other people and will provide sufficient motivation, examples, and applications.
The manuscripts or a detailed description thereof should be submitted either to one of the series editors or to the managing editor. The proposal is then carefully refereed. A final decision concerning publication can often only be made on the basis of the complete manuscript, but otherwise the editors will try to make a preliminary decision as definite as they can on the basis of the available information.
Manuscripts should be no less than 100 and preferably no more than 400 pages in length. Final manuscripts should preferably be in English, or possibly in French or German. They should include a table of contents and an informative introduction accessible also to readers not particularly familiar with the topic treated. Authors are free to use the material in other publications. However, if extensive use is made elsewhere, the publisher should be informed.
Authors receive jointly 50 complimentary copies of their book. They are entitled to purchase further copies of their book at a reduced rate. As a rule no reprints of individual contributions can be supplied. No royalty is paid on Lecture Notes in Physics volumes. Commitment to publish is made by letter of interest rather than by signing a formal contract. Springer-Verlag secures the copyright for each volume.

The Production Process

The books are hardbound, and quality paper appropriate to the needs of the author(s) is used. Yet they are modestly priced. Publication time is about ten weeks. More than twenty years of experience guarantee authors the best possible service. To reach the goal of rapid publication at a low price the technique of photographic reproduction from a camera-ready manuscript was chosen. This process shifts the main responsibility for the technical quality considerably from the publisher to the author. We therefore urge all authors to observe very carefully our guidelines for the preparation of camera-ready manuscripts, which we will supply on request. This applies especially to the quality of figures and halftones submitted for publication. Figures should be submitted as originals or glossy prints, as very often Xerox copies are not suitable for reproduction. In addition, it might be useful to look at some of the volumes already published or, especially if some atypical text is planned, to write to the Physics Editorial Department of Springer-Verlag direct. This avoids mistakes and time-consuming correspondence during the production period.
As a special service, we offer free of charge LATEX and TEX macro packages to format the text according to Springer-Verlag's quality requirements. We strongly recommend authors to make use of this offer, as the result will be a book of considerably improved technical quality. The typescript will be reduced in size (75% of the original). Therefore, e. g. any writing within figures should not be smaller than 2.5 mm.
Manuscripts not meeting the technical standard of the series will have to be returned for improvement.
For further information please contact Springer-Verlag, Physics Editorial Department II, Tiergartenstrasse 17, W-6900 Heidelberg, FRG.

Heinrich Hora

Plasmas at High Temperature and Density

Applications and Implications of Laser–Plasma Interaction

Springer-Verlag Berlin Heidelberg GmbH

Author

Prof. Dr. Dr. Heinrich Hora
CERN, CH-1211 Geneva 23
Switzerland

This book is based on the author's "Physics of Laser Driven Plasmas".

ISBN 978-3-662-13856-4 ISBN 978-3-540-88966-3 (eBook)
DOI 10.1007/978-3-540-88966-3

Originally published by Springer-Verlag Berlin Heidelberg New York in 1991
Softcover reprint of the hardcover 1st edition 1991

2153/3140-543210 - Printed on acid-free paper

Dedicated to our Grandchildren

Simon McCluskey

Barbara Hora

Alexander McCluskey

Benedikt Hora

Preface

"New physics" is an appealing new keyword, not yet devalued by the ravages of inflation. But what has this to do with such an ugly field as plasma physics, steeped in classical physics, mostly outworn, with all its unsolved and ambiguous technological problems and its messy and open ended numerical studies?

"New physics" is concerned with quarks, Higgs particles, grand unified theory, superstrings, gravitational waves, and the profound basics of cosmology and black holes. It is the field of astonishing quantum effects, demonstrated by the von Klitzing effect and high-temperature superconductors. But what can plasma physicists offer, after so many years of expensive and frustrating research to solve the problem of fusion energy?

One may suggest that the fascinating research of chaos with applications to plasma, or the achievements of statistical mechanics applied to plasmas, has something to offer and should be the subject of attention. However, this is not the aim of this book.

Complementing the traditional aim of physics, which is to *interpret the phenomena of nature by generalizing laws such that exact predictions about new properties and effects can be drawn*, this book demonstrates how new physics has been derived over the last 30 years from the state of matter which exists at high temperatures (plasma). The advent of the laser, with its very high energy densities and its concentration to extremely small volumes and to very short time periods, opened up a whole new regime for the interaction of materials and high-density plasmas, which enforced the appearance of the "rather new physics".

Here are a few examples:

- Who would have expected that optical waves in vacuum have a longitudinal component? Thomas Young discovered in 1801 the pure transversality of optical radiation. This was not understandable at that time, when it was known that mechanical waves never exist without longitudinal components. Maxwell's equation then only revealed solutions of the *purely transverse plane waves in electromagnetism*. Against all this traditional knowledge, the recent findings about the dynamics of electrons driven by laser beams by nonlinear forces led to the exact *derivation of longitudinal optical wave components*.
- The same nonlinear force interaction, in view of momentum transfer in experiments, led to the clarification of the *angular momentum of optical beams* and brought about the first substantiation of the *photon spin* by a macroscopic property.
- The conditions of very high laser intensities led to *nonlinear and relativistic generalization of the optical response* (dielectrics and absorption) with a prediction of relativistic self-focusing to understand how GeV ions are produced by laser irradiation of solid targets.
- The *quantum generalization of Coulomb collisions* in plasmas at high temperatures explains the anomalous resistivity, in agreement with observations where links are given between the simply derived Coulomb collision frequency and quantum electrodynamics, including stimulated emission.

- Against the view of the fundamentalists (that plasmas do not have internal electric fields) laser–plasma interaction enforced the derivation of *very strong electric fields and double layers inside plasmas* and unexpected properties as a new resonance to explain strange experimental results, including widespread second-harmonic emission of the plasma corona at the site of laser irradiation of solid targets.
- Many attempts to explain very complicated interaction phenomena were erroneously directed towards stimulated scattering. After it was confirmed experimentally that this scattering does not dominate the energy transfer, it was possible to understand the complicated phenomena (stuttering interaction) and how these can be overcome experimentally by random phase plates (RPP) or induced spatial incoherence (ISI).

All these developments merged into the derivation of a new principle of nonlinear physics: that very strange and unexpected phenomena can be predicted beyond well-known linear physics, and that it is necessary to increase the accuracy of linear physics more and more. Furthermore, exact treatments which avoid neglections are more important in nonlinear physics than in linear physics. This opens a fundamentally new era of predictable physics (as distinguished from chaos) and of formerly unthinkable new phenomena for technological applications.

The immediate applications of laser–plasma interaction physics are well known: the primary goal is to solve the energy problem. The increase in temperature of the atmosphere during the last 50 years by nearly one degree has been confirmed without any doubt from temperature measurements taken at locations remote from human settlements. Furthermore, there is a strong correlation between the recorded rise in temperature and the increase in the content of carbon dioxide in the atmosphere, which is linked to the excessive burning of fossil fuels in our modern, industrialised world.

Even if some energy conservation were achieved by the major industrial countries, the ever increasing need for energy by the 75% of the world's population located in developing countries – especially China and India – will result in further increases in overall energy production.

Solar energy and hydroelectric power may provide part of the answer, but experience has shown that these alternatives, while 'clean', are not likely to produce the quantities of energy required at a low enough cost.

Nuclear fission is now considered a possible alternative source for energy production, but the problems of radioactive waste disposal and the ever present threat of nuclear accidents restrict its advantages. If nuclear sources are the solution, then the answer probably rests in the alternative nuclear fusion.

Energy production from fusion reactions has been the subject of extensive and often frustrating research over many years, aimed primarily at controlling the overall reaction with magnetic confinement or inertial confinement. Fusion energy produced by magnetic confinement may be excluded because it is too expensive, based on the results of Pfirsch and Schmitter, who demonstrated that even with the most ideal assumptions, the cost of producing energy this way will be up to ten times higher than energy produced by the established light water reactors. Our attention then turns to inertial confinement fusion, especially *laser fusion*. Even at this early stage of conceptual development, it appears that by using laser pulses in the MJ energy range, *energy can be produced by inertial confinement fusion at costs similar to those of operating a light water reactor*. With intense technological development, it is conceivable that after the first laser fusion reactors become available subsequent improvements in physics and engineering knowledge will result in *energy production costing three to five times less than that produced by light water reactors.*

The application of laser fusion is in need of considerable applied research and should underline the importance of the field of physics to which this book is directed. However, the importance to physics should not merely be considered in isolation.

The problem of continued human existence is the subject of increasing debate with considerable concern being expressed over the emission of carbon dioxide from the burning of fossil fuels. The potential for *ecological catastrophe* as a result of the *'greenhouse effect'* is becoming increasingly clear and the world is in need of developing a method for the large-scale production of 'clean' low-cost energy. It might be possible in the near future to establish a world order without aggression, without suppression of ethnic groups, minorities or individuals, without police states and political prisoners, with justice and freedom and without economic crises. The solution of these problems is difficult but there is hope for a solution soon.

The subsequent problem of avoiding pollution of our planet, however, will be, by orders of magnitude, more difficult than avoiding war, suppression and other problems of humanity. The task before us is to develop a safe, inexhaustible and 'clean' source of energy for the future of our civilisation and our planet. This book therefore addresses not only an important application of science, but a key problem concerning the future of mankind.

Other less urgent but nevertheless important applications of laser–plasma interaction relate to new schemes for the acceleration of charged particles to TeV energies as alternatives to the classical accelerator schemes or advanced combinations of these schemes. The applications to material processing, welding, cutting, surface hardening and microelectronics are of very wide industrial scale. These applications, but even much more the basic physics being discovered now, and more dramatically in the future, explain the need for a presentation of this new physics. However, since the appearance of the author's book *Physics of Laser Driven Plasmas* (John Wiley, New York, 1981) ten years ago, only one further physics monograph (by W. Kruer) has appeared despite the above-mentioned exciting developments. One reason for this may be that important phenomena were not clarified before. This situation, however, may have changed just within the last few years.

As indicated in my book ten years ago, only the clarified basics of this field were presented. Now that the book is out of print, a reproduction of the established results, together with the addition of the enormous developments that have taken place during the past 10 years, form the content of this book. There is little to change from the earlier treatment: almost all findings remain valid. This is the reason why the unchanged text of *Physics of Laser Driven Plasmas* was used wherever possible with the only changes being corrections of misprints. The additional new aspects are then explained in summaries, expansions and updated comments for each section. This is the fastest and most efficient way to present this exciting field of new physics to newcomers, as well as to stimulate experts.

Thanks are due to my previous publisher (John Wiley and Sons) and to Springer-Verlag for the arrangement of this book in this special form. The preparation of this book was mainly completed at the University of New South Wales, Kensington, Sydney, Australia, and represents many years of work in this field. The continuous support by the University is gratefully acknowledged. I am further indebted to my numerous co-workers, who in recent years have helped to develop so many fundamental new insights and discoveries that only the references to each of these points in the original literature can provide an explanation. Immeasurable thanks for the preparation of this book are due to my secretary, Ms Doris Bock, and the editorial assistance in some parts by my son-in-law, Mr. Brian Minikin, is gratefully acknowledged. Further I am grateful for the support by the Gordon Godfrey Funds for Theoretical Physics at the University of New South Wales, to the Australian Research Council for continuous support as well as to overseas collaboration, especially with Prof. G.H. Miley at the University of Illinois at Urbana, and with Profs. W.C. Stwalley and G. Knorr at the University of Iowa. I thank numerous colleagues, especially Profs. A. Scharmann and W. Scheid at the University of Gießen, Prof. P. Mulser of the Technological University at Darmstadt, Dr. G. Winstel and Dr. E. Krimmel of the Siemens Research Centre Munich-Perlach, all in Germany, Prof. H.P. Weber and Drs. T. Donaldson and J. Balmer at the University of Berne, Switzerland, and the Rockford Technology Corp. in Vancouver, Canada, for their cooperation and support.

Sydney, January 1991 *H. Hora*

Contents

ONE

Aim and Scope

The study of laser produced plasmas is one of the fastest growing fields of present-day physics. It has brought about numerous innovations in materials treatment, such as quality change, welding, drilling, and related high-power beam weapons; the most exciting goal is the safe production of clean nuclear fusion energy with inexhaustive and low-cost fuel. Research projects costing several hundred million dollars have been established, some of them involving hundreds of physicists.

Despite these facts, no books have been published as monographs before 1980, apart from a short introduction [1] or a digest of the voluminous literature [2]. The first serious monograph in the classical sense was published by Hans Motz [3], while an introduction to the more basic problems of nonlinear phenomena found a preliminary formulation as lecture notes [4]. The hectic development can be seen from the fact that conference digests mostly contain short abstracts only. The Boston Conference of the Plasma Physics Division of the American Physical Society in November 1979 presented 317 papers about laser produced plasmas, while the topic conference of the Optical Society of America in San Diego, February 1980, presented--after strong refereeing--papers of 520 authors. Conferences with the intention of reviewing significant highlights using quickly printed proceedings [5] provided a way to present a synopsis of the level achieved within some set period of time.

1.1 Basic Aspects

The difficulty in finding the right motivation for presenting the new field with its obvious attractions is not so much a common denominator of the phenomena involved; it is more the question of the consciousness that really new physics has been opened up. This cannot simply be seen within the overwhelming pluralism of phenomena which individually look very trivial and without significance. There is not so much the task of advertising the fascinating applications but rather the view that one can be attracted by reasons of general physics. Singularities in this direction as for example, relativistic self-focusing, generation of high-Z GeV ions of very high density, or the way to laser induced pair production, and so on, are not the only points of significance. There is much more to report than the whole synopsis of a new physics which is on the rise.

At the beginning of this century there was one single phenomenon that was contrary to all preceding knowledge and that had been confirmed step by step in all the known world: the atomistic structure of action (quantization). This is now classical knowledge that nobody would have a doubt about.

What is dominant in the laser produced plasmas is not just one single phenomenon. It is the breakthrough of the dominance of nonlinear phenomena which, at least to some extent, have been known in part for a long time. What then, is new? It is not only the concentration of old and new nonlinear phenomena. As we shall see in detail in the simple problem of a *correct description of a laser beam and its mechanical interaction with a surrounding plasma* [6], *the nonlinear description is satisfying only if the basic physics (Maxwellian theory) is used exactly without approximations*. This fact explains the dangerous confusion that can occur with any approximate extension to nonlinearity. The more complex a nonlinear extension is, the more precise the basis must be. The desperate question on the philosophy is, then, how can we trust the basic concepts not to be embedded into a more general nonlinear background of interrelations or response?

It is the aim of this monograph to address this very basic question, which might be called *nonclassical* from the present point of knowledge. This point should appeal to the broader community of physicists who may be attracted by studying this complex and very special and apparently applied field, where nothing should be taken away from the importance of its applications for changing our basic technology, energy sources, and worldwide security.

The very complex view should not be astonishing if one remembers the painful steps in exploring the world of the high-temperature plasma. This can be seen from the description of one of the fathers of the plasma theory, Hannes Alfvén, when he wrote shortly after having received his Nobel Prize:

> The study of plasma physics developed along two parallel lines. The first one, originating about a century ago, comprised investigations into electrical dis-

> charges in gases. This approach was, to a great extent, experimental and phenomenological; only very slowly did it reach some degree of theoretical sophistication. Most theoretical physicists looked down on this field, which was complicated and awkward. The plasma exhibited striations and double layers; the electron distribution was non-Maxwellian; there were all sorts of oscillations and instabilities. In short, it was a field not at all suited for mathematically elegant theories.
>
> The other approach to plasma physics came from the highly developed kinetic theory of ordinary gases. It was thought that with a limited amount of work this field could be extended to include ionized gases as well. The theories were mathematically elegant, and the consequences of them showed that it should be possible to produce a very hot plasma and confine it magnetically. This was the starting point of thermonuclear research.
>
> However, the theories had initially very little contact with experimental plasma physics, and all the awkward and complicated phenomena that had been treated in the study of discharges in gases were simply neglected. The result of this was what has been called the 'thermonuclear crisis,' some ten years ago. It taught us that plasma physics is a very difficult field, which can only be developed by a close cooperation between theory and experiments [7].

The nonlinearities in the response of plasmas to laser irradiation will give a further magnitude of difficulties. This monograph, indeed, can only open one gate to describe this new world of physics (including applications in astrophysics or laboratory studies of matter of more than one thousand times solid-state density). This is a beginning, not a solution. It may be that we are as Winston Churchill put it- at the end of the beginning.

1.2 Limitations

The scope of the following consideration is limited against low laser intensities and low laser powers. As soon as the laser radiation *produces an irreversible process* in an irradiated material, it should be of interest in our view. No reversible processes, for example, self-focusing of laser beams in a liquid without plasma production or the frequency doubling due to the nonlinear dielectric constant in a solid, will be discussed.

The phenomena of irreversible processes with the lowest possible laser intensities are, for example, the generation or the annealing of crystal defects. This has been a very wide-open field since 1977 [8], though it could have been opened up much earlier as the necessary laser techniques were available long before 1977.

Low-intensity effects of materials processing create defects of such high density that mechanical destruction of the material occurs. These processes are well known for irradiation with electron beams, where the whole crystal lattice is deformed before breaking into parts [8].

The process of laser induced gas breakdown will be treated only marginally. This field received considerable attention in the earlier years of laser development. Besides its intrinsic importance, it was also the forerunner of the field of laser-solid interactions and stimulated many important diagnostic techniques. The first laser produced gas breakdown was achieved by Terhune et al. [9] at Ford research laboratories, followed by Meyerand et al. [10] at United Aircraft. That this field is far from a reasonable understanding has been shown by Papoular [11], who observed many complex phenomena over several orders of magnitude of laser intensity and gas parameters. This very complex situation is demonstrated by some examples: the observation of luminosity before breakdown; the generation of free electrons without breakdown; the appearance of breakdown field strengths corresponding to a multiphoton breakdown process where the ionization energy was only 5 eV, while the gases under investigation had ionization energies greater than 10 eV. A review of the field of laser induced gas breakdown has been given by Zaidel et al. [12], an early pioneer.

The processes of the laser induced gas breakdown will be touched upon in the following discussion on the self-focusing of laser radiation in plasmas, and also when the influence of low-pressure gas surrounding the target is considered. Some very significant experiments are also mentioned for example, the measurement of the polarization dependent emission of electrons from a laser produced gas breakdown experiment by Yablontovich et al. [13] and the study of the nonlinear radiation forces in the laser breakdown of extremely low-density gases [14].

Another limitation in scope is the upper limit to the available laser intensities. This is quite open due to continuing improvements in technology. The growth has been very rapid since the discovery of the laser in 1960. The available laser power of 10 kW in 1961 has grown to beyond 10 TW in 1978. Laser intensities achieved in 1978 by focusing laser beams in vacuum have reached more than 10^{18} W/cm^2, while the nonlinear interaction of such single terawatt laser beams with plasmas of sufficiently high density can self-focus in plasma to intensities exceeding 10^{21} W/cm^2 [15], as concluded indirectly from the observation of MeV ions accelerated by the fields generated when the laser beam underwent relativistic self-focusing [16].

1.3 Lasers

The more advanced lasers should be mentioned without dwelling too much on the problems of their physics and development. Presently the most commonly used laser for high-power research is the neodymium glass laser. Its wavelength is 1.06 μm and its pulse duration can be anywhere between 170 fsec (=0.17 psec) [17] and cw (continuous-wave) operation. The power in beams of 25 cm diameter, using disc laser amplifiers in the last stages, can be more than 1 TW with pulse durations of more than 10 psec (usually between 0.1 and 3 nsec [18]). One system (SHIVA) uses 20 such beams and was specially developed for laser-fusion experiments. It produces 20 TW laser pulses of 0.1 nsec duration [19]. The extension of single-beam glass amplifiers to 2.5 m diameter is under design [18], and one beam will be capable of producing 100 TW in 0.1 nsec. Another large laser system is DELFIN [20], which has 216 output beams, each of 45 mm diameter. The beams produce 50 J pulses in 0.1 nsec. A laser system using glass slabs as multipass amplifiers with a final beam cross section of 32×100 cm^2, called UMI-35, is in the design stage and should produce 10 TW in pulses between 10 psec and 1 nsec [21]. Independent designs of SHIVA-like systems are underway at LLE (University of Rochester), at the Institute of Laser Engineering (University of Osaka), at Limeil (France), and at the Shanghai Institute (Chinese Academy of Science). Other projects of similar size to the ones mentioned are at the Naval Research Labs, Washington D.C., and at the Rutherford Labs in England.

The advantage of the neodymium glass laser is the highly developed technology, which makes its use preferable to other systems despite the well-known disadvantages: nonlinear refractive index, thermal birefringence of the laser glass and the low efficiency of the laser system, which does not transfer more than 1% of electrical energy into laser energy. Exceptions with respect to the efficiency are the quasi-cw tungsten filament pumped glass laser of 3% [22], or the attempts to pump by GaAs-type semiconductor lasers [23] reaching 6%. It has been shown that an efficiency of 20% can be reached [24], if the pumping is made by laser diodes emitting at a wavelength near 900 nm, where the diodes are assumed to operate with 100% efficiency. Using erbium instead of neodymium in glass, a laser of 1.54 μm wavelength results in an efficiency of 70% if pumped by 1 μm radiation [25].

Lasers, which avoid the solid-state problems of nonlinear refractive index and thermal birefringence, and provide high repetition rates up to 1 kHz and more at high-power outputs, are based on gaseous media. Attempts have been made to keep the ideal properties of the Nd^{3+} laser action by using vapor containing molecules of neodymium compounds [26]. This development has not yet reached a state necessary for high-power laser applications.

The classical high-power gas laser is the carbon dioxide laser. The highest powers achieved with transverse electron beam discharge pumping and final stage beam diameters up to 30 cm are 3 TW in 0.5 nsec. A combination of eight beams in one laser system, at Los Alamos, reached outputs of

20 TW [27] or 40 TW [28]. Pulse lengths as short as a few psec have been reached with carbon dioxide lasers [29]. The development of high-pressure laser systems has been followed through many years [30] and has been developed independently [31] in connection with a very compact amplifier system. The only problem for several applications, including possibly laser fusion, is the long laser wavelength of 10.6 μm. Over this question, however, the argument is not finally settled. For several other applications, as for material processing, the wavelength is not of primary importance *but is of secondary importance*, as can be seen from its use in surgery for example [32], where light of a wavelength 0.5 to 1 μm is diffused by very strong scattering at the cells. This leads to much higher power thresholds for cutting tissues and results in broader cuts than with CO_2 lasers.

The future development of carbon dioxide lasers can be seen in the ANTARES project at Los Alamos [27] and similar ones elsewhere, where laser pulses of several 100 kJ in 1 nsec or less should be achieved. It should be noted, from the point of view of laser fusion, as well as material processing, that one line of development of the carbon dioxide laser could be the nuclear reactor pumped laser, where pulses of several tens of megajoules within several hundreds of nanoseconds can be expected [33].

Another important high-intensity gas laser is the photochemical iodine laser with a wavelength of 1.3 μm. This is particularly useful since its wavelength is close to that of neodymium. The high-power iodine laser was developed by Hohla [34], following the outlines given by Kompa [35]. The present design, using a final amplifier of 20 cm diameter and 10 m length, produces terawatt pulses in 0.5 nsec [36]. The most significant result is an ideal optical beam quality devoid of all the complicated lateral intensity variations due to birefringence and Fresnel diffraction. It has been shown [37], that the focused beam diameter of the completely uncorrected beam was only two times larger than the diffraction limit. One current disadvantage from the point of view of laser fusion is the relatively low efficiency of the laser, which is presently at or below 1%. The use of a specially developed UV source for dissociating the iodine molecules has resulted in higher gains [38]. Another way to achieve higher gains may be the use of exploding wires in the center of the cylindrical amplifiers, where terawatt pulses have been reached [39].

There are several further candidates for high-power lasers to be considered at present. Reed Jenssen [40] succeeded in building a working HF (hydrogen fluoride) laser, where a mixture of both H_2 and F_2 at 1 atmosphere is preexcited by an electron beam to generate more than 10^{25} free atoms/sec/cm^3. The laser consisted of a cylindrical volume of 40 cm diameter and was relatively short (1.5 m long). It produced laser pulses of 3 kJ energy and of 30 nsec duration. Another interesting laser would be one in the category of the excimer laser, which was one of the schemes originally proposed for lasers by Houtermans [41], but which was realized only after high-intensity electron beams became available [42]. With these lasers, not only wavelengths in the UV and far UV, but also laser powers above 10 GW in picosecond pulses are available (refer to Bradley [17]). By generation of high-order harmonics [42] laser wavelengths below 500 Å have been obtained at useful intensities. Lasers worked with fife times amplification at 182 Å [43] and are studied at 117 Å and less [44]. The possibility of producing short wavelengths by gamma ray lasers [45] does not seem to be outside the realms of possibility; the laser medium has to be preexcited by very intense laser beams in the optical and infrared wavelength range [46]. As a first important step, Okamoto's model [46, 47] was the basis of the first laser excitation of nuclei: Yamanaka et al. [48] succeeded in exciting ^{235}U nuclei into their isomeric state by laser excitation of uranium electrons causing a resonance with nuclear levels. A further consequence of gamma-ray lasers should be preferable for extremely high-intensity emission. Another new development of high-power lasers is the free electron laser. The first working system uses electron beams of about 40 MeV energy [49] moving through a rippled magnetic field. This mechanism of laser emission had been realized for microwaves in 1952 by Hans Motz [50]. An extensive digest of the different systems of this kind has shown [50] that the emission of coherent radiation is essentially based on second-order cyclotron radiation effects [51]. The proposed idea of using cyclotrons of 45 m radius may result in a relatively extensive apparatus for laser fusion; however, the high beam quality and the possibility of producing any magnitude of beam power may be so attractive, that the development of the cyclotron-type free electron laser could be of great importance for the future [52].

There is another free electron laser system under consideration that is basically different from the cyclotron type. It does not need any additional magnetic field to produce cyclotron-type effects and requires only a high-intensity laser beam (which has to be produced in the conventional way) that interacts with appropriate electron beams of specific energies and spatial configurations [53]. The essential mechanism is the application of the nonlinear radiation force [54]. Examples have been given [53], where a carbon dioxide laser pulse of 1 TW can be amplified to a power of 10^{15} W.

The advantage is that no materials are involved that can be damaged or ionized, because the interaction process of the electrons occurs in vacuum. The wavelength is continuously variable and, in theory, applicable to x-ray lasers. A disadvantage is that the amplification follows the square of the wavelength. However, the energetic conversion efficiency for transfer of electron beam energy into optical energy is theoretically up to 100%. This is an important point in the design of power stations or similar equipment, as the exchange of cooling energy can then be reduced.

A further basically different type of a free electron laser has been proposed by Schwarz [55], where the superposition of electron beams to produce an interference field with a quantum-modulated electron current is used. Modulation is possible by laser [56] or by the Aharonov–Bohm effect [55]. The long beating electron beam emits coherent radiation. This modulation-type laser has a higher efficiency at higher frequency and is therefore of interest for very short wavelengths (Schwarz-Hora effect type laser).

1.4 Review of Phenomena and Results

This subsection will review several significant phenomena that have been observed in experiments of laser interaction with solid targets, gases, and with plasmas usually produced by the laser itself. With the discovery of the first laser by Maiman [57] in 1960, an obvious step was to use its high-intensity radiation to study the interaction of light with solid targets. At that time the use of electron beams for grinding, drilling, welding, and other kind of material treatment was well developed, where power densities of 10^8 W/cm^2 and more had been reached [58]. It was remarkable that the first spiking ruby lasers with maximum power of 10 to 100 kW could be focused down to less than 0.05 mm diameter resulting from the beginning in comparable power densities to the electron beam. In some of the first experiments, solid targets were irradiated in vacuum and the time dependence of the emitted ions reaching a Faraday probe arranged in front of the target was measured. Time-of-flight measurements on the ions showed velocities corresponding to a few electron volts. This was in full agreement with the expected temperatures of a few ten thousand degrees centigrade for the plasma generated at the target surface [59].

These measurements were confirmed later [60], though the number of such measurements reported in the broad stream of laser plasma interaction literature has been minimal. A change occurred in the years 1977 to 1978, when solid-state physicists and semiconductor technologists began to use lasers for melting, recrystallizing, generation of crystal defects, or annealing of these defects [61]. A clear distinction of these interactions from those occurring at higher intensities, leading to evaporation and plasma generation, has to be made. It should be mentioned that the use of lasers for evaporation techniques for thin films was studied earlier [62] with the surprising result that such a complicated molecule as strontium titanate was redeposited in the initial molecular state, even after laser vaporization (or plasma generation).

The use of higher laser power than that of the spiking laser became a reality when Hellwarth [63] discovered the Q-switched laser, where the ruby laser emission was very reproducibly concentrated in pulses of 10 to 40 nsec duration. This meant that peak powers of 10 to 100 MW were attainable. Linlor was able to use such a laser to irradiate targets such as carbon, tungsten, and others in vacuum [64] and arrived at the very surprising result that the aforementioned ion energies of few eV for 100 kW irradiation were increased to more than kilovolts. It was very significant that the measurement of the ion energy as a function of the laser power P or laser peak height increased in a superlinear way. Isenor [65] measured a nearly linear increase of the ion velocity v_i on the laser power P (Fig. 1.1) corresponding to a nearly quadratic increase of the ion energy ε_i on P.

$$\varepsilon_i = \text{const}\ P^m \qquad (m = 1.8 \text{ to } 2) \tag{1.1}$$

This type of increase of the ion energy was measured by Schwarz [66] and by Namba and Schwarz [67]. It was evident that at higher powers the superlinear increase of ion energy has to undergo saturation, resulting finally in a sublinear increase [68], Fig. 1.2.

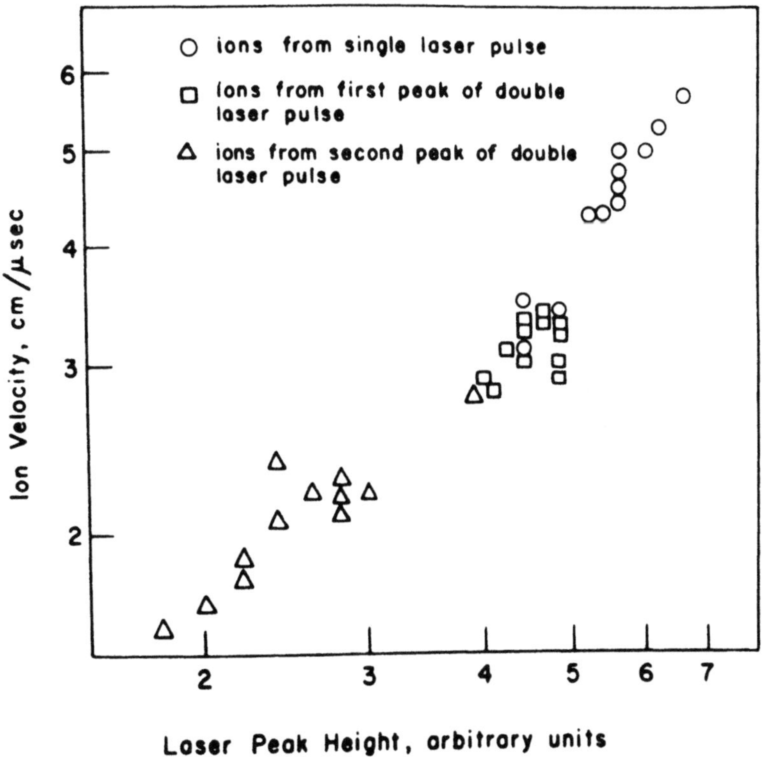

Figure 1.1 Nearly linear increase of the ion velocity with the laser power P at laser irradiation with about 10 MW (after Isenor [65]).

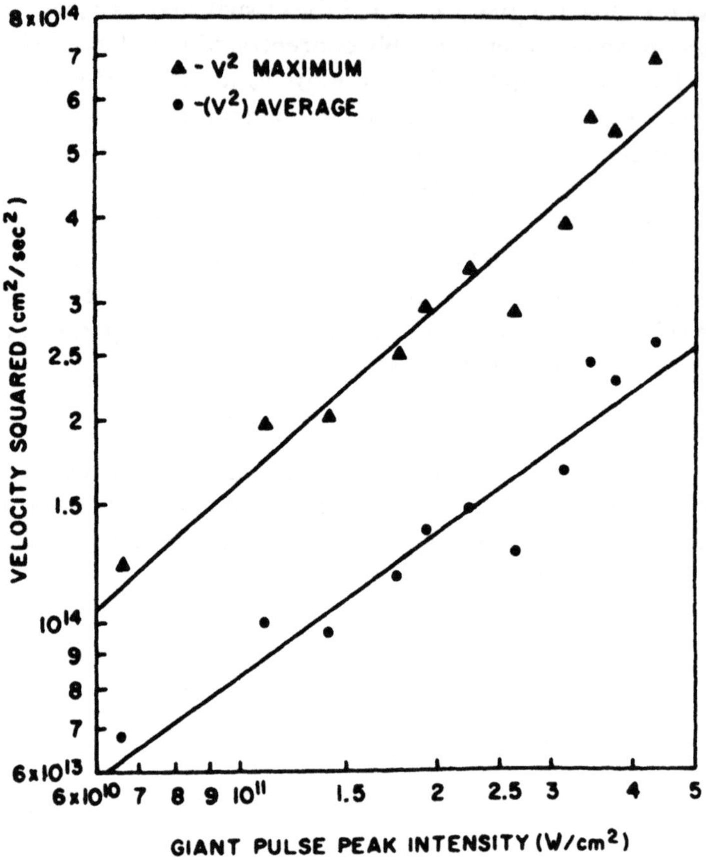

Figure 1.2 Sublinear increase of the ion energy (square of the ion velocity at 100 MW and higher laser power (after Gregg and Thomas [68]).

Another example of highly superlinear behavior with laser power using the ruby or neodymium glass wavelength around 10 MW is the measurement of the recoil exerted by the incident laser radiation on the target [69] as shown in Fig. 1.3. At higher laser powers, this increase is again saturated and merges into a sublinear slope [68, 70].

A similar mechanism between the laser powers of 1 to 10 MW for ruby or neodymium glass lasers happens for the emission of electrons. From the beginning it was an aim of workers in this field to use the interaction of laser radiation with targets for the generation of very high electron emission currents with the hope of developing "super cathodes." It was very controversial when, at the beginning of these investigations, Ready observed a fully classical behavior of electron emission with maximum currents of several hundred milliamps, completely in agreement with the space charge limitation laws of Langmuir [59]. Ready used laser powers again around,

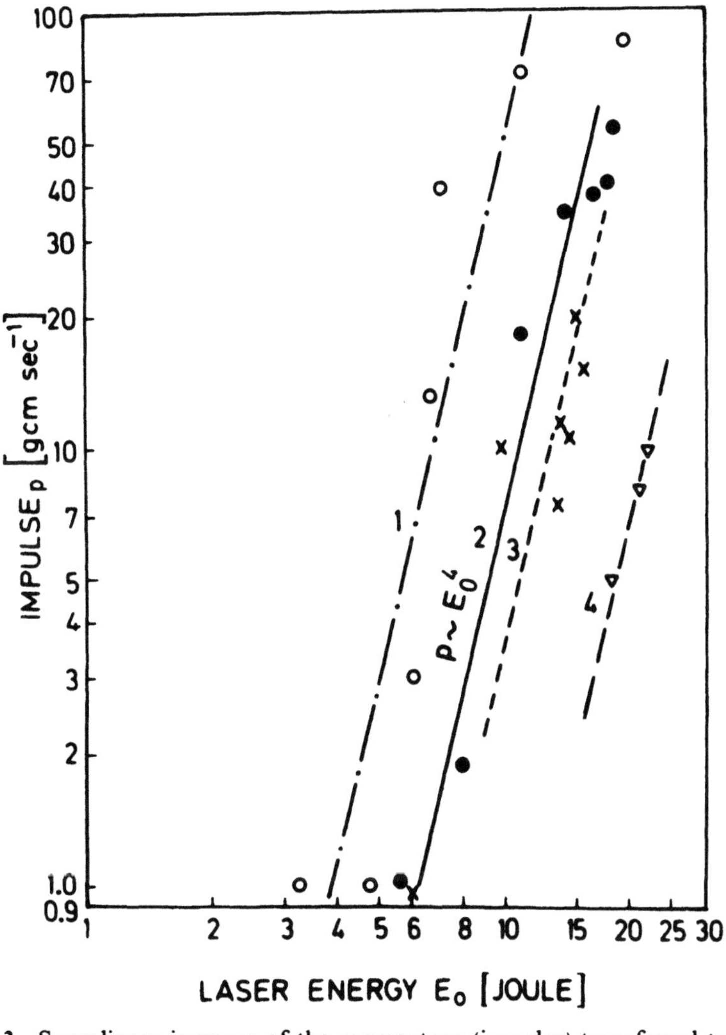

Figure 1.3 Superlinear increase of the momentum (impulse) transferred to the laser irradiated target at neodymium glass laser powers near 10 MW measured by Metz [69].

or less than, 1 MW. In contrast to this, Honig [59] measured emission currents of 100 A, which was in contradiction to any knowledge of space charge restrictions (Child–Langmuir law). This unusual result was fully reproduced [71] when using the more advanced laser techniques; emission currents of more than 1 kA were measured. It is now evident that a special mechanism takes over above powers of around 1 to 10 MW.

Another example of the complexity of the laser interaction with solids can be seen by the following examples when free spherical aluminum targets were irradiated by a laser pulse (Fig. 1.4) [72]. The framing camera

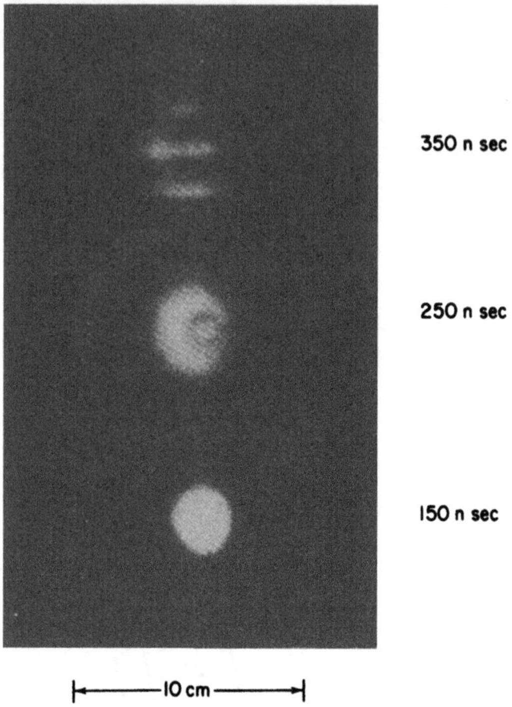

Figure 1.4 Side-on framing camera picture of a plasma produced from an aluminum sphere of 80 μm radius at the time marked. after irradiation with a 30 nsec ruby laser pulse focused to 0.4 mm diameter. The second frame shows the outer part of fast expanding plasma and an inner spherical thermally expanding part measured by Engelhardt et al. [79].

picture at 250 nsec after irradiation shows two groups of plasma: a spherical center containing 95% of the transferred laser energy expanding with a speed corresponding roughly to 10 eV temperature and a fast expanding asymmetric outer plasma with a maximum energy of 3 to 5 keV for ions moving against the laser beam direction. The fully linear or thermal behavior of the inner part was evident, while the outer part demonstrated a highly nonlinear property [72].

It appears that a special process occurs at the already mentioned ruby laser power of 1 MW, as can be seen schematically from a graph comprising the various results of measured ion energy as a function of the laser power (Fig. 1.5). This is called the Linlor effects. though Linlor has always insisted that his observations were of a fully linear nature. In the present view, including the mentioned cases of momentum transfer and of electron emission, as well as other results (e.g., [72]), the initiation of some nonlinear

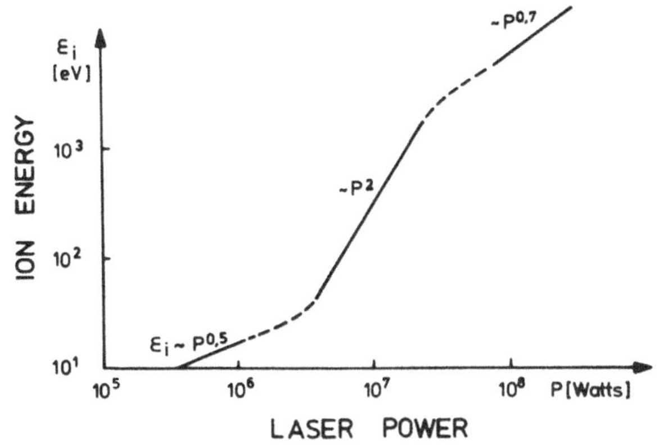

Figure 1.5 Schematic composition of measurement of the ion energy against the laser power for ruby and neodymium glass lasers with a thermal part below 1 MW, a highly superlinear part (Linlor effect) up to 10 MW, and a sublinear part above.

mechanism above a threshold somewhere in the 10 MW range of ruby laser intensities is evident.

The highly complex properties of a laser focus are another complication [73]. Figure 1.6 shows the measurement of the intensity in a laser focus where, within one small area only, the approximately Gaussian profile can be seen,

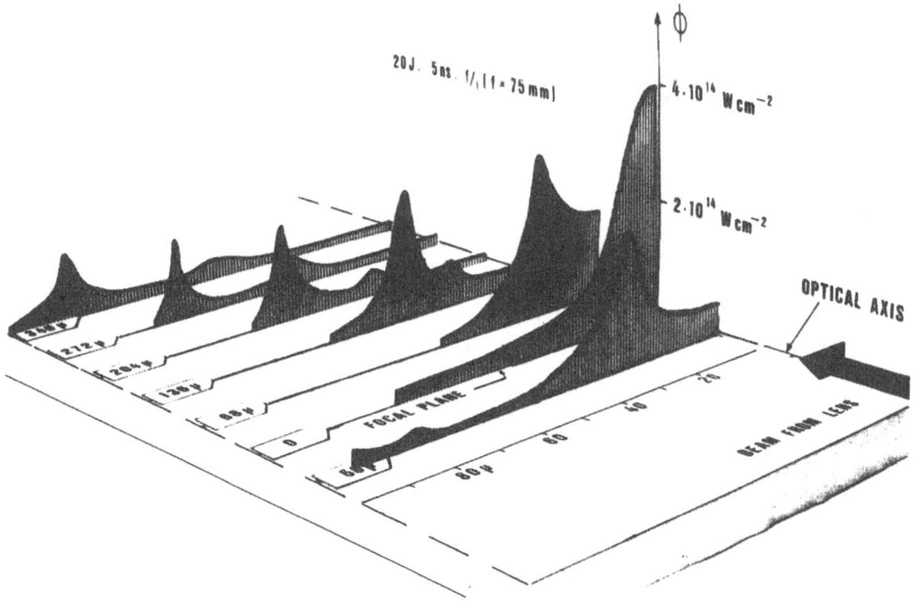

Figure 1.6 Measurement of (time integrated) spatial distribution of the laser intensity in a focus by Eidmann et al. [73].

while a complicated intensity field is observed in the off-focal regions. This consideration is even more important for laser breakdown in gases, where the whole focal region must be taken into account.

The mentioned experimental results are mostly several years old, and one may argue that they may lack accuracy when compared to present-day experiments. Even so it seems that some of the basic properties of the older measurements are often not fully taken into account when interpreting new measurements. The pluralism and curiosity of more recent experiments has still not been lessened. Comparing different measurements of the reflectivity of irradiated targets at various laser intensities, a very confusing scattering of results can be seen, Fig. 1.7 [74]. This is an example of what happens when the results from different authors working with different targets and different laser parameters are compared. It should be mentioned that, although this type of strong scatter of results may have been reduced during the last few years, a large amount of variation still exists.

To illustrate the complexity of the experimental results in the field of laser plasma interaction from more recent papers, it should be mentioned that, for neodymium glass laser intensities of 10^{13} W/cm^2 and above, several special properties have been detected. There was the observation of back-scattering of higher harmonics of the laser frequency, indicating the action of parametric instabilities, in analogy to microwave experiments. There was the observation of the half frequency radiation in the backscattered radia-

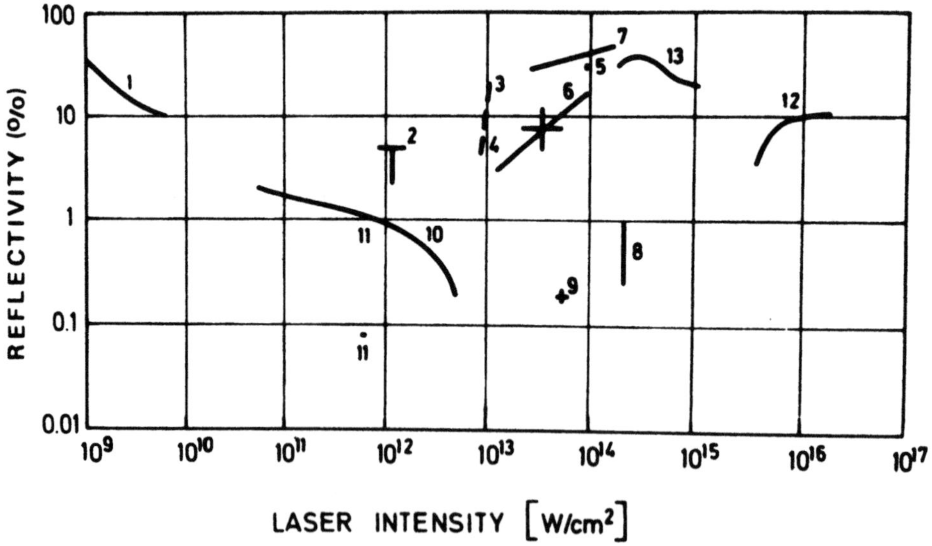

Figure 1.7 Experimental reflectivity (5) from laser produced plasma as a function of laser intensity. Data are from various authors with reference to [1, p. 4].

tion, and there was the result that no uniform temperature was present in in the focus of laser-plasma interaction. After measuring very contradictorily varying temperatures from x-ray signals, Eidmann [75] was able to analyze the emitted x-ray spectra in such a way that, apart from an expected plasma temperature of few hundred electron volts, another "temperature" of keV was shown to exist (Fig. 1.8). This was due to the intensively radiating focal region and was not an effect of different spatial properties. This elevated "temperature" turned out to be due to anomalous nonlinear processes and could reach values of 200 keV [76] and more, up to 8 MeV "temperature," appropriate for pair production [77].

Another interesting result is the fact that the initially observed different groups of expanding plasma, apart from fast ions with nonlinear behavior [72], were still seen in further observations. It was found later that the fast ions are of a few 10 keV energy [78], and that much more than 50% of the irradiated laser energy can go into the fast ions [79]. One example of the fast ions is shown in Fig. 1.9, where the probe signal indicates a clear separation of the ions by their charge number in a linear way with ion energies of several 100 keV. The detection of MeV ions was a further step [80] where, however, the inclusion of a relativistic self-focusing mechanism was necessary for an interpretation [81]. The mentioned linear dependence of the ion energy

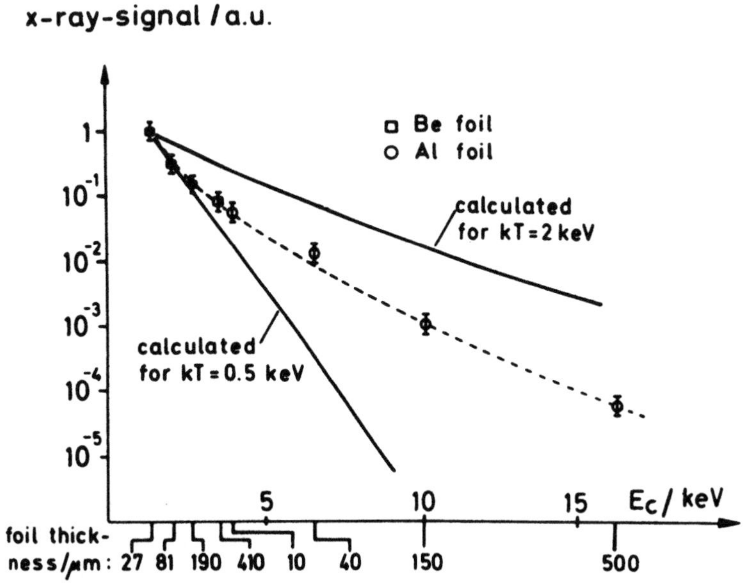

Figure 1.8 Emitted x-rays from the plasma as a function of cutoff energy of four Be and Al foils of different thicknesses. The different slopes correspond to "two different temperatures" discovered by Eidmann [75].

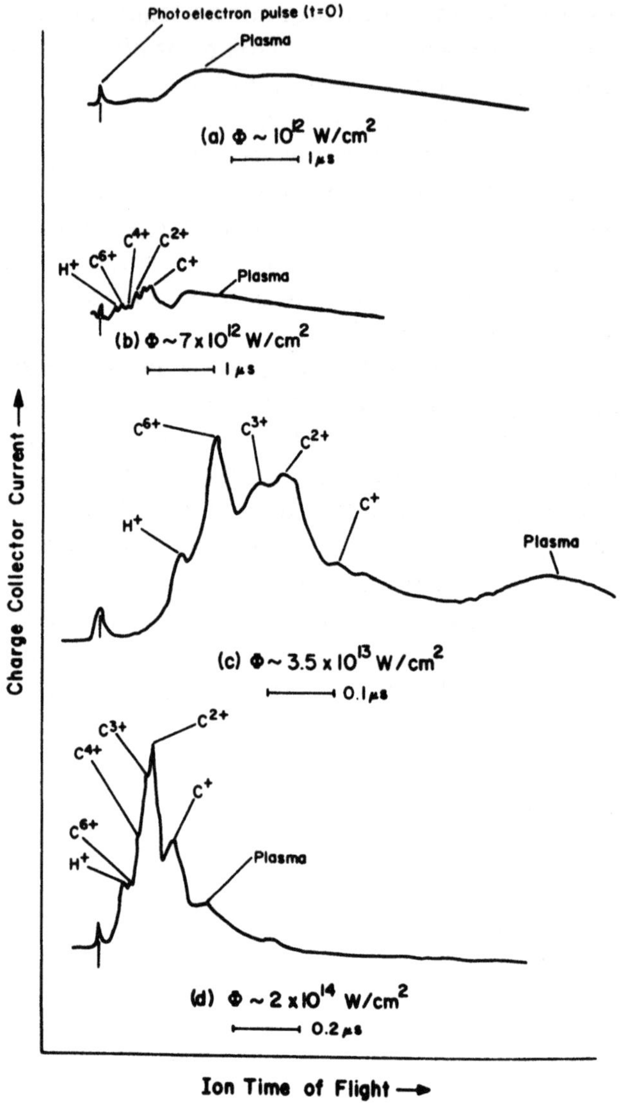

Figure 1.9 Oscillograms of probe signals for ion collection. when CO_2 laser radiation of the given intensity is incident Al and C targets. The peaks correspond to various ions of different ionization with energies of several 100 keV. measured by Ehler [76].

on the ion charge was a relation seen in several experiments, especially in the case of 100 keV ions and above. The analysis of the fast ions in the keV range, however, led to another modification if the light was incident obliquely on the plane targets [82]. It was discovered that one group of the ions behaved fully independent of the laser polarization (fast ions), while the other group showed a strong dependence on the polarization (ultrafast ions).

The most up to date experimental techniques for analyzing the properties of laser produced plasmas are able to show the most astonishing properties. It is possible to measure the spatial variation of densities of plasmas with a resolution of several micrometers and in temporal resolution down to picoseconds. One example can be seen in Fig. 1.10 [83]. Diagnostics of this kind were the basis for detecting the compression of plasma in the center of spherically irradiated gas filled glass balloons for producing genuine thermonuclear reaction in the center of the pellets [84] (see Fig. 1.11).

Another unique result is the fact that the irradiation of very thin gold foils, which are transparent to optical radiation, causes an anomalously high absorption when irradiated by very intense short laser pulses [86]. The foils are then not transparent. For oblique incidence a special behavior has been seen, the Yamanaka effect [86], and these anomalies have been confirmed by the fact [87], that the irradiation of an aluminium layer of

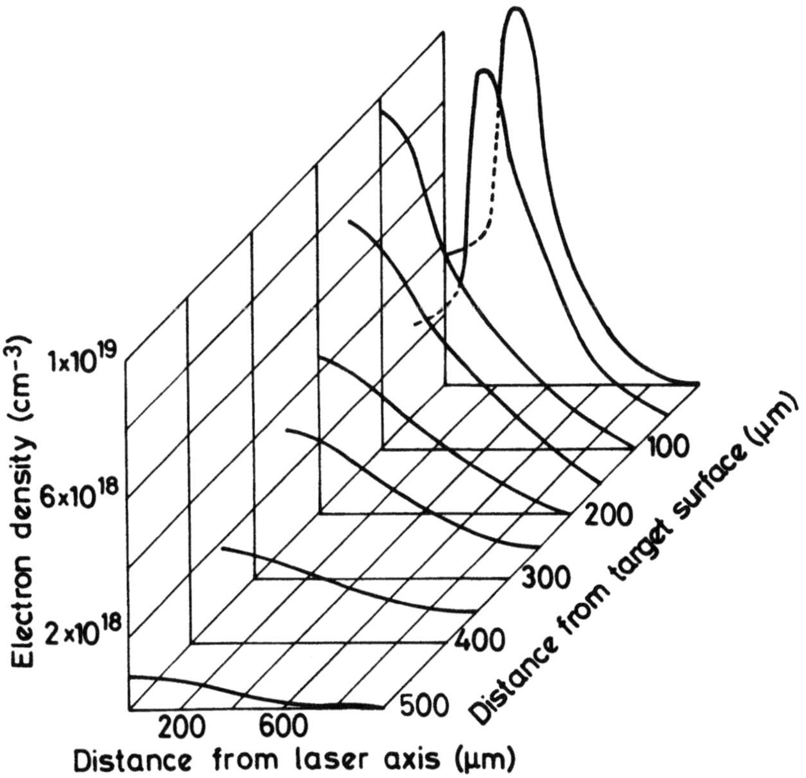

Figure 1.10 Measurement of the spatially resolved density profile of the plasma produced by a CO_2 laser pulse from a spherical target 25 psec after the start of the interaction. The generation of the density minimum was characteristic for the process. After Donaldson and Spalding [83].

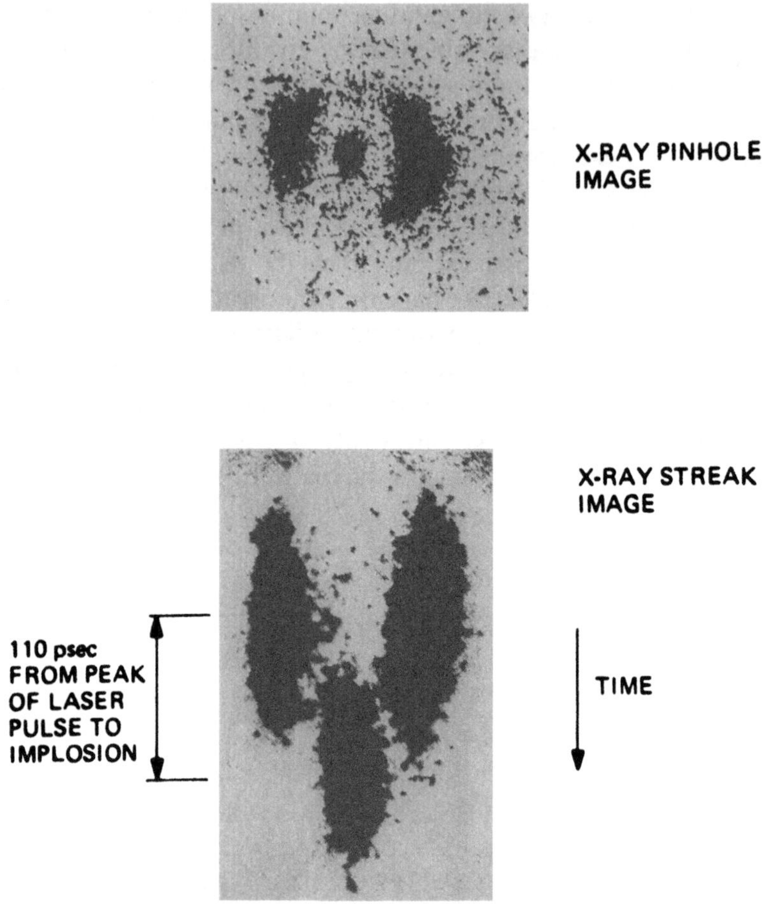

Figure 1.11 Example of high spatial and temporal resolution in the diagnostics of the diametral irradiation of two wide aperture Nd glass laser beams of 7.6 J and 100 psec duration each on a glass microballoon of 0.88 μm wall thickness and 88 μm diameter filled with 10 atm. D_2+T_2 gas. The upper picture is a time integrated x-ray pinhole photograph indicating radiation from the laser interaction at the glass surface and such from the center due to plasma compression resulting in 2×10^4 fusion neutrons. The lower picture is the x-ray streak image of a diameter showing the motion of the pellet periphery toward the center on time and the about 50 psec delayed radiation from the core after the compression has been achieved, after Evans, Key, et al. [85].

1000 Å thickness on a quartz substrate did not show silicon lines in the spectrum.

These very unexpected and very complex properties of the laser produced plasmas underscore the very complicated nonlinear processes involved. We shall nevertheless first describe the theory of the linear gasdynamics, and then plasma laser interaction processes, on the basis of which the theory for several nonlinear processes will be derived. There is definitely no complete theory possible for all the known or as yet undiscovered, unexpected, and anomalous processes. But we hope that a certain guideline can be given for a better understanding of the present developments. The complexity of the physics has to be taken into account to see how probable it will be to proceed with the already quite successful attempts to compress and heat plasma for thermonuclear reactions, as well as to understand the borderlines of low-intensity interaction for material treatment. If material treatment is taken into account the irradiation with laser intensities from CO_2 or neodymium glass lasers above 10^{12} W/cm^2 is definitely anomalous, as seen from the fact that most of the irradiated energy is transferred into a certain amount of very high energetic ions. These ions are interesting for use in accelerators or nuclear fusion. For the application of evaporation of large quantities of materials, however, these mechanisms may be disadvantageous.

1.5 Very High Power Lasers

The development of very high power laser technology since the level described in subsection 1.3 during the ten years before 1990 are quite significant. The following results are reported beginning with the longest wave lengths.

a) Free Electron Lasers (FEL)

When firing relativistic electrons through a set of static magnetic fields with alternating poles (wiggler field) and if an optical resonator cavity is used nearly coaxially to the electron beam, emission of laser radiation has been measured (Deacon et al 1977). This scheme - after overcoming enormous difficulties - arrived at schemes now where up to 20% efficiency of laser emission should be reached. In the range of millimeter waves, intensities of TW have been shown (Prossnitz 1986; Mima et al 1988). In the visible range and in the UV range, FELs with respectable conversion efficiencies have been achieved using the standing wave field of a carbon dioxide laser as wiggler field, and the use of periodic small distance magagauss fields from laser produced plasmas should arrive at intense FELs in the x-ray range (Eliezer et al 1987).

b) Carbon Dioxide Lasers

The carbon dioxide laser with the highest power was mentioned in Section 1.3, the ANTARES laser at Los Alamos. It should be added that this laser could be operated with pulses in the nsec range and in another range of 100 psec. In the latter case pulses of 100 TW power were produced. This laser was dismantled because space was needed for an experiment on magnetic confinement fusion.

Another remarkable carbon dioxide laser with an ac averaged output of 1 MW was built (Pismenny, 1990) consisting of 100 pulses of 10 microsecond duration with an energy of 10 kJ each. Another remarkable development is the achievement of carbon dioxide laser pulses of 300 fsec duration (Corkum et al 1988). The intensity of the laser pulses in the inverted laser gas can be so high in this case that optical gas discharge can occur without that the laser amplification is influenced.

c) HF lasers

While the development of cw lasers by mixing hydrogen and fluorine jets arrives at very high power cw lasers converting chemical energy into optical energy, the schemes of laser pulses with HF lasers has been proposed for laser fusion producing 100 MJ energy of several nsec duration (Phipps 1989). This laser is an extension of the highly efficient and very compact HF-laser with 3kJ pulses of 30 nsec duration from an active volume of 1.5 m length only as verified by Reed Jensen [40].

d) Iodine Lasers

The before mentioned scheme of the chemical photo dissociation laser developed by Hohla [34] has been developed to the scheme of the ISKRA-5 laser (Kyrillov et al 1990) using 12 amplifiers of 8 m length and 1 m diameter to produce 35 KJ energy in 250 psec. These 125 TW pulses have more than two times higher energy than the hitherto biggest laser in the world, the NOVA laser (see following paragraph).

e) Neodymium Glass Lasers

Following the development of the SHIVA laser at Livermore, mentioned in Section 1.3, the NOVA laser was established with 10 beams of 46 cm diameter widened to 76 cm for frequency tripling by crystals producing pulses of 125 TW of 100 psec duration, or pulses of 50 TW of 1 nsec duration (Campbell 1991.)

Lasers with glass amplifiers of square cross section of 1.5 meters are being built with output of 400 kJ in 10 nsec (Storm 1990) with the aim to produce frequency tripled laser pulses of 145 kJ and 8 nsec duration (Lowdermilk, 1991). A combination of these beams should result in pulses of 10 MJ and 10 nsec duration for the laser fusion scheme ATHENA (Manes et al 1986).

The largest laser in Europe is the PHEBUS laser at Limeil near Paris which uses three instead of the ten glass laser beams of the NOVA type. One of these beams is being converted for short pulse operation of 1 psec duration with pulses of 1 kJ producing PW (Petawatt = 1000 TW) power end of 1991 (Coutant 1990).

f) KrF Excimer Lasers

For laser fusion, the KrF excimer laser with a wavelength of 250 nm is ideal since efficiencies of 5% or even 8% have been achieved and the optical properties of these gas lasers are optimized reaching only 3 times diffraction limitation. A well developed system is AURORA at Los Alamos producing laser pulses of 1.3 kJ in 5 nsec in one beam (two beams are to be built). The energy is going to be increased to 5 kJ (Cartwright 1990). Pulses of electron beam pumped KrF gas lasers of 500 nsec duration are shortened by an optical cutting and multiplexing system. A similar but smaller system is that of SPRITE at Appleton Rutherford Laboratory where however the pulse shortening is to be done including stimulated Raman scattering (SRS) (Shaw, 1991).

The KrF lasers developed by the Troitzk Branch of the Kurchatov Institute and the Efremov Institute in the USSR produce 40 kJ pulses of 500 nsec duration in a cross section of 1 meter to the square with quartz glass plates at the exit (Baranov 1990) which are to be shortened to 5 psec duration by multiplexing (5 times) and shortening by stimulated Brillouin scattering (SBS) to produce at least 20 kJ energy. A laser for fusion application for 3 MJ pulses needs then 16 sets of beams only of 3 times 3 meter cross section consisting of nine modules each of 1 square meter cross section.

For very short pulses, F.P. Schäfer (1990) developed a scheme where a dye laser produces pulses of 250 nanometer wave length and 50 fsec duration which pulse is then amplified in a pumped KrF medium to produce pulses of 100 fsec and 50 mJ energy. A scheme is being elaborated to produce pulses of 100 J of 100 fsec duration.

g) X-Ray Lasers

Irradiating a high-Z medium cylindrically with very powerful glass (or other) lasers of 100 to 500 psec duration results in an inversion of electronic levels which produce x-ray laser emission. Wave length down to 40 Angstroms have been reached by this way. If the pumping is by the intense x-rays of nuclear explosions, wavelengths of 14 Angstrom have been reported 1982. The use of fast ionization of the K-shell electrons by intense incoherent incident x-radiation led Schäfer to the prediction of x-ray lasers with attosecond (1/1000th of a fsec) duration (Schäfer, 1986).

Pulses with fsec duration should be expected if a similar K-shell ionization is done by MeV/nucleon beams (Hora, 1988).

This overview is indeed very superficial and without the fascinating details of the dramatic developments of the continuing laser technology. It should provide only the facts of laser pulse powers and capacities which are of interest for the following discussions of the physics of the interaction of the laser beams with materials. What has been achieved at present with neodymium glass lasers and with krypton fluoride lasers are intensities in the range of 10^{19} W/cm^2 after focusing in vacuum. With the next PW-psec laser beam laser intensities in vacuum up to 10^{22} W/cm^2 can be expected and if relativistic self focussing is acting in an irradiated high density target (as explained later) in this book intensities up to 10^{24} W/cm^2 should be produced. This is 25 orders of magnitude brighter than the sunlight power density at the earth though for the whole solar spectrum of wide range frequencies.

1.6 Further Phenomena and Results

Following Section 1.4, an endorsement in support of the results reported there can be given by referring to several new results. It was underlined that there was a clearly distinguished appearance of thermal or thermokinetic processes versus typical nonthermal, mostly nonlinear and highly energetic processes. While initially this could be seen simply from the emission of ions with energies in the thermal range of several eV against the keV ions (Linlor effect), the analysis of measurements of the Fig, of 1.4 in 1968 was a clear distinguishing between the thermal and the high energetic processes.

A similar result was that of Figure 1.8 where Eidmann's measurements of the spectrum of the emitted x-rays from a laser produced plasma could be clearly separated into a thermal one corresponding to the order of hundred eV temperature and a part of energetic electrons corresponding to 20 keV energy and more. At that time, other authors also had observed the 20 keV and higher energies in laser produced plasmas, but it was the merit of K. Eidmann that these energetic electrons were to be distinguished from a thermal part with energies corresponding to the otherwise expected thermal properties of the plasma.

The energetic electrons (or sometimes very unprofessionally called "hot" electrons without a further definition what "temperature" should be considered) were indeed measured in the last years very carefully and various explanations tried. We shall return to this point in the following discussions.

It should be mentioned that measurements of the kind of Fig. 1.9 by Ehler were extended to higher laser powers and ion energies up to half GeV energy were detected by the same author.

Fig. 1.11 demonstrated one of the first sophisticated diagnostics of the spatially and time resolved x-ray emission from a laser irradiated fusion pellet in order to demonstrate the dynamics of the ablation and compression of the spherically irradiated target as seen from the x-rays through a pin hole. There was the hot plasma corona, but simultaneously the subsequent dynamical compression of the non-irradiated center of the pellet and its heating was confirmed as seen from the x-ray emission. from the center. These types of diagnostics have been elaborated in an extremely sophisticated way. An early summary was given by Ahlstrom (1982) and review papers about these advanced diagnostics were collected by Campbell (1991a).

Within the unusual and new types of x-ray diagnostics two methods should be highlighted. One is the use of spherically bent crystals such that the x-rays of a short range of wave length emitted under a very large aperture can be optically projected in the same way as by spherical mirror optics resulting in bright and up to 30 time amplified pictures of the x-ray source. This method was developed by Förster (1991) and applied extensively at the Delfin experiments (Aleksandrova et al 1985, Sklizkov 1988). Another novity was introduced by Luther-Davies 1987; Perry et al 1989) where instead of the pinhole x-ray camera, see Fig. 1.11, a large spherical aperture is used and the half shadow of the x-rays is Fourier analyzed to get a much more intensive and better resolved picture of the x-rays emitted from the pellet. This method is called penumbral technique.

A very recent development is the time resolved neutron detection from a laser produced plasma where it was possible to find out why the very high fusion neutron yields can be achieved by a stagnation free (shock free) ideal adiabatic volume compression (Yamanaka et al 1986; Nakai 1989). This type of volume compression at all could be elucidated from many hundreds of pellet fusion experiments only by this neutron diagnostics and the then resulting maximum fusion gains which were higher by many orders of magnitudes than the usual shock driven experiments. The problems of the time resolved neutron detection were explained by Lerche et al (1988). A very important extension of the x-ray penumbral method was possible for neutrons such that for the first time a very clear and detailed picture of the neutrons emitted from a laser compressed fusion pellet can be achieved (Lerche et al 1991). This technique was honoured with IR-100 prize of the 100 best inventions in the USA in 1987.

While the very wide stream of laser plasma interaction research resulted in several thousand publications in the 80th and it may be considered as impossible to give account of the experimental results, there may well be the possibility of highlighting at least a number of very unusual and pioneering results. The attention was due to the fact that there were highly confusing observations about the interaction process of the laser radiation with the plasma. Additional to the very complicated examples reported in Section 1.4, one observation by M. Lubin et al (1974) was that the reflection of neodymium glass laser light from a plane target was pulsating strongly when a very smooth laser pulse of a maximum intensity of about 10^{16} W/cm^2 in the order of 100 psec duration was incident, Fig. 1.12. There was an irregular pulsating of the reflection between few and nearly 100% with a kind of a period of 15 to 20 psec.

It is no surprize that this kind if interaction of laser light with

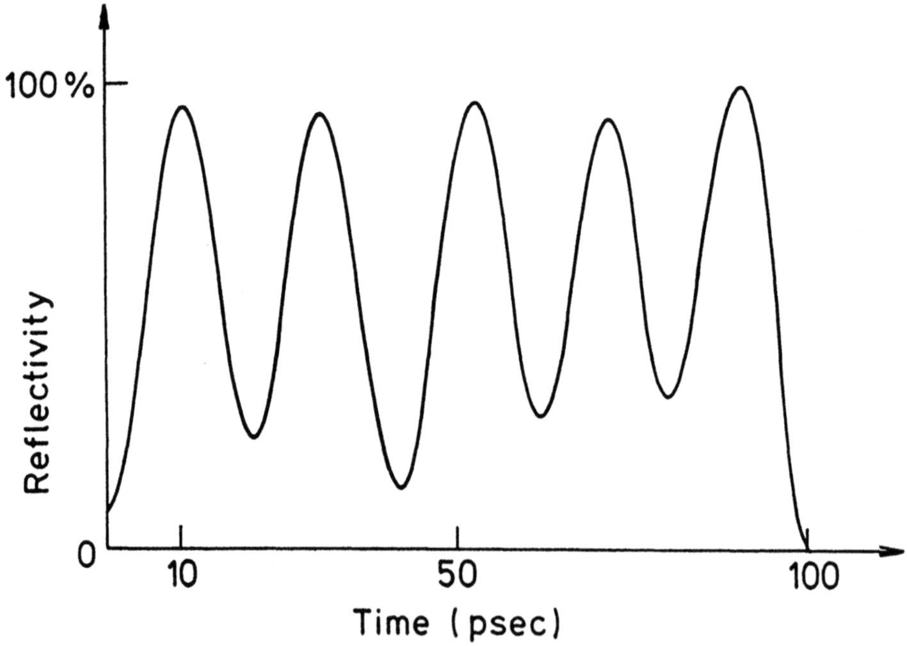

Fig. 1.12: Time dependence of the reflectivity of laser light from a target irradiated by a smooth neodymium glass laser pulse in the range close to 100 psec as observed by Lubin et al (1974).

plasma is mostly uncomfortable especially in view how to drive the plasma for laser fusion. It was assumed that this behaviour was due to the parametric instabilities especially due to stimulated Raman scattering (SRS) or stimulated Brillouin scattering (SBS). After a large number of laboratories made thousands of measurements in this direction, it was clarified that while SBS and SRS are well appearing and are most interesting for plasma diagnostic techniques, their contribution to the absorption and energy transfer process to the plasma is very low and at least not dominating. There was indeed an indication that the

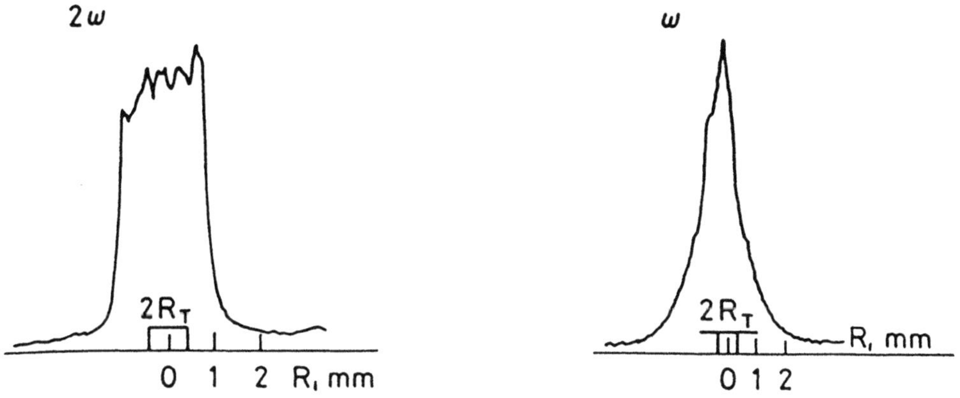

Fig. 1.13: Side-on emission intensity distribution from a pellet of a diameter 2R irradiated by a rectangular 2 nsec neodymium glass laser pulse in Delfin (upper part). The second harmonics emission is in the lower part and is nearly rectangular apart from oscillations. The narrow profile in the upper part corresponds to the strongly decreasing density in the outer plasma which, however, emits the second harmonics with unchanged strength even at very low density. (Aleksandrova et al 1985).

energy transfer from the laser to the plasma should have been of a usual absorption process and not so much by anomalous processes (Perry et al 1989). It was then confirmed from carefully elaborated experimental evidence by Drake (1988) that SRS is not dominating the absorption if not extremely pathological conditions are artificially introduced. The fact that SBS is not dominating was experimentally verified by Christine Laboune (1985). What then is the reason for the complicated interaction as shown in Fig. 1.12?
The door to open for the answer was given by the experiments of Maddever and Luther-Davies (1990) returning to the result of Lubin, Fig. 1.12, however with much more sophisticated results.

It is just the time now to understand the complex interaction mechanisms and how to achieve reproducable results and how to achieve smooth interaction. The following chapters will report about this dramatic development.

Other very strange observations have been tackled by theoretical explanation or prediction as will be shown in the following. An example of such strange result is given in Fig. 1.13. The side-on emission of the second harmonics from the corona of a laser produced plasma is much more intense for lower plasma densities than expected. This however and the observed periodicity can immediately be explained from a double layer interaction process as will be shown in the following.

TWO

Elements of the Microscopic Plasma Theory

A plasma can be defined in various ways. It has been called "the fourth state of matter" as distinct from the solid, liquid, and gaseous states. More than 99% of the cosmos consists of plasma: in the stars and to a large extent in the interstellar matter. Since there are no such sharp distinguishing marks as a melting point or boiling point, but only the fact that all matter is ionized at high temperatures (above 10,000°K all matter is ionized to some degree), the following definition can be given. *Plasma is a physical state of high electrical conductivity and mostly gaseous mechanical properties.* The term "mostly" allows for the fact that, for example, a metal or a semiconductor can have plasma properties, although its mechanical properties of compressibility and rheology are that of a solid. In the case of the high-temperature matter in the dense interior of a star, the plasma properties are evident, but the compressibility may be the same as for solids or liquids due to quantum effects (Fermi–Dirac degeneracy). The plasma may be defined alternatively as a *medium whose dielectric properties are determined only by free charges* (and not by dipoles).

In this case we neglect all bound states of electrons in atoms or molecules. We are discussing, then, the "fully ionized plasma." These are of interest at high-intensity laser interaction with plasmas. The range between the first irreversible damage in materials by lasers and the full ionization is then neglected. As self-focusing and other nonlinear effects are acting very quickly after the first damage threshold, the generated plasmas are then nearly fully ionized. This was justified also posteriorily, when Mulser included the ionization equilibrium (Saha equation) in very extensive hydrodynamic calculations [88] even at moderate neodymium glass or ruby laser intensities of 10^9 to 10^{10} W/cm^2, the dynamics of the generated plasmas was the nearly same whether the Saha equilibrium was included or not.

A fully ionized plasma is a gas consisting only of electrons and positive ions of a certain charge Z. The most basic description of such a plasma uses the single-particle equation, which are then the $3N$ equations of motion of each of the N plasma particles

$$m_n \frac{d^2 x_n}{dt^2} = f_{xn}\left(x_i, \ldots, z_N, \frac{\partial x_1}{\partial t}, \ldots \frac{\partial z_n}{\partial t}\right); \qquad n=1,\ldots,N \qquad (2.1a)$$

$$m_n \frac{d^2 y_n}{dt^2} = f_{yn}\left(x_i, \ldots, z_N, \frac{\partial x_1}{\partial t}, \ldots \frac{\partial z_n}{\partial t}\right); \qquad n=1,\ldots,N \qquad (2.1b)$$

$$m_n \frac{d^2 z_n}{dt^2} = f_{zn}\left(x_i, \ldots, z_N, \frac{\partial x_1}{\partial t}, \ldots \frac{\partial z_n}{\partial t}\right); \qquad n=1,\ldots,N \qquad (2.1c)$$

The particle coordinates x_n, y_n, and z_n and the derivatives with respect to time t are given by the masses m_n and the forces f_{xn}, f_{yn}, and f_{zn} depending on all coordinates and masses of all the N particles. In general, this task is not directly soluble though numerical solutions have been made for 50,000 or more single plasma particles by so called simulation codes [89]. However, even in such cases, the full description of the Coulomb collision forces has to be reduced to certain approximations, so that the results are not fully general and the interpretation has to be made with restrictions. The forces f will then be determined by the coordinates x_j of particles within a certain distance only for producing Coulomb forces. Particles at further distances will be ignored. Further, the first derivation of the coordinates on time had to be ignored in the forces f.

For the simulation of a plasma with 50,000 single particles, the impressive number of 150,000 differential equations had to be solved by the computer. Without going into detail, it should be mentioned that the treatment of laser plasma interaction resulted in the same properties as the following described hydrodynamic calculations resulted. Starting from a linear ramp of a plasma density, the acting laser radiation caused net forces to the plasma [90] completely equivalent to the earlier derived macroscopic nonlinear forces that will be described in the following in detail.

For the following treatment of the macroscopic hydrodynamic theory, several properties of the microscopic description will be used. This is the reason for the following subsections, which are basically related to the history of the development of plasma physics.

2.1 Plasma Frequency and Debye Length

The plasma state was discovered by Langmuir in 1920 when he attempted to explain the fact that radio waves of about 10^7 Hz frequency were totally reflected by the upper atmosphere or the ionosphere, and, in this way, were guided around the globe by the plasma shell of the ionosphere. Without having direct measurements (made later by balloons and satellites) Langmuir concluded that the upper region of atmospheric gas was ionized. He derived a characteristic frequency ω_p, the plasma frequency, for the electrostatic oscillations of the electrons in a plasma corresponding to the reflection of the waves.

Fig. 2.1 describes, in the upper part, an electron density in equilibrium given by a cell distance dx. The lower part shows a disturbance by distances $d\xi$, which causes a change of the electron density n_e of

$$\frac{dn_e}{n_e} = -\frac{d\xi}{dx} \tag{2.2}$$

Following Poisson's electrostatic potential equation ϕ, or the electric field **E**, generated by a charge density ρ in Gaussian cgs units, the equation for the displaced electrons arrives at

$$\frac{d\mathbf{E}}{dx} = -4\pi e dn_e = +4\pi e n_e \frac{d\xi}{dx} \tag{2.3}$$

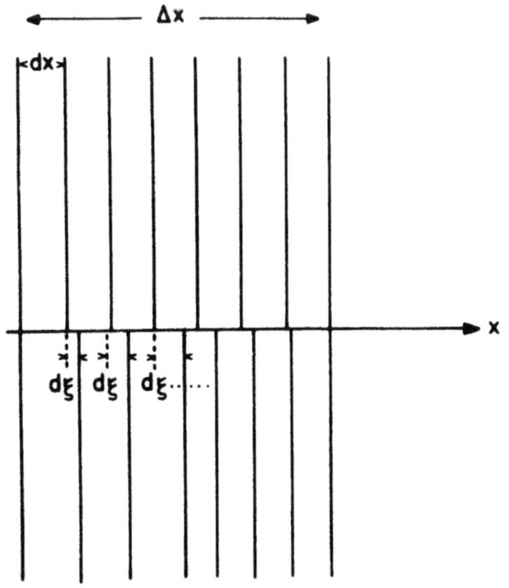

Figure 2.1 Displacements of electrons of homogeneous density (upper part) by $d\xi$ to generate electrostatic oscillations with the plasma frequency ω_p.

The electron charge is $e=4.803\times10^{-10}$ $cm^{3/2}$ $sec^{-1}g^{1/2}$ [91]. The equation of motion for the electrons is

$$m\frac{d^2\xi}{dt^2}=-e\mathbf{E}=-4\pi n_e e^2\xi(t) \tag{2.4}$$

using the electron mass $m=0.9109\times10^{-27}$ g. The solution for the differential equation (2.4), the undamped oscillation equation, is

$$\xi(t)=\text{const}\exp(i\omega_p t) \tag{2.5}$$

where n_e is given in cm^{-3}:

$$\omega_p^2=\frac{4\pi e^2 n_e}{m};\quad \omega_p=5.65\times10^4\sqrt{n_e} \tag{2.6}$$

In the case where 30 m radio waves (10^7 Hz) were totally reflected by the ionosphere, Langmuir was able to calculate an electron density $n_e=1.23\times10^6$ cm^{-3}. The process of wave reflection will be seen later, when the dielectric constant based on ω_p will be discussed. A density $n_e=10^{21}$ cm^{-3} is obtained from (2.6) for the 1.78×10^{15} Hz radiation frequency of the neodymium glass laser. The 1.78×10^{14} Hz frequency of the CO_2 laser corresponds to a "cutoff" density of $n_{ec}=10^{19}$ cm^{-3}, while an excimer laser of 1200 Å corresponds to a cutoff density $n_{ec}=7.8\times10^{22}$ cm^{-3}, which is close to the atomic density in solids.

There exists a characteristic length λ_D corresponding to the radian plasma frequency ω_p for a wave velocity equivalent to the electron thermal velocity v_e. This is distinct by the factor 2π from the usual procedure of multiplying an oscillation frequency by a wavelength to obtain the velocity of a wave or transport process. Instead of the thermal average velocity of the electrons, a value given by the average energy $KT/2$ (T=temperature; K=Boltzmann constant$=1.38\times10^{-16}$ cm^2 sec^{-2}) per degree of freedom must be used. With these minor modifications one arrives at

$$\lambda_D=\frac{v_e}{\omega_p}=\left[\frac{KT}{4\pi n_e e^2}\right]^{1/2} \tag{2.7}$$

$$\lambda_D(\text{cm})=6.9\left[\frac{T(K)}{n_e(\text{cm}^{-3})}\right]^{1/2}=743\left[\frac{T(\text{eV})}{n_e(\text{cm}^{-3})}\right]^{1/2} \tag{2.8}$$

This length is identical with that derived by Debye [92] based on Millner's work [93] for the theory of electrolytes, and is called Debye length. For the plasma state it is the limit λ_D for space charge neutrality. Only over dimensions greater than the Debye length can space charge neutrality be assumed in the macroscopic theory.

A very instructive insight into the meaning of the Debye length can be

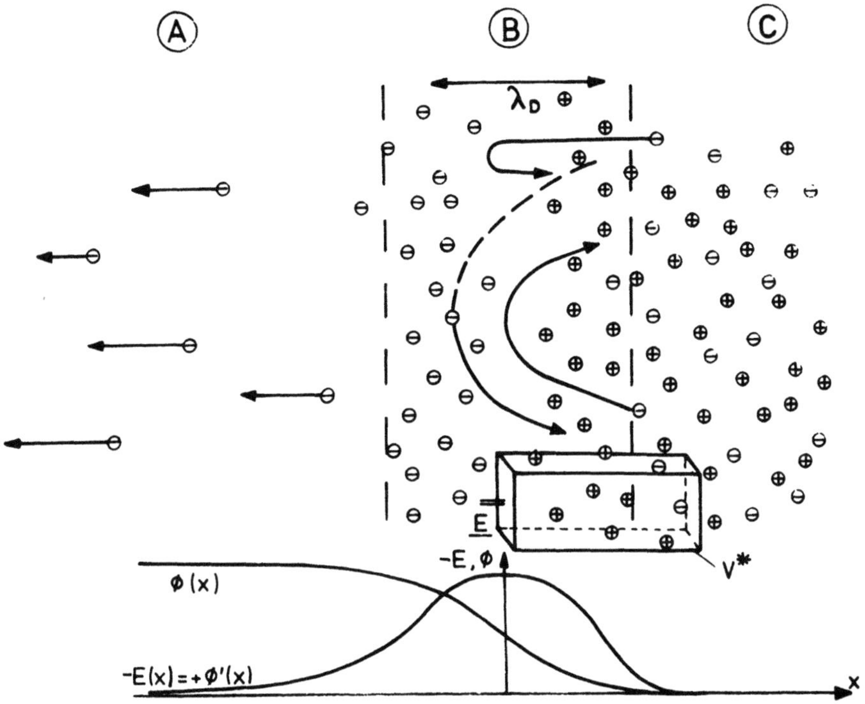

Figure 2.2 Between the vacuum range *A* and the space charge neutral plasma interior *C*, the plasma surface sheath is depleted by the escape of fast electrons until such a strong space charge is built up that the following fast electrons from the plasma *C* are electrostatically returned into *C*. The electric field **E**(x), due to the space charge separation in *B*, and its potential are given.

gained, if it is derived in another way from the surface properties of a plasma (Fig. 2.2). Between the interior of electrically neutral plasma and vacuum, a Debye sheath *B* is created due to the fact that the thermal electrons have a much higher velocity than the ions because of their much smaller masses. The number of electrons leaving the sheath is eventually limited by the equal and opposite ion charge built up, which electrostatically returns fast electrons to the plasma interior. The thermal energy KT corresponds then to an electrical potential given by an electrical field **E** times a length λ_D.

$$KT = e\phi = e\lambda_D\mathbf{E} \tag{2.9}$$

E can be derived from the charges in the interface, where the integration of the volume integral (always for 1 cm^2 cross section) results in

$$4\pi\lambda_D n_e e = \oint n^2\mathbf{E}d^2a = n^2\mathbf{E}(\times 1\ \mathrm{cm}^2) \tag{2.10}$$

Here use was made of the refractive index n, which for the case considered, can be put equal to unity. The field in C is zero and grows monotonically from the interfaces BC to AB. In the surface integral (2.10) only the value at AB is effective. The reflection of electrons from C occurs due to the negative charge in A. Expressing the right-hand side of Eq. (2.9) with the **E** from Eq. (2.10), is found

$$\lambda_D = [KT/4\pi e^2 n_e]^{1/2} \tag{2.11}$$

This is a length for the thickness of the sheath, which is equal to the Debye length of Eq. (2.7).

Considering a plasma of a certain temperature expanding in vacuum, it must be taken into account that its surface will build up an electrostatic sheath with a net space charge where the space charge neutrality equations of plasma dynamics will not be valid. The space charge will then cause an electrostatic explosion of the ions, which is called ambipolar expansion, where the ions gain energies up to a factor times thermal energy equal to the *square root of the ratio of the ion mass to the electron mass*. This generation of fast ions in laser irradiated plasmas has been detected [94] where the number of the ions accelerated by the mechanism is smaller than the number of ions in the Debye sheath of the plasma surface. If the number of fast ions (fast in the sense of much faster than thermal) is much larger than the number of ions in the Debye sheath, another acceleration process has to be assumed, and this will be discussed in the following sections.

2.2 Plasmons

The quanta of the plasma oscillations are called plasmons. Their energy, E, is given by

$$E = \hbar\omega_p = 3.73 \times 10^{-11}(n_e)^{1/2}[\text{eV}] \tag{2.12}$$

using Planck's constant $\hbar = h/2\pi$, where $h = 6.67 \times 10^{-27}$ erg·sec, and $[n_e] = \text{cm}^{-3}$. The action of these plasmons is seen in solids. Electron beams of about 50 keV energy, transmitted through a thin film of metal whose density of conduction electrons near 10^{23} cm^{-3} is equivalent to plasmon energies near 10 eV, have energy losses of the order of the plasmon energies. Historically, the energy losses were measured first by Ruthemann [95] and repeated by Möllenstedt [96] and others, while the plasmon explanation was given later by Bohm and Pines [97]. It is remarkable that this plasmon interaction also works for the electrons of insulators. One can say that the transmitted fast 50 keV electrons "see" the insulator electrons as if they were free oscillating plasma electrons.

It is worth indicating at this point that the plasmon action is present in a plasma, where the plasma densities and the well-known parameters are assumed to be fully classical quantities. The fact is that, in such a classical plasma as a stationary arc discharge, the classical Maxwellian energy distributions of the electrons can be modified quantum mechanically. Fig. 2.3 shows the spectrum of a ruby laser beam scattered in a stationary hydrogen arc plasma of 7 eV temperature and 10^{16} cm^{-3} electron density [98, 99].

Instead of the expected smooth profile of the 90° scattered light, there was a superposition of maxima and minima which correspond to distances of half the plasma frequency. Taking into account that Thomson scattering is based on that *half part* of the oscillation energy due to the quivering

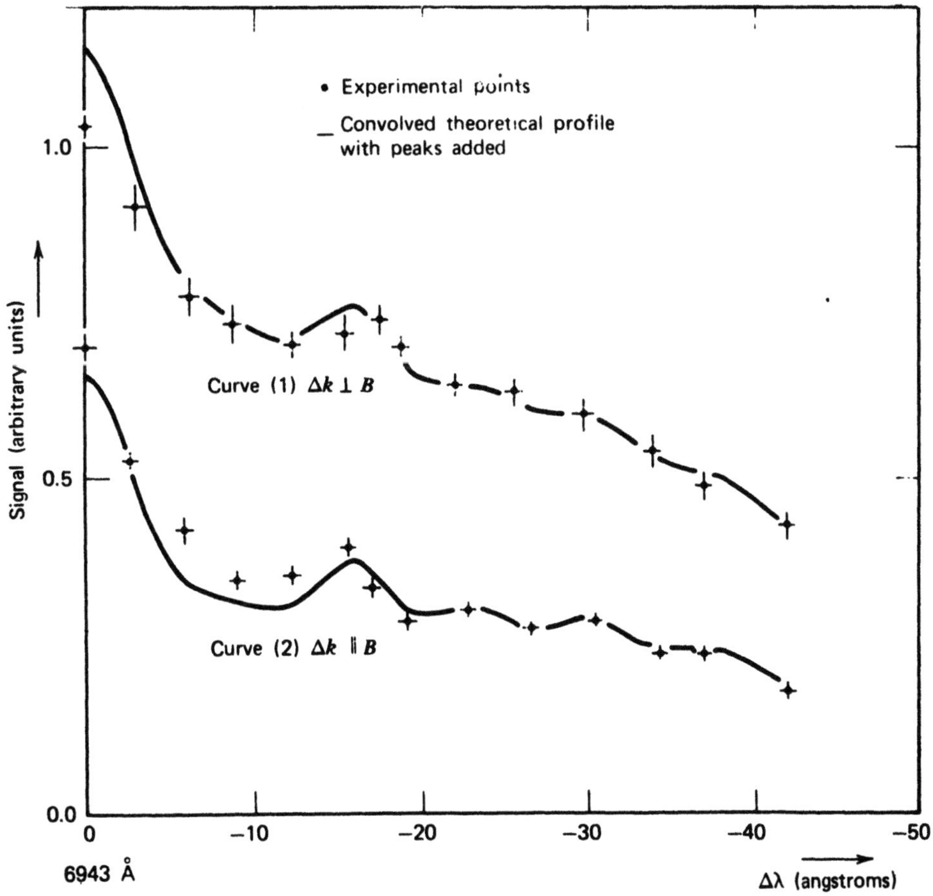

Figure 2.3 Scattering spectrum of laser light from an arc [98] where a modification by maxima and minima is due to the plasma frequency. After Ludwig and Mahn [98].

electrons in the laser field, the scattering corresponds to a leak of intensity, where the electron energy in the plasma is equal to the plasmon energy. This is very easily understood. Since the experiments on the photoemission from metals [100] confirming the volume photoeffect [101] in contrast to the earlier assumed surface photoeffect [102], it has been evident that the usual mean free path of electrons in a metal is reduced from more than 150 lattice distances to only a few if the electron has the energy of the plasmons. The same happens in the case of the arc plasma. Normally the electrons keep their Maxwellian energy distribution, unless an electron is scattered into an energy of a plasmon. Then it will lose its energy very quickly so that the Maxwellian energy distribution (Fig. 2.4) will have minima [1, p. 28].

The half-width of the minima in the energy distribution of Fig. 2.4 will be discussed. It must be related to a characteristic time Δt by a quantization relation

$$\Delta t \cdot \Delta \varepsilon = \hbar. \tag{2.13}$$

The question is what time Δt should be taken. Taking the time the electron needs, at its thermal velocity, to cross the distance of a Debye length will result in the meaningless equality of $\Delta\varepsilon$ and $h\omega_p$, the plasmon energy. A more reasonable result is obtained by starting from the remarkable fact, derived experimentally by Thomas from photoemission [100] that the mean free path of an electron of energy equal to the plasmon energy is a few ion distances. Assuming 10 distances and taking into account the spread of

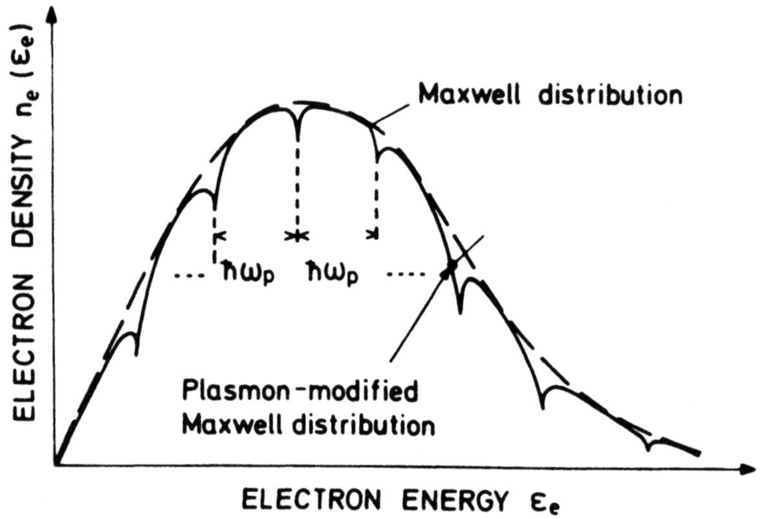

Figure 2.4 Maxwellian energy distribution for electrons and the quantum modification due to the plasmons [74].

energy in the quantum minima of the Maxwell distribution, the thermal time of flight of electrons for this condition leads to a width of the quantum minima in the Maxwellian distribution of

$$\Delta\varepsilon = h\sqrt{2KT/m\,n_e^{1/3}} \tag{2.14}$$

which, in the case of the arc result of Fig. 2.3, is 2.2 meV.

The plasmon energy is in the same case 3.74 meV, which is quite reasonable in comparison. The width of the minima in Fig. 2.3 corresponds to this relation. A further increase of accuracy leads to the assumption of a mean free path for thermal electrons with the plasmon energy of 3 to 10 ion distances, which is in quite good agreement with the experiments of Thomas [100].

This example is used to demonstrate that the quiet and stationary plasma state of an arc of low density is more complicated than was assumed before by using a classical Maxwellian energy distribution for the electrons. The experiments mentioned [98] viewed with the interpretation given here would be a clear indication of a very complicated quantum modification of the plasma state.

2.3 Polarization Shift of H-like Lines in Plasmas

We saw in subsection 2.1 how the Debye length determines the distance within which no space charge equilibrium can be assumed in a plasma, and in subsection 2.2 how the quantization of the plasma frequency can cause a modification of the Maxwellian electron distribution. This subsection considers the modification of the energy of levels of electrons bound in atoms or ions of charge Z, if these are located in plasmas. The electrostatic energy

$$\varepsilon_D = Ze^2/\lambda_D \tag{2.15}$$

within the sphere of the diameter of a Debye length will modify the energy states of the bound electrons, resulting in a shift of the spectral lines which was called plasma polarization shift by Yaakobi and Goldsmith [103].

The energy ε_D, Eq. (2.15) also appeared as the "decrease of the ionization energy" measured when a bound electron is being ionized in a plasma of electron density n_e and temperature T [104]. This decrease is not of the value e^2/λ_d, but its part to each electron in the Debye sphere. Otherwise energy could be produced by adiabatic expanding and compressing of plasma at ionization and recombination of electrons in bound states. The use of the line shift and broadening is an important tool for a direct measurement of the density of laser compressed plasmas [105].

A similar electrostatic method is used to describe the quantum state of an electron in a Coulomb potential as it was successful to describe the decrease of the ionization energy of atoms in a plasma. The aim is to compare the energy ε_B with the energy of the electron when being bound to an ion or atom. If an electron within a Coulomb field $\mathbf{E}$ of Z protons $|\mathbf{E}|=Ze/r^2$, is confined within a radius r_0, the electrostatic energy released by the electron when coming from $r=\infty$ is from the Maxwellian stress tensor

$$\varepsilon_s=\frac{Z^{-1}}{8\pi}\int_{r_0}^{\infty}\mathbf{E}^2\,d^3\tau, \qquad d^3\tau=4\pi r^2\,dr$$

$$=\frac{Ze^2}{2r_0} \tag{2.16}$$

The quantization requires that confinement of an electron to a radius r corresponds to an increase of its momentum $\mathbf{p}$ (or energy $E=\mathbf{p}^2/2m$ causing a quantum pressure) given by

$$r=\frac{n\hbar}{\sqrt{2mE}}$$

where n are integers $n=1, 2, 3, \ldots, \infty$. Apart from a factor of order one, E can be considered as a Fermi–Dirac energy:

$$E_n=\frac{n^2\hbar^2}{2mr^2} \tag{2.17}$$

The fact is that the increase of ε_s on decreasing r_0 is slower than the increase of E on r. A stationary solution is obtained when both energies are equal, $E(r_0=r_s)=\varepsilon_s(r_0=r_s)$ if

$$r_s=\frac{n^2\hbar^2}{me^2Z}=\frac{n^2r_{\text{Bohr}}}{Z} \tag{2.17a}$$

where the Bohr radius $r_s=r_{\text{Bohr}}$ results for $n=1$; $Z=1$. It is well known from the solutions of the Schrödinger equation that the radius of an electron in the state $n=2$ is $4r_s$ as given in (2.17a). Using r_s, the energy $E=E_n$ in Eq. (2.17)

$$E_n=\frac{Z^2me^4}{2\hbar n^2}=\frac{Z^2 13.6\text{ eV}}{n^2} \tag{2.18}$$

arrives in the ionization energy of hydrogen for $n=1$, $Z=1$ of 13.59 eV. In the case of an electron in the potential of a He^{2+} ion, Eq. (2.18) arrives in the ionization energy of He^+ at $n=1$ of 54.38 eV and similar for higher Z in agreement with measurements. The terms

$$E_m-E_n=Z^2R\left(\frac{1}{m^2}-\frac{1}{n^2}\right) \tag{2.19}$$

where R is the (energetic) Rydberg constant $me^4/(2\hbar^2)$, define the energy level of transitions where, for example, $E_1 - E_2 = 10.24$ eV follows from the Lyman series for hydrogen ($Z=1$) with $m=1$ and $n=2$.

This model easily answers the question why an electron does not "fall into the proton" at electrostatic contraction of the electron radius r: the rise of the "quantum energy" E or of the quantum pressure is faster than the gained electrostatic energy and is balanced at the Bohr radius. The electron has no orbital motion as in the usual quantum mechanical result (in difference to Bohr's model). This quantum pressure model has the advantage that the polarization energy ε_D per electron, Eq. (2.15) can be introduced immediately as correction to ε_s in a plasma, resulting then in the electrostatic energy

$$\varepsilon_s^{\text{plasma}} = \varepsilon_s - \varepsilon_D \tag{2.20}$$

$$= \frac{Ze^2}{2r} - \frac{Ze^2}{\lambda_D} = \frac{Ze^2}{2r}\left(1 - \frac{2r}{\lambda_D}\right)$$

The model has the advantage of distributing the energy for $n=2$ either into a nonorbiting localized state or one with one h into a localized and with the second h into an orbiting state with three possible (orthogonal) axes resulting in the 4×2 possible occupations, which equal 8 states of the L-shell. The 18 states for the M-shell are given by one localized state, three states with orbiting states and three states with orbiting states $2h$, and two states with an orbiting "ellipse" in the sense of the Sommerfeld-Bohr model where the number 2 comes from the 2 degrees of freedom for the rotation of the direction of the ellipse. The ground states are then without orbiting and cannot emit radiation, which was a difficulty in the Bohr's model and was overcome in the Schrödinger model, while the orbiting states in the electrostatic model for $n \geqslant 2$ can simply account for the dipole emission and the spontaneous transition. The Schrödinger model needed the second quantization of the field for achieving spontaneous transition.

Equating electrostatic energy in a plasma, Eq. (2.20) with the quantum energy E, Eq. (2.17), results in

$$\frac{n^2\hbar^2}{2mr^2} = \frac{Ze^2}{2r} - \frac{Ze^2}{\lambda_D}$$

and hence in the algebraic solution for

$$r = \frac{\lambda_D Z}{4}\left[1 - \left(1 - \frac{8n^2\hbar^2}{\lambda_D e^2 Z^2 m}\right)^{1/2}\right] \tag{2.21}$$

The energy state E_n for the quantum number n is then from Eq. (2.17) using Eq. (2.21):

$$E_n = \frac{Z^2}{\hbar^2} R \frac{A^2}{[1-(1-2A)^{1/2}]^2}; \qquad A = \frac{2n^2\lambda_c}{\pi Z \lambda_D \alpha} \tag{2.22}$$

where the Compton wavelength $\lambda_c = h/mc$, the fine structure constant $\alpha = 1/137$, and the mean electron velocity v_e was used. The polarization shift $E_n - E_m$ is of the same order as the low-density approximation [103]. The advantage of the result (2.22) is an immediate inclusion of the plasma temperature.

The strongest shift of the levels occurs where the limiting case $A = A^* = \frac{1}{2}$ in Eq. (2.22) is given. This means if $2A$ is close to one,

$$2A = \frac{4n^2\lambda_c}{\pi\lambda_D\alpha Z} \leqslant 1$$

The limiting case corresponds to the highest possible $\bar{n} = n^*$ in a plasma. This is the "drowning of spectral lines" (Margenau et al. [103]) which we have derived now on the basis of the Debye energy and which is temperature dependent. It limits to the Inglis–Teller continuum which is determined by the Stark broadening due to the (temperature independent) microfield (Holtzmark potential). The line shift based on the result of Eq. (2.22) is in excellent agreement with Lyman lines in difference to other models, as shown by Henry [106].

2.4 Cyclotron Frequency

An important behavior of free charges in a plasma (having an energy described by a general distribution function, or special thermal distribution (see next section) with a temperature T, or by a quantum modified distribution, is their motion in a magnetic field $\mathbf{H}$, which is assumed spatially homogeneous and temporally constant within this subsection. Without losing generality the velocity $\mathbf{v}$ of the particle of charge e and mass m can be split into one component $\mathbf{v}_p$ parallel and one component $\mathbf{v}_s$ perpendicular to the magnetic field $\mathbf{H}$. The Lorentz force leads to the following equations of motion for the particle:

$$m\frac{d\mathbf{v}}{dt} = \frac{e\,\mathbf{v} \times \mathbf{H}}{c} \tag{2.23}$$

$$m\frac{d\mathbf{v}}{dt} = 0 \tag{2.24}$$

A force-free motion of the particle parallel to the magnetic field follows. For the motion perpendicular to the magnetic field is obtained

$$m\frac{d\mathbf{v}_s}{dt} = \frac{e\,\mathbf{v}_s \times \mathbf{H}}{c} \tag{2.25}$$

Without losing generality the s component of the velocity can be expressed by an angular velocity vector $\mathbf{u}$, which has a direction parallel to $\mathbf{H}$ and a modulus (absolute value) of ω_c:

$$\mathbf{v}_s = \mathbf{u} \times \mathbf{r}; \qquad \mathbf{H} \| \mathbf{u} \| \mathbf{r} \tag{2.26}$$

Because of the constancy of $\mathbf{H}$, the angular velocity $\mathbf{u}$ is constant too, and so Eq. (2.24) results in:

$$\frac{m\mathbf{u} \times d\mathbf{r}}{dt} = \frac{e\mathbf{v}_s \times \mathbf{H}}{c} \tag{2.27}$$

$$m\mathbf{u} \times \mathbf{v}_s = \frac{e(\mathbf{u} \times \mathbf{r}) \times \mathbf{H}}{c}$$

$$m\mathbf{u} \times (\mathbf{u} \times \mathbf{r}) = \frac{e(\mathbf{u} \times \mathbf{r}) \times \mathbf{H}}{c}$$

$$m\mathbf{u}(\mathbf{u} \cdot \mathbf{r}) - m\mathbf{r}\mathbf{u}^2 = \frac{e\mathbf{r}(\mathbf{u} \cdot \mathbf{H})}{c} - \frac{e\mathbf{u}(\mathbf{r} \cdot \mathbf{H})}{c}$$

The first term on the left-hand side is zero, because $\mathbf{r}$ is perpendicular to $\mathbf{u}$, and the last term on the right-hand side is zero because $\mathbf{H}$ is perpendicular to $\mathbf{r}$. As $\mathbf{u}$ is parallel to $\mathbf{H}$, the angular frequency ω_c, the modulus of $\mathbf{u}$, is calculated to be

$$\omega_c = e|\mathbf{H}|/mc \; [\text{cgs}] = eH/m \; [\text{MKQS}] \tag{2.28}$$

This is called the cyclotron frequency, the gyro frequency, or the Larmor frequency.

The particle moves free of force along the magnetic field and rotates with a frequency ω_c around the magnetic field lines. The radius of this rotation is

$$r_L = v_s/\omega_c = v_s mc/e|\mathbf{H}| \; [\text{cgs}] = vm/e|\mathbf{H}| \; [\text{MKQS}] \tag{2.28a}$$

This is the gyration radius or the Larmor radius. The trapping of electrons and ions of a plasma on the lines of a magnetic field is exploited in the confinement of plasma for thermonuclear fusion. The problem is then to use closed magnetic fields as, for example, in a toroidal solenoid. As the initial condition of field homogeneity is then not possible, a drift can be prevented by shearing of the magnetic field. The particles would remain trapped in the field lines, if they did not undergo collisions, which makes them diffuse across the magnetic field lines. It has been discovered, however, that instead of following the classical description of diffusion, the plasma diffuses much faster [107], either according to Bohm diffusion, or the slower Pfirsch–Schlüter diffusion. The highly complex problems inherent in plasma confinement by magnetic fields for thermonuclear fusion will not be discussed here.

2.5 Collisions

An important quantity, which is used in macroscopic hydrodynamic plasma theory and which is taken from the microscopic theory, both classically and quantum mechanically, is the collison frequency of the plasma particles. It is shown now how a very primitive model for collisions is valid and how it reproduces the main properties of the collision process in quite good agreement with the most sophisticated theoretical models. Following Fig. 2.5, the Coulomb interaction of an electron with a positive ion can be described by the hyperbolic trace of the electron, where the deflection angle ϕ corresponds to an electron of initial velocity v_e which, if undeflected, passes by the positive charge at a distance r_0, which is called the "impact parameter".

The Coulomb force $\mathbf{f}$, acting between the electron and ion, is given by the distance $\mathbf{r}$ between the electron and the ion, the latter is assumed to have a charge Z

$$\mathbf{f}=\frac{-Ze^2\mathbf{r}}{r^3} \tag{2.29}$$

The main interaction between the particles is during time t

$$t=\frac{r_0}{v} \tag{2.29a}$$

when a change of electron momentum

$$\Delta(mv)=|\mathbf{f}t|=\frac{Ze^2}{r_0 v} \tag{2.30}$$

occurs. It is the aim to calculate the interaction for a 90° scattering event. The change of the momentum is then equal to the initial momentum of the electron.

$$\Delta(mv)\simeq mv\simeq\frac{Ze^2}{r_0 v}; \qquad r_0=\frac{Ze^2}{mv^2} \tag{2.31}$$

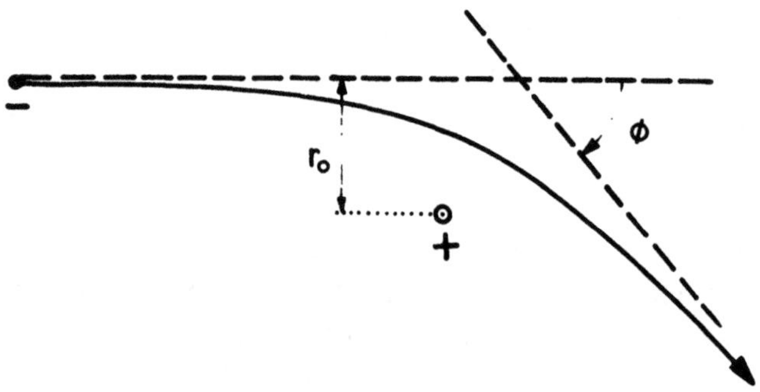

Figure 2.5 Coulomb collision of an electron and an ion.

The resulting cross section is

$$\bar{p}=\pi r_0^2=\frac{Z^2\pi e^4}{m^2v^4} \tag{2.32}$$

The electron-ion collision frequency ν_{ei} is then for a plasma with an ion density $n_i=n_e/Z$:

$$\nu_{ei}=n_i\bar{p}v=\frac{Zn_e\pi e^4}{m^2v^3} \tag{2.33}$$

The average velocity v of the electrons can be expressed by the electron temperature T_e leading to a collision frequency

$$\nu_{ei}=\frac{Zn_e\pi e^4 3^{-3/2}}{m^{1/2}(KT)^{3/2}} \tag{2.34}$$

For the collision process the most probable velocity v_m, $(mv_m^2/2=3KT/2)$, and not the velocity corresponding to the average energy ($E=KT$), has to be used. Though a very crude description of the 90° scattering has been used, the calculated collision frequency of the electrons with the ions is in very good agreement with the exact classical calculation of Spitzer, and Härm [108], which takes into account small angle scattering. These authors included also the collisions for electrons with electrons, which resulted in a correction factor $\gamma_e(Z)$ varying between 0.5 for hydrogen and 1 for large Z (so that this correction will be neglected mostly in the following). Their calculation arrived at an electron collision frequency

$$\nu_e=\frac{n_e}{(KT)^{3/2}}\frac{Z\pi^{3/2}e^4\ln\Lambda}{m^{1/2}2^{5/2}\gamma_e(Z)} \tag{2.35}$$

where the Coulomb logarithm ln Λ is used, given by

$$\Lambda=\frac{\lambda_D}{r_0(90°)}=\frac{3}{2Ze^3}\left(\frac{K^3T^3}{\pi n_e}\right)^{1/2} \tag{2.36}$$

The Coulomb logarithm varies between 5 and 20, so that the difference between (2.35) and the primitive calculation (2.34) is really surprisingly small. The numerical value of Spitzer's electron collision frequency is

$$\nu_e=8.513\times10^{-7}\,\frac{Zn_e}{\gamma_e(Z)T_e^{3/2}}\ln\left(1.55\times10^{10}\,\frac{T_e^{1/2}}{Zn_e^{1/2}}\right) \tag{2.37}$$

where the electron temperature T_e is given in electron volts and the electron density n_e is given in cm^{-3}.

The electrical resistivity $1/\sigma$ of the plasma can now be calculated from the collison frequency ν_e. Starting from the mean free path l of the electrons, their thermal velocity v is

$$l=\frac{v}{\nu_e} \tag{2.38}$$

the velocity $\mathbf{v}_D$ gained by drift in an electric field $\mathbf{E}$

$$\frac{d\mathbf{v}_D}{dt}=\frac{e\mathbf{E}}{m} \tag{2.39}$$

The average drift velocity between two collisions is then

$$\langle \mathbf{v}_D \rangle=\frac{e\mathbf{E}}{2m\nu_e} \tag{2.40}$$

which can be used to calculate the current density $\mathbf{j}$

$$\mathbf{j}=en_e\mathbf{v}_D=\sigma\mathbf{E} \tag{2.41}$$

$$\mathbf{j}=\frac{n_e e^2\mathbf{E}}{2m\nu_e}=\sigma\mathbf{E}$$

The definition of the electrical conductivity σ from Ohm's law and (3.40) are used. The Ohmic conductivity in a plasma, using the simplified collision frequency (2.34), is then

$$\sigma=\frac{3(KT)^{3/2}}{2Z\sqrt{m/3\pi}e^2} \tag{2.42}$$

which is numerically

$$\sigma=\frac{T^{3/2}}{Z}\times 1.93\times 10^8 \text{ cgs}$$

$$\sigma=\frac{T^{3/2}}{Z}\times 2.14\times 10^{-4}\Omega^{-1}\text{ cm}^{-1} \tag{2.43}$$

In both cases, the temperature is in degrees Kelvin. If the temperature is given in electron volts, the numerical value of the conductivity is

$$\sigma=\frac{T^{3/2}}{Z}\times 268\ \Omega^{-1}\text{ cm}^{-1}$$

The electrical resistivity is given by

$$1/\sigma=\frac{Z}{T^{3/2}}\times 4.66\times 10^3\ \Omega\text{ cm}$$

or (2.44)

$$1/\sigma=\frac{Z}{T^{3/2}}\frac{\ln\Lambda}{\gamma_e(Z)}\ 3.08\times 10^3\ \Omega\text{ cm} \qquad ([T]=K)$$

where in the last equation the Spitzer collision frequency is used, Eq. (2.35).

It is remarkable that the electrical conductivity in fully ionized plasmas agrees with these values, confirming the correctness of the Coulomb collision frequency derived above. A comparison with the quantum mechanically derived collision frequencies is possible by considering the high collision frequency given by the optical constants from inverse bremsstrahlung theory. This will be discussed later in connection with the theory of the refractive index of plasmas.

The electrical resistivity of plasma, Eq. (2.44), does not depend on the electron density. Comparing with the electrical resistivity of metals, for example, of aluminium as a good conductor with $\sigma = 36 \times 10^4\ \Omega^{-1}\ \mathrm{cm}^{-1}$, the plasma conductivity reaches that of metals when the electron temperature is in the region of 20 to 100 eV (corresponding to about 200,000 to 1,000,000 °K).

At this point the case of metals is briefly mentioned. In metals, the nearly free conduction electrons will undergo Coulomb collisions. When at the beginning of this century Drude and Lorentz were developing their electron theory of metals, they calculated the Coulomb collision frequency in a similar way as shown here for plasmas. Using the room temperature of the metals in Eq. (3.42), the conductivities were about 10^6 times less than measured. This disappointing result was due to the fact that the quantum mechanical properties of the electrons had not been taken into account. Planck's discovery of the atomistic nature of all quantities which have the dimension of a unit of action (quantization) leads to the result that, if an electron is concentrated within a volume of length x, its momentum p along each direction has to be so large to result in an energy E, as given in Eq. (2.17).

The exact Fermi energy E_F, which is the corrected energy E of Eq. (2.17) with respect to spherical geometry using Pauli's double occupation

$$E_F = \frac{h^2}{2m} n_e^{2/3} \frac{(3/\pi)^{2/3}}{4} \tag{2.45}$$

This energy is valid also for plasmas of high density. If any plasma is compressed to densities such that the Fermi energy of the electrons is larger than the thermal energy, the electrons are called Fermi–Dirac degenerate (see Appendix A). The compression of a degenerate plasma is adiabatic with respect to an increase of the Fermi energy. The compression is against their Fermi pressure. It results in a compressibility that is equal to the compressibility of a solid within an order of magnitude. The fact that an electron in an atom or in a solid cannot be compressed as easily as in a dilute gas is due to the Fermi pressure.

The Fermi energy of electrons in a metal is about 6 to 20 eV. Thus it is understandable that, when conduction electrons interact with the electric field, the electrons act as if they have a "temperature" of about 100,000°K.

If in this way the electrical conductance based on the Coulomb collisions, Eq. (2.42), is still about 10 times less than measured values for the best metals, this can be ascribed to the so-called effective mass (see Appendix A) of the conduction electrons, that differs from the mass of the electrons in a dilute plasma that is equal to the vacuum electron mass m. This effective mass indeed varies within the order of magnitude necessary to give high conductance. Another point is that the Coulomb logarithm in the more accurate equation (2.42) must be determined. The theory restricts the validity of the Coulomb logarithm to values between 2 and 3 as a lower limit. For very high electron densities in metals and temperatures T~10eV, the Coulomb logarithm has to be corrected within one order of magnitude.

2.6 Anomalous Resistivity, Quantum Collisions and Tokamak Experiments

We just mentioned some minor corrections of the values of the resistivity of fully ionized plasmas which however are not changing the exponents of the power laws of the just derived values nor its order of magnitude for agreement of measured values with those derived from these theoretical models. The detailed agreement was not easy to be measured with the classical plasmas especially if there were atoms of higher charge number and where the plasma was partially ionized only. In this case, the collisions of electrons with neutral atoms or with partially ionized ions has to be taken into account, which cross sections could not be treated in the rather simplified way as described before for Coulomb collisions.

The situation was transparent only if using plasma of hydrogen or of its isotopes where at higher temperature, one had the transparent conditions of a fully ionized plasma for comparison of the Ohmic resistivity with the formulas derived here. Since the last number of years it was then possible to study these hydrogen plasmas at much higher temperatures than before, e.g. at 100,000 K (corresponding to 10 eV) and at several 100 eV. It was then observed that the Ohmic resistivity was higher than the classical value derived in the preceding subsections and it was vaguely said that this is due to anomalous resistivity without specifying its properties.

We are presenting here a model for this anomalous resistivity which at least for a number of cases show agreement with experiments. We are not arguing that other types of anomalous resistivity due to turbulence, chaos or other phenomena may exist but simply restrict to the following model and types of experiments.

According to Fig. 2.5, the classical Coulomb collision was described by the hyperbolical motion of an electron around a positive ion. We neglected in this case that the acceleration of the electron is causing the emission of radiation, the so-called bremsstrahlung to which question we shall come back in connection with the optical constants in Chapter 6. The neglecting of the bremsstrahlung in our preceding model for the dc-resistivity of plasma was at least justified by the good agreement with the experiments.

A quantum mechanical change of the classical hyperbolic motion of Fig. 2.5 was discussed early by Bethe for the case that the de Broglie wave length

$$\lambda_{dB} = \frac{h}{\sqrt{2mE}} \tag{2.46}$$

using Planck's constant h, the electron mass m and the electron energy E, is of the order of magnitude or larger than the impact parameter r_o of Eq. (2.31). Bethe (1934) concluded verbally that then the electron will perform a wave mechanical diffraction around the ion. Instead of the impact parameter r_o of the classical hyperbolic motion, the 90° wave mechanical diffraction results then in an impact parameter

$$r_o = \frac{h}{2mv} \tag{2.47}$$

Both impact parameters are equal at an electron velocity determining a plasma temperature as first pointed out by Marshak (1941)

$$T^* = 4Z^2mc^2\alpha^2 /(3K) = Z^2 4.176\times10^5 (^oK) \tag{2.48}$$

and both branches of the impact parameter of Eq. (2.31) and (2.47) can be combined by the following formula

$$r_o = \frac{r_{Bohr}}{2Z}\frac{1}{(1+4T/T^*)^{1/2}-1} =$$

$$= \begin{cases} r_{Bohr}T^*/(4ZT) = Ze^2/(3KT); \text{ if } T < T^* \\ \\ \dfrac{r_{Bohr}}{4Z}\sqrt{\dfrac{T^*}{T}} = \dfrac{\hbar}{2\sqrt{3KT/m}}; \text{ if } T > T^* \end{cases} \tag{2.49}$$

where the Bohr radius $r_{Bohr} = \hbar^2/me^2$ and the fine structure constant $\alpha = e^2/\hbar c$ was used. It is remarkable that even in this case of a fully ionized atom, the Bohr radius plays an important role branching two different ranges with dfferent powers at lower plasma temperature (classical range) and causing a higher impact temperature, i.e. a higher resistivity, at higher temperatures than T^*, the quantum mechanical range.

Taking the same steps as from Eq. (2.31) to (2.35), the electron ion collision frequency is

$$\nu_{ei} = \frac{\pi r_{Bohr}^2 n_e}{4Z^3}\sqrt{\frac{3kT}{m}}\,[(1+4T/T^*)^{1/2}-1]^{-2} =$$

$$= \begin{cases} \dfrac{Z\pi e^4}{3^{3/2}\,m^{1/2}}\dfrac{n_e}{(kT)^{3/2}} = \nu_{ei} & \text{if } T \ll T^*, \\[2ex] \dfrac{\pi\hbar^2}{3^{1/2}\,Zm^{3/2}}\dfrac{n_e}{(kT)^{3/2}} = \nu_{ei}\dfrac{T}{T^*} & \text{if } T \gg T^*, \end{cases} \tag{2.50}$$

The electrical conductivity is then in a similar way as derived before

$$\sigma = \frac{en_e}{2m\nu_{ei}} = \frac{eZ^3}{\sqrt{3mkT}\,r_{Bohr}}\,[(1+4T/T^*)^{1/2}-1]^2 =$$

$$= \begin{cases} \dfrac{3^{3/2}(kT)^{3/2}}{2Zm^{1/2}e^2} = \sigma_{cl} & \text{if } T \ll T^*, \\[2ex] \dfrac{\sigma_{cl}}{T}T^* & \text{if } T \gg T^*, \end{cases} \tag{2.51}$$

showing the anomalous resistivity at higher temperature than T^*, being increased by the value T/T^* above the classical value.

This result is not generally accepted yet. While Spitzer [107] used this within his Coulomb logarithm, he did not accept it for the total value of the cross section. In his case the discussion about a factor two with respect to small angle and wide angle classical scattering traces in contrast to a model of Chandrasekhar [108] was so much more important that it was not taken into account to determine the whole cross section changes by quantum mechanics.

At this stage it is a philosophical question whether the classical or the quantum cross section is valid. One may argue that the classical cross section is always valid even if the wave mechanical diffraction would require a larger cross section. The wave mechanical scattering would then be a Mie scattering as known from optics and the shorter classical cross section would survive. Philosophically the situation, however, may be different to the optical Mie scattering analogy. One may argue that always that cross section is valid which is larger. In this case (2.50) has to be taken with the different branches.

Since thermal equilibrium plasmas at temperatures above 100 eV are not long available it may be understandable that the quantum branch was not noticed in the past experimentally. In the following we are presenting experimental examples which confirm the quantum properties of plasmas at high temperatures.

Just at the time when the quantum collisions became evident (Hora, 1981), the experiments with the stellarator plasmas were achieved at high temperatures. These experiments of magnetic confinement fusion with zero current plasmas for producing fusion energy were rather transparent and understandable. However the temperatures achieved earlier were very low, about 50 to 80 eV only for a long time. Most of these experiments were closed down, e.g., in Princeton, where L. Spitzer et al had initiated this type of magnetic confinement fusion, after the high current torroidal magnetic field experiments of the Kurchatov Institute in Moscow initiated by Yawlenski and pushed forward finally by Artsimovich, turned out to be successful. The excellent heating of such tokamak plasmas to rather high temperatures due to ohmic resistivity should be mentioned later again.

The zero current stellarator experiments were nevertheless continued - despite broad criticisms - by G. Grieger and his team at the Max-Planck-Institute for plasma physics in Garching, Germany based on the high class experimental technics and craftsmanship, and motivated by the attempts to understand the Pfirsch-Schlüter diffusion of the plasma across the magnetic field contrary to the disappointing Bohm diffusion which was considered to be so disadvantageous in Princeton and the other stellarator experiments. Grieger et al (1980) succeeded - just before they would have been forced to dismantle their experiments under the pressure of the success of tokamaks and of other personal arguments - that their deuterium stellarator was producing long confinement times and high temperatures up to keV in rather transparent conditions of ideally fully ionized deuterium plasma. Nuclear fusion gains were respectable and confirmed the achievements of the high temperatures.

The diffusion across the magnetic field was nevertheless much faster than that of classical diffusion. This diffusion is due to the collisions in plasma otherwise the ideal confinement of plasma in magnetic fields would be rather close to be realized in collisionless plasmas trapping the electrons and ions by the magnetic fields. The collisions disturb this trapping and a diffusion across the magnetic field appears. Using the quantum collision branch (2.50) the diffusion branches into the following equation

$$D = \begin{cases} \dfrac{\pi^{3/2} m_e^{1/2} n_e Z e^2 c^2 \ln\Lambda}{2^{5/2} \gamma_e(Z)(kT)^{1/2} B^2} = D_{cl} & \text{if } T \ll T^*, \\ D_{cl} T/T^* & \text{if } T \gg T^*, \end{cases} \quad (2.52)$$

What was measured in a rather transparent and precise way by Grieger et al (1980) was that the diffusion across the magnetic field was 20 times faster than the classical value for plasma temperatures of 800 eV. Taking into account Eqs. (2.48) and (2.52), the transition temperature between the low temperature classical case and the quantum collision is just about 40 eV. The classical diffusion is changed in the same way as the collision frequency by the factor T/T^* and for the plasma temperature of $T = 800$ eV is then just 20 higher exactly as measured. This immediately indicates the mechanism of the quantum collisions.

Another example is taken from the measurement of the Ohmic conductivity of plasmas. The measurements are highly complex and very rare cases supply the sufficiently accurate information. Taking the case of a hydrogen tokamak at the center line parallel to the magnetic field (Hugill, 1983, Fig. 5, analyzing Dimock's results), the Ohmic conductivity according to Eq. (2.51) (Hora, 1981) with the use of Spitzer's Ohmic conductivity value for the classical branch, Eq. 5.35 of Ref. [107],

$$\sigma_{cl}^{sp} = T^{3/2}/(3.80 \times 10^3 Z \ln\Lambda) \quad (2.53)$$

one finds from Eq. (2.51) the value of 2.805×10^4 Ohm^{-1} cm^{-1} which is about 30% below the experimental value. However the classical Spitzer value (2.53) is 1.27082×10^6 Ohm^{-1} cm^{-1} and therefore more than 30 times higher than the measured value. This is a convincng result in favour of the quantum collisions. In order to save the classical collisions for explaining the resistivity in plasmas, one assumes that there are low concentrations of highly charged heavy ions as contamination in the plasma to reproduce the measured conductivities. Though the whole complicated analysis of the experiments is still not transparent easily and any differentiation is to be discussed, it may have been shown that plasmas at high temperature have an anomalous resistivity given by the quantum collisions, and that this quantum property of the plasma may be the reason for the achievement of the high temperature by Ohmic heating in tokamaks.

Another example of tokamak plasmas is given to explain the quantum collision at high plasma temperatures from the measurements of the thermal conductivity parallel to the magnetic field. This direction avoids the complication of the magnetic field and one can then relate simply to conditions of homogeneous plasmas free from the difficulties of double layer effects well known from the reduction of the thermal conductivity by double layers (Cicchitelli et al 1984). Under these conditions of the experiments the plasmas were homogeneous and no double layers possible (Eliezer et al 1989).

Taking the measurements of Razumova (1984 see Fig. 5), for thermal conduction parallel to the magnetic field of the mentioned rare and difficult case with a high beta value (of the inner part of the tokamak only for $r<a/2$), Fig. 2.6 contains these experimental values of the thermal electron conductivity depending on the electron temperature. This is compared (drawn plots) with the absolute values of the anomalous branch of the generalized thermal conductivity κ of the following equations and the absolute values of the classical κ_c.

$$\kappa = \kappa_c \ \{(T^*/2T)\ [(1 + 4T/T^*)^{1/2} - 1]\}^2 \tag{2.54}$$

which can be approximated by

$$\kappa = \begin{cases} \kappa_c & \text{if } T < T^* \\ \\ \kappa_c\, T^*/T, & \text{if } T > T^* \end{cases} \tag{2.55}$$

with the classical thermal conductivity

$$\kappa_c = 4.67 \times 10^{-12}\, \varepsilon\delta_T \frac{T^{5/2}}{Z\ln\Lambda} \ \text{cal/sec deg cm} \tag{2.56}$$

where Spitzer's Coulomb logarithm $\ln\Lambda$ was used. Other measurements for low beta or for the turbulent range are basically different with a decreasing dependence on the temperature and other much different behaviour.

The comparison in Fig. 2.6 shows that the high-beta tokamak is fitting the exact theory within 40%, Eq. (9), quite well in absolute numbers and it well represents the slope of a $T^{3/2}$ law of equation (2.54). The classical branch (intentionally extended as a plot in Fig. (2.54) to temperatures above T^* is higher than the measurements by a factor 71 at 1 keV and

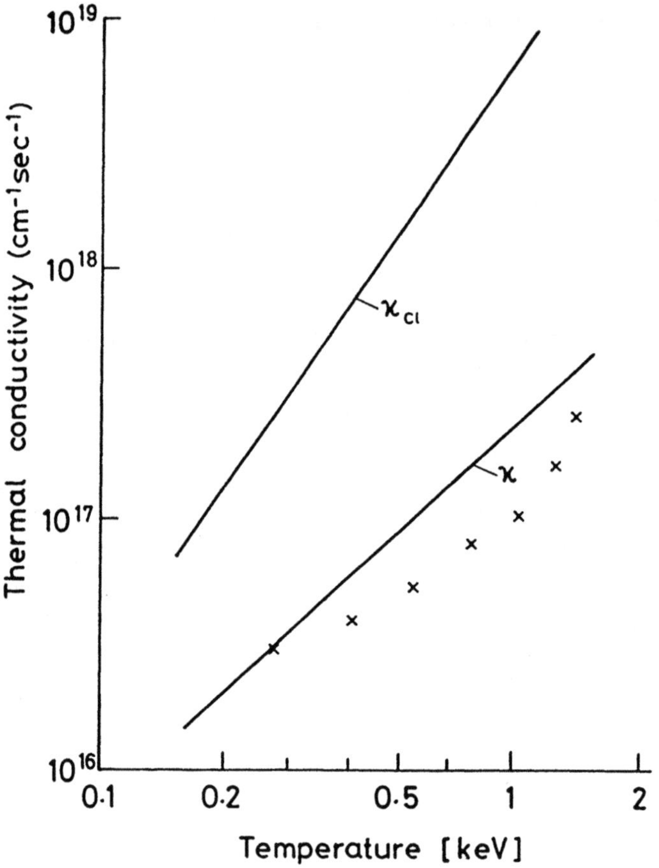

Fig. 2.6: Measured electron thermal conductivity in a high q(a) = 1.3 tokamak parallel to the magnetic field in the nonturbulent central range (r <a/2) depending on the electron temperature (Razumova, 1983) compared with the quantum generalized thermal conduction κ, formula (9b), in its anomalous conductivity branch. This is in contrast to the strongly deviating classical conductivity κ_c with a clearly different slope and more than 10 times higher values.

generally by one or two orders of magnitude and has a different slope of $T^{5/2}$. Though there is no total fit of the experimental values, the difference in the slope and in the absolute values clearly confirms the quantum collisions, if one takes into account the difficulty of the experiments.

It is remarkable that the scaling law of the energy confinement time τ_E for tokamaks derived from the T-11 group (Razumova, 1984) apart from a purely geometric factor with the dimensions of the tokamak and a weak factor q (a) was reported (using the electron density n_e)

$$\tau_E \approx n_e/T^{1/2} \approx \nu \tag{2.57}$$

which corresponds to the collision frequency ν taking into account its quantum modified anomalous branch, Eq. (2.50). This is contrary to the classical branch given by $n_e/T^{3/2}$. The quantum branch Eq. (2.56) is valid only for electron temperatures $T > T^*$. The result of Eq. (2.57) would very directly represent the expectation that the confinement time is proportional to the collision frequency ν responsible for the drift of the plasma across the magnetic field, but given at high temperatures by the anomalous values of equation (2.50).

These examples indicate how experiments - definitely uninfluenced by any theoretical prejudice - may prove the prediction of a change of the classical Coulomb collision of electrons with fully ionized ions into the quantum branch at higher electron energies. The electron energy of this change is given by the electron rest mass and the square of the fine structure constant, Eq. (2.48) (Hora, 1981). These results are providing a connection of the dc-current plasma resistivity to the results of quantum electrodynamics with or without stimulated emission as will be explained in Sect. 6.6.

THREE

Kinetic Theory

Despite the fact that the following discussion of the laserplasma interaction follows the macroscopic (hydrodynamic) description, this section will sketch the kinetic theory of plasmas for some background understanding. The reader looking more for the experimental or engineering aspects may skip this section.

The kinetic theory or the theory of kinetic equations is a link between the single-particle description of the microscopic theory and of the macroscopic continuum theory of hydrodynamic quantities such as particle densities, net velocities, and temperatures. Just the definition of temperature as a macroscopic quantity involves complications if no thermal equilibrium has been achieved and the microscopic state shows several "temperatures" of plasma components.

3.1 Distribution Functions

For a straightforward understanding, an ad hoc description of the elements of the kinetic theory is given in this subsection. We first have to underscore that distribution functions do not necessarily relate—in a philosophical sense—to probabilities, uncertainties, or inaccuracies. If one has to find the average value of an ensemble of N integers a_i

$$M = \frac{\sum_{i=1}^{N} a_i}{N} \tag{3.1}$$

If there are several of the integers of the same numerical value, one can order the numbers in microensembles $j=1,\ldots,n$, where f_j determines the number of elements in each microensemble. The task of finding the average value (3.1) is then solved by

$$M=\frac{\sum_{i=1}^{n} f_i a_i}{\sum_{i=1}^{n} f_i} \tag{3.2}$$

Therefore a set of f_i or, continuously, a distribution function $f(x)$ can then determine the average value of a quantity $a(x)$

$$\langle a\rangle=\frac{\int f(x)a(x)\,dx}{\int f(x)\,dx} \tag{3.3}$$

The same procedure with distribution functions is used to reduce the microscopic N-particle problem with its $3N$ differential equations (2.1) by the following way: The configuration space is subdivided into cells of a volume $d^3r=dx_1\,dx_2\,dx_3$ about $\mathbf{x}$ and the velocity space is subdivided into cells $d^3w=dw_1\,dw_2\,dw_3$ about $\mathbf{w}$. Instead of describing each of the N particles by their differential equation (2.1) in time, one can ask, how many particles are present in the mentioned cells, depending on time t, given by the distribution function

$$f(x_1,x_2,x_3,v_1,v_2,v_3,t)d^3x\,d^3v \tag{3.4}$$

The density of the particles (number of particles per cubic centimeter) is then from this definition with the index e or i (electrons or ions)

$$n_{e,i}(\mathbf{r},t)=\iiint_{-\infty}^{+\infty} f(\mathbf{r},\mathbf{w},t)\,d^3w \tag{3.5}$$

Similar to Eq. (3.3) we find the value of any physical quantity $Q(\mathbf{r},\mathbf{w},t)$ depending on the velocity $\mathbf{w}$ at a point $\mathbf{r}$ at a time t as

$$\overline{Q(\mathbf{r},t)}=\frac{\iiint_{-\infty}^{+\infty} Q(\mathbf{r},\mathbf{w},t)f(\mathbf{r},\mathbf{w},t)\,d^3w}{\iiint_{-\infty}^{+\infty} f(\mathbf{r},\mathbf{w},\ )\,d^3w} \tag{3.6}$$

The average velocity of the N particles at $\mathbf{r}$ and t is then

$$\mathbf{v}(\mathbf{r}, t)=\frac{\displaystyle\iiint_{-\infty}^{+\infty} \mathbf{w} f(\mathbf{r}, \mathbf{w}, t)\, d^3 w}{\displaystyle\iiint_{-\infty}^{+\infty} f(\mathbf{r}, \mathbf{w}, t)\, d^3 w} \tag{3.7}$$

If $Q(\mathbf{w})$ does not depend on $\mathbf{r}$ and t, we find the following relations using Eq. (3.5):

$$\begin{aligned}\int Q(\mathbf{w}) \frac{\partial f}{dt}\, d^3 w &= \frac{\partial}{\partial t} \int Q(\mathbf{w}) f\, d^3 w \\ &= \frac{\partial}{\partial t}(n\bar{Q})\end{aligned} \tag{3.8}$$

$$\int Q(\mathbf{w}) \mathbf{w} \cdot \nabla f\, d^3 w = \nabla \cdot \int Q(\mathbf{w}) \mathbf{w} \mathbf{f}\, d^3 w = \nabla(n\overline{\mathbf{w} Q}) \tag{3.9}$$

where the vector operator $\nabla=(\partial/\partial x_1;\ \partial/\partial x_2;\ \partial/\partial x_3)$ is called "del" or nabla. Using a general vectorial function $\mathbf{F}$ and the operator $\nabla_w=(\partial/\partial w_1;\ \partial/\partial w_2;\ \partial/\partial w_3)$

$$\int Q(\mathbf{w}) F(\mathbf{r}, \mathbf{w}) \cdot \nabla_w f\, d^3 w = -\int f \nabla_w \cdot \{\mathbf{F}(\mathbf{r}, \mathbf{w}) Q(\mathbf{w})\}\, d^3 w$$

and using the fact at partial integration of $f(\pm\infty)=0$ [for reasons of convergence of (3.5)],

$$\int Q(\mathbf{w}) F(\mathbf{r}, \mathbf{w}, t) \cdot \nabla_w f\, d^3 w = -n \nabla_w \cdot (\overline{FQ}) \tag{3.10}$$

We arrive at a kinetic equation for the distribution function f by the very trivial statement that, if there are no changes, there is no explicit dependence on time, or—in other words—the total derivative

$$\frac{d}{dt} f=0 \tag{3.11}$$

has to be zero. Taking into account all seven variables of f, Eq. (3.4), we find by partial differentiation from (3.11)

$$\begin{aligned}\frac{d}{dt} f = {} & \frac{\partial}{\partial t} f+\frac{\partial}{\partial x_1} f \frac{\partial x_1}{\partial t}+\frac{\partial}{\partial x_2} f \frac{\partial x_2}{\partial t}+\frac{\partial}{\partial x_3} f \frac{\partial x_3}{\partial t} \\ & +\frac{\partial}{\partial w_1} f \frac{\partial w_1}{\partial t}+\frac{\partial}{\partial w_2} f \frac{\partial w_2}{\partial t}+\frac{\partial}{\partial w_3} f \frac{\partial w_3}{\partial t}\end{aligned} \tag{3.12}$$

or using the velocity vector $\mathbf{w}=(\partial x/\partial t; \partial y/\partial t; \partial z/\partial t)$, and the force density $\mathbf{F}$, where

$$\mathbf{F}=m\frac{\partial}{\partial t}\mathbf{w} \tag{3.13}$$

if all particles have the same (or averaged) mass m, we arrive at

$$\frac{\partial f}{\partial t}+\mathbf{w}\cdot\nabla f+\frac{\mathbf{F}}{m}\cdot\nabla_w f=0 \tag{3.14}$$

This equation is a *kinetic equation* and is called the *Vlasov equation*, if the force is due to electrical quantities ($\mathbf{E}$=electric and $\mathbf{H}$=magnetic field strength, c=velocity of light)

$$\mathbf{F}=e\left(\mathbf{E}+\frac{1}{c}\,\mathbf{w}\times\mathbf{H}\right) \tag{3.14a}$$

resulting in

$$\frac{\partial f}{\partial t}+\mathbf{w}\cdot\nabla f+\frac{e}{m}\left(\mathbf{E}+\frac{1}{c}\,\mathbf{w}\times\mathbf{H}\right)\cdot\nabla_w f=0 \tag{3.15}$$

If f does depend explicitly on the time, the total differentiation in Eq. (3.11) is not zero but equal to the change due to collisions (index c):

$$\frac{\partial f}{\partial t}+\mathbf{w}\cdot\nabla f+\frac{\mathbf{F}}{m}\nabla_w f=\left(\frac{\partial f}{\partial t}\right)_c \tag{3.16}$$

which is the Boltzmann equation. The whole problem is then concentrated in the collision term on the right-hand side. An approximation for collisions with neutral atoms is the Krook collision term

$$\left(\frac{\partial f}{\partial t}\right)_c=\frac{f_n-f}{\tau} \tag{3.17}$$

where f_n is the distribution function of the neutral atoms and τ is the averaged collision time.

For Coulomb collisions, Eq. (3.16) can be approximated using binary collisions by the Fokker–Planck equation

$$\frac{df}{dt}=\sum_{n=1}^{\infty}\frac{(-1)^n}{n!}\,\frac{\partial^n}{\partial_{vi_1}\partial v_{i_2}\cdots\partial_{vi_n}}(\bar{\alpha}_{(m)}f) \qquad (i_n=1, 2, 3, \ldots, n) \tag{3.18}$$

where the $\bar{\alpha}$'s are the Fokker–Planck coefficients [109].

Without discussion we mention here Liouville's theorem: for a conservative system, f is constant along a dynamic trajectory. This means that the volume $d^3r\, d^3w$ which is taken by a number of particles, does not change in time if the particles are not interacting.

While the kinetic equations provide the possibility of treating a plasma without having reached thermal equilibrium, a central meaning has the distribution function for equilibrium. If one normalizes f at each point of certain particle density $n(\mathbf{r}, t)$, the factor f_n in

$$f(\mathbf{r}, \mathbf{v}, t) = n(\mathbf{r}, t)\hat{f}_M(\mathbf{r}, \mathbf{v}, t) \tag{3.19}$$

is called the Maxwellian distribution (or Maxwell–Boltzmann distribution), see Appendix B,

$$\hat{f}_M = \left(\frac{m}{2\pi KT}\right)^{3/2} \exp(-\mathbf{w}^2/v_{\text{th}}^2) \tag{3.20}$$

using Boltzmann's constant K and the average thermal velocity $v_{\text{th}} = (2KT/m)^{1/2}$. Equation (3.20) takes into account positive and negative velocity components. Based on the scalar magnitude of the velocity w one defines another distribution $g(w)$ by

$$\int_0^{\infty} g(w)\, dw = \int_{-\infty}^{+\infty} f(\mathbf{w})\, d^3w \tag{3.21}$$

where

$$g(w) = 4\pi n(\mathbf{r}, t)\left(\frac{m}{2\pi KT}\right)^{3/2} w^3 \exp(-w^2/v_{\text{th}}^2) \tag{3.22}$$

to be distinguished from f.

3.2 Loss of Information

The description of the kinetic equations in the preceding subsection is sufficient to understand the derivation of the macroscopic equations in the next following subsection. In this subsection, some basic problems of the kinetic theory should be mentioned. Boltzmann's basic question was the description of the collisions even in the most simplified way where the collision time can be neglected. The derivation of the Boltzmann equation can be based on the more general Liouville distribution function for the N particles

$$F(\mathbf{r}_1, \mathbf{r}_2 \cdots \mathbf{r}_N, \mathbf{w}_1, \mathbf{w}_2 \cdots \mathbf{w}_N, t) \tag{3.23}$$

which gives the joint probability of finding particle 1 at $\mathbf{r}_1$ with velocity $\mathbf{w}_1$, and particle 2 at $\mathbf{r}_2$ with velocity $\mathbf{w}_2$, and particle N at $\mathbf{r}_N$ with $\mathbf{w}_N$. A single-particle distribution function f defines the probability of finding a particle at $\mathbf{r}$, with a velocity $\mathbf{w}$, at a time t from integrating (3.23) over all but the first particle coordinate:

$$f(\mathbf{r}, \mathbf{w}, t)=\int d^3r_2 \cdots d^3r_N d^3w_1 \cdots d^3w_N F(\mathbf{r}_1 \cdots \mathbf{r}_N, \mathbf{w}_1 \cdots \mathbf{w}_N) \qquad (3.24)$$

If the particles move independently, a factorization is possible:

$$F(\mathbf{r}_1 \cdots \mathbf{r}_N, \mathbf{w}_1 \cdots \mathbf{w}_N, t)=f(\mathbf{r}_1, \mathbf{w}_1, t) \cdots f(\mathbf{r}_N, \mathbf{w}_N, t)=f^N(\mathbf{r}, \mathbf{w}, t) \qquad (3.25)$$

F satisfies the Liouville equation

$$\frac{\partial F}{\partial t}+[F, H]=0 \qquad (3.26)$$

where the Poisson bracket of F with the Hamiltonian H of the complete system

$$[F, H]=\sum_{i=1}^{N} \frac{\partial H}{\partial p_i}\frac{\partial F}{\partial q_i} - \frac{\partial H}{\partial q_i}\frac{\partial F}{\partial p_i} \qquad (3.27)$$

remembering Hamilton's equations

$$\dot{q}_i=\frac{\partial H}{\partial p_i}; \qquad \dot{p}_i=-\frac{\partial H}{\partial q_i}$$

and stepping back from the generalized coordinates q_i to the Cartesians $\mathbf{x}_i$ and from the generalized momenta p_i to velocities $\mathbf{w}$; or accelerations or forces (as in Eq. (3.13)), Eq. (3.27) can be written

$$[F, H]=\sum_{i=1}^{N} \mathbf{w}_i \cdot \nabla F+\frac{\mathbf{F}}{m} \cdot \nabla_{w_i} F \qquad (3.28)$$

rewriting (3.26) in

$$\frac{\partial}{\partial t} F+\mathbf{v}_i \cdot \nabla F+\frac{\mathbf{F}}{m} \cdot \nabla_{w_i} F=0 \qquad (3.29)$$

Integration over $N-1$ coordinates as in Eq. (3.24), results in the collisionless Boltzmann equation

$$\frac{\partial f}{\partial t}+\mathbf{v} \cdot \nabla f+\frac{\mathbf{F}}{m} \cdot \nabla_v f=0 \qquad (3.30)$$

Following the result of Blatt and Opie [110], there are fundamental insufficiencies about the loss of information about the initial state when the system is described by the kinetic equations. The approximation (3.25) is an especially strong restriction. But even there are considerable doubts as to whether the Liouville equation (3.26) with a general F represents any real thermal system adequately [111]. The careful derivation of the kinetic model [110] is further made on the basis of minor approximations appropriate to a very dilute gas, ignoring effects of ternary and higher order collisions.

In laser produced plasmas, the ternary process at the usual high densities are of importance (see, e.g., J. M. Dawson [112]) and the predominance of three-body recombinations [113] was observed even in the early experiments [114] of laser produced plasmas. The basic problems involved are described within a consequent derivation of the Langrangean and Hamilton theory on the basis of the d'Alembert's principle [115].

These arguments should be taken into account if doubts are considered against the macroscopic hydrodynamic models: the kinetic models still are not free of doubts, especially for the high densities of laser produced plasmas. A comeback to the single-particle description of the microscopic models based on Eq. (2.1) is difficult because of computer capacity, the necessary neglection of low-range interactions, an approximative description of the Coulomb collisions with respect to small-angle scattering, and the approximation of collective effects.

3.3 Derivation of Macroscopic Equations

In this subsection macroscopic equations will be derived from the integration of the Boltzmann equation (3.16)

$$\frac{\partial f}{\partial t}+\mathbf{w}\cdot\nabla f+\frac{\mathbf{F}}{m}\nabla_w f=\left(\frac{\partial f}{\partial t}\right)_c \tag{3.31}$$

Each term will be integrated over the whole velocity space d^3w, if we assume that there are no velocity dependent forces. The forces are then holonomic following the definition of H. Hertz [115].

Integration from $-\infty$ to $+\infty$ of the first term of (3.31) results in

$$\int\frac{\partial f}{\partial t}\,d^3w=\frac{\partial}{\partial t}\int f\,d^3w=\frac{\partial}{\partial t}n \tag{3.32}$$

following Eq. (3.5). Integration of the second term of (3.31) based on Eq. (3.9) with a quantity $Q=1$ results in

$$\int\mathbf{w}\cdot\nabla f dw=\nabla\cdot(\overline{n\mathbf{w}})=\nabla\cdot\overline{n\mathbf{v}} \tag{3.33}$$

where the velocity of the particles was split in the drift velocity $\mathbf{v}$ and the random (thermal) velocity $\mathbf{u}$

$$\mathbf{w}=\mathbf{v}+\mathbf{u} \tag{3.34}$$

with

$$\bar{\mathbf{u}}=0 \tag{3.35}$$

Integration of the third term of (3.16) using Eq. (3.10) with $Q=1$ arrives at

$$\int \frac{\mathbf{F}}{m} \cdot \nabla_w f \, d^3w = -\int f \nabla_w \cdot \mathbf{F} \, d^3w = -n\overline{\nabla_w \cdot \mathbf{F}} = 0$$

which is zero because we assume velocity independent (holonomic) forces $\mathbf{F}$. The right-hand side of Eq. (3.31) results in the integration

$$\int_{-\infty}^{+\infty} \left(\frac{\partial f}{\partial t}\right)_c d^3w = 0 \tag{3.36}$$

because collisions cannot change the total number n of particles per cubic centimeter in average.

Summarizing Eqs. (3.32) to (3.36) results in the velocity integral of the Boltzmann equations

$$\frac{\partial}{\partial t} n + \nabla \cdot n\mathbf{v} = 0 \tag{3.37}$$

which is the hydrodynamic equation of continuity (equation of conservation of mass).

Another integration of the Boltzmann equation over the velocity space after multiplying with the momentum ($m\mathbf{w}$) will lead to the hydrodynamic equation of motion (equation of conservation of momentum). From Eq. (3.31) we arrive then at

$$\int m\mathbf{w} \frac{\partial}{\partial t} f \, d^3w + \int m\mathbf{w}\mathbf{w} \cdot \nabla f \, d^3w + \int m\mathbf{w} \frac{\mathbf{F}}{m} \cdot \nabla_w f \, d^3w = \int m\mathbf{w} \left(\frac{\partial f}{\partial t}\right)_c d^3w \tag{3.38}$$

We use, as before, Eqs. (3.7) to (3.10), where

$$Q(\mathbf{w}) = m\mathbf{w} \tag{3.39}$$

The first term A is from Eq. (3.7)

$$A = \int m\mathbf{w} \frac{\partial}{\partial t} f \, d^3w = \frac{\partial}{\partial t}(\overline{nm\mathbf{w}}) \tag{3.40}$$

where the separation into drift velocity $\mathbf{v}$ and random velocity $\mathbf{u}$, Eq. (3.34) leads to

$$A = mn \frac{\partial}{\partial t}(\overline{\mathbf{v}+\mathbf{u}}) + (\overline{\mathbf{v}+\mathbf{u}}) \frac{\partial}{\partial t} mn \tag{3.41}$$

and using $\bar{\mathbf{u}}=0$ to

$$A = mn \frac{\partial}{\partial t} \mathbf{v} + \mathbf{v} \frac{\partial}{\partial t} mn \tag{3.42}$$

The second term B from Eqs. (3.8), (3.34), and (3.39) arrives at

$$\begin{aligned} B &= \int m\mathbf{w}\mathbf{w}\cdot\nabla f\, d^3w \\ &= \nabla\cdot n\overline{\mathbf{w}m\mathbf{w}} = \nabla\cdot nm\overline{(\mathbf{v}+\mathbf{u})(\mathbf{v}+\mathbf{u})} \\ &= \nabla\cdot nm[\overline{\mathbf{v}\mathbf{v}}+\overline{\mathbf{u}\mathbf{u}}+\overline{\mathbf{v}\mathbf{u}}+\overline{\mathbf{u}\mathbf{v}}] \end{aligned} \tag{3.43}$$

where the last two dyadic terms are zero ($\bar{\mathbf{u}}=0$).

Furthermore,

$$B=\overline{\mathbf{v}\mathbf{v}}\cdot\nabla mn + mn\overline{\mathbf{v}\cdot\nabla\mathbf{v}} + nm\overline{\mathbf{v}\nabla\cdot\mathbf{v}} + \nabla\cdot mn\overline{\mathbf{u}\mathbf{u}} \tag{3.44}$$

For the following addition of A and B it is important that the second terms of (3.42) and the first term of (3.44) are zero

$$\mathbf{v}\frac{\partial}{\partial t}mn+\mathbf{v}\mathbf{v}\cdot\nabla mn=\mathbf{v}\frac{d}{dt}mn=0 \tag{3.45}$$

because of the conservation of mass within one volume element at substantial motion (total differentiation for explicit time).

The third term on the right-hand side of Eq. (3.44) of B is

$$mn\mathbf{v}\nabla\cdot\mathbf{v}=-\mathbf{v}\frac{\partial}{\partial t}mn-m\mathbf{v}\mathbf{v}\cdot\nabla n \tag{3.46}$$

using (3.37) which describes adiabatic heating after compensating the first term of B (3.44) against the last term of (3.46). The fourth term in Eq. (3.44) is

$$\nabla\cdot mn\mathbf{u}\mathbf{u}=\nabla\cdot 2mn\tfrac{1}{2}\mathbf{u}^2\mathbf{1}=\nabla nKT \tag{3.47}$$

using the unity tensor

$$\mathbf{1}=\mathbf{i}_1\mathbf{i}_1+\mathbf{i}_2\mathbf{i}_2+\mathbf{i}_3\mathbf{i}_3 \tag{3.48}$$

and interpreting the energy of random motion $nm\mathbf{u}^2/2$ as inner energy.

The third term C in Eq. (3.38) is derived using Eqs. (3.9), (3.10), and (3.39):

$$C=\int m\mathbf{w}\frac{\mathbf{F}}{m}\cdot\nabla_w f\, d^3w=-n\nabla_w\cdot\mathbf{F}\mathbf{w}=-n\mathbf{F}\cdot\nabla_w\mathbf{w}-n\mathbf{w}\nabla_w\cdot\mathbf{F} \tag{3.49}$$

The second term in the last expression vanishes because the forces $\mathbf{F}$ should depend on the velocity $\mathbf{w}$ as given by Eq. (3.14a), where $\nabla_w\times\mathbf{w}=0$,

$$\nabla_w\cdot\mathbf{F}=0 \tag{3.50}$$

The tensor of the first term in the last expression of Eq. (3.49) is

$$\begin{aligned} \nabla_w\mathbf{w} &= \mathbf{i}_1\mathbf{i}_1\frac{\partial}{\partial w_1}w_1+\mathbf{i}_1\mathbf{i}_2\frac{\partial}{\partial w_1}w_2+\mathbf{i}_1\mathbf{i}_3\frac{\partial}{\partial w_1}w_3 \\ &+\mathbf{i}_2\mathbf{i}_1\frac{\partial}{\partial w_2}w_1+\mathbf{i}_2\mathbf{i}_2\frac{\partial}{\partial w_2}w_2+\mathbf{i}_2\mathbf{i}_3\frac{\partial}{\partial w_2}w_3 \\ &+\mathbf{i}_3\mathbf{i}_1\frac{\partial}{\partial w_3}w_1+\mathbf{i}_3\mathbf{i}_2\frac{\partial}{\partial w_3}w_2+\mathbf{i}_3\mathbf{i}_3\frac{\partial}{\partial w_3}w_3 \end{aligned}$$

where all nondiagonal terms vanish for Cartesian coordinates; therefore, using Eq. (3.48)

$$\nabla_w \mathbf{w} = \mathbf{1}$$

and from (3.49) we arrive at

$$C = -n\mathbf{F} \cdot \mathbf{1} = -n\mathbf{F} \tag{3.51}$$

the forces in plasmas should be

$$\mathbf{F} = Ze\mathbf{E} + \frac{Ze}{c} \mathbf{v} \times \mathbf{H} + \mathbf{F}_g \tag{3.52}$$

where Z is the number of charges for the equation of ions, while $Z=1$ for electrons, and $\mathbf{F}_g$ are forces of gravitation, Coriolis forces, and others.

The last term D in Eq. (3.38)

$$D = \int m\mathbf{w} \left(\frac{\partial f}{\partial t}\right)_c d^3w \tag{3.53}$$

expresses the net momentum per volume transferred to the ions by collisions with the electrons

$$D = \mathbf{P}_{ie} \tag{3.54}$$

If there are no asymmetric velocity distributions of the electrons,

$$\mathbf{P}_{ie} = 0 \tag{3.55}$$

This is not the case, for example, if an electron beam is fired into the plasma.

Putting together the results of A, B, C, and D, we arrive at

$$mn \frac{\partial}{\partial t} \mathbf{v} + mn\mathbf{v} \cdot \nabla \mathbf{v} = \nabla (nKT) + nmZe \left[\mathbf{E} + \frac{1}{c} \mathbf{v} \times \mathbf{H} \right] + mn\mathbf{F}_g \tag{3.56}$$

where the second term of A, (3.42), and the first term of B, (3.44), canceled because of (3.45) and the first terms and second term of B, (3.44). The macroscopic (hydrodynamic) equation of motion is then

$$mn \frac{\partial}{\partial t} \mathbf{v} + mn\mathbf{v} \cdot \nabla \mathbf{v} = -\nabla nKT + mnZe \left[\mathbf{E} + \frac{1}{c} \mathbf{v} \times \mathbf{H} \right] + mn\mathbf{E}_g \tag{3.57}$$

or

$$mn \frac{d}{dt} \mathbf{v} = \mathbf{f} \tag{3.58}$$

where $\mathbf{f}$ is the force density.

From Eq. (3.56) we derive immediately the Bernoulli equation for a vortex-free stationary ($\partial/\partial t = 0$) motion without external forces $\mathbf{F}_g$ and fields $\mathbf{E}$ and $\mathbf{H}$:

$$\nabla \left(nKT + \frac{mn}{2} \mathbf{v}^2 \right) = 0 \tag{3.59}$$

or using the (static) pressure $p=nKT$ and the density $\rho=nm$, integration of Eq. (3.59) results in Bernoulli's equation

$$p+\frac{\rho}{2}\mathbf{v}^2=\text{const} \tag{3.60}$$

3.4 Landau Damping

Landau [116] studied the behavior of a distribution function of electrons in a plasma when being changed slightly from the thermal equilibrium. Without any dissipation or energy transfer, a damped oscillation will occur for returning into the undisturbed state. This can be seen in the following way. The plasma is assumed to be field free ($\mathbf{E}_0=\mathbf{H}_0=0$) apart from small fields due to the perturbation. The equilibrium thermal electron distribution should be a Maxwellian $f_0(\mathbf{v})$, Eq. (3.20), and the perturbation should be described by the linear approximation f,

$$f(\mathbf{r},\mathbf{v},t)=f_0(\mathbf{w})+f_1(\mathbf{r},\mathbf{w},t) \tag{3.61}$$

No collisions should be present; therefore, Eq. (3.61) will result in a first-order Vlasov equation from (3.14)

$$\frac{\partial f_1}{\partial t}+\mathbf{v}\cdot\nabla f_1-\frac{e}{m}\mathbf{E}_1\cdot\nabla_v f_0=0 \tag{3.62}$$

where the electric field $\mathbf{E}_1$ is due to the perturbation, and f_1 has been neglected compared with f_0 in the last term of (3.62). If f_1 and $\mathbf{E}_1$ are described as plane waves in the x direction

$$f_1=f_{10}\exp i(kx-\omega t);\qquad E_x=E_{x0}\exp i(kx-\omega t) \tag{3.63}$$

using a wave number $k=\omega/c$ and a radian frequency ω where c is the velocity of the wave. Eq. (3.62) with Eq. (3.63) becomes

$$-i\omega f_1+ikv_x f_1=\frac{e}{m}E_x\frac{\partial f_0}{\partial v_x} \tag{3.64}$$

or

$$f_1=\frac{ieE_x}{m}\frac{\partial f_0/\partial x}{\omega-kv_x} \tag{3.64a}$$

Using Poisson's equation

$$\nabla\cdot\mathbf{E}=ikE_x=-4\pi en_1=-4\pi e\int f_1\,d^3v \tag{3.65}$$

with the disturbation n_1 of the density. Substituting (3.64) and E_x from

(3.65) in the integral in (3.65) results in

$$1=-\frac{4\pi e^2}{km}\int_{-\infty}^{+\infty}\frac{\partial f_0/\partial v_x}{\omega-kv_x}\,d^3v \tag{3.66}$$

The integral has to be finite, but as any real ω would cause a divergence, the only possibility is to assume a complex ω whose imaginary part will contribute to a damping of an oscillation.

The evaluation of the integral (3.66) with Cauchy's residue theorem is not very easy, as the integral along the half-circle for infinite $|v|$ of complex values v_x is not zero. Only for a small imaginary part of ω, can the following approximation be achieved from (3.66):

$$1=\frac{\omega_p^2}{\omega^2}+i\pi\frac{\omega_p^2}{k^2}\frac{\partial f_0}{\partial v}\bigg|_{v=v\phi} \tag{3.67}$$

where v_ϕ is given from $\mathrm{Re}(\omega)-k\cdot\mathrm{Re}(v_x)=0$ and ω_p is the plasma frequency. The resulting frequency is then

$$\omega=\omega_p\left[1+i\frac{\pi}{2}\frac{\omega_p^2}{k^2}\left(\frac{\partial f_0}{\partial v}\right)_{v=v\phi}\right] \tag{3.68}$$

The result shows that a disturbance of the distribution of the electrons results in an oscillation with a frequency ω_p (as we have seen in Section 2.1 in a very direct way), and, if the perturbation has been generated once, its oscillation will be damped by the plasma without an exchange of energy. A plasma therefore has a self-stabilizing property

If collisions are included, the Landau damping may modify the usual absorption process. These effects are strong only if the wavelength of the electromagnetic wave $\lambda=\omega/2\pi c$ is close to the Debye length λ_D of the plasma. Thinking of laser wavelengths near the visible range and plasmas of solid-state density and above, even for extremely high temperatures of 10^5 eV, the Debye length is 100 times less than λ.

Considering this section about the kinetic theory on which the macroscopic theory can be based, we can summarize as follows. The most general (microscopic) description of a plasma as an N-particle system is limited by computing capacity and/or restricting simplifications on the range of forces and so on. The kinetic theory using distribution functions results in an irreversible loss of information. Even with the general Liouville distribution, the Liouville equation (3.28) contains basic insufficiencies as has been shown by Blatt in 1959 [111]. Taking these reservations into account, the derivation of macroscopic (hydrodynamic) equations (Section 3.3) is convincing but limited. The Landau damping was an example of how deviations from an average value are automatically damped down to the equilibrium state, showing how the equilibrium is stable. There might be difficulties at strong deviations. Meixner [117] finds from the Chapman-Enskog solution [118] of Boltzmann's equation (3.31) with a restriction of the local equilibrium of a system that temperature variations ΔT must be less than the average temperature in a cell of the mean free path l_c of the particles.

There was a discussion on whether the kinetic theory limits a derivation of the macroscopic hydrodynamics if conditions of irreversible thermodynamics appear. This could lead to criticism of hydrodynamic computer results for very short times of laser interaction with plasma. However, it must be remembered that highly irreversible processes can be described by hydrodynamic codes in full agreement with experiments. A splendid example is the compression of thermonuclear plasma in a theta-pinch. The irrelevance of the mentioned criticism is perhaps due to the well-known fact that the limit for reversible processes can be very much extended into the irreversible range. The role of fluctuations on entropy production and the formulation of the optical absorption by Onsager coefficients [119] should be mentioned in this connection.

The problems of Landau damping is by far not exhausted theoretically. A more recent paper by Alexeff et al (1990) presented an alternative new derivation of Landau damping where the error by a factor two in Landau's initial derivation was elaborated in a very transparent way.

3.5 Concluding Remarks on Microscopic Theory

Since the following discussion of laser interaction with plasma will be presented nearly exclusively by macroscopic, hydrodynamic theory, some more reference should be given to the alternative work on microscopic theory. In the preceding Chapters 2 and 3, only some first steps of microscopic theory have been shown.

The advent of more powerful computers permitted an extension of the direct calculation of the N-particle model of a plasma as mentioned with equations (2.1) where the numerical follow-up of each particle under the influence of its surrounding or collectively interacting plasma particles is treated very simply by computation. The initial work by Kruer [89] or Biskamp and Welter [90] was extended now to one million particles. An introduction into plasma physics with strong attention to this treatment was presented by Kruer (1988). By technical reasons, the mechanism of collisions, especially of Coulomb collisions was neglected in most cases. Nevertheless, numerous collective effects were very transparently reproduced, as the generation of higher harmonics if a plasma is irradiated by a laser of basic frequency ω. The generation of electron waves (Langmuir oscillations and waves) and of ion-acoustic waves could be seen. At the critical density [$\omega = \omega_p$, see Eq. (2.6)], the strong resonances were seen and the generation of electron oscillations with very high energy and of translative electrons motion with high energy have been reproduced immediately showing one of the possible phenomena how the observed energetic electrons can be generated.

The N-particle simulation is the ideal tool to study the interpenetration of high temperature electrons or ions from one area when touching another area of cold plasma. Contrary to the very primitive hydrodynamic picture of a shock front and all the otherwise very successful models of shock wave theory, the N-particle simulation does cover the interpenetration of the plasma particles at the interface between hot and cold plasma. The hydrodynamics confines the particles to idealized partial layers which do not interpenetrate.

N-particle simulation also covers effects of plasma inhomogeneities automatically and it is therefore most predestined for studying the extremely inhomogeneous plasmas at laser interaction. Just these inhomogeneities imply many phenomena which were not covered before by the established plasma theory based on Langmuir's definition that plasmas are space-charge quasi neutral [apart from ranges less than the Debye length, see Eq. (2.11)]. This fundamentalistic view is correct only for homogeneous plasmas and it prevented the realization of many phenomena, e.g. of internal electric fields inside of inhomogeneous plasmas and their rich dynamic behaviour. Hannes Alfven (1981; 1989) had good reasons to consider these electric fields based on the knowledge of earlier plasma physics, especially from results of Birkeland (Peratt, 1988, 1989; Fälthammer 1988). But it is not surprising that fundamentalists denounced that "Alfven's electric fields are intuitively not clear" (Kulsrud, 1983).

The following extensive discussion of the hydrodynamic (macroscopic) foundation of plasma physics arrived at a clear confirmation of these dynamic internal electric fields in inhomogeneous plasmas as a result of numerical studies and subsequent analytical derivations to explain several unexpected observations in laser produced plasmas in a straight forward way. These phenomena are second harmonics emission from a low density plasma corona, surface tension of plasmas or of the interior of inhomogeneous plasmas and the action of this surface tension against bending of interfaces (Rayleigh-Taylor instabilities).

These phenomena automatically will appear in N-particle simulations. It can be concluded that the output of these computations of 1980 do have all these phenomena of internal electric fields, their dynamics, and stabilization. However, nobody looked for them or arranged a printout of these values in long time existing computer results. Only after the results of the hydrodynamic theory-described in the following sections - were gained against the prejudice of the before mentioned fundamentalists, a richer evaluation of the N-particle simulations can be predicted.

The just mentioned phenomena (not yet all exhausted from the simulations) were mostly working without the collisions of the electrons with the ions. For other phenomena the neglection of collisions in the simulations is highly dangerous. We shall learn about cases of the optical response of plasmas, especially near the poles of the relevant functions how the omission of collisions results in minus infinity while adding a very minor part of collisions changes the values to nearly plus infinity.

The N-particle simulation without collisions is indeed of interest and may even reveal more interesting facts as mentioned when internal fields will be evaluated, but very insufficient results may be gained in other cases if no way will be found, how the complete Coulomb collisions with their far ranging forces and with their quantum modification at higher energies (see end of the preceding chapter) are included.

Collisions, however involve enormous difficulties for N-particle simulations. Though several rough inclusions were tried in the earlier stages, even with the latest generation of supercomputers, only an improvement of the approximations, no final slution could be reached. And these attempts needed enormous computer times (Nishihara, 1988; Yabe 1988). This handicap with the collisions is just the reason why the otherwie ideal treatment of plasma interpenetration in fusion detonation waves could not succeed in the past since the energy transfer and equipartition between the various particles, given by collisions, is just the essential process.

After discussing the difficulties of the N-particle simulation, the other very wide range of microscopic plasma theory is based on using the kinetic equations, the Boltzmann (3.16) or the collisionless Boltzmann (Vlasov) (3.30) equation as given by treating the N-particle model via distribution functions (statistical mechanics). In the preceding subsections of this Chapter 3, only a very initial use of this kinetic equation was demonstrated when deriving the macroscopic hydrodynamic equations by integration of the Boltzmann equation after multiplying with momenta.

The traditional study of plasmas with this kinetic theory is indeed very rich and caused reactions as reflected in Alfven's statement (see citation on pages 2 and 3 [7]). It is remarkable what all can be derived from minor deviations of the distribution function, Eq. 3.4, from its equilibrium value given by the Maxwell-Boltzmann distribution (3.22) and (B.30) (Nicholson

1983). Similarly to the derivation of the Landau damping, see Subsection 3.4, all kinds of plasma waves, general dispersion laws, the dielectric response (optical constants) and many other properties of microscopic instabilities can be derived. Especially the study of deviations of electron distributions from a temperature determined equilibrium distribution, e.g. of many maxima (expressing multiple temperatures or even much more general situations) can be studied preferably by this kinetic theory. Electrons can be changed into two stream conditions at laser irradiation causing then instabilities before equilibrating.

While the impressive textbook treatment of this kinetic theory is restricted to the most significant and elegant results which astonishingly can be gained despite of drastic simplifications, neglections and linearizations, one has to take into account how much richer the theory is if less approximative assumptions are used. These more difficult and more general derivations are mostly unexplored at the moment and one can be sure that enormous new knowledge may be derived in the future. However, the complexity of the treatment will require to decide what properties still have to be neglected and what have to be considered as being important. As we shall show in this book only from examples of macroscopic theory, there is the fact how easy it is to arrive at totally wrong results in nonlinear physics if neglections and approximations are used. This warning explains why the exploration of the much richer kinetic theory in future may be by several orders more difficult than the macroscopic theory.

One big problem in kinetic theory is again, as in the case of the N-particle model, the correct inclusion of collisions. While well the Boltzmann collisional term, Eq. 3.16, expresses the general description of the collisions, its detailed evaluation is a problem and some well known approximations have been mentioned before. If one does not permit linearizations, simplifications, and neglections as usual, the general treatment of the collisional term arrives at more than one thousand terms and their algebraic treatment can be done only by using advanced methods of cybernetics (Drska, 1987).

Needless to say that laser interaction with plasmas requires the careful inclusion of collisions in most cases if not significant changes of results will appear, and that there is one of the weak points of kinetic theory. On the other hand, if the following emphasis on macroscopic theory is given, one should not miss the fact that this is based on the very severe simplification that the electrons and ions are assumed to have an equilibirum distribution given by temperatures each. One advantage of macroscopic theory is that the results can be presented in a quite transparent way and can even be treated numerically in real time such that reconsideration with complicated scaling factors and reductions of dimensionalities can be avoided.

FOUR

Hydrodynamics

Macroscopic plasma theory is a combination of electromagnetic theory with the hydrodynamic properties of plasma. There are various hydrodynamic descriptions of plasma in literature, yet starting in the 1940s with the one-fluid model of Alvén, very general macroscopic equations for fully ionized plasma are the two-fluid equations derived by Schlüter, which will be discussed in Section 6. To make the reader more familiar with the background of hydrodynamics, this section will consider some significant properties.

4.1 Euler's Equation of Motion

The hydrodynamic equation of motion is the field-theoretical generalization of Newton's single-particle equation of motion:

$$\bar{m}\mathbf{a} = \mathbf{F} = -\nabla\phi \tag{4.1}$$

The product of mass $\bar{m}$ times the acceleration $\mathbf{a}$ of a body equals the force $\mathbf{F}$, which can be expressed by the gradient of a potential ϕ. In the case of a fluid, there is a velocity field $\mathbf{v}(x, y, z, t)$, of which the temporal derivative corresponds to the acceleration; a mass density field $\rho(x, y, z, t)$, which corresponds to the mass m; and a force density, which is given by the gradient of the pressure field $p(x, y, z, t)$. Consider the fluid as being composed of electrons of mass m, density n_e, temperature T_e, and ions of mass m_i, density n_i, and charge Z. Assuming space charge neutrality $n_i = n_e/Z$, the mass density field is given by

$$\rho(x, y, z, t) = m_i n_i(x, y, z, t) + m n_e(x, y, z, t) \tag{4.2}$$

The pressure field (also as function of x, y, z, t) is

$$p = n_e K T_e + n_i K T_i \approx (1+Z) n_i K T_e, \qquad \text{if } T_e \approx T_i \tag{4.3}$$

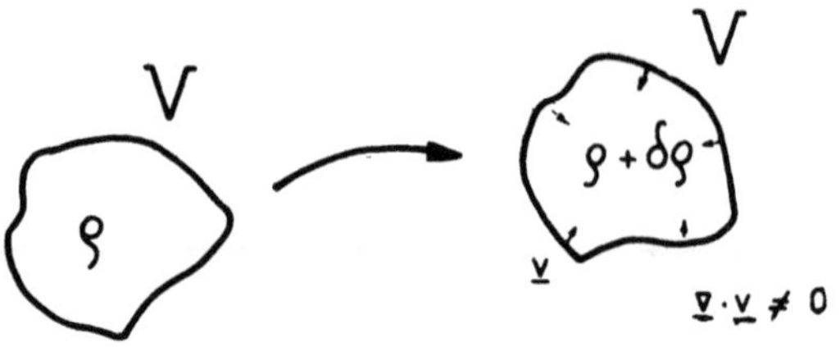

Figure 4.1. Geometrical variation with conservation of mass for the derivation of the equation of continuity.

The equation of motion corresponding to Newton's Eq. (4.1), the Euler equation, is

$$\rho \frac{d\mathbf{v}}{dt} = -\nabla p + \eta \nabla^2 \mathbf{v} \tag{4.4}$$

The last term on the right-hand side is added to the original Euler equation and is called the Navier–Stokes term. This includes the hydrodynamic viscosity η, which determines the internal friction of the fluid. The operation on the left-hand side of Euler's equation (4.4) is

$$\rho \frac{d\mathbf{v}}{dt} = \rho \frac{\partial \mathbf{v}}{\partial t} + \rho \frac{\partial \mathbf{v}}{\partial x}\frac{dx}{dt} + \rho \frac{\partial \mathbf{v}}{\partial y}\frac{dy}{dt} + \rho \frac{\partial \mathbf{v}}{\partial z}\frac{dz}{dt}. \tag{4.5}$$

We combine the last three terms to $\rho \mathbf{v}\cdot\nabla\mathbf{v}$, which is a nonlinear term. Remember that the spatial "del" operator is $\nabla = \mathbf{i}_x \partial/\partial x + \mathbf{i}_y \partial/\partial y + \mathbf{i}_z \partial/\partial z$, where the unit vectors $\mathbf{i}_x$, $\mathbf{i}_y$, and $\mathbf{i}_z$ are of the modulus 1 and are in the direction of the Cartesian coordinates x, y, and z, respectively. Using this way of writing, the Euler equation is then

$$\rho \frac{\partial \mathbf{v}}{\partial t} + \rho \mathbf{v}\cdot\nabla\mathbf{v} = -\nabla p \tag{4.6}$$

where the viscosity has been dropped.

4.2 Bernoulli's Stationary Solution

From Euler's equation (4.6), Bernoulli's equation can be derived as a stationary solution. Because of the then necessary time independence:

$$\frac{\partial}{\partial t} = 0 \tag{4.7}$$

the Euler equation (4.6) reduces to

$$\nabla \mathbf{v} + \nabla p = 0 \tag{4.8}$$

By using the vector identity:

$$\mathbf{v}\cdot\nabla\mathbf{v}=\tfrac{1}{2}\nabla\mathbf{v}^2-\mathbf{v}\times(\nabla\times\mathbf{v}) \tag{4.9}$$

The last term can be dropped for vortex-free motion, giving:

$$\frac{\rho}{2}\nabla\mathbf{v}^2+\nabla p=0 \tag{4.10}$$

and under the further special assumption that the spatial variation of the density ρ is zero,

$$\rho=\text{const} \tag{4.11}$$

Eq. (4.10) can be written as:

$$\nabla\left(\frac{\rho}{2}\mathbf{v}^2+p\right)=0 \tag{4.12}$$

which, when integrated, yields the Bernoulli equation

$$\frac{\rho}{2}\mathbf{v}^2+p=\text{const} \tag{4.13}$$

Note that the Bernoulli equation is a very particular case of the general Euler hydrodynamic equation, not only because of the *stationary* condition, Eq. (4.7), but also because of the neglect of the last term in Eq. (4.9) (*vortex-free motion*) and by condition (4.11), which corresponds to a chemical homogeneity and *incompressibility* of the fluid. Instead of this general derivation, which is appropriate for a theoretical method of deduction, Bernoulli's equation (4.13) can be derived primitively by compensating the kinetic pressure or the first term in Eq. (4.13), by the hydrostatic pressure, or second term on the left-hand side of Eq. (4.13).

4.3 Equation of Continuity

The Euler equation is one of the three basic equations of hydrodynamics and corresponds to the conservation of momentum. The next basic equation is the conservation of mass, which is sometimes called the equation of continuity. It can be derived from the following geometrical consideration. Take a constant volume V with a density ρ, which is moved during a time interval dt along the large arrow of Fig. 4.1. After the time dt the volume V has changed its density to the value $\rho+\delta\rho$. The increase of mass in the volume V can only be due to material streaming into the volume V; therefore it has a negative divergence of the velocity field (expressed by $-\nabla\cdot\mathbf{v}$)

$$\frac{d\rho}{dt}=-\rho\nabla\cdot\mathbf{v} \tag{4.14}$$

Gauss' law, expressing the converging or diverging velocity field at a closed surface by the divergence $(\nabla \cdot \mathbf{v})$ of a volume integral within a closed area, gives

$$\oiint \mathbf{v}\, d^2\mathbf{a} = \iiint \nabla \cdot \mathbf{v}\, d^3\tau$$

The left-hand side of equation (4.14) can be expressed by the partial differentiation:

$$\frac{\partial \rho}{\partial t} + \mathbf{v} \cdot \nabla \rho = -\rho \nabla \cdot \mathbf{v} \tag{4.15}$$

and using the differentiation relation:

$$\nabla \cdot \rho \mathbf{v} = \rho \nabla \cdot \mathbf{v} + \mathbf{v} \cdot \nabla \rho \tag{4.16}$$

Eq. (4.15) can be rewritten as:

$$\frac{\partial \rho}{\partial t} + \nabla \cdot (\rho \mathbf{v}) = 0 \tag{4.17}$$

which is the equation of continuity [see Eq. (3.37)]. For the special case of incompressible fluids, Eq. (4.11), the special formulation of the equation of continuity is obtained:

$$\frac{\partial \rho}{\partial t} + \rho \nabla \cdot \mathbf{v} = 0 \tag{4.18}$$

which is related to a fast change of the density within a very large volume. It will be seen in the following that hydrodynamic waves are specially related to this process. It should be noted that a large number of problems in hydrodynamics and aerodynamics can be studied on the assumption of an incompressible flow.

4.4 Compressibility

For the following derivation of acoustic waves (in the next subsection) compressibility will be now discussed. By definition, compressibility is the relation of a change δV in the volume V of a medium which is generated by the change ∂p of the pressure p. With an initial volume V_0,

$$V(p) = V_0 + \delta V \tag{4.19}$$

the change expressed by a variation δp of the pressure p is

$$V(p) = V_0 - \frac{\partial V}{\partial p} \delta p$$

or the relative change of the volume V is

$$\frac{\delta V}{V} = -\frac{1}{V}\frac{\partial V}{\partial p}\delta p \tag{4.20}$$

The definition of compressibility is the proportionality factor between the relative change of the volume against a variation δp of the pressure:

$$\kappa = -\frac{1}{V}\frac{\partial V}{\partial p} \tag{4.21}$$

This is nothing more than a definition for which a reasonable explanation was given by the preceeding equations. The connection must now be found with the thermodynamic quantity of the adiabatic compression. Referring to thermodynamics, the connection between pressure p and volume V in an adiabatic compression (i.e., for a change without exchange of heat or energy with any other medium outside) is

$$pV^{\gamma} = \text{const} \tag{4.22}$$

where the exponent $\gamma = c_p/c_v$ is the ratio of the specific heat c_p for constant pressure to that for constant volume c_v. Thermodynamics derives this ratio from the number F of the degrees of freedom of particles in the medium:

$$\gamma = \frac{c_p}{c_v} = \frac{F+2}{F} \tag{4.23}$$

For the case of a fully ionized plasma there are $F=3$ degrees of freedom, as in the case of rare gases, where the particles are single atoms.

From Eq. (4.22) is found:

$$\frac{\partial V}{\partial p} = -\frac{1}{\gamma}\frac{\text{const}^{1/\gamma}}{p^{(1/\gamma)+1}} \tag{4.24}$$

and from (4.21) by using Eqs. (4.22) to (4.24), finally:

$$\kappa = -\frac{1}{V}\frac{\partial V}{\partial p} = \frac{1}{\gamma p} \tag{4.25}$$

the compressibility for an adiabatic change of state, expressed in term of the pressure.

4.5 Acoustic Waves

The description of acoustic waves uses the Euler equation (4.4) in its linearized form:

$$\frac{\partial \mathbf{v}}{\partial t} = -\frac{1}{\rho}\nabla p \tag{4.26}$$

and the equation of continuity (4.17) with the assumption, that $|\mathbf{v}\cdot\nabla\rho| \ll |\nabla\cdot\mathbf{v}|$ (quasi incompressible)

$$\frac{\partial\rho}{\partial t} = -\rho\nabla\cdot\mathbf{v} \tag{4.27}$$

where the relation of the density ρ to particle densities is given by Eq. (4.2). The variation of the density ρ with the pressure p

$$\rho = \rho_o + \frac{\partial\rho}{\partial p}\delta p \tag{4.28}$$

can be expressed by definition through the compressibility κ from Eq. (4.21)

$$\rho = \rho_0[1+\kappa(p-p_0)] \tag{4.29}$$

The variations $\delta p = p - p_o$ and $\delta\rho = \rho - \rho_o$ can be expressed by their initial values (index o) and their instantaneous values (without index), so that:

$$\delta p = \frac{1}{\rho_0\kappa}\delta\rho \tag{4.30}$$

or by differential description:

$$\nabla p = \frac{1}{\rho_0\kappa}\nabla\rho \tag{4.31}$$

Substituting this into the Euler equation (4.26), it is found:

$$\frac{\partial\mathbf{v}}{\partial t} = -\frac{1}{\rho}\frac{1}{\rho_0\kappa}\nabla\rho = -\frac{1}{\rho_0\kappa}\nabla\ln\rho \tag{4.32}$$

From the equation of continuity, Eq. (4.27), is found the approximation:

$$\frac{1}{\rho}\frac{\partial\rho}{\partial t} = \frac{\partial}{\partial t}\ln\rho = -\nabla\cdot\mathbf{v} \tag{4.33}$$

Differentiation of Eq. (4.32) by t and of Eq. (4.33) by ∇ leads to:

$$\frac{\partial^2\mathbf{v}}{\partial t^2} = -\frac{1}{\rho_0\kappa}\nabla\frac{\partial}{\partial t}\ln\rho = +\frac{1}{\rho_0\kappa}\nabla^2\mathbf{v} \tag{4.34}$$

or

$$\nabla^2\mathbf{v} - \rho_o\kappa\frac{\partial^2\mathbf{v}}{\partial t^2} = 0 \tag{4.35}$$

The solutions of the wave equation (4.35) are, for example, plane waves of a radiation frequency ω

$$\mathbf{v} = \mathbf{v}_o\exp(\pm i\mathbf{k}\cdot\mathbf{r} - i\omega t)$$

where the wave vector k determines the direction of the wave propagation and $|\mathbf{k}| = \omega/c_s$ gives a result for the phase velocity c_s:

$$c_s^2 = \frac{1}{\kappa\rho_0} \tag{4.36}$$

This velocity is the velocity of sound or the ion acoustic velocity. Substituting the compressibility from Eq. (4.25) leads to

$$c_s = \sqrt{\frac{\gamma p}{\rho_o}} \tag{4.37}$$

The same sound equation could have been obtained if the expression of ln ρ had been eliminated from the equation of motion and the equation of continuity to give:

$$\nabla^2 \ln \rho - \kappa\rho_o \frac{\partial^2 \ln \rho}{\partial t^2} = 0 \tag{4.38}$$

where again the same wave equation is reproduced as for the velocity with the same wave velocity c_s.

4.6 Equation of Energy

In addition to the conservation equations of momentum (Euler's equation of motion) and of mass (equation of continuity), the equation of energy conservation is needed to arrive at the complete set of differential equations for uniquely solving hydrodynamic problems. This equation is of the type:

$$\frac{\partial}{\partial t}\frac{\rho}{2}\mathbf{v}^2 = -\frac{\partial}{\partial t} n_i KT(1+Z) - \nabla\cdot(\kappa_T \nabla T) + W \tag{4.39}$$

where the left-hand side describes the temporal change of the kinetic energy of the fluid to be compensated by the change of internal energy (first term on the right-hand side), by thermal conduction, characterized by the thermal conductivity κ_T, and by any power density W of energy exchange by radiation and so on.

The expression for the power density W for energy transfer to the plasma will include the linear or nonlinear absorption constant. This is derived from the optical linear, nonlinear, or relativistic refractive index or from an effective collision frequency due to parametric instabilities or by an effective dynamic nonlinear absorption process. These steps will be discussed later in the contents of specific applications.

Any additional potentials are not to be included in Eq. (4.39), if these are independent of the time. This related to gravitational potentials or such of other static fields. In the case of laser produced plasmas, electrodynamic net potentials can change in time. These components will then have to be included in Eq. (4.39).

FIVE

Self-Similarity Model

The hydrodynamic equations of the preceding section can be used to analyze the gasdynamic expansion of a laser produced plasma of spherical shape into a vacuum. The transfer of the laser energy to the plasma is assumed to be fully symmetric, given by the spatially constant power density $W(t)$ in the energy equation (4.39). All specific conditions as to how this fast power transfer and equilibration can be realized are neglected and will be the topic of subsequent sections. Agreement with experiments at not too high laser intensities and not too short pulses for a wide range of parameters will justify the assumptions.

The expansion of a spherical plasma into vacuum can be described by a relatively simple model where the radius R is found as a function of time. The plasma temperature T has, at a time t_0 an initial value T_0, the radius R of the plasma an initial value R_0 and the velocity of expansion $\partial R/\partial t$ an initial value $\dot{R}_0$. During the expansion, an adiabatic transfer of thermal energy into kinetic energy of expansion will occur. The complete radial symmetric hydrodynamic calculation, for example, by Fader [120], with any initial radial velocity profile $v(r, t=0)$ and an ion density profile $n_i(r, t=0)$, resulted after some time t into a solution where n_i became a Gaussian density profile, while the velocity became a linear profile

$$v(r, t)=\frac{v_0(t)r}{R} \tag{5.1}$$

and the temperature dropped adiabatically. The fact that the density and velocity profiles remained similar is the reason for using 'self-similar' expansion. This has nothing to do with the similarity laws of hydrodynamics, for example, the Reynold's number, which are relations of dimensions and quantities of a fluid characteristic for the hydrodynamic motion.

Historically, the self-similarity model was used in the expansion of the universe [121] where (5.1) was derived [122]. The relation to the case of laser produced plasmas was underlined by Lengyel and Salvat [123]. This model was used for thermonuclear plasmas, for example, by Zeldovich and Raizer [124], especially for the calculation of laser plasmas by Basov and Krokhin [125], Dawson [112], and for the optimization of nuclear fusion gains [126].

Instead of the very global derivation of the self-similarity model [112, 125], here a general derivation from the hydrodynamic equations [127] is followed to show the limitations and restrictions of the model and to discuss a classical error because the global consideration was not based on hydrodynamics. Finally, some applications to laser produced plasmas will be considered.

5.1 Hydrodynamic Derivation

We use the definitions (4.2) and (4.3) and hydrodynamic equations of conservation (4.6), (4.17), and (4.39) in the following way. The equation of continuity;

$$\frac{\partial n_i}{\partial t}+\nabla\cdot(n_i\mathbf{v})=0 \tag{5.2}$$

the equation of motion:

$$\frac{d}{dt}n_i\mathbf{v}\left(1+Z\frac{m_e}{m_i}\right)m_i=-\nabla p \tag{5.3}$$

and the equation of energy conservation:

$$\frac{\partial}{\partial t}\frac{n_i m_i}{2}\left(1+Z\frac{m_e}{m_i}\right)\mathbf{v}^2=-\frac{\partial}{\partial t}(1+Z)n_i KT+W \tag{5.4}$$

where the terms Zm/m_i are very small compared with unity. Equations (5.2), (5.3), and (5.4) are the general hydrodynamic equations to solve n_i, v, and T in space and time for given initial conditions.

In the case of a spherical plasma with radial symmetry resulting in only a spatial dependence on the radial coordinate r with the radial velocity component v_r, we get the equation of continuity from Eq. (5.2):

$$\frac{\partial}{\partial t}n_i+\frac{\partial}{\partial t}n_i v_r+\frac{2n_i}{r}v_r=0 \tag{5.5}$$

Using the equation of state

$$p=n_i(1+Z)KT \tag{5.6}$$

in Eq. (5.3) one gets the equation of motion for the spherical case:

$$\frac{d}{dt} n_i m_i \left(1+Z\frac{m_e}{m_i}\right) \mathbf{v}_r = -\frac{\partial}{\partial r} n_i (1+Z) \geq T \tag{5.7}$$

and finally the equation of energy conservation from Eq. (5.4):

$$\frac{\partial}{\partial t}\frac{n_i m_i}{2}\left(1+Z\frac{m_e}{m_i}\right) v_r^2 = -\frac{\partial}{\partial r}(1+Z) n_i KT - \kappa \frac{1}{r^2}\frac{\partial}{\partial r}\left(r^2 \frac{\partial}{\partial t} T\right) + W \tag{5.8}$$

Then, the three basic equations (5.5), (5.7), and (5.8) have to be solved together with the initi al conditon at a time $t=t_0$

$$T(r, t_0); \qquad n_i(r, t_0); \qquad v_r(r, t_0) \tag{5.9}$$

to find solutions for the three equations T, n_i, and v_r as a function of r and t.

Now the formulas of the self-similarity model, which were used by Dawson [112], are derived starting from the general radially symmetric hydrodynamic equations. The equation of motion (5.7) is multiplied with v_r

$$\frac{1}{2}\frac{d}{dt} n_i m_i \left(1+Z\frac{m_e}{m_i}\right) \mathbf{v}_r^2 = -v_r \frac{\partial}{\partial r} p \tag{5.10}$$

and integrated over the whole volume of the spherical plasma with the radius R of the plasma surface. This leads to

$$-\int_0^R v_r \left(\frac{\partial p}{\partial r}\right) 4\pi r^2 dr = \frac{1}{2}\frac{d}{dt}\int_0^R n_i m_i \left(1+Z\frac{m_e}{m_i}\right) v_r^2 4\pi r^2 dr \tag{5.11}$$

It is essential to point out that the procedure of integration in Eq. (5.11) and the following equations give a loss of information. Instead of details of $n_i(r, t)$, only averaged values of the functions under the integrals can be expected.

If the pressure within the spherical plasma is assumed to be a constant value p from $r=0$ until $r=R-\varepsilon$ and decreases to $p=0$ at $r=R$, then the left-hand side of Eq. (5.11) can be written

$$\begin{aligned} -\lim_{\varepsilon\to 0} v_r 4\pi R^2 \int_{R-\varepsilon}^{R} \frac{\partial p}{\partial r} dr &= -v_r 4\pi R^2 (p(R) - p(R-\varepsilon)) \\ &= 4\pi R^2 p v_r(R) \end{aligned} \tag{5.12}$$

Constant p at constant T gives an averaged constant n_i in the inner of the sphere. Assuming additionally a linear velocity profile,

$$v_r(r) = v_{r0}\frac{r}{R} \qquad (0<r<R) \tag{5.13}$$

which expresses the properties of a similarity expansion. then the right-hand side of Eq. (5.11) leads to

$$\frac{1}{2}\frac{d}{dt}\int_0^R n_i m_i\left(1+Z\frac{m_e}{m_i}\right)v_r^2 4\pi r^2 dr=\frac{1}{2}\bar{M}\frac{d}{dt}v_r(R)^2 \tag{5.14}$$

with the abbreviation of an averaged mass $\bar{M}=\frac{3}{5}n_i m_i[1+Z(m/m_i)]$ with respect to the spherical expansions. Combining Eqs. (5.14) and (5.12) leads to

$$4\pi R^2 p\frac{dR}{dt}=\frac{1}{2}\bar{M}\frac{d}{dt}v_r(R)^2=\frac{1}{2}\bar{M}\frac{\partial}{dt}\left(\frac{dR}{dt}\right)^2 \tag{5.15}$$

A further integration over the volume is of interest when similar assumptions to those for n_i are used as in Eq. (5.15), making n_i constant in the averaged sense within the plasma sphere and using a linear velocity profile Eq. (5.13). In this case, one gets from the left-hand side of Eq. (5.8)

$$\frac{\partial}{\partial t}\int_0^R\frac{n_i m_i}{2}\left(1+Z\frac{m_e}{m_i}\right)v_r^2 4\pi r^2\,dr=\frac{\partial}{\partial t}\frac{n_i m_i}{2}\left(1+Z\frac{m_e}{m_i}\right)\frac{v_0}{R^2}4\pi\int_0^R r^4\,dr \tag{5.16}$$

Using

$$\frac{3}{5}n_i m_i\left(1+Z\frac{m_e}{m_i}\right)v_0=\bar{M}v_0=\frac{2}{3}p \tag{5.17}$$

which by definition led to the pressure p.

One finds for the integral of the left-hand side of Eq. (5.8)

$$\frac{\partial}{\partial t}\frac{2}{3}4\pi p R^3 \tag{5.18}$$

Integrating the right-hand side of Eq. (5.8) over the plasma volume and assuming a constant temperature $[(\partial/\partial t)T=0]$, one obtains an equation of energy conservation with the total power W absorbed in the plasma after subtraction of the radiation loss

$$4\pi p R^2\frac{\partial}{\partial t}R=-\frac{3}{2}K\frac{dT}{dt}\left[\frac{4\pi}{3}R^3(1+Z)n_i\right]+W \tag{5.19}$$

Equations (5.6), (5.15), and (5.19) are the same that Dawson derived from the basis of a phenomenological combination of gas kinetic laws and are the basic equations of the self-similarity model. The equations determine the time dependence of the plasma radius $R(t)$ and the power transfer $W(t)$ to the plasma. The two essential points in the derivation from the hydrodynamic equations are first that the energy transfer from the radiation has to be assumed in such a way that the plasma temperature can be assumed spatially constant at each instant, and second that a linear velocity profile is valid, Eq. (5.13). The properties of an averaged density n_i, an averaged mass M, and a steep pressure decrease at the surface are consequences of the integration procedures; these are mathematically correct but they involve a loss of details of mostly unnecessary information. Therefore, the self-similarity model contains the constant density profile as an averaged value only.

The problem is different from that of the preceeding result when *details* of the density profile $n_i(r)$ at varying times must be evaluated. This leads to a mathematical contradiction. To reach a statement of the actual density profile, Haught and Polk [128] started from a linear (not Gaussian) decrease of the density $n_i(r)$ from the center of the plasma to the surface. Simultaneously, a linear velocity profile, as given by Eq. (5.13) was erroneously assumed to be conserved. It is easy to show that both linear profiles are not conservable in time. Substituting $n_i = n_{i0}(1 - r/R)$ with the normalization $n_{i0} \sim R(t)^{-3}$ with a spatially invariant T in the equation of motion one finds:

$$\frac{dv_r}{dt} = \frac{(1+Z)KT}{[1+Z(m_e/m_i)]m_i R(1-r/R)}; \tag{5.20}$$

this gives an acceleration profile varying nonlinearly with r even with a pole at $r/R = 1$. Should, at any time t_0, the velocity profile be linear and the density profile is linear for $t > t_0$, then the velocity profile must change its linearity at $t > t_0$ due to Eq. (5.20).

The motivation for the controversial view of Haught [128] about the self-similarity model is not trivial if one looks at the question of the pressure profile from the ad hoc assumptions of the not easily understandable self-similarity model [112]. The analysis given here [127] of a derivation of Eqs. (5.15) and (5.19) from the general hydrodynamic equations has the advantage—as always in general theory—of demonstrating how the mentioned pressure problem has been solved by the justified assumptions of the averaging procedure at Eq. (5.12).

One way in which a linear velocity profile (self-similarity motion) can be observed under the condition of spatially constant (but temporally varying) temperatures, can be verified if the density profile is Gaussian [129] and [130]. Such a profile

$$n_i(r, t) = \frac{n_0}{\pi^{3/2} l^3} \exp\left(-\frac{r^2}{l^2}\right) \tag{5.21}$$

where the length $l(t)$ is only a function of t, produces an acceleration in the equation of motion (5.7):

$$\frac{\partial^2 r}{\partial t^2} = -\frac{KT(1+Z)}{m_i[1+Z(m_e/m_i)]n_i}\frac{\partial}{\partial r}n_i = \frac{KT(1+Z)}{m_i[1+Z(m_e/m_i)]}\frac{2r}{l^2} \tag{5.22}$$

If the starting velocity $v_r(r, t=0)$ is zero or linear with r, the acceleration conserves this property. The Gaussian density profile expands similarly. The limitation of the model with the Gaussian profile is given by the fact that the real plasma has a definite surface and a finite expansion velocity, while the Gaussian profile distributes the plasma to any distance. It is interesting that the Gaussian density profile was observed by interferometry of plasma, produced by lasers from thin foils [131], and that a direct numerical calculation, based on the hydrodynamic equations of an initial linear density profile, assimilated a Gaussian profile after a certain time [120]. The function $l(t)$ behaves like $R(t)$ from Eqs. (5.15) and (5.19).

5.2 Laser Irradiation with Varying Pellet Radius

As an example of the application of the self-similarity model of Eqs. (5.6), (5.15), and (5.19) some experimental results [132] are interpreted. For the model some generalizations with respect to the energy transfer from the radiation to the plasma were used. A further step is to include the special condition of the experiment of a varying diameter for a spherical plasma, due to the heating and expansion during irradiation within a constant laser focus. One can go to analytical expressions, demonstrating immediately the physical properties. A numerical application is based on a numerical stable iteration procedure.

As evaluated by Dawson [112], the treatment of Eqs. (5.6), (5.15), and (5.19) for an energy input $W_1 = W$ from a laser of frequency ω starting at a time t_0 and remaining constant gives the solutions, derived also by Basov and Krokhin [125], for the time dependence of the laser heated sphere of the radius R and temperature T

$$R^2 = \left[R_0^2 + \frac{10}{9} G\right] \tag{5.23}$$

$$KT = \frac{W_1 t}{3N_i(1+Z)} \frac{2R_0^2 + \frac{5}{9}G}{R_0^2 + \frac{10}{9}G} \tag{5.24}$$

where R_0 is the initial radius of the target before t_0 and

$$G = \frac{W_1 t^3}{N_i m_i} \tag{5.25}$$

The total number of ions N_i was used

$$N_i = n_i(t_0) \frac{4\pi}{3} R_0^3 \tag{5.26}$$

The model used, [112, 125], implies a Z constant in time, which is quite reasonable.

A modification in the time variation of the input power W is made, taking into account that the plasma sphere expands and changes its cross section for energy transfer. A power density in the focus, which is spatially constant, has a time dependence as a step function and is constant for $t > t_0$ as assumed. The further assumption that all energy incident within the cross section of the plasma is transferred to the plasma as long as the plasma is overdense, $\omega_p > \omega$ [see Eq. (3.13)], with the ion density n_i, leads to

$$\omega_p^2 = \frac{4\pi e^2}{m_e} Z n_i \tag{5.27}$$

The details of the energy transfer from the irradiated plasma corona to the whole plasma are assumed to be fast enough, and all laser energy within the cross section of the target is taken as contributing to W. The power input then gives the formula:

$$W(t) = \frac{R(t)^2}{R_0^2} W_1 \tag{5.28}$$

Here W_1 is the constant power input of the initial cross section, if the laser power density is constant in time and within the focal area. Using this time dependent, $W(t)$, one cannot solve Eqs. (5.6), (5.15), and (5.19) immediately. An iteration is used as done in Eqs. (5.23) and (5.24), a first iteration $R_1(t)$, $T_1(t)$ using $W = W$ and a second iteration $R_2(t)$, $T_2(t)$ using $W = R(t)W/R$, and so on. The second iteration is then (avoiding the index 2):

$$R^2 = R_0^2 + \frac{10}{9} G \left(1 + \frac{1}{18R_0^2} G\right) \tag{5.29}$$

$$KT = \frac{W_1}{3N_i(1+Z)} \frac{2R_0^2 + 5G/3 + (80G^2/81R_0^2) - (200G^3/1458R_0^4)}{R_0^2 + (10G/9)(1 + G/18R_0^2)} \tag{5.30}$$

The difference of this solution compared with the solutions (5.23) and (5.24) is obvious when we evaluate the time t_{TP} at which the plasma becomes transparent ($\omega_p = \omega$). In the case of ruby laser radiation ($\omega = 2.7 \times 10^{15}$ Hz) and an initial density $n_0 = 6 \times 10^{22}$ cm^{-3}, as is the case for solid hydrogen or solid aluminium, from Eq. (5.23) with Eq. (5.25) such an R is found, for which $(R/R_0)^3 = n_0/n_{c0}$. With the cutoff density, given by Eq. (5.27) at $n_i = n_{c0}$ for $\omega_p = \omega$,

$$t_{TP}^{(1)} = \left(\frac{7.78 r_0^2 N_i m_i}{W_1}\right)^{1/3} \tag{5.31}$$

For the solution with varying cross section, from Eq. (5.29) there follows in the same way

$$t_{TP} = \left(\frac{7.15 r_0^2 N_i m_i}{W_1}\right)^{1/3} \tag{5.32}$$

The higher energy input in this case makes the plasma transparent a little earlier.

A greater difference can be seen in plasma temperature. The maximum temperature of the plasma is reached when the total energy input $W_1(t)$ is so fast that the expansion of the plasma is negligible. In this case, the temperature is [112]:

$$T = T_{max} = \frac{2}{3KN_i(1+Z)} \tag{5.33}$$

Considering the temperature T at the time $t = t_{TP}$, one finds in the case of Eq. (5.24) the first-order temperature at the time of transparency

$$T^{(1)} = 0.32 T_{max} \tag{5.34}$$

and in the case of varying diameter

$$T = 0.583 T_{max} \tag{5.35}$$

The temperature is differing nearly by a factor of two and is larger, although $t_{TP} < t_{TP}^{(1)}$.

In order to evaluate the time dependence of the plasma parameters of self-similarity expansion, a numerical program to solve $R(t)$ and $T(t)$ for a more general input power W in Eqs. (5.6), (5.15), and (5.19) is based on the following assumptions.

Evaporation, ionization, and recombination of the plasma were neglected. The focal region is approximated by a boxlike intensity profile constant in space. The intensity has a time dependence of variable forms (rectangular, triangular, symmetric, or with steepened rise time, tailored with triangular initial prepulse and triangular main pulse), expressed by $W_1(t)$ given by the laser power within the area of the cross section of the plasma $t = t_0$. The geometry is given by a factor

$$f = \begin{cases} R^2/R^2 & \text{for } R < R_F \\ R_F^2/R^2 & \text{for } R \geqslant R_F \end{cases} \tag{5.36}$$

with the focus diameter R_F taking into account, that the plasma can reach a larger R than the focus radius R_F before $\omega_p < \omega$ is reached. The energy transfer due to absorption, given by the optical absorption constant K, is approximated with respect to the spherical geometry by

$$g = \begin{cases} 1 & \text{for } n_i \geqslant n_{c0} \\ 1 - \dfrac{1-(1+2KR)\exp(-2KR)}{2K^2R^2} & \text{for } n_i < n_{c0} \end{cases} \tag{5.37}$$

The power input into the plasma with negligible radiation losses is then

$$W(t) = W_1(t) g(t) f(t) \tag{5.38}$$

The time dependent functions $R(t)$ and $T(t)$ are solved from the following system of equations, where the total mass of the plasma is $M=4\pi R_0^3 n_0 m_i/3$

$$R^2(t)=R_0^2+\frac{20}{3M}\int_0^t d\tau\int_0^\tau d\tau'\int_0^{\tau'} d\tau'' W\left[R(\tau''), T(\tau'')\right] \qquad (5.39)$$

and

$$KT(t)=\frac{M}{5N_i(1+Z)}\left\{-\left(\frac{dR}{dt}\right)^2+\frac{10}{3M}\int_0^t d\tau W[R(\tau), T(\tau)]\right\} \qquad (5.40)$$

The solution is verified by iteration with the first step $R_1(t)T_1(t)$ as described before, put into $W(R_1(t), T_1(t))$ and Eqs. (5.39) and (5.40) solved to find the second iteration $R_2(t)$, $T_2(t)$ and with these values $W(R_2(t), T_2(t))$ is used to solve the third iteration $R_3(t)$, $T_3(t)$, and so on. The iteration was performed until R- and T-values differ by less than 10^{-4} from the values of the previous iteration.

In the following examples, the necessary number of iterations of this procedure is given by the result that the temperature $T(t)$ decreased to zero at large t compared with the laser pulse length t_L, defining $W_1=0$ at $t>t_L$ in Eq. (5.38)

5.3 Numerical Example

The results of the self-similarity model in the formulation described here will be compared with experiments. A first comparison is performed with experiments [132] where single aluminum balls of 50 to 150 μm radius were irradiated by focused laser pulses of 2 to 5 J energy and 15 to 35 nsec pulse length. A numerical solution of the time dependence of the ball radius $R(t)$ and the temperature $T(t)$ for a case measured [132] is shown in Fig. 5.1. The velocity of the plasma surface $v_{max}^{th}(t)=dR/dt$ is evaluated and also the maximum ion energy $\varepsilon_{max}^{th}(t)=m_i v^2/2$. In addition, the amount of the totally absorbed energy was evaluated, taking into account the geometry of the growing plasma within the laser focus and the absorption constant. The results show that the calculated absorption of energy is equal to the measured values within 15%, and the dependence of the amount of the absorbed energy on the ball radius and the pulse length shows the same systematic variation as seen from the measurements.

The final maximum ion energy of the expanding plasma has been measured from side-on framing camera pictures [132]. These values fit very well the theoretical slope with a triangular pulse shape, the parameters of which were determined by the measured pulse shape (Fig. 5.2). The index "th" in ε_{max}^{th} expresses the inner thermal part of the created plasma studied,

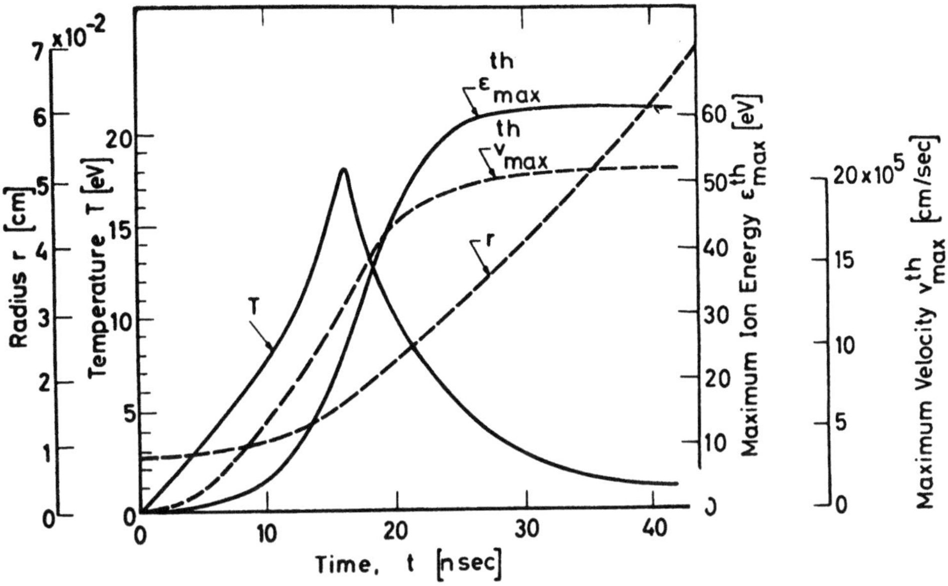

Figure 5.1 Numerical calculation by iteration of Eqs. (5.39) and (5.40) to evaluate the time dependence of the plasma radius R, temperature T, maximum velocity v_{max} and ion energy of the plasma surface for an aluminium ball of 80 μm radius, irradiated by a laser pulse of 3.4 J energy and a rectangular pulse length of 16 nsec.

while an outer part of the plasma [132] has properties of a nonlinear surface mechanism (see Fig. 1.4), which is due to nonlinear processes and will be discussed later.

The result of Eqs. (5.34) and (5.35) indicates a higher temperature of the plasma if the self-similarity model is applied in the way described. This increase of the temperature was observed by Thomson scattering experiments [132]. Another possibility is that recombination mechanisms increase the electron temperature.

5.4 Application to Foils

It should not surprise us that the self-similarity model is so successful in explaining the experiments and the gasdynamics for interaction of medium intensity laser radiation (10^{10} to 10^{12} W/cm^2, ruby or Nd glass lasers) with spherical targets. The intensities are so low that nonlinear effects, if present, are at least not dominant, and the conditions of a fully gasdynamic behavior are realized to a large extent. Nevertheless, it is not clear from the beginning, whether the energy transfer to the whole overdense pellet can occur fast enough (for about 10 nsec total irradiation time) to fulfill the conditions of

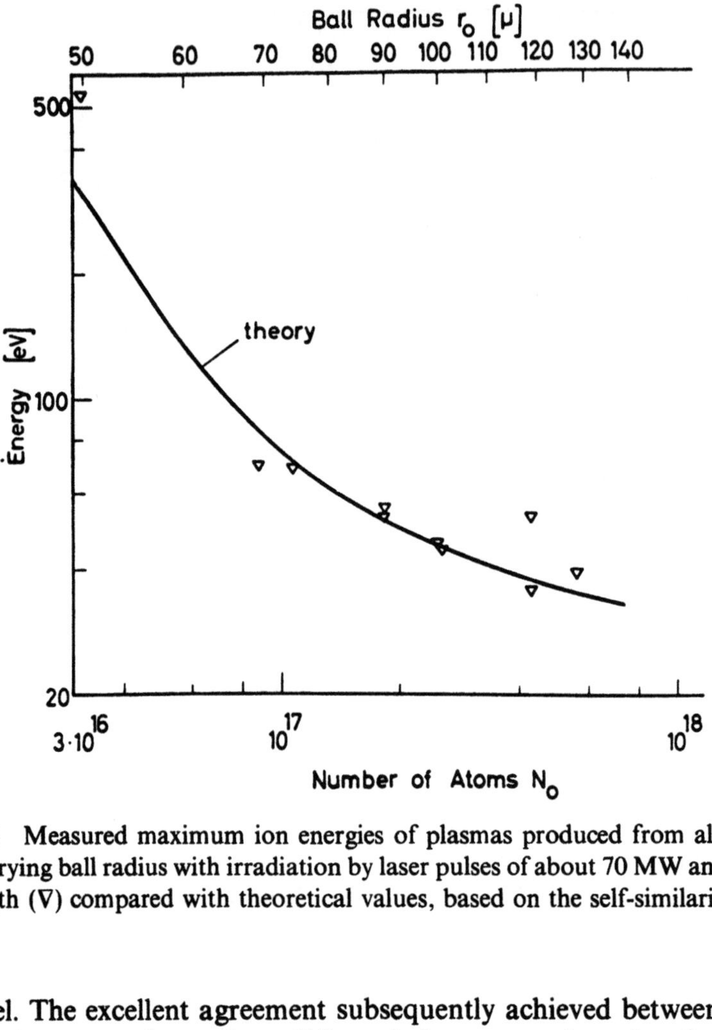

Figure 5.2 Measured maximum ion energies of plasmas produced from aluminium balls of varying ball radius with irradiation by laser pulses of about 70 MW and 30 nsec pulse length (∇) compared with theoretical values, based on the self-similarity model (curve).

the model. The excellent agreement subsequently achieved between theory and experiment confirms the validity of the assumptions for the energy deposition.

A greater surprise is the fact that the self-similarity model reproduces the thermal expansion properties of plasmas produced by laser irradiation of thin foils. Here, a complication of the energy transfer should be expected, due to the interactions of the plasma with the nonirradiated cold foil material. Nevertheless, a reasonably good agreement with the self-similarity model is possible.

The experiments consist of the production of thin foils of solid hydrogen for measuring the transmitted ruby laser pulse [133]. Initially, light passes through the solid target, but is then blacked out sharply by the generated plasma. When the plasma is assumed to expand according to the self-similarity model to smaller electron densities, at a certain time t_{TP}, the plasma becomes transparent [when $\omega_p(n_e(t_{TP}))\leqslant\omega$], which can be measured from the onset of the transmission of light.

For solid hydrogen and ruby laser radiation, from Eq. (5.31) the transparency time t_{TP} is evaluated, using r_0 for the thickness of the foil we have

$$W_1=\frac{W_0 r_0^2}{r_F^2}=I\pi r_0^2 \tag{5.41}$$

W_0 is the input laser power at the front of the layer, where the laser beam is focused to a radius r_F, and I is the laser intensity. Using the density $\rho=0.1$ g/cm^3 of solid hydrogen and Eq. (5.31) leads to

$$t_{TP}=\left(\frac{7.78(4/3)r_0^3\rho}{I}\right)^{1/3} \tag{5.42}$$

In the experiment [133] with $I=2.4\times10^{12}$ W/cm^2, Eq. (5.42) results in

$$t_{TP}=3.50\times10^{-7}r_0 \qquad \text{for } W_0=200 \text{ MW} \tag{5.43}$$

$$t_{TP}=4.75\times10^{-7}r_0 \qquad \text{for } W_0=500 \text{ MW} \tag{5.44}$$

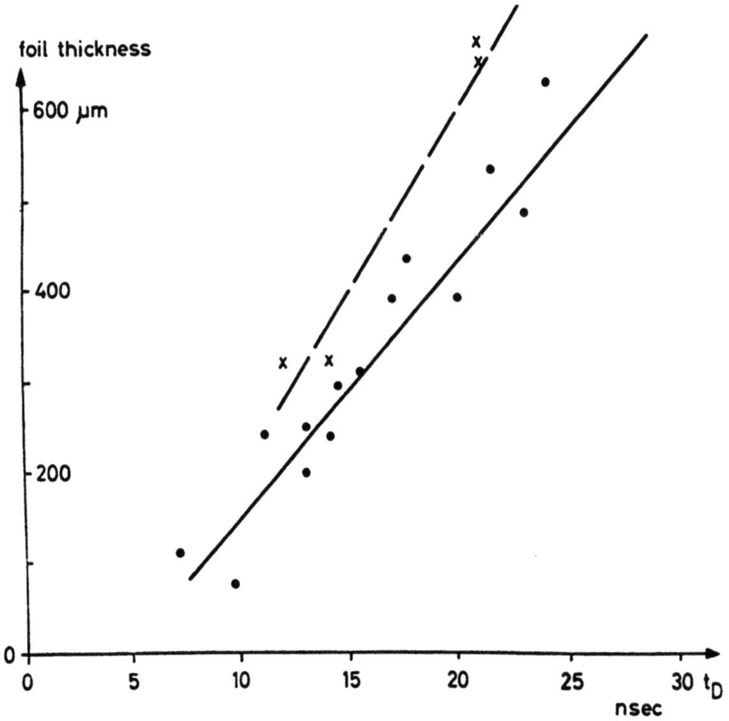

Figure 5.3 Measured values (•. ×) [133] of the delay time of transparency t_{TP} for solid hydrogen foils of given thickness compared with the calculated (lines) transparency time t_{TP} by the self-similarity model [127], when t_D-t_{TP} is 5 nsec. The laser intensities were 200 MW (———) and 500 MW (- - - - - -).

measuring foil thicknesses r_0 in centimeters and transparency time t_{TP} in seconds. To compare these values with the measured transparency time, we have to add to t_{TP} the time between the beginning of the laser pulse and the creation of the plasma, which is about 5 nsec (Fig. 5.3). In agreement with analogous experiments, Eqs. (5.43) and (5.44) fit the measurements as shown in Fig. 5.3 very well. At this point, nothing has been said about the details of the process achieving the very fast deposition of the laser energy to the plasma for justifying the self-similarity model. It has to be taken into account that self-focusing and related complicated dynamics may be responsible for the fast energy transfer.

It is indeed a surprise that the self-similarity model with homogeneous heating fits the experimental values of t_{TP} so well. The numerical calculation using a shockfront type heating [134] of plasma arrived at 30 times longer transparency times. This result induces scepticism about several shock wave models developed for explaining the gasdynamic laser plasma interaction for laser intensities of 10^{10} to 10^{12} W/cm^2 of about 1 μm wavelength, pulses of 1 to 30 nsec duration and targets up to 0.4 mm diameter or characteristic size. While the plane shock wave calculation will be correct for such experimental conditions, the experiments might provide different conditions of a fast energy transfer due to self-focusing and other mechanisms [135].

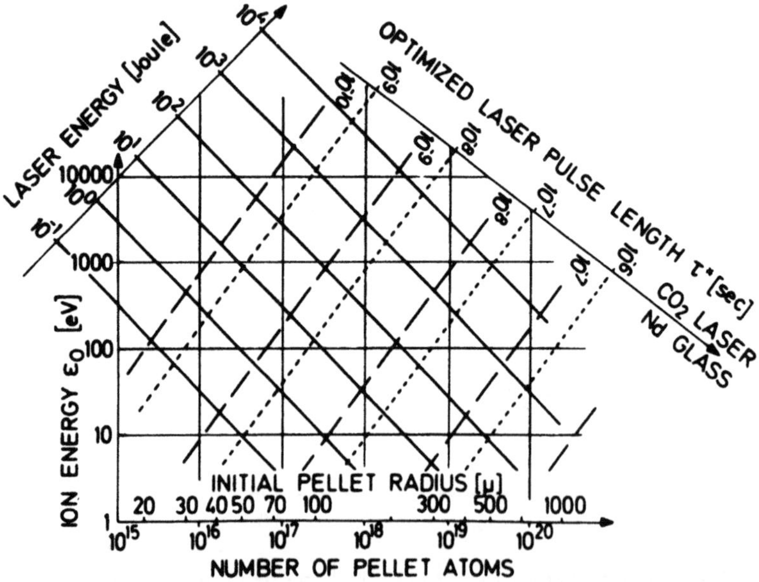

Figure 5.4 Energy and pulse length τ^* of neodymium-glass and CO_2 laser radiation for heating a solid deuterium pellet of given initial radius to averaged ion energies ε_0, derived from the self-similarity model of homogeneous heating [74, p. 37].

Finally, a diagram has been derived from the self-similarity model for constant pellet radius [74,Fig. 5.3] for calculating maximum temperature, Eq. (5.24) corresponding to the ion energy ε_o after subsequent free expansion if t=t* is the time of achieving transparency t_{TP} = t*, Eq. (5.31) at irradiation by a constant power W, onto a solid D_2 pellet of initial radius R_o (corresponding to a number N_i of atoms). $Wt^*\pi R_o^2$ is then equal to the laser energy E_L for the irradiation up to the time t*. The transparency depends on the wavelength (cutoff density) and results in different optimized times τ^* for CO_2 laser or Nd glass irradiation. The nomogram in Fig. 5.4 is read in the following way as an example. One may start from the basic diagram by using a pellet radius of, for example, 350 μm (10^{19} atoms) for a maximum temperature (averaged ion energy) of 100 eV. From this print one can read a necessary laser energy of 3 kJ for 10 nsec pulse length for Nd glass or for 45 nsec pulse length for CO_2 laser irradiation. The laser intensity is then 1.02 x 10^{15} W/cm^2 for Nd glass and 2.2 x 10^{14} W/cm^2 for CO_2.

Finally it should be mentioned that the self-similarity model is being used in many applications. In section 13, it is a key formulation of the hydrodynamic motion of spherical pellets after laser irradiaiton, i.e., using Eq. (5.19) for W=0, after laser energy has been transferred into the pellet and the subsequent ideal adiabatic compression or expansion is followed up and the nuclear fusion gain is summed up during this process. It is then an essential mechanism for understanding "volume compression" in inertial confinement fusion, as will be explained in Sect. 13.

Despite the rather straightforward theory of the self-similarity model, it was reconsidered and discussed generally during the recent years as can be seen by the work of Schmalz (1986), Liu Renhong and Tan Weihan (1990), or Boiko (1990). One should note that the use of the self-similarity model including the fuel depletion and energy deposition by alpha reheat and energy loss by bremsstrahlung is a rather strong modification. During the numerical calculation of the time steps each a re-definition of the initial conditions is included continuously, a rather overstressing of the usual meaning of "initial condition".

5.5 Introductory Remarks to the Following Three Chapters.

It should be noted that the hydrodynamics described in Chapter 4 and its extension to the self-similarity model in this chapter, are well on the level of classical hydrodynamics textbooks or their applications. What was important for the hydrodynamics of plasmas was the inclusion of forces by magnetic fields (initially excluding very artistically electric fields) to the charged particles of electrons and ions, and the formulation of this "magneto-hydrodynamics" for the two fluids of electrons and ions.

While the inclusion of the magnetic forces in a single fluid model was surprisingly successful and led Alfvén (1942) to the discovery of the "magnetohydrodynamic" or Alfvén waves [see the following Eq. (12.49)] for which discovery he received the Nobel Prize similar to Langmuir whose discovery of the "electrostatic" electron oscillations in a plasma, Eq. (2.4) with the discovery of the plasma frequency (2.6), was honored by the Nobel Prize.

Langmuir's subsequent derivation of waves however is restricted. We shall see that these Langmuir waves make sense only in purely homogeneous plasmas, while inhomogeneous plasmas produce a process of energy transport which is in the best way only a "Pseudo"-Langmuir wave since its main characteristic of a wave for having everywhere the same frequency, is being lost. The local Langmuir's plasma frequency is changing locally on the density change. Nevertheless we shall report on the numerical observation of energy transport by longitudinal electron oscillations, a typical property of waves.

A more consequent but not yet the final-macroscopic magnetohydrodynamic description of plasmas using the electrons and ion fluids was achieved by Schlüter [136] as will be explained in the following Section 6. The enormous success of the Schlüter equations is given by reducing the motion to one equation of momentum conservation, and a kind of a diffusion equation which appears as a highly generalized Ohm's law of plasmas.

The rather complicated derivation of Schlüter's equations is shown in Appendix C, which is the only printed presentation of this derivation. The complexity, however, consists in the fact that it arrives at more general formulations than Schlüter [136] achieved. This generalization was indeed the result of the study of laser-plasma interacton as it is proved and explained from momentum conservation in Chapter 8. Therefore after knowing this correct general result with necessary further nonlinear terms on top of Schlüter's initial derivation, the derivation in Appendix C was possible only, motivated and directed from our solution of the momentum conservation treatment at laser-plasma interaction as shown in Chapter 8.

What can be done after starting from the initial formulation of Schlüter in Section 6 without all the other competing nonlinear terms, is substituting Schlüter's formulation for plasmas of Ohm's law in Maxwell's equations by which procedure the dielectric response or - in other words -

the optical constants of plasma including its dissipation or absorption by collisions are derived. When the author of this book derived this using the direct current electric resistivity of plasma, this was a surprise, even for Schlüter himself, but it was confirmed - as will be shown in more detail - that this result agreed (apart from a minor multiplicator of the order of one) with the optical constants derived from the quantum electrodynamic theory of bremsstrahlung.

This success of the optical constants of plasmas within macroscopic hydrodynamic theory enables us then to summarize the problems of the propagation of electromagnetic waves in inhomogeneous plasmas, as presented in Chapter 7. This is not only a condensation of the main results of Ginzburg's book [150], it "shows surprising completeness for such a brief introduction" (Herbst, 1981).

After this exhaustion of Schlüter's diffusion equation (Ohm's law) simultaneously as providing a preparation for the later studies of laser propagation in the highly inhomogeneous plasmas produced by the lasers, we shall present in Chapter 8 how the initial equation of motion of Schlüter had to be generalized by adding further nonlinear terms as required for confirming momentum conservation. Only after these steps the derivation of the two-fluid equations in Appendix C has to be considered in retrospect for this whole discussion of Chapters 6 to 8.

The result of this two-fluid model, however, is not complete. It is based on space charge quasi-neutrality. This means that for dimensions larger than for microscopic fluctuations given by the Debye length, Eq.(2.11), there should be no electric field within plasmas and motion of plasma is driven (apart from hydrodynamic pressures) by magnetic fields only. This, however, is correct only for homogeneous plasmas and is one of the basic properties of plasmas postulated by Langmuir.

Contrary to these fundamentalistic views, the experience with the inhomogeneous plasmas in space (see Alfvén and Fälthammer [102, 137] and with the strongly inhomogeneous laser produced plasmas led to contradictions. The conclusion of electric fields (and double layers) inside of plasmas in ranges much larger than the Debye length became evident also from the forces in laser produced plasmas as explained in the second half of Chapter 8 leading then to the formulation of a "genuine two-fluid model" whose properties are consequently evaluated and highlighted by analytical and numerical results. This model drops any of the earlier simplifying neglection with respect to space charge neutrality and arrived at a complete macroscopic plasma model. Only after these results we shall follow up further consequences of the magnetohydrodynamic theory as a preparation to understand why laser-plasma interaction is so highly complicated, how these difficulties can be overcome, and how one can provide the physics basis for laser fusion for low cost, clean, and inexhaustive large scale energy production.

SIX

Plasma Dynamics and Lorentz Theory

In the two preceding sections the three basic plasma hydrodynamic equations have been described. These have been used in simplified form to determine the expansion of a spherical plasma with the self-similarity model. In this section, the mechanical response of the plasma to electric and magnetic fields **E** and **H** will be discussed on the basis of the two-fluid equations [136]. These basic equations show that mechanical equations can lead to Ohm's law, which is an electrodynamic equation.

6.1 The Two-Fluid Equation of Motion

Schlüter [136] started from the Euler equations for the electron and ion fluids in a plasma.* The indices e and i denote the electron and ion parameters, respectively.

$$m_i n_i \frac{d\mathbf{v}_i}{dt} = Z n_i e\mathbf{E} + n_i \frac{Ze}{c}\, \mathbf{v}_i \times \mathbf{H} - \nabla n_i K T_i - m n_e \nu_{ei}(\mathbf{v}_i - \mathbf{v}_e) + \mathbf{K}_i \tag{6.1}$$

$$m n_e \frac{d\mathbf{v}_e}{dt} = -n_e e\mathbf{E} - n_e \frac{e}{c}\, \mathbf{v}_e \times \mathbf{H} - \nabla n_e K T_e + m n_e \nu_{ei}(\mathbf{v}_i - \mathbf{v}_e) + \mathbf{K}_e \tag{6.2}$$

The force densities on the right-hand side of Eqs. (6.1) and (6.2) arise from the electric field **E**, the Lorentz force $\mathbf{v} \times \mathbf{H}$, and the pressure $p = n_{i,e} K T$.

*A generalization to a three-fluid model with remaining neutral atoms was a subsequent step [137]. These plasmas of partial ionization are of marginal importance for laser produced plasmas.

The ultimate term corresponds to the viscosity, where ν_{ei} is the electron ion collision frequency given by Eqs. (3.34) and (3.35). For any additional forces, such as, for example, the gravitational force $\mathbf{K}_i$ and $\mathbf{K}_e$ are used. The net velocity $\mathbf{v}$, as defined by Schlüter [136], is (see Appendix C)

$$\mathbf{v}=\frac{m_i\mathbf{v}_i+Zm v_e}{m_i+Zm} \tag{6.3}$$

and the current density $\mathbf{j}$ is

$$\mathbf{j}=e(n_i\mathbf{v}_i-n\mathbf{v}_e) \tag{6.4}$$

Addition of the Eqs. (6.1) and (6.2) and substitution of (6.3) and (6.4) and rearranging terms leads to an *equation of motion*, given by a force density $\mathbf{f}$

$$\mathbf{f}=m_i n_i\frac{d\mathbf{v}}{dt}=-\nabla p+\frac{1}{c}\mathbf{j}\times\mathbf{H}+\frac{1}{4\pi}\left(\frac{\omega_p}{\omega}\right)^2\mathbf{E}\cdot\nabla\mathbf{E} \tag{6.5}$$

where p represents the total gasdynamic pressure in the plasma. The additional force densities due to gravitation and so on, $\mathbf{K}_i$ and $\mathbf{K}_e$, are neglected in Eq. (6.5). The last term was written originally [136] as:

$$\frac{1}{4\pi}\left(\frac{\omega_p}{\omega}\right)^2\mathbf{E}\cdot\nabla\mathbf{E}=\mathbf{j}\cdot\nabla\frac{1}{\omega^2}\frac{\partial\mathbf{E}}{\partial t} \tag{6.6}$$

For this substitution see Ref. [138]. The importance of generalizing the equation of motion (6.5) by more nonlinear terms to describe the laser-plasma interaction, will be shown in Section 8. The derivation of an equation of motion for a plasma without the nonlinear term in (6.5) has been shown by Spitzer [107], starting from kinetic theory (Boltzmann equation), see Section 3.

6.2 The Diffusion Equation (Ohm's Law)

In order to obtain an equation for the motion of the electrons relative to the ions, Schlüter [136] subtracted Eq. (6.1) from Eq. (6.2) to obtain (see Appendix C)

$$\frac{m}{e^2 n_e}\left(\frac{d\mathbf{j}}{dt}+\nu\mathbf{j}\right)=\mathbf{E}+\frac{1}{c}\mathbf{v}\times\mathbf{H}+\frac{1}{en_e c}\mathbf{j}\times\mathbf{H}+\frac{c}{en_e}\frac{\nabla p}{1+1/Z} \tag{6.7}$$

Schlüter called this the "diffusion equation", which is a *generalized Ohm's law*, containing a relation between the current density $\mathbf{j}$ and the electrical field $\mathbf{E}$, which—in the sense of an acting electric field—must be extended by the Lorentz term $\mathbf{v}\times\mathbf{H}$, the Hall term $\mathbf{j}\times\mathbf{H}$, and an electron pressure term. Neglecting these last terms, a form of Ohm's law is obtained:

$$\frac{d\mathbf{j}}{dt}+\nu\mathbf{j}=\frac{\omega_p^2}{4\pi}\mathbf{E} \tag{6.8}$$

This is how it was originally formulated for a plasma by Langmuir, so the purely mechanical Euler equations (6.1) and (6.2) lead to the electrical relation (6.8), known as Ohm's law, where automatically the plasma frequency ω_p of the electrostatic plasma oscillations were determined [see Eq. (2.16)].

6.3 Electrodynamic Equations

The electric and magnetic fields **E** and **H** in the equations of motion (6.1) and (6.2) obey the electrodynamic equations derived by Maxwell. The integral formulation of Faraday's induction law, including the magnetic permeability μ, is

$$\oint \mathbf{E}\cdot ds = \frac{1}{c}\frac{\partial}{\partial t}\iint \mu\mathbf{H}\cdot d^2\mathbf{a} \tag{6.9}$$

The induction of an electric field **E** along a closed loop is created by a temporal change of the magnetic flux **H** through this loop. With Stokes' law,

$$\oint \mathbf{E}\cdot ds = \iint \nabla\times\mathbf{E}\cdot d^2\mathbf{a}$$

Faraday's law (6.9) in differential form results in the first Maxwellian equation:

$$\nabla\times\mathbf{E} = -\frac{1}{c}\frac{\partial}{\partial t}\mu\mathbf{H} \tag{6.10}$$

The integral formulation of Ampere's law

$$\oint \mathbf{H}\cdot ds = \frac{4\pi}{c}\iint \mathbf{j}\cdot d^2\mathbf{a} \tag{6.11}$$

expresses the magnetic field **H**, generated along a closed loop, within which an electric current, given by the current density **j**, is produced. It can be formulated again by the use of Stokes' law in differential form, to which Maxwell added the dielectric displacement current, given by the dielectric constant ε, and arrived at his second equation:

$$\nabla\times\mathbf{H} = \frac{4\pi}{c}\mathbf{j} + \frac{1}{c}\frac{\partial}{\partial t}\varepsilon\mathbf{E} \tag{6.12}$$

The source equation of the electric field, given by a charge density ρ_e, is

$$\oiint \varepsilon\mathbf{E}\cdot d^2\mathbf{a} = 4\pi\iiint \rho_e\, d^3\tau \tag{6.13}$$

The integration of ε**E** is along the closed area determining the integration volume of ρ_e. Using Gauss' law

$$\oiint \varepsilon \mathbf{E} \cdot d^3\mathbf{a} = \iiint \nabla \cdot (\varepsilon \mathbf{E})\, d^3\tau$$

the differential form of (6.13) can be written as

$$\nabla \cdot (\varepsilon \mathbf{E}) = 4\pi\rho_e = 4\pi e(Zn_i - n_e) \tag{6.14}$$

With the electron charge $e = 4.803 \times 10^{-10}$ cgs, charge densities are the source of the **E** fields. Magnetic monopoles, as the source of magnetic fields, have not yet been observed, though Dirac's theory of 1933 [139] shows that their existence is possible. With the exclusion of this possibility, the source of the magnetic field is zero.

$$\nabla \cdot \mathbf{H} = 0 \tag{6.15}$$

It was of the ingenious discoveries of Maxwell to differentiate Eq. (6.10) by ∇ and Eq. (6.12) by $\partial/\partial t$ and eliminate **E** or **H** (assuming $\varepsilon = \mu = 1$ and $\mathbf{j} = 0$) to arrive at

$$\nabla^2 \mathbf{E} - \frac{1}{c^2}\frac{\partial^2}{\partial t^2}\mathbf{E} = 0 \tag{6.16}$$

This is a wave equation with the speed $c = 299{,}796$ m/sec, the speed of light. An equivalent can be written for **H**,

It was O. D. Chwolson, who in 1905 introduced Maxwellian theory into the textbooks for a *deductive* description of the electrodynamics: firstly to establish the Maxwellian equations, and secondly to treat electrostatics or magnetostatics separately for $\partial/\partial t = 0$, the quasistatic case for slowly varying **j** and the wave fields as the general case. It is surprising that this basic methodological scheme has not yet been introduced into all physics textbooks for university students even 100 years after Maxwell's death.

The treatment of media in the Maxwellian theory uses ε or μ as material constants. Lorentz described the phenomena with $\varepsilon = \mu = 1$, as in vacuum, and described all material phenomena by charge densities ρ_e and current densities **j**. The dielectric properties of insulators are then due to electric dipoles and currents. This way of description is preferable for plasmas. In the microscopic plasma theory, only currents **j** are present, while any space charge is balanced to zero for spatial dimensions exceeding the Debye length, Eq. (2.8), due to the good electrical conductivity of the plasma. Additionally, electric dipoles determine the dielectric properties. The only exception is higher frequency electromagnetic fields, where high-frequency oscillations of charge densities can be influenced.

The Lorentz theory of plasma uses the Maxwellian equations for vacuum and a current density, given by Ohm's law (6.8), derived from the equations of mechanics

$$\nabla\times\mathbf{E}=-\frac{1}{c}\frac{\partial}{\partial t}\mathbf{H} \tag{6.17}$$

$$\nabla\times\mathbf{H}=\frac{4\pi}{c}\mathbf{j}+\frac{1}{c}\frac{\partial}{\partial t}\mathbf{E} \tag{6.18}$$

If the quantities **E**, **H**, and **j** are of periodic time dependence with a frequency ω, we have

$$\begin{aligned}\mathbf{E}&=\mathbf{E}_r\exp(i\omega t)\\ \mathbf{H}&=\mathbf{H}_r\exp(i\omega t)\\ \mathbf{j}&=\mathbf{j}_r\exp(i\omega t)\end{aligned} \tag{6.19}$$

where $\mathbf{E}_r$, $\mathbf{H}_r$, and $\mathbf{j}_r$ only depend on spatial coordinates. Integration of (6.8) leads to

$$\mathbf{j}=\frac{\omega_p^2}{4\pi i\omega(1-i\nu/\omega)}\mathbf{E} \tag{6.20}$$

and the time-independent Maxwellian equations

$$\nabla\times\mathbf{E}_r=-\frac{i\omega}{c}\mathbf{H}_r \tag{6.21}$$

$$\nabla\times\mathbf{H}_r=-\frac{i\omega_p^2}{c\omega(1-i\nu/\omega)}\mathbf{E}_r+\frac{1}{c}\omega\mathbf{E_r} \tag{6.22}$$

With the operation $\nabla\times$ on (6.22) and the substitution of $\nabla\times\mathbf{E}$ from (6.21), the following equation is obtained:

$$\nabla^2\mathbf{H}_r+\frac{\omega^2n^2}{c^2}\mathbf{H}_r-i\frac{\omega}{c}\mathbf{E}_r\times\nabla n^2=0 \tag{6.23}$$

with Eq. (6.15) and resubsititution according to Eq. (6.19) a wave equation is obtained.

$$\nabla^2\mathbf{H}-\frac{n^2}{c^2}\frac{\partial^2}{\partial t^2}\mathbf{H}-\frac{1}{n^2}(\nabla\times\mathbf{H})\times\nabla n^2=0 \tag{6.24}$$

or

$$\nabla^2\mathbf{H}-\frac{n^2}{c^2}\frac{\partial^2}{\partial t^2}\mathbf{H}+2(\nabla\mathbf{H})\cdot\nabla\ln n-2(\nabla\ln n)\cdot\nabla\mathbf{H}=0 \tag{6.25}$$

with the phase velocity

$$c_\phi=c\sqrt{\mathrm{Re}(n)} \quad \cdot \tag{6.26}$$

The complex constant n is the time independent complex refractive index and is related to the complex dielectric constant ε by

$$\varepsilon = n^2 = 1 - \frac{\omega_p^2}{\omega^2(1 - i\nu/\omega)} \tag{6.27}$$

The first-order term of **H** in Eq. (6.24) is zero in a homogeneous plasma with $\nabla \cdot n^2 = 0$. For an inhomogeneous plasma, where n is a function of x, y, z due to the spatial dependence of n_e or T [Eqs. (2.33) or (2.34) with (2.35)], Eq. (6.24) can be written as:

$$\nabla^2 \mathbf{H} + \frac{\omega^2 n^2}{c^2}\mathbf{H} + 2(\nabla \mathbf{H}) \cdot \nabla \ln n - 2(\nabla \ln n) \cdot \nabla \mathbf{H} = 0 \tag{6.28}$$

This equation will be important in the later chapter on resonance absorption.

It should be noted that all the steps from Eqs. (6.2) to (6.28) presumed time independent n_e and T. When these quantities are time dependent, as is possible in plasma, the derivation is much more complex. In most cases of the study of transient behavior of laser plasma interactions, the time independence of n_e and T in the Maxwellian equations (N.B.—not in the mechanical equations!) is a reasonable approximation. However, there comes a point where a more general treatment is necessary, such as to treat the case of very short time interactions.

Again, with the same assumptions of a time independent n, $\nabla \times$ is operated on Eq. (6.21) and $\nabla \times \mathbf{H}$ is substituted from Eq. (6.22). The resulting equation is

$$\nabla^2 \mathbf{E} + \frac{\omega^2 n^2}{c^2}\mathbf{E} - \nabla\nabla \cdot \mathbf{E} = 0 \tag{6.29}$$

By using $\rho_e = 0$ in Eq. (6.14)

$$\varepsilon \nabla \cdot \mathbf{E} + \mathbf{E} \cdot \nabla \varepsilon = 0$$

Equation (6.29) can be written as:

$$\nabla^2 \mathbf{E} + \frac{\omega^2 n^2}{c^2}\mathbf{E} + \nabla \frac{2}{n}\mathbf{E} \cdot \nabla n = 0 \tag{6.30}$$

The wave equation results again by resubstitution from Eq. (6.19) to

$$\nabla^2 \mathbf{E} + 2(\nabla \mathbf{E}) \cdot \nabla \ln n - \frac{1}{c^2}\left[n^2 + 2\left(\frac{c}{\omega}\right)^2 \nabla^2 \ln n\right]\frac{\partial^2}{\partial t^2}\mathbf{E} = 0 \tag{6.31}$$

This wave equation is still specialized for monochromatically oscillating fields (6.19) and for the time independent n only. On the other hand, it is more general than Eq. (6.16), first due to a first-order derivation, a spatial damping term which is determined by $\nabla \ln n$, and second due to a refractive index n, which is modified by a second-order term $\nabla^2 \ln n$. If for any further study, the time dependence of n has to be included, the time periodic dependence (6.19) has to be revised, as any Fourier superposition of single-frequency solutions would have to be further generalized.

6.4 Refractive Index of Plasma and Its Relation to Absorption

A discussion of the complex refractive index n, Eq. (6.24), will be given in this subsection representing the dispersion of electromagnetic waves in plasma by using the dependence of ω_p upon the electron density n_e, Eq. (2.6), and the collision frequency $\nu(n_e, T)$, Eq. (2.37) including the following nonlinear generalizations.

The complex optical refractive index n is given as the dispersion relation of electromagnetic waves in plasma, Eq. (6.27), where the real part n' and the imaginary part κ are evaluated algebraically:

$$n = n' + i\kappa = \left[1 - \frac{\omega_p^2}{\omega^2(1 + i\nu/\omega)}\right]^{1/2} \tag{6.32}$$

$$n' = \frac{1}{\sqrt{2}}\left[\sqrt{\left(1 - \frac{\omega_p^2}{\omega^2 + \nu^2}\right)^2 + \left(\frac{\nu}{\omega}\frac{\omega_p^2}{\omega^2 + \nu^2}\right)^2} + \left(1 - \frac{\omega_p^2}{\omega^2 + \nu^2}\right)\right]^{1/2} \tag{6.33}$$

$$\kappa = \frac{1}{\sqrt{2}}\left[\sqrt{\left(1 - \frac{\omega_p^2}{\omega^2 + \nu^2}\right)^2 + \left(\frac{\nu}{\omega}\frac{\omega_p^2}{\omega^2 + \nu^2}\right)^2} - \left(1 - \frac{\omega_p^2}{\omega^2 + \nu^2}\right)\right]^{1/2} \tag{6.34}$$

Sometimes n', the real part only, is called the refractive index. For a collisionless plasma ($\nu = 0$), both values are equivalent

$$n = n' = (1 - \omega_p^2/\omega^2)^{1/2} \qquad (\text{if } \nu = 0) \tag{6.35}$$

The cutoff density n_{ec} is that electron density n_e where the collisionless refractive index vanishes [$\omega_p = \omega$, and Eq. (2.6)]

$$n_{ec} = \frac{m\omega^2}{4\pi e^2} \tag{6.36}$$

Since $c_\varphi = 0$, Eq. (6.26), no propagation of transversal electromagnetic waves is possible. A plasma of this density causes total reflection.

The imaginary part of n, κ, is called the absorption coefficient. Its meaning is seen immediately from its relation to the absorption constant $\bar{K}$, which determines the attenuation of a laser intensity I at some depth x; if I_0 is the intensity at $x = 0$

$$I = I_0 \exp(-\bar{K}x)$$

The absorption constant is then

$$\bar{K}=\frac{2\omega}{c}\kappa \tag{6.37}$$

As demonstrated by the preceding equations, the optical properties of a plasma depend on the plasma frequency ω_p, Eq. (2.6), and, therefore, on the electron density n_e, the electron mass m (to the extent it can be changed relativistically), and the collision frequency ν. Here it is important to recognize in what sense the collision frequency is defined. In Eqs. (2.34) and (2.35) it was defined by changes in the motion of the electrons and ions due to Coulomb interaction, in which energy is exchanged. This kind of collision leads to equipartition and characterizes the thermal conductivity and the friction in plasmas, Eqs. (6.1) and (6.2). It also characterizes the exchange of energy, given by an equipartition within one component (e.g., in the electrons if for some reason, the velocity distribution is non-Maxwellian), or between electrons and ions, if there is no thermal equilibrium for some reason (e.g., if only the electrons are heated by incident laser radiation). The following steps show that this collision frequency can be identified with collisions for high-frequency processes.

In a numerical evaluation of n, $\bar{K}$, and n' [140] values for neodymium glass and CO_2 laser radiation were calculated. The agreement with plasma experiments can be considered as a simple proof of the equivalence of the dc and HF collision frequencies. Furthermore, the very simple derived dc collision frequency in Section 2, now used for the HF optical constant, is essentially the same—apart from a minor factor—because the quantum mechanical derivation of the absorption constant of a plasma is based on inverse bremsstrahlung. Indirectly, therefore, a quantum mechanical justification of the very primitive 90° collision frequency, Eq. (2.34), has been established.

The numerical evaluation of $\bar{K}$ and n' leads to the curves in Figs. 6.1 to 6.4. For lower plasma temperatures, the Coulomb logarithm reaches $\ln \Lambda = 1$, $\Lambda = 2.718$. Below this point, the collision theory is *not valid*. This restriction means, from Eq. (2.36), that

$$n_e \leqslant \frac{9K^3T^3}{4Z^2e^6\pi} = 2.42\times 10^{20}\,T^3; \qquad [n_e]=\mathrm{cm}^{-3}; \qquad [T]=\mathrm{eV} \tag{6.38}$$

If the computation is extended to lower temperature, neglecting (6.38), all curves of $\bar{K}$ merge together. This expresses a density independent absorption. This case has to be excluded on the plots [140]. Another restriction is that the calculations assume a classical Boltzmann-type energy distribution for the electrons. The curves of Figs. 6.1 to 6.4 are drawn as lines for temper-

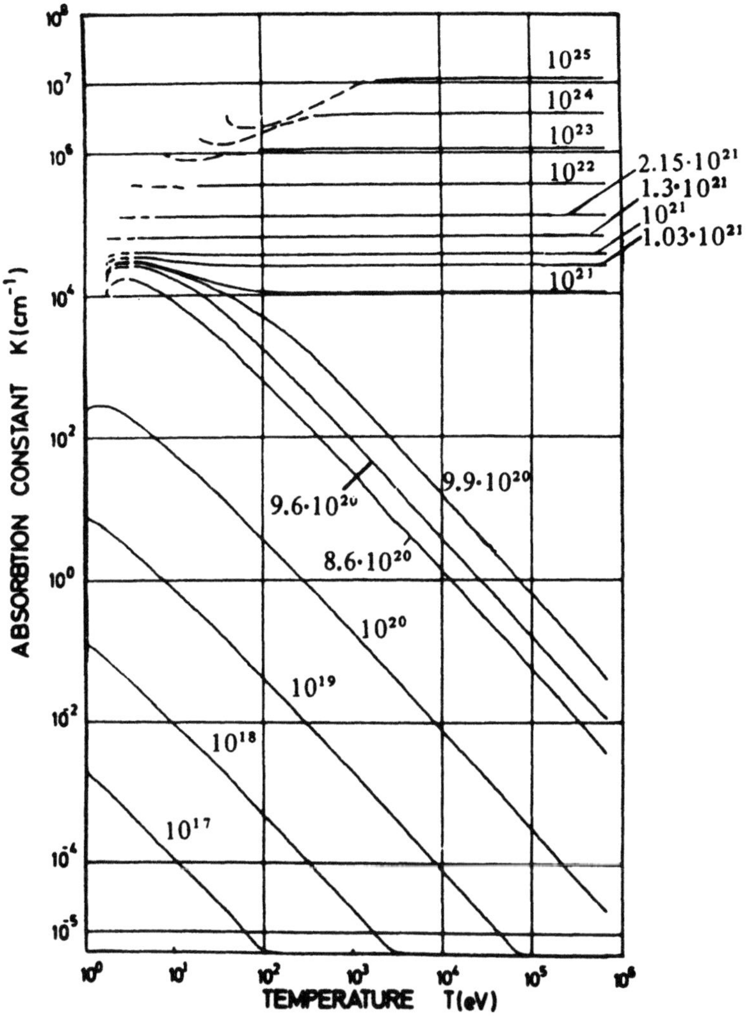

Figure 6.1 Absorption constant $\tilde{K}$ (cm^{-1}) from Eqs. (6.35) and (6.38) with Coulomb collision for neodymium-glass laser radiation in plasma with temperature T(eV) and density n_e (cm^{-3}) [140].

atures equal to or above 10 times the Fermi energy E_F (*Eq.* 2.47)

$$T > 10E_F = 10\frac{h^3}{2m}\left(\frac{3n_e}{8\pi}\right)^{2/3} = 3.65 \times 10^{-14} n_e^{2/3} \tag{6.39}$$

$[T] = \text{eV}$; $[n_e] = \text{cm}^{-3}$. For lower temperatures, the curves are dashed because it is not certain that this model is valid for Fermi–Dirac degenerate plasmas.

In quantum mechanical terms, the motion of an electron within a Coulomb field of an ion has continuous energy eigenvalues, between which

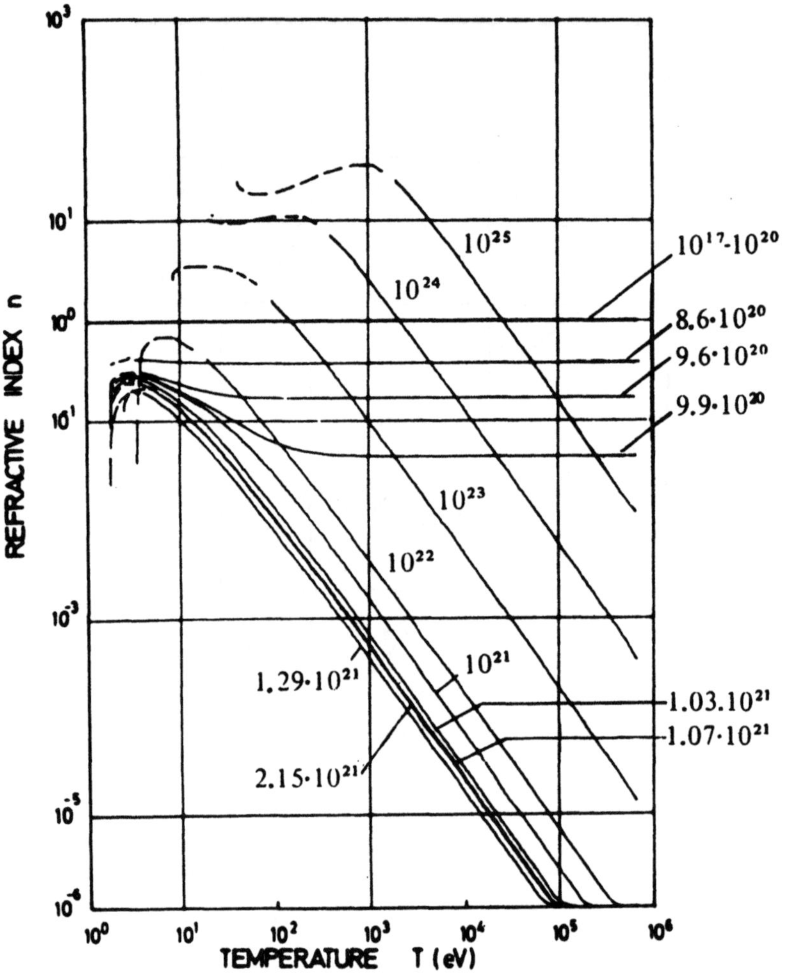

Figure 6.2 Variation of refractive index n' (real part), Eq. (6.34), for neodymium-glass laser radiation in plasma with temperature T(eV) and density n_e (cm^{-3}) [140].

the electron changes to a higher energy state by absorbing the energy of a photon. The exact quantum mechanical description [141] results in an absorption constant (index B for bremsstrahlung):

$$\bar{K}_B = \frac{8\pi^2}{3^{1/2}} \frac{Z^2 n_e n_i e^6 g(T)}{c\omega^2 (2\pi m k T)^{3/2}} \tag{6.40}$$

T is the plasma temperature and $g(T)$ is the "Gaunt factor" [142], a value that corrects the point-mechanical description by a factor between 0.1 and 10. The comparison of the quantum mechanical process of inverse bremsstrahlung and that defined by the collisions is justified by the ratio of $\bar{K}_B$ to $\bar{K}$. This ratio is derived from Eq. (6.37) using the plasma collision frequency

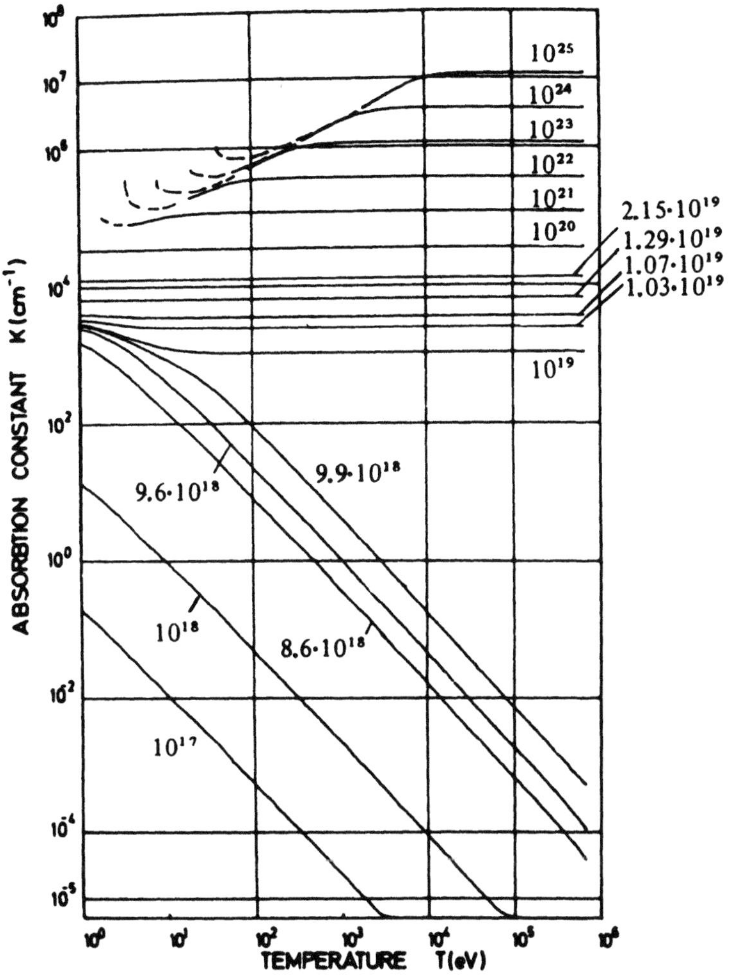

Figure 6.3 Dependence of absorption constant $\bar{K}$ (cm^{-1}) for CO_2 laser radiation in plasma on temperature T(eV) and density n_e (cm^{-3}) [140].

Eq. (2.35):

$$\frac{\bar{K}}{\bar{K}_B}=0.324\,\frac{\ln\Lambda}{\gamma_E(Z)g} \tag{6.41}$$

The validity of $\bar{K}$ in

$$\bar{K}=\frac{4\pi^2}{\gamma_E(Z)} \tag{6.42}$$

is restricted to plasma densities below the cutoff density, $\omega \gg \omega_p$, $n_e \ll n_{ec}$ (Eq. 2.6) and to relatively low collision frequencies $\nu \ll \omega$. For the following representation of the Gaunt factor, the validity is limited in a similar way

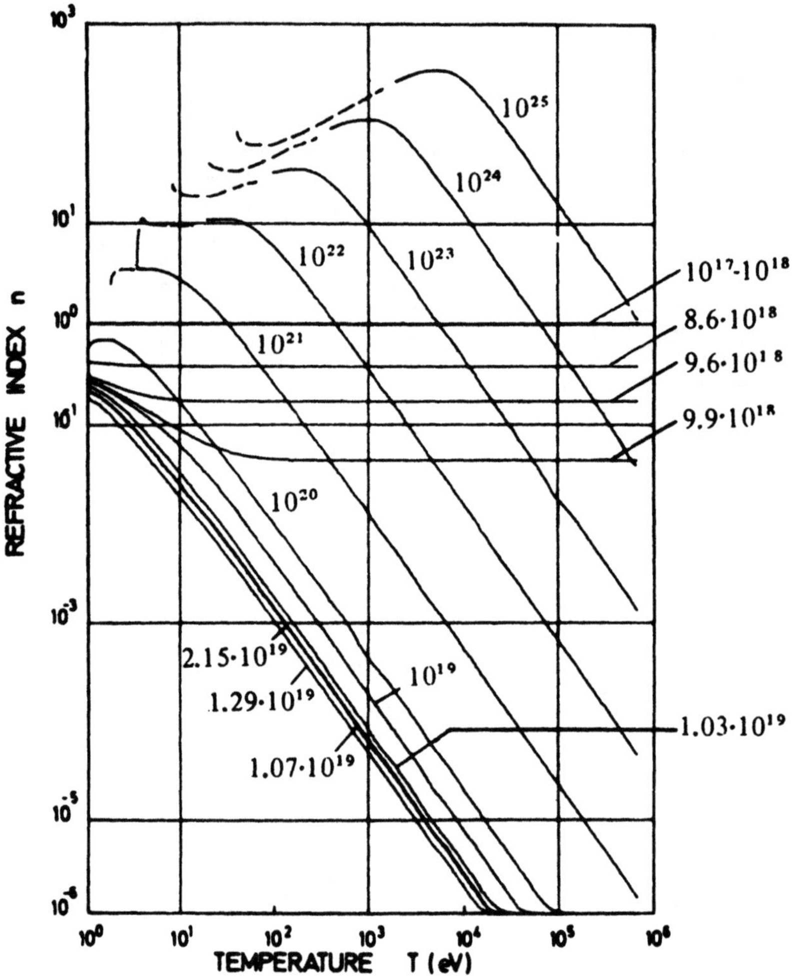

Figure 6.4 Refractive index (real part) n' for CO_2 laser radiation as a function of plasma temperature T (eV) and density n_e (cm^{-3}) [140].

(T in eV and n_e in cm^{-3}) [143]

$$g(T, n_e) = 1.2695(7.45 + \ln T - \tfrac{1}{3}\ln n_e) \tag{6.43}$$

If the value of the Coulomb logarithm is

$$\ln \Lambda = 3.45(6.69 + \ln T - \tfrac{1}{3}\ln n_e - \tfrac{2}{3}\ln Z) \tag{6.44}$$

the ratio becomes

$$\frac{\bar{K}}{\bar{K}_B} = \frac{0.88}{\gamma_E(Z)}\left(1 - \frac{0.66 + \tfrac{2}{3}\ln Z}{7.45 + \ln T - \tfrac{1}{3}\ln n_e}\right) \tag{6.45}$$

The agreement of $\bar{K}$ derived from Spitzer's value [140] with K_B, derived from quantum mechanics, can be seen for $Z=1$, $T=10^4$ eV, $n_e=10^{21}$ cm^{-3}, where $\bar{K}/\bar{K}_B=1.06$.

The absorption of laser radiation in a plasma was considered by Dawson and Oberman [144]. The plasma was described by the collisionless Vlasov equation (3.7) and included the interaction of the particles by phase mixing. This theory [140] was justified by a comparison with quantum mechanical theory in the same way as shown previously for the theory of absorption due to Coulomb collisions. For the absorption constant derived by Dawson and Oberman K_{DO}, the ratio is quite close to unity. Comparing this result with Eq. (6.41), the absorption theory based on Coulomb collisions is found to be closer to the inverse bremsstrahlung theory (e.g., for deuterium by a factor of 2.02).

The quantum mechanical absorption constants may be considered as the most probable values. The disadvantage of the theory is its limitation to low-density and high-temperature plasmas (low collision frequency). However, the laser-plasma interaction is important for plasmas near and above the cutoff densities. For these cases, the refractive index n' (real part) and the absorption constant $\bar{K}$ have been evaluated numerically for several interesting wavelengths (ruby, neodymium-glass, and CO_2 lasers and their second harmonics [140]).

The optical constants are very similar to those of metals and semiconductors. The low-density majority carriers follow very well the oblique lines of $\bar{K}$ in Figs 6.1 and 6.3. The high absorption constants for nearly visible radiation of 10^5 cm^{-3} and more are similar with that of metals for the superdense case ($\omega<\omega_p$). It is remarkable that, for high electron densities, the real part of n can grow to values of 10 and more [145].

For some purposes it is important to evaluate the absolute minimum value of the refractive index n, which can be much less than unity in a hot plasma near the cutoff density. From the exact value

$$|n|=\left[\left(1-\frac{\omega_p^2}{\omega^2+\nu^2}\right)^2+\left(\frac{\nu}{\omega}\frac{\omega_p^2}{\omega^2+\nu^2}\right)^2\right]^{1/4} \tag{6.46}$$

the minimum is found at

$$|n|_{\min}=\left[\left(\frac{\nu}{\omega}\right)^4+\left(\frac{\nu}{\omega}\right)^2\right]^{1/4} \quad \left[\text{at } \omega_p^2=\frac{\omega^2+\nu^2}{1+(\nu/\omega)^2}\right] \tag{6.47}$$

If (as is usual) $\nu \ll \omega$, this can be written as

$$|n|_{\min}=\left(\frac{\nu}{\omega}\right)^{1/2}=\frac{a'}{T^{3/4}} \tag{6.48}$$

The constant a' in the preceding equation can be derived from Eqs. (2.6) and (2.35)

$$a' = \frac{\omega_p \pi^{3/4} m^{1/2} Z^2 \ln \Lambda}{8\pi\gamma_E (Z)(2K)^{3/2}} \tag{6.49}$$

for example, for deuterium ($Z=1$) and $\ln \Lambda = 10$

$$a = 3.25\ (\mathrm{eV})^{3/4}, \qquad \text{for Nd-glass lasers}$$

$$a = 1.03\ (\mathrm{eV})^{3/4}, \qquad \text{for } CO_2 \text{ lasers}$$

6.5 Nonlinear and Relativistic Absorption

The assumption in the preceding discussion of the optical constants concerns n_e and T only. Now, derivations based on the fact that the oscillation energy ε_{osc} of the electrons (coherent quiver motion) in a laser field can exceed the thermal energy KT will be discussed. The absorption is dependent on the laser intensity and the whole Coulomb collision induced absorption process becomes nonlinear. A further generalization is necessary if the quiver motion arrives at such high energies, that relativistic effects (such as, e.g., changes of the electron mass) have to be taken into account.

The equation of motion of a single electron in an electromagnetic wave field is given in the nonrelativistic case ($|\mathbf{v}| \ll c$) by

$$m \frac{d}{dt} \mathbf{v} = e\mathbf{E} + \frac{e}{c} \mathbf{v} \times \mathbf{H} \tag{6.50}$$

Owing to $|\mathbf{v}| \ll c$ the Lorentz term can be neglected. If the elongation of the oscillation, induced by the real part of the electric field $\mathbf{E}$ of a laser varying with frequency ω (Eqs. 6.19), is much less than the wavelength $\lambda = 2\pi c/\omega$,

$$\mathbf{v} = \frac{e}{m} \frac{\mathbf{E_v}}{i\omega} \cos \omega t \tag{6.51}$$

The maximum value of the energy of the quiver motion is

$$\frac{m}{2} \mathbf{v}^2 = \frac{e^2 \mathbf{E_v}^2}{2m\omega^2} = \varepsilon_{osc} \tag{6.52}$$

This is equal to the total oscillation energy of the electrons. The average value of the kinetic energy at quivering is

$$\varepsilon_{osc}^{kin} = \tfrac{1}{2}\varepsilon_{osc} = \frac{e^2}{4m} \frac{\mathbf{E_v}^2}{\omega^2} \tag{6.53}$$

Inserting the plasma frequency ω_p, Eq. (2.6), and the cutoff density n_{ec},

Eq. (6.36), results in

$$\varepsilon_{osc} = \frac{\mathbf{E}_V{}^2}{8\pi n_{ec}}; \qquad \varepsilon_{osc}^{kin} = \frac{\mathbf{E}_V{}^2}{16\pi n_{ec}} \tag{6.54}$$

The definition of the energy density in electromagnetic fields leads to the relation of the field $\mathbf{E}_r$ to the laser intensity I:

$$|\mathbf{E}_V|(\text{cgs}) = \sqrt{8\pi I/c} = 2.91 \times 10^{-5}\sqrt{I}\,(\text{erg/cm}^2\,\text{sec}) \tag{6.55}$$

$$|\mathbf{E}_V|(\text{V/cm}) = \sqrt{2I/\Omega_0} = 27.4\sqrt{I}\,(\text{W/cm}^2)$$

Ω_0 is the resistivity of vacuum (377 Ω). Using this, (6.54) in cgs units can be written as

$$\varepsilon_{osc} = \frac{I}{cn_{ec}}; \qquad \varepsilon_{osc}^{kin} = \frac{I}{2cn_{ec}} \tag{6.56}$$

To arrive at the optical constant n for $\varepsilon_{osc}^{kin} \gtrsim KT_{th}$ we have to take into account that the Coulomb collision frequency ν_{ei} must be modified to include the velocity due to the coherent quiver motion. Very simply, this means that we have to use an effective temperature T^*

$$T^* = T_{th} + \varepsilon_{osc}^{kin}/K \tag{6.57}$$

instead of the temperature T_{th} representing the mean energy of random thermal motion only. The absorption constant and collision frequency ν_{ei} (2.35) is then calculated to [146]

$$\bar{K}_{NL} = \frac{4\pi^4 e^6 n_e^2 n_i^2 Z^2 \ln \Lambda}{\sqrt{2}\, 16(KT_{th} + \varepsilon_{osc}^{kin})^{3/2}} \tag{6.58}$$

where

$$\nu_{ei} = \frac{(m\pi)^{1/2}\omega_p^2 e^2 Z \ln \Lambda}{\sqrt{2}\, 16(KT_{th} + \varepsilon_{osc}^{kin})^{3/2}} \tag{6.58a}$$

The index NL is used for "nonlinear". The Spitzer factor is then unity, as there is no electron–electron collision for coherent motion of the electrons in the laser field. This simple generalization is fully justified by the fact that the general quantum mechanical calculation, including the quiver motion, arrives at a very similar result (Rand [147])

$$\bar{K}_{NL}^{Rand} = \frac{\sqrt{\pi c}}{4} n_e^2 n_i^2 Z^2 e^3 \omega m^{1/2} \ln\left(\frac{32e^2 I}{mKT_{th}c\omega^2}\right)/I^{3/2} \tag{6.59}$$

In both cases, (6.58) and (6.59),

$$\bar{K} \sim (\varepsilon_{osc}^{kin})^{-3/2} \sim I^{-3/2}$$

The ratio of the two nonlinear absorption constants

$$\frac{\bar{K}_{\mathrm{NL}}^{\mathrm{Rand}}}{\bar{K}_{\mathrm{NL}}} = \frac{4 \ln(16\varepsilon_{\mathrm{osc}}^{\mathrm{kin}}/KT_{\mathrm{th}})}{\sqrt{2\pi}\,\pi \ln \Lambda} \tag{6.60}$$

is a constant, which is close to unity within one order of magnitude.

The relativistic generalization of the optical constants has to consider the single-particle motion of an electron in an electromagnetic field

$$\frac{d}{dt} m\mathbf{v} = e\mathbf{E} + \frac{e}{c}\mathbf{v} \times \mathbf{H} \tag{6.61}$$

Assuming a linear polarized laser field propagating along the x-direction ($\mathbf{E} = \mathbf{i}_y E_y$, $\mathbf{H} = \mathbf{i}_z E_z$), the following equations are obtained:

$$\frac{d}{dt}\frac{m_0 v_y}{[1-(v_x^2+v_y^2)/c^2]^{1/2}} = eE_y \cos \omega t - \frac{e}{c} v_x H_z \cos(\omega t + \phi) \tag{6.62}$$

$$\frac{d}{dt}\frac{m_0 v_x}{[1-(v_x^2+v_y^2)/c^2]^{1/2}} = \frac{e}{c} v_y H_z \cos(\omega t + \phi) \tag{6.63}$$

m_0 is the rest mass of the electrons and ϕ describes a possible phase between **E** and **H** in a medium. The spatial dependence of **E** and **H** is neglected. Furthermore, the second term on the right-hand side of Eq. (6.62) is neglected for the moment, but this point will be discussed later again. Using

$$\gamma = [1-(v_x^2+v_y^2)/c^2]^{1/2} = (1-\mathbf{v}^2/c^2)^{1/2} \tag{6.64}$$

Eq. (6.62) can be integrated using the wave number $k = \omega/c$

$$v_y = \frac{eE_y\gamma}{m_0\omega} \sin(-kx + \omega t) \tag{6.65}$$

With Eqs. (6.64) and (6.65) there follows from Eq. (6.63)

$$\frac{d}{dt}\{c^2 - \gamma^2[c^2 + e^2E_y \sin^2(-kx+\omega t)/m_0^2\omega^2]\}^{1/2}\gamma^{-1}$$

$$= \gamma \frac{e^2\mathbf{E}_\mathbf{v}^2}{m_0^2 c\omega}\{\tfrac{1}{2}\sin 2(-kx+\omega t)\cos\phi - \sin^2(-kx+\omega t)\sin\phi\} \tag{6.66}$$

Obviously the expression under the radical cannot become negative, therefore γ can vary only between

$$0 < \gamma^{*2} \leq \gamma^2 \leq 1 \tag{6.67}$$

where

$$\gamma^{*2}\left\{1 + \frac{e^2\mathbf{E}_\mathbf{v}^2}{c^2 m_0^2 |n| \omega^2}\right\} = 1 \tag{6.68}$$

The minimum value of γ is

$$\gamma^* = (1 - v_{max}^2/c^2) = 1 \Big/ \left(1 + \frac{e^2 |\mathbf{E}_V|^2}{m_0^2 \omega^2 c^2 |n|}\right)^{1/2} \tag{6.69}$$

This can be used to evaluate the maximum value of the kinetic energy ε_{kin} directly from the oscillation (which can be identified by the oscillation energy ε_{osc} of the electron) without solving the equation of motion [Eqs. (6.62) and (6.63)]

$$\varepsilon_{osc} = m_0 c^2 \left(\frac{1}{\gamma} - 1\right) = m_0 c^2 \left[\left(1 + \frac{e |\mathbf{E}_V|^2}{m_0^2 \omega^2 c |n|}\right)^{1/2} - 1\right] \tag{6.70}$$

For $|\mathbf{E}_r| < E^r$ the nonrelativistic value of the oscilation energy, (Eq. 6.52), is reproduced

$$\varepsilon_{osc} = \frac{1}{2} \frac{e_2 |\mathbf{E}_V|^2}{m_0 \ \omega^2} \qquad (|\mathbf{E}_r| < E^r) \tag{6.71}$$

E^r is the relativistic limit where, in Eq. (6.70) $\varepsilon_{osc} = mc^2$ given by a relativistic threshold intensity I_r

$$E^r = \sqrt{3} m_0 \omega c/e; \qquad I_r = \frac{3 m_0^2 \omega^2 c^3}{8 \pi e^2} \tag{6.72}$$

This limit for neodymium glass lasers [with a wavelength $\lambda = 2\pi c/\omega = 1.06\ \mu m$ (Nd)] and for CO_2 lasers ($\lambda = 10.6\ \mu m$) amounts to

$$E^r = \begin{cases} 5.17 \times 10^{10}\ \text{V/cm} & \text{(Nd)} \\ 5.17 \times 10^{9}\ \text{V/cm} & (CO_2) \end{cases} \tag{6.73}$$

The corresponding laser intensities $I = I_r$ in vacuum are

$$I_r = \begin{cases} 3.68 \times 10^{18}\ \text{W/cm}^2 & \text{(Nd)} \\ 3.68 \times 10^{16}\ \text{W/cm}^2 & (CO_2) \end{cases} \tag{6.74}$$

For the pure relativistic case at very high laser intensities I and amplitudes $E_r > E^r$, Eq. (6.70) leads to

$$\varepsilon_{osc} = \frac{ec |\mathbf{E}_V|}{\omega \sqrt{|n|}} \propto I^{1/2} \qquad (|\mathbf{E}_r| \gg E^r) \tag{6.75}$$

This result demonstrates that the oscillation energy of relativistic particles does not depend on their rest mass m_0 and that the increase is proportional only to the square root of the intensity I.

In Eq. (6.62) the last term on the right-hand side was neglected. The complete solution results in an expression for the oscillation energy with a correction function $A(I)$.

$$\varepsilon_{osc} = m_0 c^2 \left\{\left[1 + \frac{A(I) e^2 |\mathbf{E}_r|^2}{m^2 \omega^2 c^2 |n|}\right]^{1/2} - 1\right\} = m_0 c^2 \left\{\sqrt{1 - 3A(I) \frac{I}{I_r}} - 1\right\} \tag{6.76}$$

The function $A(I)=1$ for $I \ll I^r$ and grows monotically to the limit of $\pi/2^{3/2}$ according to Max et al. [148] or $3/2^{3/2}$ according to Schwarz and Talenski [149]. The correction of ε_{osc} due to A is therefore less than 5.3%. This permits the use of Eq. (6.70) or (6.76) with $A=1$ as a good approximation in most applications.

Owing to the relativistic change of the rest mass of the electron, the plasma frequency ω_p has to be corrected to:

$$\omega_p^2=\frac{4\pi e^2 n_e}{m_0}\sqrt{1-v^2/c^2}=\frac{4\pi e^2 n_e}{m_0}\frac{1}{[1+3A(I)I/I_r]^{1/2}} \tag{6.77}$$

The relativistic cutoff density is then

$$n_{ec}=\frac{\omega^2 m_0}{4\pi e^2}[1+3A(I)I/I_r]^{1/2} \tag{6.78}$$

The collision frequency (6.58) of the electrons in a plasma at relativistic conditions changes from the value of Eq. (2.34) if $(T \ll \varepsilon_{osc}/K)$ to become:

$$\nu=\frac{n_e Z\pi^{3/2}e^4 \ln \Lambda^4}{m_0^2 c^3\{[1+3A(I)I/I_r]^{1/2}-1\}^{3/2}2^{5/2}[1+3A(I)I/I_r]^{1/2}} \tag{6.79}$$

or

$$\nu=\begin{cases}\dfrac{n_e Z\pi^{3/2}e^4 \ln \Lambda^*}{m_0^2 c^3 2^{5/2} A(I)I/I_r}, & \text{if } I \gg I_r\\[2ex] \dfrac{n_e Z\pi^{3/2}e^4 \ln \Lambda}{(3I/I_r)^{3/2}2m_0^{1/2}}, & \text{if } I \ll I_r\end{cases} \tag{6.80}$$

The expression $\ln \Lambda^*$ is the relativistic Coulomb logarithm for $\varepsilon_{osc} \gg KT$ with

$$\Lambda^*=\frac{3}{2Ze^3(\pi n_e)^{1/2}}(m_0 c)^{3/2}\{[1+3A(I)I/I_r]^{1/2}-1\}^{3/2} \tag{6.81}$$

Finally, it has to be noted that, for absorption in plasmas, the only Coulomb collision induced processes have been considered. Their extension to superdense conditions $(\omega<\omega_p)$, the nonlinear extension $(\varepsilon_{osc} \gg KT)$ and the relativistic extension (neglecting spatial generalizations of Eq. (6.61) leading to harmonic generation) are given here. Another type of absorption is due to instabilities, or due to the macroscopic acceleration of plasma by the nonlinear forces, described below, or due to soliton generation. The direct transfer of optical energy without collision into kinetic energy of plasma has to be distinguished from the usual Coulomb collision absorption as a purely macroscopic nonlinear absorption mechanism.

6.6. Absorption Constant and QED Theory

This subsection is going to describe a problem for which no final solution has been found yet. This problem is related to fundamental questions of quantum electrodynamics (QED) in the sense that the Einstein coefficients of spontaneous emission can be described only after the second quantization of the electromagnetic field had been introduced by Dirac (1927). The problem is connected to the quantum correction of the classical Coulomb collision frequency, discussed in subsection 2.6.

The quantum mechanical derivation of optical constants including its imaginary part of absorption was based on the theory of bremsstrahlung, i.e. the emission of radiation when electrons are being slowed down in any material, in our case: in plasma. This case is most important for astrophysics. Despite intensive attempts (this problem was urgent since Röntgen's discovery of the x-rays) it was not solved before Gaunt used second quantization theory [142]. Extending this to the inverse bremsstrahlung, the optical absorption constant K_B, Eq. 6.40, was derived. However, the dependence on the plasma temperature of the

$$K_B \sim T^{-3/2} \tag{6.82}$$

law was achieved only if the quantum electrodynamic theory for the free-free transitions of electrons in the fields of fully ionized ions of a plasma included stimulated emission (see Appendix B).

The comparison of our result derived from Spitzer's collision frequency [140] with that of Gaunt's theory, Eq. (6.46), was indeed not immediately derived from Gaunt's work but only after its extension as shown e.g. by Allen [141] including some helpful comments by F.R. Ribe. When others attempted to derive the optical absorption in plasmas from Gaunt's theory [147] they arrived at

$$K_B \sim T^{-1/2} \tag{6.83}$$

The discrepancy to Eq. (6.82), agreeing with Spitzer's model [140] and agreeing with experiments, consists in the fact that in Eq. (6.83), the stimulated emission has not been used.

It is remarkable that the agreement of the absorption based on the classical collision frequency and that based on the quantum electrodynamic derivation at inclusion of the stimulated emission enforced the conclusion that photons of less energy than kT of a plasma at thermal equilibrium take out photons from the plasma by stimulated emission. This is a laser process without any inversion of states happening in all plasmas. This may be of interest after conditions have been found recently (Scully, 1990) that laser emission indeed is possible without an inversion of states.

While this discrepancy was understood soon, the victory of Gaunt's model with stimulated emission was not the last answer. There could well be conditions that absorption of the kind of Eq. (6.83) could occur. This is the case, e.g. if x-ray photons, e.g. of 1 keV energy, are propagating in a plasma of 50 eV temperature. The keV photons cannot remove photons of the same energy from the plasma since these are nearly not present at such temperature, therefore, no stimulated emission is possible and Eq. (6.83) will appear. The results are drawn in Fig. 6.5 as full plot.

It is remarkable that the branches of Eq. (6.82) and (6.83) are of the same powers as the branches of classical and the quantum collision frequency ν, Eq. (2.50). Since the absorption constant $\bar{K}$ is proportional to ν, there should be a relation between quantum collisions and the non-stimulated inverse bremsstrahlung absorption model. However, if one would draw the absorption constant with a collision frequency of the kind of Eq. (2.50), the dashed plot in Fig. 6.5 would result, defining the $T^{-1/2}$ range at higher temperature inversely to the case with the inverse bremsstrahlung theory (fully drawn plot).

In a physics decision of the problem given in Fig. 6.5, one may argue that always the strongest interaction or absorption process will dominate, such that finally an absorption constant of the dashed line may survive. This fact however needs to be checked experimentally. Just the laser interaction in hydrogen plasma at temperatures above 100 eV should provide an example for this proof. One should be able to directly measure the absorption constant by lasers or microwaves in hydrogen plasma and check, whether there is a kink in the slope when the temperature changes from below 40 eV to above. Indirect measurements may relate to observations of stronger collisional absorption of laser light in hydrogen plasmas at temperatures above 50 eV than expected.

It is indeed an intriguing question (Hora, 1982) to what extent the simple dc-collision frequency ν of a plasma is related to inverse bremsstrahlung and to the question of the use or to the suppression of stimulated emission. Since this question could not be solved yet convincingly, we were using the classical collisions only in our evaluaton of the absorption constants in Fig. 6.1 and 6.3. While it is essential in most of the following theoretical models and in numerical evaluations not to neglect collisions, we checked many numerical cases of optical interaction with or without a quantum branch of the collision frequency, and no drastic changes of the

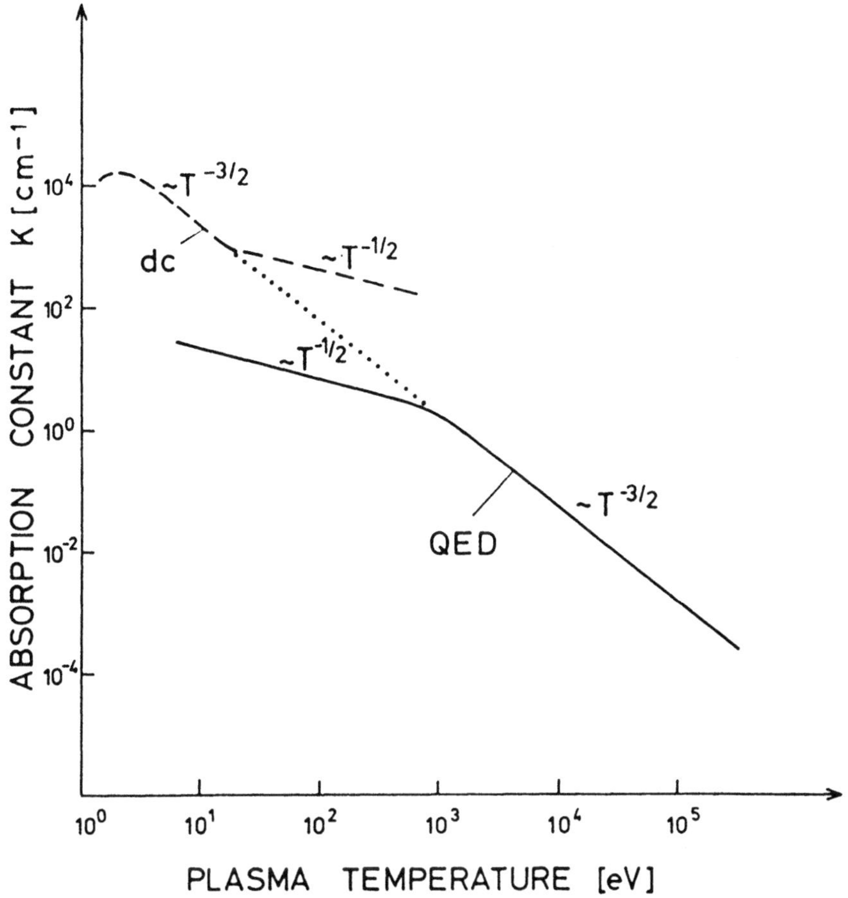

Fig. 6.5. Absorption constant K for optical (x-ray) radiation of 1 nm wavelength in plasma of density $n_e = 10^{24}$ cm^{-3} and at varying temperature T based on QED (___) or using the dc classical and quantum collision frequency (------). For $\hbar\omega<kT$, the gap of the $T^{-1/2}$ branches vanishes and QED and classical models agree apart from a modification factor of the order of one.

results occurred. However, for thermal conduction, thermal equilibration by equipartition, viscosity and electric conduction, the quantum branches are essential and may have then to be included correctly.

SEVEN

Waves in Inhomogeneous Plasma

Some properties of electromagnetic waves in plasmas with spatially varying refractive index (inhomogeneous plasma) will be discussed in this section before discussing the two-fluid equation of motion. Equation (6.5) and the necessary nonlinear generalization for laser-plasma interaction and some properties of electromagnetic waves in plasmas with spatially varying refractive index (inhomogeneous plasma) will be discussed in this section. It is important to note that the temporal change of the refractive index n is not taken into account. Only the wave equations are discussed,

$$\nabla^2\mathbf{H}_r+\frac{\omega^2}{c^2}n^2\mathbf{H}_r=0 \tag{7.1}$$

and

$$\nabla^2\mathbf{E}_r+\frac{\omega^2}{c^2}n^2\mathbf{E}_r=0 \tag{7.2}$$

These follow from Eqs. (6.24) and (6.30) if the terms with spatial derivatives of the refractive index n are neglected. The coupling between **E** and **H** is given by one of the Maxwell equations (6.10) or (6.12). The spatial dependence of the plasma frequency ω_p on the electron density n_e and the nonlinear and/or relativistic dependence on the magnitude of **E** and **H** as well as the dependence of the collision frequency ν on n_e, **E**, **H**, electron temperature T, and ion charge Z have all been described in the preceding chapter. This results in a spatial dependence of the refractive index n:

$$n^2=1-\frac{\omega_p^2}{\omega^2(1-i\nu/\omega)} \tag{7.3}$$

The solutions of the general stationary equations (6.24) and (6.29) including the spatial dependent term are neglected in this section and will be discussed in Section 11. The restriction to the simplified wave equations (7.1) and (7.2) is useful for understanding the essential behaviour of electromagnetic waves propagating in inhomogeneous media. The mathematical problem consists of the discussion of differential equations of the type

$$\frac{\partial^2}{\partial x^2} f(x) + \bar{a}(x) f(x) = 0 \tag{7.4}$$

$\bar{a}(x)$ is a given function and $f(x)$ has to be determined. This is an extensive problem for differential equations. A simple case is with $\bar{a}=\text{const}$ given by the general solutions with elementary functions.

$$\begin{aligned} f(x) &= C_1 \cos(\bar{a}^{1/2}x) + C_2 \sin(\bar{a}^{1/2}x), &&\text{if } (\bar{a} \geqslant 0) \\ &= C_3 \exp(\sqrt{-\bar{a}}x) + C_4 \exp(-\sqrt{-\bar{a}}x), &&\text{if } (\bar{a} \leqslant 0) \end{aligned}$$

with integration constants $C_1 \cdots C_4$. For general $\bar{a}(x)$, higher functions are determined where in most cases a very special algebraic form of $\bar{a}(x)$ defines a whole class of functions, for example, the Bessel functions, Hermite functions, confluent hypergeometric functions and others, between which, as well as with the elementary functions, relations exist and solutions in the form of series of higher functions exist. Since the computer era has been established, the solution of (7.4) for any reasonable $\bar{a}(x)$ is a relatively easy task for a direct numerical treatment (using Runge–Kutta difference schemes).

To understand the general properties of waves in inhomogeneous media, first approximate solutions of Eqs. (7.1) and (7.2) are discussed. They are called WKB solutions.

The properties of the waves can also be seen from the special case of an inhomogeneous medium with refractive index

$$n(x) = \frac{1}{1+\alpha x} \qquad (\alpha = \text{const}) \tag{7.5}$$

The corresponding differential equation (7.4) is Euler's differential equation. This has solutions with elementary functions which have been discussed by Rayleigh. The solutions for a linearly increasing $n(x)$ leads to the more complicated Airy functions.

7.1 WKB Approximation for Perpendicular Incidence

The approximate solution of a wave equation with spatially variable refractive index n [general $\bar{a}(x)$ in Eq. (7.4)] was known in the last century. It is called WKB approximation, after Wentzel, Kramers, and Brillouin, who used this approximation for the wave equation of electron waves (Schrödinger equation).

A linear polarized plane electromagnetic wave ($\mathbf{E}=\mathbf{i}_y E_y$ and $\mathbf{H}=\mathbf{i}_z H_z$) is assumed to be perpendicularly incident on a stratified plasma, whose n is only dependent on the x coordinate (the general elliptic polarized light can be considered as superposition of linear polarized waves of different **E**-directions and different temporal phase). Equations (7.1) and (7.2) are then

$$\frac{\partial^2}{\partial x^2} E_y + \frac{\omega^2}{c^2} n^2(x) E_y = 0 \tag{7.6}$$

and

$$\frac{\partial^2}{\partial x^2} H_z + \frac{\omega^2}{c^2} n^2(x) H_z = 0 \tag{7.7}$$

The WKB approximation is the ansatz

$$E_y = \frac{E_v}{|n|^{1/2}} \exp\left(i\frac{\omega}{c} \int^x \mathrm{Re}(n)\, d\xi - \frac{\omega}{c} \int^x \mathrm{Im}(n)\, d\xi \right) \tag{7.8}$$

or

$$E_y = \frac{E_v}{|n|^{1/2}} \exp iF; \qquad F = i\frac{\omega}{c} \int^x n\, d\xi \tag{7.9}$$

E_v is the vacuum amplitude of the incident field. This solution is valid in Eq. (7.6) only as long as certain terms are sufficiently small.
Differentiating Eq. (7.8), for $\nu \ll \omega (\rightarrow |n| \approx n)$

$$\frac{\partial}{\partial x} E_y = -\frac{E_v}{2|n|^{3/2}} (\exp iF) \frac{\partial |n|}{\partial x} + \frac{E_v}{\sqrt{n}} (\exp iF) i \frac{\omega}{c} |n|$$

$$\frac{\partial^2}{\partial x^2} E_y = \frac{3E_v}{4|n|^{5/2}} (\exp iF) \left(\frac{\partial |n|}{\partial x} \right)^2 - \frac{E_v}{2|n|^{3/2}} \frac{\partial^2 |n|}{\partial x^2} (\exp iF)$$
$$- \frac{E_v}{|n|^{3/2}} \frac{\partial |n|}{\partial x} i\frac{\omega}{c} |n| (\exp iF) + i\frac{\omega}{c} \frac{E_v}{|n|^{1/2}} (\exp iF) \left(\frac{\partial |n|}{\partial x} \right)$$
$$- \frac{\omega^2}{c^2} |n|^2 \frac{E_v}{|n|^{1/2}} (\exp iF)$$

and inserting into Eq. (7.6)

$$-\frac{\omega^2}{c^2} |n^2| \frac{E_v}{\sqrt{|n|}} (\exp iF) + \frac{3E_v}{4|n|^{5/2}} (\exp iF) \left(\frac{\partial |n|}{\partial x} \right)^2$$
$$- \frac{E_v}{2|n|^{3/2}} \frac{\partial^2 |n|}{\partial x^2} (\exp iF) + \frac{\omega^2}{c^2} |n|^2 \frac{E_v}{\sqrt{|n|}} (\exp iF) = 0$$

shows that Eq. (7.8) is a good approximation of Eq. (7.6), if

$$\frac{\sqrt{3}}{2|n|}\left|\frac{\partial n}{\partial x}\right| \ll \frac{\omega}{c}|n| \quad \text{or} \quad \Theta = \frac{\sqrt{3}}{2}\frac{c}{\omega|n|^2}\left|\frac{\partial n}{\partial x}\right| \ll 1 \tag{7.10}$$

and

$$\frac{1}{2|n|}\left|\frac{\partial^2 n}{\partial x^2}\right| \ll \frac{\omega^2}{c^2}|n^2| \quad \text{or} \quad \psi = \frac{1}{2|n|^3}\frac{c^2}{\omega^2}\left|\frac{\partial^2 n}{\partial x^2}\right| \ll 1 \tag{7.11}$$

The second-order equation (7.11) is usually not taken into account. The WKB condition (7.10) restricts the refractive index to a relatively small spatial variation. Using the definitions (7.8) and (6.32) and the geometry for plane waves perpendicularly incident on a stratified plasma leads to

$$\mathbf{E} = \mathbf{i}_y \frac{E_v}{|n|^{1/2}} \cos\left(\frac{\omega}{c}\int^x \mathrm{Re}(n)\, d\xi - \omega t\right) \exp\left(\frac{-\bar{k}x}{2}\right) \tag{7.12}$$

$\bar{k}$ is an averaged absorption constant

$$k = 2\frac{\omega}{xc}\int^x \mathrm{Im}(n)\, d\xi \tag{7.13}$$

The real part of **E** is given for further use in calculating forces. The denominator for E_v is used approximatively by the absolute value of n. This approximation causes the neglect of terms of the third order, as will be shown below from a discussion where the complex denominator of **E** is used.

The solution of Eq. (7.7) can be derived in a similar way. To derive the correct phase between E_y and H_z, the value of H_z is obtained by using **E** from Eq. (7.12) in the Maxwell equation (6.10) and by spatial differentiation and temporal integration. The spatial differentiation produces the following term with the derivative of n.

$$H_z = -\frac{ic}{2\omega}\frac{E_v}{|n|^{3/2}}\frac{dn}{dx}(\exp iF) + E_v\sqrt{|n|}\,(\exp iF) \tag{7.14}$$

or with an approximation of n by $|n|$

$$\mathbf{H} = -\mathbf{i}_z \frac{c}{2\omega}\frac{E_v}{|n|^{3/2}}\frac{d|n|}{dx}\sin\left(\frac{\omega}{c}\int^x \mathrm{Re}(n)\, d\xi - \omega t\right)\exp(-\tfrac{1}{2}\bar{k}x)$$
$$+\mathbf{i}_z E_v |n|^{1/2}\cos\left(\frac{\omega}{c}\int^x \mathrm{Re}(n)\, d\xi - \omega t\right)\exp(-\tfrac{1}{2}\bar{k}x) \tag{7.15}$$

The second terms in the last two equations correspond to the usual form of the plane waves in vacuum. Decreasing n leads to decreasing **H** and increasing **E** in Eqs. (7.15) and (7.12), respectively. This increase or swelling of the amplitude of **E** in a plasma will be of essential importance in the following sections. The product $|\mathbf{E}\times\mathbf{H}|$, however, is not influenced by $|n|$. Thus, the

Poynting vector or the electromagnetic energy flow is unchanged. The first terms in Eqs. (7.14) and (7.15) correspond to the phase shift between **E** and **H**. Similar properties are shown below from the discussion of the exact mathematical solution for the Rayleigh case. The phase shift term will be essential for the generation of forces in plasma due to the electromagnetic radiation. Decreasing Re(n) in a plasma causes an increase of the wavelength in the oscillating factors in Eqs. (7.12) and (7.15).

7.2 Oblique Incidence and WKB Solution

If plane waves are obliquely incident on a stratified plasma, they have to be connected analytically with the plane waves in vacuum outside the plasma. For vacuum, $n=1$, Eq. (7.2) is

$$\frac{\partial^2}{\partial x^2}\mathbf{E}_r+\frac{\partial^2}{\partial y^2}\mathbf{E}_r+\frac{\partial^2}{\partial z^2}\mathbf{E}_r+\frac{\omega^2}{c^2}\mathbf{E}_r=0 \tag{7.16}$$

Each vector component of $\mathbf{E}_r$ can be separated, for example, from the y-component, by a product ansatz.

$$E_y(x, y, z)=E_{yx}(x)E_{yy}(y)E_{yz}(z) \tag{7.17}$$

Substituting into Eq. (7.16) and dividing by E_y leads to

$$\frac{1}{E_{yx}}\frac{\partial^2}{\partial x^2}E_{yx}+\frac{1}{E_{yy}}\frac{\partial^2}{\partial_y^2}E_{yy}+\frac{1}{E_{yz}}\frac{\partial^2}{\partial_z^2}E_{yz}+\frac{\omega^2}{c^2}=0$$

In this way, the solution of the partial differential equation (7.16) is reduced to a solution of the ordinary differential equations

$$\frac{d^2}{dx^2}E_{yx}+(-k_x^2)E_{yx}=0$$

$$\frac{d^2}{d_y^2}E_{yy}+(-k_y^2)E_{yy}=0$$

$$\frac{d^2}{dz^2}E_{yz}+(-k_z^2)E_{yz}=0$$

The constant k_x, k_y, and k_z are called eigenvalues. The eigenfunctions E_{yx}, E_{xy}, and E_{yz} can be combined by Eq. (7.17) to give a solution

$$E_y=E_r\exp(i\mathbf{k}\cdot\mathbf{r}) \tag{7.18}$$

k is the propagation vector of the plane wave, whose components

$$k_x=\frac{\omega}{c}\cos u_x;\qquad k_y=\frac{\omega}{c}\cos u_y;\qquad k_z=\frac{\omega}{c}\cos u_z \tag{7.19}$$

determine the angle **u** between the direction of **k** and the x-, y-, or z-axis of a Cartesian coordinate system, respectively. Obviously

$$k_x^2+k_y^2+k_z^2=\frac{\omega^2}{c^2}$$

Without losing generality, the plane of incidence is assumed to be the x–y plane (Fig. 7.1). The plasma is still stratified so that n is only dependent on x.

The plane waves have an angle of incidence u_0 in vacuum. The solution of the Maxwell equations for a plane wave with an electric vector $\mathbf{E}_p$ oscillating in the plane of incidence (p-polarization) are then, from the preceding steps,

$$\mathbf{E}_p=E_v(-\mathbf{i}_x \cos u_0+\mathbf{i}_y \sin u_0) \cos\left[\frac{\omega}{c}(\cos u_0)x+\frac{\omega}{c}(\sin u_0)y-\omega t\right] \tag{7.20}$$

$$\mathbf{H}_p=E_v\mathbf{i}_z \cos\left[\frac{\omega}{c}(\cos u_0 x+\frac{\omega}{c}(\sin u_0)y-\omega t\right] \tag{7.21}$$

and, for the perpendicular polarization (s-polarization)

$$\mathbf{E}_s=E_v\mathbf{i}_z \cos\left[\frac{\omega}{c}(\cos u_0)x+\frac{\omega}{c}(\sin u_0)y-\omega t\right] \tag{7.22}$$

$$\mathbf{H}_s=E_v(\mathbf{i}_x \cos u_0-\mathbf{i}_y \sin u_0) \cos\left[\frac{\omega}{\tau}(\cos u_0)x+\frac{\omega}{c}(\sin u_0)y-\omega t\right] \tag{7.23}$$

The linear combination of the two cases generate the general polarization in vacuum.

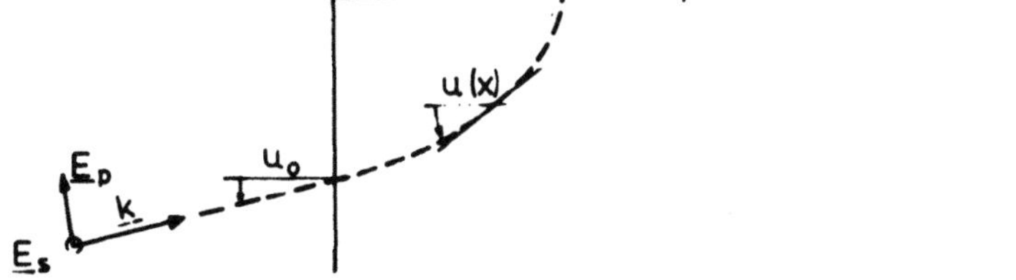

Figure 7.1 Linearly polarized plane waves with p- or s-polarization obliquely incident at an angle u_0 in vacuum onto a stratified plasma. This angle $u(x)$ varies in the plasma along the x-direction.

In this subsection, the formulation of the WKB approximation is given for the case of a collisionless plasma only ($\nu=0$), where the refractive index n is real. The substitution of the following solutions into the wave equations leads to first- and second-order restrictions of validity of the WKB approximation in a similar way to that of the preceding subsection for perpendicular incidence. These restrictions are

$$\Theta(u)=\frac{\sqrt{3}}{2}\frac{c}{\omega}\frac{1}{n^2\cos^2 u}\left|\frac{d(n\cos u)}{dx}\right|\ll 1 \tag{7.24}$$

$$\psi(u)=\frac{1}{2}\frac{c^2}{\omega^2}\frac{1}{n^2\cos^2 u}\left|\frac{d^2(n\cos u)}{dx^2}\right|\ll 1 \tag{7.25}$$

The angle $u(x)$ of the direction of wave propagation in plasma follows Snell's law:

$$n^2(x)\sin^2 u(x)=\sin^2 u_0 \tag{7.26}$$

This can be seen by analytical connection of the solution in vacuum with that in plasma.

First to be resolved is the question whether the WKB conditions (7.24) and (7.25) are met in plasmas. The general linearly polarized, obliquely incident plane wave is written as a sum of two parts, one with the oscillating **E** vector parallel to the plane of incidence (subscript p) and one with the **E** perpendicular to this plane (subscript s). It is well known that this separation is not generally possible [150]. In order to obtain the electric vector in the perpendicular case, the following equation has to be solved

$$\left(\frac{\partial^2}{\partial x^2}+\frac{\partial}{\partial y^2}+\frac{\omega^2}{c^2}n^2(x)\right)\mathbf{E}_s=0 \tag{7.27}$$

The WKB solution gives

$$\mathbf{E}_s=\mathbf{i}_z\frac{E_v\sqrt{\cos u}}{(n(x)-\sin^2 u_0)^{1/4}}\cdot\cos G \tag{7.28}$$

where

$$G=\pm\frac{\omega}{c}\left(\int_0^x\sqrt{n^2(\xi)-\sin^2(u_0)}\,d\xi+n(x)y\sin u(x)\right)+\omega t \tag{7.29}$$

The upper sign is chosen for an electromagnetic wave whose propagation direction has a component into the plasma. The lower sign is for the opposite direction. Applying the Maxwell equations, Eq. (7.28) leads to

$$\begin{aligned}\mathbf{H}_s=&\pm\frac{E_v\sqrt{n\cos u_0}}{\sqrt{\cos u(x)}}[\mathbf{i}_x\sin u(x)-\mathbf{i}_y\cos u(x)]\cos G\\ &-\mathbf{i}_y\frac{c}{\omega}\frac{E_v\sqrt{\cos u_0}}{2n^{3/2}(x)\cdot\cos^{5/2}u(x)}\frac{dn}{dx}\sin G\end{aligned} \tag{7.30}$$

In the case of E_p the electric vector is no longer perpendicular to the direction of propagation as it was for the general case, and the electric field is no longer divergenceless. The components of the electric field strength **E** are given by Eq. (6.31)

$$\left(\frac{\partial^2}{\partial x^2}+\frac{\partial^2}{\partial y^2}+\frac{\omega^2}{c^2}\, n^2(x)\right)E_{px}-\frac{\partial}{\partial x}\,\nabla\cdot\mathbf{E}_p=0 \tag{7.31}$$

and

$$\left(\frac{\partial^2}{\partial x^2}+\frac{\partial^2}{\partial y^2}+\frac{\omega}{c^2}\, n^2(x)\right)E_{py}-\frac{\partial}{\partial y}\,\nabla\cdot\mathbf{E}_p=0 \tag{7.32}$$

The last terms in Eqs. (7.31) and (7.32) couple all components of the electric field so that a representation as a sum of the two differently polarized components is not possible. In the case of the WKB condition, however, the last terms in Eqs. (7.31) and (7.32) are negligible, as will be shown. Using the relation

$$n(x)\nabla\mathbf{E}_p=-2E_{px}\frac{dn}{dx} \tag{7.33}$$

which can be derived from Ginzburg [150, Eq. (19.19)] from (7.31) and (7.32)

$$\left(\frac{\partial^2}{\partial x^2}+\frac{\partial^2}{\partial y^2}+\frac{\omega^2}{c^2}\, n^2(x)\right)E_{px}+2\frac{\partial}{\partial x}\left(E_{px}\frac{d\ln n(x)}{dx}\right)=0 \tag{7.34}$$

and

$$\left(\frac{\partial^2}{\partial x^2}+\frac{\partial^2}{\partial y^2}+\frac{\omega^2}{c^2}\, n^2(x)\right)E_{py}+2\frac{\partial}{\partial y}\left(E_{py}\frac{d\ln n(x)}{dx}\right)=0 \tag{7.35}$$

Comparison with the WKB conditions (7.24) and (7.25) shows that the time averaged last term in Eq. (7.34) is always $(\Theta+2\psi)$ times the $n^2(\mathbf{x})$ term in the first bracket of Eq. (7.34). Therefore, neglect of the last term in Eq. (7.35), compared with n^2, is possible in the same way when

$$2\tan(x)\sin u(x)\leqslant 1 \tag{7.36}$$

that is, when $u\leqslant 40°$. So, with the conditions given by Eqs. (7.24), (7.25), and (7.36), one can solve Eqs. (7.34) and (7.35) separately without coupling terms to get

$$\mathbf{E}_p=\frac{E_v\cos^{1/2}u_0}{(n(x)\cos u(x))^{1/2}}\left[-\mathbf{i}_x\sin u(x)+\mathbf{i}_y\cos u(x)\right]\cos G \tag{7.37}$$

Substitution of Eq. (7.37) into Maxwell's equations gives

$$\mathbf{H}_p=\frac{E_v(n\cos u_0)^{1/2}}{\cos^{1/2}u(x)}\,\mathbf{i}_z\left[\cos G-\frac{c}{\omega}\frac{\frac{1}{2}-\sin^2 u(x)}{n\cos u(x)}\frac{dn}{dx}\sin G\right] \tag{7.38}$$

These solutions for the s- and p-polarization will be used in the following evaluations.

7.3 The Rayleigh Profile

The Rayleigh profile of the refractive index $n(x)$ corresponds to a very special mathematically defined electron density profile of a plasma. It looks very artificial in the first instance. However, it is important for the electrodynamic forces in the plasma. Furthermore, this profile has the advantage of leading without any restriction to an exact solution of the Maxwell equations which can be described with elementary functions. This is used for a very basic study of the wave structure in inhomogeneous media and is a tool for comparison with numerical approximations.

The case of perpendicular incidence on a collisionless, stratified plasma will be considered. For a linearly polarized plane wave with

$$n=\frac{1}{1+\alpha x} \qquad \alpha\geqslant 0 \tag{7.39}$$

$$\mathbf{E}_r=i_yE_y; \qquad \mathbf{H}_r=i_zH_z \tag{7.40}$$

the complete wave equation (6.29) can be written as

$$\frac{\partial^2}{\partial x^2}E_y+\frac{\omega^2}{c^2}n^2E_y=0 \tag{7.41}$$

Equation (6.28) neglecting the second spatial derivative of the refractive index is

$$\frac{\partial^2}{\partial x^2}H_z-2\left(\frac{\partial}{\partial x}\ln n\right)\frac{\partial}{\partial x}H_z+\frac{\omega^2}{c^2}n^2H_z=0 \tag{7.42}$$

Using the Rayleigh profile of Eq. (7.39) for the refractive index in Eq. (7.42), for real $\alpha\geqslant 0$ one obtains

$$\frac{\partial^2}{\partial x^2}E_y+\frac{\omega^2}{c^2}\frac{1}{(1+\alpha x)^2}E_y=0 \tag{7.43}$$

The corresponding electron density, according to the definition of the plasma frequency ω_p [Eq. (2.6)] and of the cutoff density n_{ec} [Eq. (6.37)], is

$$n_e(x)=n_{ec}[1-(1+\alpha x)^{-2}] \qquad (x\geqslant 0) \tag{7.44}$$

As can be seen, a plasma with $n\leqslant 1$ is possible only for $\alpha\geqslant 0$. For $x\rightarrow\infty$, n_e is increasing from $n_e=0$ at $x=0$ monotonically to the cutoff density. No higher densities than the cutoff density are possible for the Rayleigh case without collisions.

For the solution of Eq. (7.43), the following substitution is used

$$\xi = 1 + \alpha x \tag{7.45}$$

An Euler differential equation is obtained

$$\xi^2 \frac{\partial^2}{\partial \xi^2} E_y + \frac{4\omega^2}{c^2\alpha^2} E_y = 0 \tag{7.46}$$

where

$$E_y = E_v \xi^{1\ 2 \pm 1/2[1 - 4\omega^2/(c^2\alpha^2)]^{1/2}} \tag{7.47}$$

After resubstitution of (7.45), the result is

$$E_y = (1+\alpha x)^{1/2} E_v \exp\left[\frac{i}{\mp 2} \left(\frac{4\omega^2}{c^2\alpha^2} - 1 \right)^{1/2} \ln(1+\alpha x) \right] \tag{7.48}$$

or using Eqs. (7.39), (7.40), and (6.20)

$$\mathbf{E} = \mathbf{i}_y \frac{E_y}{|n|^{1/2}} \exp\left\{ \frac{i}{\mp 2} \left[\frac{4\omega^2}{c^2\alpha^2} - 1 \right]^{1/2} \ln(1+\alpha x) - i\omega t \right\} \tag{7.49}$$

The solution for H_z can be derived from Eq. (7.42) or by substitution of E_y into a Maxwell equation and subsequent differentiation and integration exactly, for $\mu = 0$:

$$\mathbf{H} = \mp \left(\sqrt{1 - \frac{\alpha^2 c^2}{4\omega^2}} - i \frac{\alpha c}{2\omega} \right) \mathbf{i}_z E_v n \exp\left[\mp \frac{i}{2} \sqrt{\frac{4\omega^2}{c^2\alpha^2} - 1}\ \ln(1+\alpha x) - i\omega\iota \right] \tag{7.50}$$

The increase of the absolute value of **E** with z due to the decreasing refractive index n in Eq. (7.49) is exactly compensated by a decrease of **H** very similar to the WKB approximation. The signs in the exponential functions indicate a wave propagating to the $+x$ and $-x$ direction, respectively. The linear combination of both solutions is the complete solution of the differential equations. The change in the purely oscillating exponential factor of Eqs. (7.49) and (7.50) into that of the case of vacuum ($\alpha = 0$) is evident by Taylor expansion of the ln function near $x = 1$:

$$\lim_{\alpha \to 0} \exp[\cdots] = \exp\left(\pm \frac{i}{2} \sqrt{\frac{4+\omega^2}{c^2\alpha^2} - 1}\ \alpha x \right) = \exp\left[\mp i \frac{\omega}{c} x \right]$$

The phase between **E** and **H**, given by the complex first factor of **H** in (7.50), is responsible for reflection. If a medium with refractive index 1 (vacuum) for $x \leqslant 0$ is connected with a Rayleigh medium with a density distribution given by Eq. (7.44), then the refractive index and the density slope are continuous. If for $x > 0$ only a wave propagating to $+x$ is permitted, necessarily a reflected wave in the vacuum ($x \leqslant 0$) is necessary to fit the phase shift of

H. This reflection is very small, as long as

$$\alpha \ll \frac{2\omega}{c} \tag{7.51}$$

but it is definitely different from zero, despite the continuous connection between vacuum and inhomogeneous plasma.

Calculation of the reflection R [151] shows that $R < 10\%$ for $\alpha = \omega/c$. However, a quick rise appears. Total reflection occurs for

$$\alpha \geqslant \frac{2\omega}{c} \tag{7.52}$$

This phenomenon is curious again. For perpendicular incidence of a wave onto a *discontinuity* from vacuum to a nonabsorbing medium of a certain refractive index n the Fresnel formula

$$R = \frac{(1-n)^2}{(1+n)^2} \tag{7.53}$$

results in $R = 1$ only for $n = 0$ or $n = \infty$. The *continuous* connection of vacuum with a Rayleigh plasma, with a slow increase in electron density where $\alpha = 2\omega/c$, results in $R = 1$.

There was a basic discussion about the question of whether an "internal reflection" is generated in an inhomogeneous plasma or not. The treatment with ray-optical approximations or with an approximation using a sequence of stepwise homogeneous regions with exact Fresnel-type calculations of the reflectivity at each interface resulted in the standard opinion of an "internal reflection." The Rayleigh solutions, however, did not show an internal reflexion. There were simply the two penetrating waves, one to $+x$, the other to $-x$, and a reflection was produced only at a discontinuity of the derivatives of the refractive index. Osterberg [152] found the same very generally from a medium with an analytical function of $n(x)$: no internal reflection.

The solution is that the stepwise homogeneous approximation is a tool to determine the "local reflection," if the condition of a purely propagating wave should follow for $x \to +\infty$. This "Ausstrahlungsbedingung" for plane waves, in analogy to the Sommerfeld conditions for spherical waves, must be used in the analytical case as well as in the step approximation. The plane wave approximation of the two possible exact solutions is probing from point to point. This is illustrated by the following numerical examples, which, however, lead to a paradox, as shown by V. F. Lawrence [153].

The following media are considered (Fig. 7.2): vacuum ($x \leqslant 0$) is continuously connected with a Rayleigh medium ($0 \leqslant x \leqslant D$), which becomes a homogeneous plasma for $x \geqslant D$. The refractive index is real owing to the

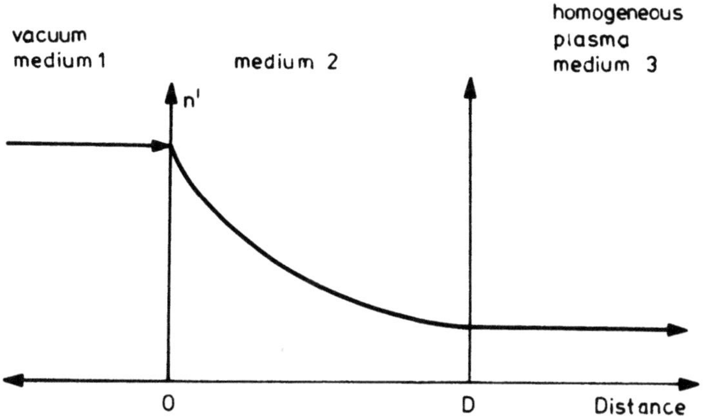

Figure 7.2 The schematic dependence on x of the refractive index n' for connections of homogeneous media 1 and 3 with an inhomogeneous Rayleigh-like density profile [153].

assumption of no absorption.

$$\begin{aligned} &n_1 = 1 = n_v && \text{medium 1} \quad x \leqslant 0 \\ &n_2 = \frac{1}{1+\alpha x} && \text{medium 2} \quad 0 \leqslant x \leqslant D \\ &n_3 = \frac{1}{1+\alpha D} = \text{const} && \text{medium 3} \quad D \leqslant x \end{aligned} \tag{7.54}$$

The components of the electric and magnetic fields can be described by plane waves

$$E_3 = C_{3+} \exp\left[i\left(n_3 \frac{\omega}{c} x - \omega t\right)\right] \tag{7.55}$$

$$H_3 = C_{3+} n_3 \exp\left[i\left(n_3 \frac{\omega}{c} x - \omega t\right)\right] - C_{3-} n_3 \exp\left[-i\left(n_3 \frac{\omega}{c} x + \omega t\right)\right] \tag{7.56}$$

The subscript denotes the medium and the constants C_+, C_-, the amplitudes of the transmitted and reflected waves, respectively. In medium 3 only the transmitted waves and no reflected waves are postulated. This is specified at $x = D$ by $C_{3+} = 1.0$, $C_{3-} = 0.0$. Then for medium 3 results at $t = 0$

$$E_3 = \exp\left(i n_3 \frac{\omega}{c} D\right) \qquad H_3 = n_3 \exp\left(i \frac{\omega}{c} n_3 D\right) \tag{7.57}$$

In the Rayleigh medium 2 from Eqs. (7.49) and (7.50) the components of the E and H fields are ($t = 0$)

$$E_{2+} = C_{2+} \sqrt{1+\alpha x} \exp\left[i \sqrt{\frac{\omega^2}{c^2 \alpha^2} - \frac{1}{4}} \ln(1+\alpha x)\right] \tag{7.58}$$

$$E_{2-}=C_{2-}\sqrt{1+\alpha x}\,\exp\left[-i\sqrt{\frac{\omega^2}{c^2\alpha^2}-\frac{1}{4}}\ln(1+\alpha x)\right] \tag{7.59}$$

$$H_{2+}=C_{2+}\frac{1}{\sqrt{1+\alpha x}}\left[\sqrt{1-\frac{\alpha^2}{4\omega^2c^2}}-\frac{i\alpha}{2\omega c}\right]\exp\left[i\sqrt{\frac{\omega^2}{c^2}-\frac{1}{4}}\ln(1+\alpha x)\right] \tag{7.60}$$

$$\mathbf{H}_{2-}=C_{2-}\frac{1}{\sqrt{1+\alpha x}}\left[-\sqrt{1-\frac{\alpha^2}{4\omega^2c^2}}+\frac{i\alpha}{2\omega c}\right]\exp\left[-i\sqrt{\frac{\omega^2}{c^2}-\frac{1}{4}}\ln(1+\alpha x)\right] \tag{7.61}$$

The indices y and z are dropped. The boundary conditions at the interfaces can be derived for $t=0$ at the junction of media 2 and 3, $x=D$:

$$E_{2+}+E_{2-}=E_3 \tag{7.62}$$

$$H_{2+}+H_{2-}=H_3 \tag{7.63}$$

The reflection and transmission coefficients of medium 2, C_{2+} and C_{2-}, can be derived algebraically.

The same can be written for waves in medium 1 where $n=1$

$$E_1=C_{1+}\exp\left(i\frac{\omega}{c}n_1x-i\omega t\right)+C_{1-}\exp\left(-i\frac{\omega}{c}n_1x-i\omega t\right) \tag{7.64}$$

$$H_1=C_{1+}n_1\exp\left(i\frac{\omega}{c}n_1x-i\omega t\right)-C_{1-}n_1\exp\left(-i\frac{\omega}{c}n_1x-i\omega t\right) \tag{7.65}$$

At the junction of medium 1 and medium 2 ($x=0$) at $t=0$, the continuity of E_y and H_z requires

$$E_1=C_1+C_{1-} \tag{7.66}$$

$$H_1=C_{1+}n_1-C_{1-}n_1 \tag{7.67}$$

$$E_2=C_{2+}+C_{2-} \tag{7.68}$$

$$H_2=(C_{2+}-C_{2-})\left[\left(1-\frac{\alpha^2}{4\omega^2c^2}\right)^{1/2}-\frac{i\alpha}{2\omega c}\right] \tag{7.69}$$

Equating

$$E_1=E_2;\qquad H_1=H_2 \tag{7.70}$$

the reflection coefficient R_A of the amplitudes as an exact solution depending on α and D is obtained:

$$R_A=\frac{E_2-H_2}{E_2+H_2} \tag{7.71}$$

Now the case of Fig. 7.2 is approximated by a series of homogeneous media with a stepwise decreasing refractive index n in the Rayleigh medium 2. Equations (7.55) and (7.57) remain the same for medium 3. In medium 2, however, using plane wave approximations at $t=0$, the first-order single-step approximation with $n_2=(n_1+n_3)/2$ leads to

$$E_2=C_{2+}\exp\left(in_2\frac{\omega}{c}x\right)+C_{2-}\exp\left(-in_2\frac{\omega}{c}x\right) \tag{7.72}$$

$$H_2=C_{2+}n_2\exp\left(in_2\frac{\omega}{c}x\right)-C_{2-}n_2\exp\left(in_2\frac{\omega}{c}x\right) \tag{7.73}$$

For higher approximations $n_2^i=(n_2^{i+1}+n_2^i)/2$ are used. i denotes the slab number depending on the number of steps. It is then possible to calculate the reflectivity from the conditions of medium 3 back to medium 1, where the refractive index $n_2(x)$ in medium 2 is given by the Rayleigh case.

It is necessary to note that in using step approximations the refractive index varies, depending on the number of steps used. For each step beginning at distance D, the values of C_{2+} and C_{2-} are calculated These constants are used as a basis for the next step approximation until $x=0$. The analytical result of the first approximation is derived here. Eqs. (7.63) and (7.64) can be written for $n=1$, $t=0$ as

$$E_1=C_{1+}\exp\left(i\frac{\omega}{c}x\right)+C_{1-}\exp\left(-i\frac{\omega}{c}x\right) \tag{7.74}$$

$$H_1=C_{2+}\exp\left(i\frac{\omega}{c}x\right)-C_{1-}\exp\left(-i\frac{\omega}{c}x\right) \tag{7.75}$$

At the junction of medium 1 and medium 2 ($x=0$), $n=1$, and $t=0$

$$C_{1+}+C_{1-}=C_{2+}+C_{2-} \tag{7.76}$$

$$C_{1+}-C_{1-}=n_2C_{2+}-n_2C_{2-} \tag{7.77}$$

so that

$$R=\frac{C_{1+}}{C_{1-}}=\frac{(1+n_2)C_{2+}+(1-n_2)C_{2-}}{(1-n_2)C_{2+}+(1+n_2)C_{2-}} \tag{7.78}$$

The following calculations are performed for a set of cases, where the Rayleigh parameter α is set constant (for a wavelength $\lambda=1.06$ μm corresponding to the wavelength of neodymium glass laser radiation) and the thickness D of the Rayleigh medium is varied. Figure 7.3 shows the results, where the abscissa gives D in micrometers and the ordinate the reflectivity. It is obvious that an oscillation of the reflectivity R is found with zero values at such thicknesses D, where the phases of the incident and reflected waves in the Rayleigh medium are just canceling the reflection, as known from the interference for transmission of light through parallel plates. Evidently, a higher

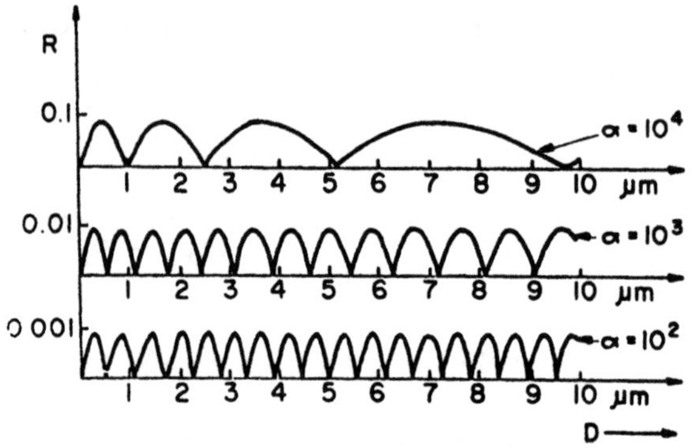

Figure 7.3 Reflectivity R depending on the distance D of the Rayleigh-like plasma interface between homogeneous plasmas described in Fig. 7.2. The parameter α determines the refractive index n from Eq. (7.39). The computed values of analytical solutions for the reflectivity, Eq. (7.7), is shown for the decreasing refractive index [153].

α leads to a higher reflectivity. At $\alpha = 2\omega/c = 1.18 \times 10^5$ cm^{-1}, total reflection at the discontinuity between the profile and the vacuum occurs, although the waves are perpendicularily incident as discussed above.

The maxima of R are of the same height. This is because no absorption is assumed and the reflection is determined only at $x=0$ and $x=D$. As known from the Rayleigh case [151], in agreement with the general result of Osterberg [152], the reflection is only determined by α. Therefore, the maxima are of constant value, although the refractive index of medium 3 is monotonically decreasing with increasing D.

The stepwise approximation gives for a small number of steps corresponding minima, but increasing maxima with medium thickness D, by orders of magnitudes larger than the exact case. This can be understood from the crude approximation of the refractive index n, which shows the insufficiency of the approximation with few steps. This is hardly surprising as large numbers of steps imply the use of small mesh sizes, which increases the accuracy of calculations.

The numerical calculations with large numbers of steps (100 or 1000) Fig. 7.4 converges to the exact case. The same reflectivity as in the exact case, the same constancy of the maxima and the corresponding distances of zero reflectivities are found. On closer observation, however, there is a slight difference: zero reflectivity distances increase for the stepwise approximation. Such a "wavelength" effect is of a basic nature. Any numerical inaccuracies or instabilities were excluded, there is a definite paradox in the difference between the exact solution and the asymptotic value of the stepwise model.

Leaving aside the aforementioned paradox, the Osterberg problem can be discussed in the following manner. It is a mathematical fact that there are two exact linearly independent solutions in the homogeneous medium,

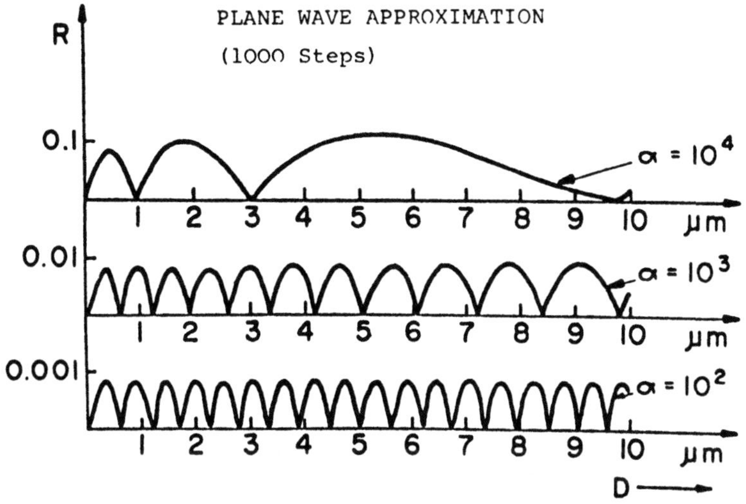

Figure 7.4 The plane wave stepwise approximation using 1000 steps for the corresponding calculations of reflectivity R as in Fig. 7.3. Both figures have varying refractive indices depending on x [153].

whose ratios are the reflectivity determined by the boundaries to the homogeneous media. The condition of only penetrating waves, that is, only transmitted waves in medium 3 (no standing waves), determines the phase between **E** and **H** in media 1 and 2. The phase depends on the thickness D and on α. This is originally reproduced by the stepwise approximation

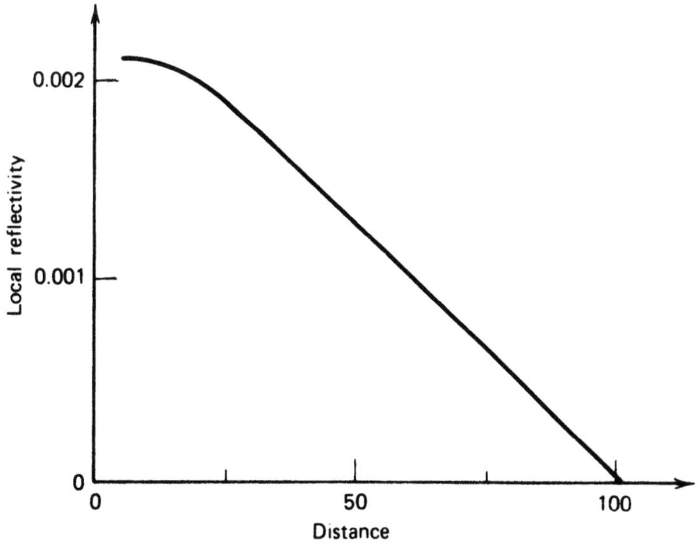

Figure 7.5 Variation of the reflection on the coefficient R of the plane wave approximation with distance in medium 2, representing a local reflection.

(apart from the paradox). However, in the case of approximation, the reflection coefficient of the plane waves within the steps decreases from medium 1 to 3, Fig. 7.5. Therefore, the results of Osterberg "nonreflectivity" and the plane waves "local reflectivity" do not contradict each other.

The problem is how the solutions for the inhomogeneous medium 2 have to be determined. The only condition is similar to that of Sommerfeld's spherical radiation condition (at large distance r, the amplitude has to decrease as $1/r$) expressed here for the plane waves: when $x \to +\infty$, the solution is approximately that of a homogeneous medium with only forward propagating waves and no standing waves, while any approximation of the exact case by fine steps for other x will produce, in effect, internal reflection or standing waves. The exact solution does not exhibit reflection properties and the stepwise approximation is, so to speak, a probe for mathematically detecting local reflectivity.

7.4 The Airy Profiles

One example of an exact solution of the wave equation for an inhomogeneous plasma with a special refractive index profile will be mentioned, which is solved by higher functions. The refractive index n should have a linear spatial dependence

$$n^2 = -ax + is \tag{7.79}$$

where

$$a = \frac{2\Theta\omega}{c}; \qquad s = \frac{\nu}{\omega} \ll 1 \tag{7.80}$$

using Θ from Eq. (7.10). The wave equation for plane waves perpendicularly incident for the E_y component is then [154]

$$\frac{\partial^2 E_y}{\partial x^2} + \frac{\omega^2}{c^2}(-ax+is)E_y = 0 \tag{7.81}$$

Using the substitution

$$\zeta = \left(\frac{\omega}{ca}\right)^{2/3}(-ax+is) = \rho^{2/3}(-ax+is); \qquad \rho = \frac{\omega}{ca} \tag{7.82}$$

Eq. (7.81) reduces to

$$\frac{d^2 E_y}{d\zeta^2} + \zeta E_y = 0 \tag{7.83}$$

The solution of this differential equation is the well-known Airy function a_i, which can be expressed by Bessel functions J and I (see Watson [155]) of the order $\frac{1}{3}$

$$E_y = 3Aa_i(-\zeta) = \begin{cases} A\zeta^{1/2}[J_{1/3}(\frac{2}{3}\xi^{3/2}) + J_{-1/3}(\frac{2}{3}\xi^{3/2})]; & \text{if Re } \xi > 0 \\ A(-\zeta)^{1/2}[I_{-1/3}(\frac{2}{3}(-\xi)^{3/2}) + I_{1/3}(\frac{2}{3}(-\xi)^{3/2})]; & \text{if Re } \xi < 0 \end{cases} \tag{7.84}$$

where

$$A = \tfrac{2}{3}\pi^{1/2}\left(\frac{\omega}{ca}\right)^{1/6} E_v \exp\left[\frac{i\omega}{c}\int_{-i/a}^{+is/a} n^{1/4}\, dz - \frac{i\pi}{4}\right] \tag{7.85}$$

The solution of the magnetic field is derived from the Maxwellian equations

$$H_z = i\zeta^{-1/3}\frac{dE_y}{d\zeta} \tag{7.86}$$

These solutions for E_y and H_z will be used below when the results derived by Lindl and Kaw [154] are discussed.

EIGHT

Equation of Motion

This section is devoted to the central question of the basic equations for the nonclassical treatment of laser produced plasmas, the equation of motion including its nonlinear properties. The equation of motion contains the nonlinear forces of which one part is the high-frequency (optical) extension of the ponderomotive forces, and of which another part, however, is of non-ponderomotive nature.

One should remember the steps of the preceding discussions: after the introduction to the very complex phenomena of laser-plasma interaction with fast ions, nonlinearities, and other unusual observations (Section 1), basic microscopic plasma parameters as the Debye length and the plasma and collision frequency were derived (Section 2). The properties of the kinetic theory were then discussed showing the derivation of macroscopic equations from the Boltzmann equation, the nondissipative return into equilibrium (Landau damping), but also indicating basic problems with respect to the Liouville equation and irreversibility (Section 3). After introducing and evaluating hydrodynamics (Section 4), the first global description of laser produced plasmas based on the self-similarity model was possible (Section 5) where the necessary fast thermalization for times of several nanoseconds of interaction was confirmed by agreement with experiments. Section 6 derived the basis of the macroscopic plasma theory (two-fluid model of magnetohydrodynamics) including the electrodynamics, on which basis the optical properties as refractive index and absorption (nonlinear and relativistic) were deduced. The wave-optical properties of the plasma with respect to its inhomogeneities, propagation problems, and reflection were described in Section 7.

The following discussion of the equation of motion (conservation of momentum) for plasmas including electrodynamics is essentially related to the combination of mechanical and electrodynamic phenomena, which are basically of nonlinear nature, as for example, electrostriction, magnetostriction, or the Maxwellian stress tensor, to mention the long-known phenomena. In Section 6, the plasma dynamics were based on the two Euler equations, one for ions and the other for electrons. By adding both equations, an equation of motion (first Schlüter equation) for the net plasma was achieved [136]; whereas by subtracting the Euler equations, the second Schlüter equation was found: the generalized Ohm's law, with which the Lorentz type electrodynamics of plasma was formulated and the wave properties in inhomogeneous media were derived.

This section will discuss the equation of motion, the first Schlüter equation (6.5), and its generalization as far as it is needed for the nonlinear processes of laser plasma interactions. To take it in advance, the following general equation of motion reproduces all known results of subrelativistic plasma dynamics at interactions with plane electromagnetic waves. The treatment of laser beams of finite diameter involves the solution of a basic question including the need for exactness in the presumed linear theories.

If plasma physics is based on kinetic equations, see Section 3, Spitzer derived an equation of motion given by a force density **f** [107]

$$\mathbf{f}=m_i n_i \frac{\partial \mathbf{v}}{\partial t}+m_i n_i \mathbf{v}\cdot\nabla\mathbf{v}=-\nabla p+\frac{1}{c}\mathbf{j}\times\mathbf{H} \qquad \text{(Spitzer)} \qquad (8.1)$$

Several simplifications were made and terms neglected in the derivation. In Schlüter's derivation [136] of the equation of motion, (6.5), from adding the Euler equations for electrons and ions,

$$\mathbf{f}=-\nabla p+\frac{1}{c}\mathbf{j}\times\mathbf{H}-\frac{1}{4\pi}\frac{\omega_p^2}{\omega^2}\mathbf{E}\cdot\nabla\mathbf{E} \qquad \text{(Schlüter)} \qquad (8.2)$$

one nonlinear term appeared, see Eq. (6.6). The most general equation of motion, satisfying all cases of plane optical wave interaction with plasma, is [138]

$$\begin{aligned}\mathbf{f}=&-\nabla p+\frac{1}{c}\mathbf{j}\times\mathbf{H}+\frac{1}{4\pi}\mathbf{E}\nabla\cdot\mathbf{E}\\ &-\frac{1}{4\pi}\frac{\omega_p^2}{\omega^2+\nu^2}\left(1+\frac{\nu}{\omega}\right)\mathbf{E}\nabla\cdot\mathbf{E}-\frac{1}{4\pi}\frac{\omega_p^2}{\omega^2+\nu^2}\left(1+\frac{\nu}{\omega}\right)\mathbf{E}\cdot\nabla\mathbf{E} \qquad (8.3)\\ &-\frac{1}{4\pi}\mathbf{E}\mathbf{E}\cdot\nabla\ \frac{\omega_p{}^2}{\omega^2+\nu^2}\left(1+i\frac{\nu}{\omega}\right)\end{aligned}$$

where the derivation is based on the exact connection with the Maxwellian stress tensor, by which way the problems of dispersion and dissipation were solved automatically (to be described). It should be mentioned that Dragila

[156] derived an additional term in Eq. (8.3), $(\partial\varepsilon/\partial t)\mathbf{E}\times\mathbf{B}/(4\pi c)$, which, however, only contributes to relativistic plasma dynamics. While we shall include microscopic relativistic effects, for example, relativistic electron quivering (Section 6.5), we shall not consider effects where the net macroscopic plasma velocities or currents are becoming relativistic. We can drop therefore the Dragila term and other terms of relativistic dynamics.

In Eq. (8.3) "nonlinear terms" follow the thermokinetic force

$$\mathbf{f}_{th} = -\nabla p \tag{8.4}$$

These remaining terms in Eqs. (8.2) and (8.3) can be seen for the case of electromagnetic waves, to introduce quantities **j**, **E**, and **H** which are fast oscillating at least with the frequency ω, or as a result of a product, with higher harmonics. As binary products of these quantities appear, oscillations with 2ω are produced by these "nonlinear terms."

Historically, forces are called ponderomotive forces if they are generated by electric or magnetic fields. It will be shown by a special example that, apart from the purely field generated ponderomotive terms, mixed terms exist. These nonponderomotive terms are characterized by dissipative effects and are proportional to the collision frequency. Due to these mixed terms the terminology "nonlinear force" f_{NL} will be preferred:

$$\mathbf{f}_{NL} = \mathbf{f} - \mathbf{f}_{th} \tag{8.5}$$

This terminology was accepted by several authors [156–157].

8.1 Equivalence to Maxwellian Stress Tensor

The derivation of the most general formulation up to the present stage of the equation of motion (8.3) or of the nonlinear force (8.5)

$$\mathbf{f}_{NL} = \frac{1}{c}\mathbf{j}\times\mathbf{H} + \frac{1}{4\pi}\mathbf{E}\nabla\cdot\mathbf{E} + \frac{1}{4\pi}\nabla\cdot(n^2-1)\mathbf{E}\mathbf{E} \tag{8.6}$$

is connected with a physical criterion: the forces due to obliquely incident (plane wave) laser radiation on a stratified collisionless plasma must produce a net force only toward or against the plasma surface, because no momentum transfer is possible perpendicular to the surface. This criterion of oblique incidence is a tool to check whether the nonlinear force (8.6) is complete, whether it contains all and not too many terms. This was shown in 1969 [138]. In the meantime, many authors could not suppress their excitement in discovering several nonlinear terms, for which only one measure exists: these terms are those that exist only since 1969 (the rederivation nevertheless is appreciable and can lead to new insights) or they are incomplete or over-determining and therefore wrong.

The direct derivation of Eq. (8.6) from the two-fluid theory is given in Appendix C. With regard to the assumptions involved there, especially on the interpretation of (oscillating) space charges and high frequency dielectric displacement, one should be skeptical about their correctness; but we have the tool for correctness, the criterion of oblique incidence. This can be shown by algebraic rewriting of Eq. (8.6) into an expression using the Maxwellian stress tensor that we discuss in this subsection.

Historically, the steps leading to this formulation were quite complicated and required prior steps. For perpendicular incidence, the Schlüter formulation, Eq. (8.2), was sufficient, arriving at a nonlinear force of laser plasma interaction which indicated the basic mechanism of the gradient of the refractive index [158]. The formula agreed at low density approximation with the force-formula derived at microwave interaction with plasmas [159–161], but in cases of absorption, first discussed 1969 [138], interpretation of the basic formulation (8.6) led to interpretations of nonponderomotive terms of the nonlinear force [this has nothing to do with the thermokinetic force (8.4)] [162–164] and higher order terms [165].

The motivation to use the formulation of the Maxwellian stress tensor [138] came from the expression of Landau and Lifshitz [166] where the problems of variable electromagnetic fields had not yet been solved, however. These authors underlined that the time averaged stress tensor for generating forces in matter in a variable electromagnetic field had not been derived then. Landau and Lifshitz used only nondispersive and nonabsorbing media. Pitaevski found a generalization to dispersive fluids with high frequency electric fields and only static magnetic fields [167]. Our formal use of this very restricted Landau–Lifshitz formula [166] as equation (6) in reference [138] was therefore for the completely different conditions of high frequency electromagnetic waves in dispersive plasma with absorption. This was justified later only by algebraic identity to the general plasma equation (8.3) derived from the two-fluid model and was proved only by the previously mentioned criterion of oblique incidence [138].

Following these outlines, the mathematical transformation of the general equation of motion, Eq. (8.3), including all nonlinear terms, into one, using the Maxwellian stress tensor is given. For this derivation, vector algebraic and vector analytic identities as well as the Maxwell equations are used, where, finally, the Poynting vector **S** will be included.

$$\frac{\partial}{\partial t}\left(\frac{\mathbf{E}\times\mathbf{H}}{4\pi c}\right)=\frac{\partial}{\partial t}\frac{1}{c^2}\mathbf{S} \tag{8.7}$$

Compared with quantities of the nonlinear force, this is less by the ratio

$2\pi/\omega\tau_R$. τ_R is the rise time of the laser pulse. For neodymium glass lasers $\tau_R \simeq 10^{-12}$ sec and $\omega/2\pi = 2.8 \times 10^{14}$ sec^{-1}. A similar ratio is valid for CO_2 lasers. In the following sections, the Poynting term can be neglected, but the derivation here is without any neglect. Rewriting Eq. (8.5) from Eq. (8.3) as

$$\mathbf{f}_{NL} = \mathbf{A} + \mathbf{B} \tag{8.8}$$

where

$$\mathbf{A} = \frac{1}{c}\mathbf{j} \times \mathbf{H} \tag{8.9}$$

$$\mathbf{B} = \frac{1}{4\pi}\mathbf{E}\nabla\cdot\mathbf{E} + \frac{1}{4\pi}\nabla\cdot(n^2-1)\mathbf{E}\mathbf{E} \tag{8.10}$$

The refractive index n is defined by Eq. (6.27). From Eq. (6.18) $\mathbf{j}$ is eliminated

$$\mathbf{j} = \frac{c}{4\pi}(\nabla \times \mathbf{H}) - \frac{1}{4\pi}\frac{\partial}{\partial t}\mathbf{E} \tag{8.11}$$

and inserted in Eq. (8.9)

$$\begin{aligned}\mathbf{A} &= \frac{1}{4\pi}(\nabla \times \mathbf{H}) \times \mathbf{H} - \frac{1}{4\pi c}\left(\frac{\partial}{\partial t}\mathbf{E}\right) \times \mathbf{H} \\ &= -\frac{1}{4\pi}\mathbf{H} \times (\nabla \times \mathbf{H}) - \frac{1}{4\pi c}\left(\frac{\partial}{\partial t}\mathbf{E}\right) \times \mathbf{H}\end{aligned} \tag{8.12}$$

Using the vector identity

$$\mathbf{H} \times (\nabla \times \mathbf{H}) = \tfrac{1}{2}\nabla\mathbf{H}^2 - \mathbf{H}\cdot\nabla\mathbf{H} \tag{8.13}$$

$\mathbf{A}$ can be written as

$$\mathbf{A} = -\frac{1}{4\pi}\left[\frac{1}{2}\nabla\mathbf{H}^2 - \mathbf{H}\cdot\nabla\mathbf{H}\right] - \frac{1}{4\pi c}\left(\frac{\partial}{\partial t}\mathbf{E}\right) \times \mathbf{H} \tag{8.14}$$

Turning to $\mathbf{B}$, Eq. (8.10)

$$\mathbf{B} = \frac{1}{4\pi}\mathbf{E}\nabla\cdot\mathbf{E} + \frac{1}{4\pi}\nabla\cdot n^2\mathbf{E}\mathbf{E} - \frac{1}{4\pi}\mathbf{E}\nabla\cdot\mathbf{E} - \frac{1}{4\pi}\mathbf{E}\cdot\nabla\mathbf{E} \tag{8.15}$$

can be changed to

$$\mathbf{B} = \frac{1}{4\pi}n^2\mathbf{E}\cdot\nabla\mathbf{E} + \frac{1}{4\pi}\mathbf{E}\nabla\cdot(n^2\mathbf{E}) - \frac{1}{4\pi}\mathbf{E}\cdot\nabla\mathbf{E} \tag{8.16}$$

From Eq. (6.27),

$$\nabla\cdot(n^2\mathbf{E}) = 4\pi\rho_e \tag{8.17}$$

the time average of which is zero in a space charge neutral plasma [136]. We note that the space charge density ρ_e cannot be neglected here, as oscillating space charges of the frequency ω can be generated by the electromagnetic waves. This process is the driving of electrostatic (Langmuir) waves in a plasma and is possible in a plasma with incident electromagnetic waves of frequency ω. These electrostatic waves will be induced even at plasma densities n_e where no reaonance exists with the electrostatic oscillation at the plasma frequency, see Fig. 2.1.

Adding zero by the following additional two last terms and combining the first terms in Eq. (8.16)

$$\mathbf{B}=\frac{1}{4\pi}\left[\nabla\cdot n^2\mathbf{EE}-\mathbf{E}\cdot\nabla\mathbf{E}+\tfrac{1}{2}\nabla\mathbf{E}^2-\tfrac{1}{2}\nabla\mathbf{E}^2\right] \tag{8.18}$$

the second and the third term are combined to

$$\frac{1}{4\pi}\mathbf{E}\times(\nabla\times\mathbf{E})=\frac{1}{8\pi}\nabla\mathbf{E}^2-\frac{1}{4\pi}\mathbf{E}\cdot\nabla\mathbf{E}. \tag{8.19}$$

Using the Maxwell equation (6.18) in Eq. (8.19) leads to

$$\mathbf{B}=\frac{1}{4\pi}\left[\nabla\cdot n^2\mathbf{EE}-\tfrac{1}{2}\nabla\mathbf{E}^2\right]-\frac{1}{4\pi c}\mathbf{E}\times\frac{\partial}{\partial t}\mathbf{H} \tag{8.20}$$

Writing **B** from Eq. (8.20) and **A** from Eq. (8.14), Eq. (8.8) is

$$\mathbf{f}_{\mathrm{NL}}=\frac{1}{4\pi}\nabla\cdot\left[\mathbf{EE}+\mathbf{HH}-\tfrac{1}{2}(\mathbf{E}^2+\mathbf{H}^2)\mathbf{1}+(n^2-1)\mathbf{EE}\right]-\frac{1}{4\pi c}\frac{\partial}{\partial t}\mathbf{E}\times\mathbf{H} \tag{8.21}$$

where use was made of

$$\mathbf{H}\cdot\nabla\mathbf{H}=\nabla\cdot\mathbf{HH}-\mathbf{H}\nabla\cdot\mathbf{H} \tag{8.22}$$

with a negligible last term because of Eq. (6.15). The unity tensor **1** is in Cartesian coordinates defined by

$$\mathbf{1}=\mathbf{i}_x\mathbf{i}_x+\mathbf{i}_y\mathbf{i}_y+\mathbf{i}_z\mathbf{i}_z \tag{8.23}$$

Eq. (8.20) can be written in the form of

$$\mathbf{f}_{\mathrm{NL}}=\nabla\cdot\left[\mathbf{T}+\frac{n^2-1}{4\pi}\mathbf{EE}\right]-\frac{1}{4\pi c}\frac{\partial}{\partial t}\mathbf{E}\times\mathbf{H}. \tag{8.24}$$

T is the Maxwellian stress tensor

$$\mathbf{T}=\frac{1}{4\pi}\left[\mathbf{EE}+\mathbf{HH}-\tfrac{1}{2}(\mathbf{E}^2+\mathbf{H}^2)\mathbf{1}\right] \tag{8.25}$$

The components **T** are given by the scalar components of **E** and **H**

$$4\pi\mathbf{T}=\begin{pmatrix} 0.5(E_x^2-E_y^2-E_z^2+H_x^2-H_y^2-H_z^2) & E_xE_y+H_xH_y & E_xE_z+H_xH_z \\ E_xE_y+H_xH_y & 0.5(-E_x^2+E_y^2-E_z^2-H_x^2+H_y^2-H_z^2) & E_yE_z+H_yH_z \\ E_xE_z+H_xH_z & E_yE_z+H_yH_z & 0.5(-E_x^2-E_y^2+E_z^2-H_x^2-H_y^2+H_z^2) \end{pmatrix} \tag{8.26}$$

Formula (8.24) is very similar for the formulation of the force density in a dielectric medium without dispersion following Landau and Lifshitz [166]. The only problem is that a plasma is a medium with dispersion, and an appropriate interpretation of the factor of **EE** in the brackets in Eq. (8.24) is necessary [162]. If no absorption is present, the force is purely of a ponderomotive nature, while collisions cause an extension (see the following Section 8.3) to nonponderomotive terms of the nonlinear force.

For the very special case of plane waves of linear polarization (**E** in y-direction), perpendicularly incident on a stratified, inhomogeneous plasma with a depth along the x-axis, is from Eq. (8.24), given by

$$\mathbf{f}_{\mathrm{NL}} = -\frac{\mathbf{i}_x}{8\pi}\frac{\partial}{\partial x}(E_y^2 + H_z^2) \tag{8.27}$$

where the Poynting term has been neglected for the reasons mentioned before Eq. (8.8). This formula will be used as the most general expression for describing nonlinear forces in stratified plasma with perpendicularly incident waves. One curiosity should be pointed out at this stage. When deriving the nonlinear force for perpendicularly incident plane waves from the ($\mathbf{j}\times\mathbf{H}$) expression, Eq. (8.3), without collisions [158], the term with $\partial n/\partial x$ in the WKB solution of **H**, Eq. (7.15), was essentially necessary. When evaluating Eq. (8.27), this term is not necessary to arrive at the same formula of the nonlinear force for the collisionless case. A refractive index n, following the WKB condition, is used, Eq. (7.10),

$$\Theta = \frac{c}{2\omega}\frac{\sqrt{3}}{|n|^2}\frac{\partial |n|}{\partial x} \ll 1 \tag{8.28}$$

The complex electric and magnetic field strengths are then

$$\mathbf{E} = \mathbf{i}_2 \frac{E_v}{|n|^{1/2}} \exp iF_0 \exp[\mp \bar{k}(x)x/2] \tag{8.29}$$

and approximately)

$$\mathbf{H} = \mathbf{i}_3 E_v |n|^{1/2} \exp iF_0 \exp[\mp \bar{k}(x)x/2] \tag{8.30}$$

where

$$F_0 = \omega\left(t \mp \int^x (\mathrm{Re}\, n(\xi)/c)\, d\xi\right) \tag{8.31}$$

Terms with derivatives of n in space are neglected here, for these only give terms of second and higher order in the following time-averaged expressions. By selecting only the real parts, from Eq. (8.29) is found

$$E_y^2 = E_v^2\left[\left(\frac{1}{\mathrm{Re}|n|^{1/2}}\right)^2 \cos^2 F_0 + \left(\mathrm{Im}\frac{1}{|n|^{1/2}}\right)^2 \sin^2 F_0 \right. \tag{8.32}$$
$$\left. + \tfrac{1}{2}\mathrm{Re}|n|^{-1/2}\,\mathrm{Im}|n|^{-1/2} \sin 2F_0\right] \exp(\bar{k}(x)x/2)$$

This leads to the time-averaged value

$$\overline{E_y^2}=\frac{E_v^2}{2|n|}\exp(\mp\bar{k}(x)x/2) \tag{8.33}$$

In the same way, from Eq. (8.30) is obtained

$$\overline{H_z^2}=\pm\tfrac{1}{2}E_v^2|n|\exp(\mp\bar{k}(x)x/2) \tag{8.34}$$

The time-averaged nonlinear force density (8.27) is then

$$\overline{\mathbf{f}_{\mathrm{NL}}}=-\mathbf{i}_x\frac{E_v^2}{16\pi}\frac{\partial}{\partial x}\left(\frac{1}{|n|}+|n|\right)\exp(\mp\bar{k}(x)x/2) \tag{8.35}$$

or

$$\begin{aligned}\overline{\mathbf{f}_{\mathrm{NL}}}=&\,\mathbf{i}_x\frac{E_v^2}{16\pi}\frac{1-|n|^2}{|n|^2}\exp(\mp\bar{k}(x)x/2)\frac{\partial|n|}{\partial x}\\ &\pm\mathbf{i}_x\frac{E_v^2}{16\pi}\frac{1+|n|^2}{|n|^2}\frac{2\omega}{c}\,\mathrm{Im}(n)\exp(\mp\bar{k}(x)x/2)\end{aligned} \tag{8.36}$$

In the collisonless case with $k=0$ and n taken from Eq. (6.36), the second term of Eq. (8.36) vanishes and the following result is obtained

$$\overline{\mathbf{f}_{\mathrm{NL}}}=i_x\frac{E_v^2}{16\pi}\frac{\omega_p^2}{\omega^2 n^2}\frac{\partial n}{\partial x} \tag{8.37}$$

It is important to note that this result, as it was first derived in 1969 [138], produces a collisional term and gives the basis of the approximation for the nonlinear force in the WKB condition by

$$\overline{\mathbf{f}_{\mathrm{NL}}}=-\mathbf{i}_x\frac{E_v^2}{16\pi}\frac{\partial}{\partial x}\left(\frac{1}{|n|}+|n|\right)\exp(\mp kx) \tag{8.38}$$

The exponential factor in Eq. (8.36), with the absorption constant $\bar{k}$, is close to one in the ranges where WKB is valid and where $\omega_p \leqslant \omega$.

8.2 Obliquely Incident Plane Waves

The insufficiency of a nonlinear force in Eq. (8.2) is seen when the force density in a stratified plasma for obliquely incident plane waves is calculated. Using only Eq. (8.2), net forces along the plasma surface appear. This cannot be accepted, because there is no recoil possible for the necessary momentum transfer. Therefore, the nonlinear force from the equation (8.3) is used

$$\mathbf{f}_{NL}=\frac{1}{c}\mathbf{j}\times\mathbf{H}+\frac{1}{4\pi}\mathbf{E}\nabla\cdot\mathbf{E}+\frac{1}{4\pi}\nabla\cdot(n^2-1)\mathbf{E}\mathbf{E} \tag{8.39}$$

or from the equivalent Eq. (8.24)

$$\mathbf{f}_{NL} = \nabla \cdot \left[(\mathbf{EE} + \mathbf{HH} - (\mathbf{E}^2 + \mathbf{H}^2)\mathbf{1}/2)4\pi + \frac{n^2 - 1}{4\pi} \mathbf{EE} \right] - \frac{1}{4\pi c} \frac{\partial}{\partial t} \mathbf{E} \times \mathbf{H} \tag{8.40}$$

Both are reconfirmed by the following procedure, valid for plane electromagnetic waves.

The calculation of the time-averaged net force for obliquely incident plane waves will be done for a collisionless plasma $\nu = 0$, for a WKB approximation, see Section 7. With the same geometry of the p-polarized plane waves, Eqs. (7.37) to (7.38), this results in time-averaged values of

$$\frac{\partial}{\partial y} \mathbf{EE} = \frac{\partial}{\partial z} \mathbf{EE} = \frac{\partial}{\partial y} \mathbf{HH} = \frac{\partial}{\partial z} \mathbf{HH} = 0. \tag{8.41}$$

The time-averaged nonlinear force density (equation of motion) from Eq. (8.40) for a collisionless plasma ($\nu = 0$) is then

$$\begin{aligned} \overline{\mathbf{f}_{NL}} = {} & \frac{1}{8\pi} \mathbf{i}_x \frac{\partial}{\partial x} [(2n^2 - 1)(\overline{E_{px}^2} - \overline{E_{py}^2})\cos\beta - \overline{E_{sx}^2} \sin^2\beta \\ & + (\overline{H_{sx}^2} - \overline{H_{sy}^2})\sin^2\beta + \overline{H_{px}^2} \cos^2\beta] \\ & + \frac{1}{4\pi} \mathbf{i}_y \frac{\partial}{\partial x} [n^2 \overline{E_{px} E_{py}} \cos^2\beta + \overline{H_{sx} H_{sy}} \sin^2\beta] \\ & + \frac{1}{8\pi} \mathbf{i}_z \frac{\partial}{\partial x} [n^2 \overline{E_{px} E_{sx}} + \overline{H_{px} H_{sx}}] \sin 2\beta \end{aligned} \tag{8.42}$$

where β is the angle between the electrical vector $\mathbf{E}$ and the plane of incidence. The $\mathbf{i}_z$ component shows a coupling between parallel and perpendicular polarization. To evaluate this component, one finds from Eqs. (7.28), (7.30), (7.37), and (7.38)

$$n^2 E_{px} E_{sx} = -\frac{n E_v^2 \cos\alpha_0}{\cos\alpha(x)} \sin\alpha \sin^2 G \tag{8.43a}$$

$$H_{sx} H_{px} = \frac{n E_v^2 \cos\alpha_0}{\cos\alpha(x)} \sin\alpha \cos^2 G \tag{8.43b}$$

As a result, the time-averaged bracket in the $\mathbf{i}_z$ component of Eq. (8.42) vanishes identically.

Because of the last result, one can construct an expression for the case of a general β by a simple addition of the two expressions valid for each polarization. For $E_s(\beta = \pi/2)$ the expression of the time-averaged perpendicular component $\overline{\mathbf{f}}_s$ of the force density is

$$\overline{\mathbf{f}}_s = \frac{1}{8\pi} \mathbf{i}_x \frac{\partial}{\partial x} (-E_{sz}^2 + H_{sx}^2 - H_{sy}^2) + \frac{1}{4\pi} \mathbf{i}_y \frac{\partial}{\partial x} (H_{sx} H_{sy}) \tag{8.44}$$

From (7.30) is found

$$H_{sx}H_{sy} = -\mathbf{n}E_v \cos\alpha_0 \sin\alpha \cos^2 G$$
$$+\frac{c}{k\omega}\frac{E_v^2 \cos\alpha_0 \sin\alpha}{|n|^{1/2}\cos^3\alpha}\frac{dn}{dx}\sin 2G \tag{8.45}$$

From Eq. (7.29) the time averaged value of Eq. (8.45) is

$$\overline{H_{sx}H_{sy}} = -\tfrac{1}{2}E_v^2 \cos\alpha_0 \sin\alpha_0 = \text{const} \tag{8.46}$$

and as a result, the $\mathbf{i}_y$ component of $\mathbf{f}_s$ vanishes in (8.44). From (7.28) and (7.30) one finds

$$\overline{-E_{sz}+H_{sx}-H_{sy}} = -E_v^2 \cos\alpha_0 \frac{\cos^2\alpha_0 + n^2\cos\alpha}{2n\cos\alpha} + \frac{\tilde{A}}{2} \tag{8.47}$$

No second-order term was found, but a third-order term was given by

$$\tilde{A} = \frac{c^2}{4\omega^2}\frac{1}{n^3\cos^5\alpha}\left(\frac{dn}{dx}\right)^2 \tag{8.48}$$

The force density from (8.45) is then

$$\mathbf{f}_s = \mathbf{i}_x \frac{E_v^2\cos\alpha_0}{16\pi}\left(\frac{1}{\cos^3\alpha}\frac{\omega_p^2}{\omega^2}\frac{1}{n^2}\frac{dn}{dx} - \frac{d}{dx}\tilde{A}\right) \tag{8.49}$$

In the case of $E_p(\beta=0)$ one finds in (8.42)

$$\mathbf{f}_p = \mathbf{i}_x \frac{1}{8\pi}\frac{\partial}{\partial x}\left[\overline{E_{px}^2\left(1-2\frac{\omega_p^2}{\omega^2}\right) - E_{py}^2 - H_{pz}^2}\right] + \mathbf{i}_y \frac{1}{4\pi}\frac{\partial}{\partial x} n^2 \overline{E_{px}E_{py}} \tag{8.50}$$

The last term vanishes because of the constance in space of

$$\overline{n^2 E_{px}E_{py}} = \frac{E_v^2}{2}\cos\alpha_0 \sin\alpha_0 \tag{8.51}$$

Only the general formulas (8.39) or (8.40) give the correct result of a vanishing force in the plane of the plasma surface. Thus, the importance of the non-linear terms of Eq. (8.39) is demonstrated. Using Eq. (8.48) and n with $\nu=0$ leads to

$$\overline{(2n^2-1)E_{px}^2 - E_{py}^2 - H_{pz}^2} = -E_v^2\cos\alpha_0 + \frac{\cos^2\alpha_0 + n^2\cos^2\alpha}{2n\cos\alpha}$$
$$+\frac{\tilde{A}}{2}\cos^2\alpha(1-2\sin 2\alpha)^2 \tag{8.52}$$

and finally from Eqs. (8.50) and (8.57)

$$\bar{\mathbf{f}}_p=\mathbf{i}_x\frac{E_v^2\cos\alpha}{16\pi}\left[\frac{1}{\cos^3\alpha}\frac{\omega_p^2}{\omega^2}\frac{1}{n^2}\frac{dn}{dx}-\frac{d\tilde{A}}{dx}\cos^2\alpha(1-2\sin^2\alpha)^2\right.$$
$$\left.+2\tilde{A}\frac{dn}{dx}\frac{\sin^3\alpha}{\sin\alpha_0}(2\sin^2\alpha-1)(3\sin 2\alpha-2\cos^4\alpha-1)\right] \quad (8.53)$$

A comparison with the collisionless case of perpendicular incidence, Eq. (8.37), shows in the first order from Eqs. (8.49) and (8.53) that

$$\bar{\mathbf{f}}(\alpha)=\overline{\mathbf{f}(0)}\frac{\cos\alpha_0}{\cos^3\alpha} \quad (8.54)$$

The force density of this analysis is only in the negative direction, that is, toward decreasing electron density, up to the third-order in the spatial variation of the refractive index in all polarizations and is independent of the propagation direction of the light. The magnitude of the force density, however, is weakly dependent on the polarization of the light in the third-order terms. The consideration of these third-order terms is also justified in the WKB approximation, because it can be shown that deviations of the WKB approximation from the exact solution are exponentially small [168].

8.3 Nonponderomotive Collisional Term of the Nonlinear Force

Perpendicular incidence of linearly polarized electromagnetic waves on an inhomogeneous plasma along the x-direction is assumed within this subsection.

$$\mathbf{E}=\mathbf{i}_y E_y; \qquad \mathbf{H}=\mathbf{i}_z H_z \quad (8.55)$$

The generated force density in the irradiated inhomogeneous plasma can be calculated from Eq. (8.1). Because all other nonlinear terms in (8.3) vanish ($\mathbf{E}$ is perpendicular to ∇n_e and $\nabla\cdot\mathbf{E}=0$) the following equation is obtained:

$$\mathbf{f}_{\mathrm{NL}}=\frac{1}{c}\mathbf{j}\times\mathbf{H} \quad (8.56)$$

If the complex refractive index n of the plasma is given by Eq. (6.27) and if the inhomogeneous plasma of a certain temperature T has an electron density profile fitting the WKB condition Eq. (7.10), the (complex) electric field is

$$E_y'=\frac{E}{\sqrt{n}}\exp iF \quad (8.57)$$

where

$$F=\frac{\omega}{c}\int^{x} n(\xi)\,d\xi+\omega t \tag{8.58}$$

$\bar{k}$ is given by Eq. (7.13). The (complex) magnetic field is

$$H'_z=E_v|n|^{1/2}\exp(-\bar{k}x/2)\exp iF-\frac{icE_v\exp(\bar{k}x/2)\exp iF}{2\omega n^{3/2}}\frac{dn}{dx} \tag{8.59}$$

The temporal integration constants are neglected since f_{NL} will be time averaged later.

For calculating the time-averaged force, real quantities are required:

$$\mathbf{H}=\mathrm{Re}(\mathbf{H}')=\tfrac{1}{2}(\mathbf{H}'+\mathbf{H}'^{*}) \tag{8.60}$$

Thus

$$\begin{aligned}\mathbf{H}=\mathbf{i}_y\Bigg\{&E_v\exp(-\bar{k}x/2)[\mathrm{Re}(n)^{1/2}\cos F-\mathrm{Im}(n)^{1/2}\sin F]\\&-\frac{cE_v\exp(-\bar{k}x/2)}{2\omega}\left[Im\left(\frac{1}{|n|^{3/2}}\frac{dn}{dx}\right)\cos F+\mathrm{Re}\left(\frac{1}{|n|^{3/2}}\frac{dn}{dx}\right)\sin F\right]\Bigg\}\end{aligned} \tag{8.61}$$

The crucial point for deriving the collisional term is to use the complete diffusion equation with respect to collisions has to be used to find $\mathbf{j}$ for the plasma, still with neglect of the Hall and Lorentz terms

$$\frac{\partial \mathbf{j}'}{\partial t}+\nu\mathbf{j}'=\frac{\omega_p^2}{4\pi}\mathbf{E}' \tag{8.62}$$

The importance of ν was suggested [127]. Both ν and ω_p are assumed to be constant in time. In the numerical evaluations the nonlinear generalization [146] of Spitzer's collision frequency, Eq. (6.59) is used. This shows that heating of plasma by electromagnetic waves and other dynamic effects have to be neglected. Using the WKB solution for $\mathbf{E}$, Eq. (8.57) in Eq. (8.62), the following equation is found

$$\mathbf{j}'=\mathbf{i}_y\left[\frac{iG'F_0}{|n|^{1/2}(\nu-i\omega)}\right] \tag{8.63}$$

where

$$G'=\frac{\omega_p^2E_v^2}{4\pi} \tag{8.64}$$

is real and Eq. (8.58) has to be used. Again the temporal integration constant has been neglected for the same reason as in earlier papers.

The real part of $\mathbf{j}$ [remembering that $\exp(iF_0)=\exp(-\bar{k}x/2)\exp iF$],

(8.63) can be written as

$$\mathbf{j}=\mathbf{i}_y\left\{\frac{G'\exp(-\bar{k}x/2)}{|n|(\nu^2+\omega^2)}[(\mathrm{Re}|n|\nu+\mathrm{Im}|n|\omega)\cos F\right.$$
$$\left.-(\mathrm{Re}|n|^{1/2}\omega-\mathrm{Im}|n|^{1/2}\nu)\sin F]\right\} \tag{8.65}$$

Now, the time-averaged general nonlinear force f_{NL} in Eq. (8.56), is calculated for the considered plane waves. Eq. (8.56) and the expressions in Eqs. (8.61) and (8.65) for **j** and **H** are used. The only time-dependent terms in these equations are cos F and sin F terms since noting $\langle\cos^2 F\rangle=\langle\sin^2 F\rangle=\frac{1}{2}$ and $\langle\cos F\sin F\rangle=\langle\cos F\rangle=\langle\sin F\rangle=0$ the following equation is obtained for the nonlinear force

$$\overline{\mathbf{f}_{NL}}=\mathbf{i}_x\frac{\omega_p^2E_v^2\exp(-\bar{k}x)}{8\pi c(\nu^2+\omega^2)|n|}[(\mathrm{Re}^2|n|-\mathrm{Im}^2|n|^{1/2})\nu+2\omega\,\mathrm{Im}|n|\mathrm{Re}|n|^{1/2}]$$
$$-\mathbf{i}_x\frac{\omega_p^2E_v^2\exp(-\bar{k}x)}{16\pi\omega|n|(\omega^2+\nu^2)}\left[\mathrm{Im}\left(\frac{1}{|n|^{3/2}}\frac{dn}{dx}\right)\left(\nu\,\mathrm{Re}|n|^{1/2}+\omega\,\mathrm{Im}|n|^{1/2}\right)\right. \tag{8.66}$$
$$\left.-\mathrm{Re}\left(\frac{1}{|n|^{3/2}}\right)\left(\omega\,\mathrm{Re}|n|^{1/2}-\nu\,\mathrm{Im}|n|^{1/2}\right)\right]$$

This equation holds under the following assumptions are made. A plane electromagnetic wave travelling in the x-direction is incident perpendicularly on a plasma with an inhomogeneity in the x-direction only. The WKB condition is satisfied, and ν and ω_p are approximately constant over the time of interaction. Eq. (8.66) consists basically of two terms each with a square bracket. To arrive at a more transparent discussion, one reduces Eq. (8.66) for $\bar{k}\approx 0$ and

$$\bar{\mathbf{f}}_{NL}=\mathbf{i}_x\left(\frac{\omega_p^2E_v^2}{8\pi c\omega^2}\nu+\frac{\omega_p^2E_v^2}{16\pi\omega^2n^2}\frac{dn}{dx}\right) \tag{8.67}$$

where now $n^2=1-(\omega_p^2/\omega^2)$. The second term of Eq. (8.67) is recognized as the usual ponderomotive force [158],

$$\bar{\mathbf{f}}_{NL}=-i_x\frac{E_v^2\omega_p^2}{16\pi\omega^2}\frac{d}{dx}\frac{1}{|n|} \tag{8.68}$$

which, for $\nu=0$ and for the WKB formulation of the electric field **E**, Eqs. (8.55) and (8.57), is

$$\bar{\mathbf{f}}_{NL}=-\frac{\omega_p^2}{8\pi\omega^2}\overline{\frac{\partial}{\partial x}\mathbf{E}^2} \tag{8.69}$$

The factor $\frac{1}{2}$ is generated by temporal avaraging of **E**. This formula (8.69) is formally identical with that derived by Boot [159], Gapunov and Miller

[160], and by Weibel [161] for the forces generated by microwaves in plasmas, if the motion of low-density collisionless plasma towards the nodes of a standing electromagnetic wave is considered. The difference to the derivation in the case of lasers was the force in high-density plasma, when a propagating wave is incident. In this case the spatial variation of the dielectric properties, given by $(d/dx)(1/n)$ in Eq. (8.68) [158], is the essential driving mechanism. The general case of high-plasma densities includes the low-density case of microwaves discussed earlier. A further generalization was to include collisions [162] and to generalize the equation of motion for oblique incidence, as was discussed in Section 8.2.

The description of the collision property by a separate term, namely the first in Eq. (8.67) (in contrast to the use of a correction factor only [162]) was given by Stamper [163], and, in the more general form of Eq. (8.66) by Miller et al. [164]. For relating the first term of Eq. (8.67) to Stamper's term his expression is discussed.

$$\mathbf{f}_{\mathrm{NL}} = \frac{\mathrm{Re}(\varepsilon) - 1}{8\pi} \nabla \overline{\mathbf{E}^2} + \left(\frac{\omega_p}{\omega}\right)^2 \frac{\nu I}{c^2} \frac{k}{|\mathbf{k}|} \tag{8.70}$$

The first term is the usual nonlinear (ponderomotive) force, and the second term is referred by Stamper to the collisional force. This can be seen from the energy flux density

$$I = \frac{c}{4\pi} |\mathbf{E} \times \mathbf{H}|$$

Absorption in the WKB solutions for **E** and **H** is neglected.

$$\mathbf{E} = \mathbf{i}_y \frac{E_v}{|n|^{1/2}} \cos F_N$$

$$\mathbf{H} = \mathbf{i}_x \left\{ E_v |n|^{1/2} \cos F_0 - \frac{c}{2\omega} \frac{E_v}{|n|^{3/2}} \frac{d|n|}{dx} \sin F_0 \right\}$$

Inserting these last equations in the second term for the collisional force $\mathbf{f}_c$ in Eq. (8.70) leads to the nonponderomotive part of the nonlinear force

$$\mathbf{f}_c = \mathbf{i}_x \frac{\omega_p^2 E_v^2 \nu}{8\pi c \omega^2} \tag{8.71}$$

This is the first term of Eq. (8.67). The first term of Eq. (8.66) can be considered as a generalization of the collisional force.

The importance of Stamper's term in the special formulation of Eq. (8.67) can be seen in the following examples of a parabolic electron density

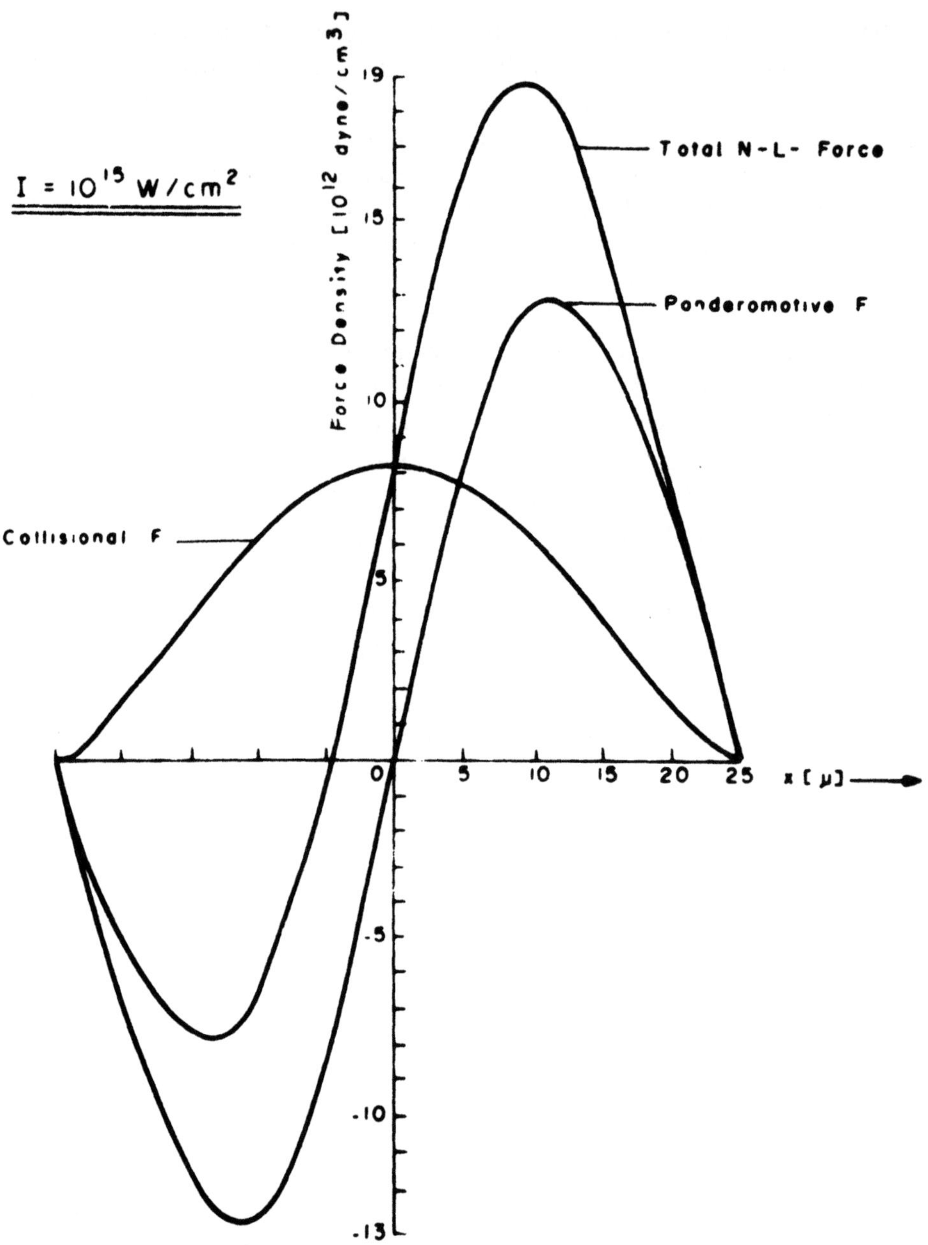

Figure 8.1 Total nonlinear force. collisional (Stamper) term and usual ponderomotive force for a 50 μm thick parabolic density profile [Eq. (22)] with a maximum of $n_e = n_{ec}/2$ for hydrogen plasma of 100 eV temperature at plane wave perpendicular neodymium glass laser irradiation of 10^{15} W/cm^2 intensity [164].

profile of a plasma:

$$n_e(x) = \begin{cases} n_0\left(1 - \frac{x^2}{b^2}\right) & \text{if } |x| < b \\ 0, & \text{elsewhere} \end{cases} \tag{8.72}$$

n_0 is chosen to be $n_0 = n_{ec}/2$ to see a strong effect. For higher densities, the dielectric swelling $(1/|n|)$ will be more dominant over the Stamper term, and the restriction to the WKB condition may be lost. b is chosen to be $b = 25$ μm, to allow fcr some reasonable conditions for irradiation by neodymium glass laser radiation ($\lambda = 1.06$ μm; $n_{ec} = 10^{21}$ cm). The plasma is a Hydrogen plasma ($Z = 1$) at a temperature of 100 eV. Two vacuum intensities of 10^{15} W/cm^2 and 10^{16} W/cm^2 are used. The light (plane waves) is incident from the left in the Figs. 8.1, 8.2 and 8.3.

The profile of Eq. (8.72) satisfies the WKB condition, Eq. (7.10), resulting in

$$\Theta \leqslant 1.9 \times 10^{-3} \tag{8.73}$$

In Figs. 8.1 and 8.2 the ponderomotive force and collisional force of Eq. (8.18)

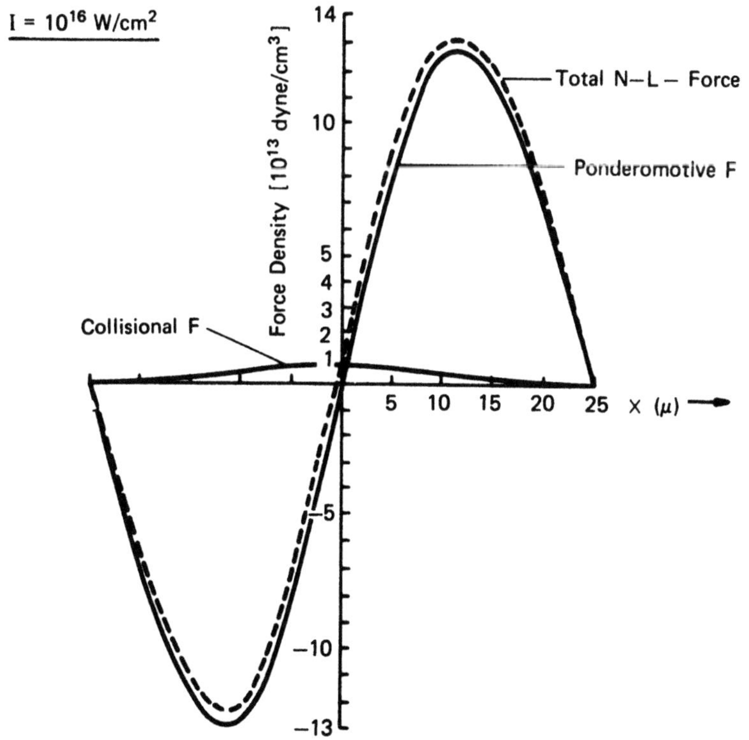

Figure 8.2 Same as Fig. 8.1 for 10^{16} W/cm^2.

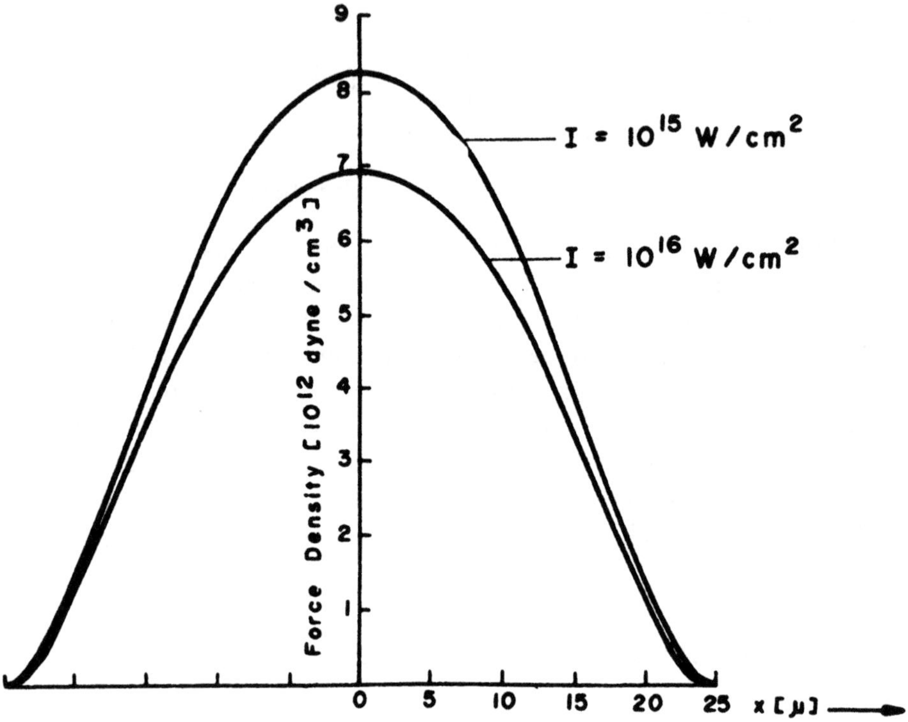

Figure 8.3 The collisional force for the case of Figs. 8.1 and 8.2 decreases for higher intensity due to the nonlinear intensity of the collision frequency.

(second and first term, respectively) are plotted for intensities of 10^{15} and 10^{16} W/cm^2, which are added to obtain the overall shape of the nonlinear force f_{NL}. Fig. 8.3 shows the Stamper's collisional force for both intensities. At an intensity of 10^{15} W/cm^2, which is very close to the threshold for the predominance of the nonlinear force, the collisional force is relatively large. At higher intensities, the nonlinear force is much less influenced by the collisional force. Their intensity dependence, Fig. 8.3, corresponds to the fully nonlinear description of the intensity dependence of the collision frequency.

The collisional part diminishes with increasing temperature and intensity, when the fully nonlinear dependence of the collision frequency (6.59) is used. It is less than about 5% of the total nonlinear force for $T = 100$ eV at neodymium glass laser intensities of 10^{16} W/cm^2 for a maximum density of one-half of the cutoff density.

8.4 Additional Third-Order Terms for Perpendicular Incidence

It is remarkable that higher order terms appear if the generally valid formula (8.27) for perpendicularly incident plane waves is evaluated for the WKB approximation. Using E_y, Eq. (8.57), and the first terms of Eq. (8.61) for H_z (neglecting then the phase between E_y and H_z), the nonlinear force without collisions (8.68) and the nonponderomotive Stamper term [first in bracket of (8.67)] had been derived in 1969 [138], while the phase was essential in Section 8.3 for the derivation from the $\mathbf{j}\times\mathbf{H}/c$ expression. If the phase is used in the formula (8.27) we should find additional third-order terms as will be shown now.

Using the same expression for the WKB approximation (8.57) and (8.61) as before, from Eq. (8.27), we arrive at the nonlinear force [169, 170]

$$\mathrm{Re}\,\overline{\mathbf{f}_{\mathrm{NL}}}=-i\frac{1}{4\pi}\left\{\overline{\mathrm{Re}(E_y)\,\mathrm{Re}\left(\frac{\partial E_y}{\partial x}\right)}-\overline{\mathrm{Re}(H_z)\,\mathrm{Re}\left(\frac{\partial H_z}{\partial x}\right)}\right\} \tag{8.74}$$

From Eqs. (8.57) and (8.61) we find

$$\mathrm{Re}(E_y)=\frac{E_v}{|n|^{1/2}}\exp(-\bar{k}\omega x/2)\left[\mathrm{Re}\frac{1}{n^{1/2}}\cos F-\mathrm{Im}\left(\frac{1}{n^{1/2}}\right)\sin F\right] \tag{8.75}$$

and

$$\begin{aligned}\mathrm{Re}\left(\frac{\partial E_y}{\partial x}\right)=&-\frac{E_v}{2}\exp(-k(x)x/2)\left[\mathrm{Re}\left(\frac{1}{n^{3/2}}\frac{dn}{dx}\right)\cos F-\mathrm{Im}\left(\frac{1}{n^{3/2}}\frac{dn}{dx}\right)\cos F\right]\\&-\frac{E_v\omega}{c}\exp(-\bar{k}(x)x/2)\,[\mathrm{Re}(n^{1/2})\sin F+\mathrm{Im}(n^{1/2})\cos F],\end{aligned} \tag{8.76}$$

also

$$\begin{aligned}\mathrm{Re}(H_z)=&E_v\exp(-\bar{k}(x)x/2)[\mathrm{Re}(n^{3/2}\cos F-\mathrm{Im}(n^{3/2})\sin F]\\&-\frac{E_v\omega}{c}\exp(-\bar{k}(x)x/2)\left[\mathrm{Im}\left(\frac{1}{n^{3/2}}\frac{dn}{dx}\right)\cos F+\mathrm{Re}\left(\frac{1}{n^{3/2}}\frac{dn}{dx}\right)\sin F\right]\end{aligned} \tag{8.77}$$

and

$$\begin{aligned}\mathrm{Re}\left(\frac{\partial H_z}{\partial x}\right)=&-\frac{E_v\omega}{c}\exp(-\bar{k}(x)x/2)[\mathrm{Re}(n^{2/3})\sin F+\mathrm{Im}(n^{3/2})\cos F]\\&-\frac{E_v c}{2\omega}\exp(-\bar{k}(x)x/2)\left\{\left[\mathrm{Re}\frac{1}{n^{3/2}}\frac{d^2n}{dx^2}-\frac{3}{2}\frac{1}{n^{3/2}}\left(\frac{dn}{dx}\right)^2\right]\sin F\right.\\&\left.+\mathrm{Im}\left[\frac{1}{n^{3/2}}\frac{d^2n}{dx^2}-\frac{3}{2}\frac{1}{n^{5/2}}\left(\frac{dn}{dx}\right)^2\right]\cos F\right\}\end{aligned} \tag{8.78}$$

Hence, Eq. (8.74) becomes

$$
\begin{aligned}
\operatorname{Re}\{\bar{f}_{\mathrm{NL}}\} = & \frac{E_v^2}{16\pi} \exp(-\bar{k}(x)x) \left[\operatorname{Re}\left(\frac{1}{n^{1/2}}\right) \operatorname{Re}\left(\frac{1}{n^{3/2}}\frac{dn}{dx}\right) \right. \\
& \left. + \operatorname{Im}\left(\frac{1}{n^{1/2}}\right) \operatorname{Im}\left(\frac{1}{n^{3/2}}\frac{dn}{dx}\right) \right] \\
& - \frac{E_v^2}{16\pi} \exp(-\bar{k}(x)x) \left[\operatorname{Re}\left(\frac{1}{n^{3/2}}\frac{dn}{dx}\right) \operatorname{Re}(n^{3/2}) + \operatorname{Im}\left(\frac{1}{n^{3/2}}\frac{dn}{dx}\right) \operatorname{Im}(n^{3/2}) \right] \\
& - \frac{E_v^2 \omega}{8\pi c} \exp(-\bar{k}(x)x) \left[\operatorname{Im}\left(\frac{1}{n^{1/2}}\right) \operatorname{Re}(n^{1/2}) - \operatorname{Im}(n^{1/2}) \operatorname{Re}\left(\frac{1}{n^{1/2}}\right) \right] \\
& + \frac{E_v^2 \omega}{8\pi c} \exp(-k(x)x) [\operatorname{Re}(n^{1/2}) \operatorname{Im}(n^{3/2}) - \operatorname{Im}(n^{1/2}) \operatorname{Re}(n^{3/2})] \\
& - \frac{E_v^2 c}{16\pi\omega} \exp(-\bar{k}(x)x) \left\{ \operatorname{Re}\left[\frac{1}{n^{3/2}}\left(\frac{d^2n}{dx^2}\right) - \frac{3}{2n^{5/2}}\left(\frac{dn}{dx}\right)^2 \right] \operatorname{Im}(n^{1/2}) \right. \\
& \qquad \left. - \operatorname{Im}\left[\frac{1}{n^{3/2}}\left(\frac{d^2n}{dx^2}\right) - \frac{3}{2n^{5/2}}\left(\frac{dn}{dx}\right)^2 \right] \operatorname{Re}(n^{1/2}) \right\} \\
& - \frac{E_v^2 c}{32\pi c} \exp(-\bar{k}(x)x) \left\{ \operatorname{Re}\left[\frac{3}{n^{3/2}}\left(\frac{d^2n}{dx^2}\right) - \frac{3}{2n^{5/2}}\left(\frac{dn}{dx}\right)^2 \right] \operatorname{Re}\left(\frac{1}{n^{3/2}}\frac{dn}{dx}\right) \right. \\
& \left. + \operatorname{Im}\left[\frac{1}{n^{3/2}}\left(\frac{d^2n}{dx^2}\right) - \frac{3}{2n^{5/2}}\left(\frac{dn}{dx}\right)^2 \right] \operatorname{Im}\left(\frac{1}{n^{3/2}}\frac{dn}{dx}\right) \right\}
\end{aligned} \tag{8.79}
$$

This is the hitherto most extensive evaluation of the general nonlinear force [Eq. (8.3)] or (8.24) for plane waves perpendicularly incident on a stratified plasma for the WKB approximation. If we consider the case in which collisions are neglected, then $\nu = 0$ and the imaginary terms vanish, and the above expression then yields

$$
\mathbf{f}_{\mathrm{NL}} = \frac{E_v^2 \omega_p^2}{16\pi\omega^2 n^2} \frac{dn}{dx} - \frac{E_v^2 c^2}{32\pi\omega^2} \left[\frac{1}{n^3} \left(\frac{d^2n}{dx^2}\right) \left(\frac{dn}{dx}\right) \right] + \frac{3E_v^2 c^2}{64\pi\omega^2} \frac{1}{n^4} \left(\frac{dn}{dx}\right)^3 \tag{8.80}
$$

The first term in Eq. (8.80) is the one obtained before [138], Eq. (8.68) neglecting the phase term between E_y and H_z. It would be reasonable to assume that the higher order terms in Eq. (8.80) are due to the phase terms. This is not as can be seen by spatially differentiating our expression (8.57) for the magnetic field strength **H**. When the phase term is differentiated, one of its terms cancel with one of the derivatives of the first term of Eq. (8.57). Hence the nonlinear force arises from the phase term alone. This is in agreement with the model of quivering motion [54]. As described before [138], the first term of Eq. (8.80) indicates a deconfining collisionless acceleration, because the direction of the force density is toward decreasing plasma densities, which is clearly independent of the polarization of the incident laser light from symmetry considerations. The first of the higher order terms is a confining one, while the other is a deconfining force term. These third-order terms may contribute to the momentum transferred to the homogeneous interior [138]. With collisions, the other terms can be interpreted as collision produced radiation pressure of the light, within the inhomogeneous plasma an analogy with the usual radiation pressure of homogeneous media, where, however, the very complex influence of the refractive index is included now.

8.5 The General Non-Transient Nonlinear Force

The most general equation of motion of a plasma consists of a thermokinetic term, Eq. (8.5), and the following nonlinear terms derived 1969 [138]

$$f_{NL} = \frac{1}{c}\,\mathbf{j}\times\mathbf{H} + \frac{1}{4\pi}\,\mathbf{E}\nabla\cdot\mathbf{E} + \frac{1}{4\pi}\,\nabla\cdot(n^2-1)\mathbf{E}\mathbf{E} \tag{8.81}$$

which is identical with the formulation using the Maxwellian stress tensor **T**, Eq. (8.25)

$$f_{NL} = \nabla\cdot(\mathbf{E}\mathbf{E}+\mathbf{H}\mathbf{H}-\tfrac{1}{2}(\mathbf{E}^2+\mathbf{H}^2)\mathbf{1}+(n^2-1)\mathbf{E}\mathbf{E})/4\pi - \frac{1}{4\pi c}\frac{\partial}{\partial t}\mathbf{E}\times\mathbf{H} \tag{8.82}$$

This formulation is formally identical with the nondispersive Landau–Lifshitz expression [166] if a special formulation of the density is used, where our derivation of the algebraic identity of Eq. (8.82) with Eq. (8.81) proves the general validity for dispersive media of a plasma with dissipation. Only expressions (8.81) or (8.82) result in nonlinear forces for oblique incidence of plane wave on stratified collisionless plasma which have no wrong time-averaged component in the plasma surface.

For perpendicular incidence (x-direction) of plane waves, the forces in the plasma are especially (from 8.81)

$$f_{NL} = \frac{1}{c}\,\mathbf{j}\times\mathbf{H} \tag{8.83}$$

or (from 8.82)

$$f_{NL} = -\frac{\partial}{\partial x}\,\frac{\mathbf{E}^2+\mathbf{H}^2}{8\pi} \tag{8.84}$$

These equations are valid for any general density profiles (differences to the WKB approximation have been discussed by Lindl and Kaw [154]).

For the simplification of the WKB approximation, the special result is for a plasma with collisions

$$f_{NL} = i_x \frac{E_v^2}{16\pi} \frac{\omega_p^2}{\omega^2} \frac{1}{|n|^2} \frac{d|n|}{dx} + i_x \frac{E_v^2}{16\pi} \frac{\omega_p^2}{\omega^2} \frac{2\omega v}{c} \frac{}{\omega} \tag{8.85}$$

where the second term is a non ponderomotive dissipative part of the nonlinear force, first derived 1969 [138] and derived later in another way by Stamper [163].

For nondissipative (collisionless) plasmas, the nonlinear force (for the very special case of plane waves perpendicularly incident on a WKB-like plasma) reduces to

$$f_{NL} = - i_x \frac{E_v^2}{16\pi} \frac{\omega_p^2}{\omega^2} \frac{\partial}{\partial x} \frac{1}{n} = -i_x \frac{1}{16\pi} \frac{\omega_p^2}{\omega^2} \frac{\partial}{\partial x} \tilde{E}^2 \tag{8.86}$$

The last relation can simply be seen from the fact that the WKB approximation results in $\tilde{E} = E_v/(n)^{1/2}$.

The terminology for f_{NL} is not unique. While the expression "nonlinear force" was used [3, 156, 163, 171, 172, 173, 174], the expressions "nonlinear radiation force" [175], "electrostrictive force" [176], "j x B-force" [178, 178], or an unspecified "ponderomotive force" have also been used. The fact that additional (nonponderomotive) terms appear in the nonlinear force apart from those of ponderomotive forces, may be a convincing reason for remaining with the name "nonlinear force."

8.6 The Transient Nonlinear Force

The results in the preceding subsection 8.5 are the outcome of the extensive steps with the aim to derive the correct and finally complete formulation of the force density, or equation of motion, of a plasma solving the problem of the equations 6.1 to 6.6 as explained in Appendix C and in the whole context of this Chapter Eight up to this point. We repeat: The force density - apart from the thermokinetic part, Eq. (8.4) - is the nonlinear force (8.81) or (8.82) as confirmed from the fact that **this and only this formulation** results in the fulfilment of the **momentum conservation** when plane electromagnetic waves are obliquely incident on a planely striated plasma if there are no collisions. For the case of collisions, the energy transfer results in differential form of radiation pressure of which the integrated value is identical with the ordinary radiation pressure in absorbing media (eg. in metals) [165].

It is to be noted that this result still has limitations as it is covering only:

a) the case of a space-charge quasi-neutral plasma, as a basic assumption of Schlüter's (and here [135] extended) two-fluid model, see Eq. C.5, and
b) the case of stationary (non-transient) electromagnetic fields, excluding any transient switching-on or off, or any temporal change of the electromagnetic radiation, hitting the plasma.

While we shall discuss the case a) in the following subsection, this subsection is discussing and summarizing the transient behaviour.

It is to the merit of Klima and Petrzilka [184] that the transient case of the nonlinear force was treated for the very first time. While our initial treatment [158] before going to the general non-transient form of the nonlinear force [138], was for the first time ever treating the problem of plasmas inhomogeneity for the pondermotive problem, Klima and Petrzilka [184] considered a homogeneous plasma or electron gas and studied the force density there due to the propagation of a plane optical wave packet. It turned out that the same nonlinear force Eq. (8.37) was the result for this special geometry. We shall come back to the work of Klima and Petrzilka in Chapters 9.4 and 12.3.

A more aggressive discussion of the transient case was started by Kono et al (1981), where, however only one special shearing term of the force was considered. Finally six different models were published (Karpman et al 1982; Tskhakaya, 1981; Stratham et al, 1983; Lee et al, 1983; Mulser et al, 1983; Zeidler et al, 1985) of which each was considered as being the solution. The rather controversial situation was discussed in many details to explain the merits of each model, by Zeidler et al (1985). This work was nearly complete, but since all these theories were using the transient process as an approximation only for slowly varying laser intensity, it is no surprize that this did not arrive at the final formulation.

By carefully analyzing the problem it was discovered (Hora, 1985) that a further logarithmic term was missing which obviously was so small that it was lost due to the approximations. The final result was that the complete transient nonlinear force has the form (Hora, 1985)

$$\mathbf{f}_{NL} = \frac{1}{c}\mathbf{j}\times\mathbf{H} + \frac{1}{4\pi}\mathbf{E}\nabla\cdot\mathbf{E} + \frac{1}{4\pi}\left(1 + \frac{1}{\omega}\frac{\partial}{\partial t}\right)\nabla\cdot\mathbf{E}\,\mathbf{E}\,(n^2-1) \qquad (8.87)$$

or in tensioral formulation

$$\mathbf{f}_{NL} = \nabla\cdot\left(\mathbf{E}\,\mathbf{E} + \mathbf{H}\mathbf{H} - \frac{1}{2}(\mathbf{E}^2 + \mathbf{H}^2)\,\mathbf{1} + \left(1 + \frac{1}{\omega}\frac{\partial}{\partial t}\right)(n^2 - 1\quad \mathbf{E}\,\mathbf{E})/4\pi\right.$$
$$\left. - \frac{1}{4\pi c}\frac{\partial}{\partial t}\mathbf{E}\times\mathbf{H}\right) \qquad (8.88)$$

where ω is the central frequency of the irradiated wave packet.

The results (8.87) and (8.88) contrary to all the other results discussed by Zeidler et al (1985) are very well understandable: the dielectric response of the plasma with respect to the incident electromagnetic radiation is not only due to the refractive index $\tilde{n}^2$ and its inhomogeneities (as seen from the gradients or - as our very first discovered result 1967 [158] - from the spatial differentiation of the form of Eq. (8.37), but furthermore, the transient process is just due to the temporal differentiation (time gradient, see B. Philbert et al, 1982), of the terms including the refractive index. This just expresses the feeding in or taking out of the electromagnetic energy to or from the plasma.

A proof of the general correctness of Eqs. (8.87) and (8.88) cannot be done simply from the conservation of momentum as it was possible for the non-transient case [138]. The general validity, however, was confirmed by Rowlands (1990) by establishing that the formulation (8.88) in the covariant form satisfies the Lorentz and the gauge invariance. A further result is (Rowlands, 1991) that a general study of the covariant formulation including a material (dielectric and magnetic) response for a collisionless plasma, given by a tensor of the fourth grade, arrives at a magnetic permeability $\mu = 1$ fully generally for all plasmas.

The discussion of the transient shear force term of Kono (1981) arrived at a rather remarkable result (Kono, 1989; Chernikov et al 1989). It turned out that the plasma state goes into chaos at very high laser intensity. This analysis, however is non-relativistic and the transition appears at such laser intensities where the electron oscillation is getting values of 0.3 times mc^2 (see Eq. 6.70). This experience may have a fundamental meaning, since Kono's treatment was simply subrelativistic: the limit ation of all velocities is given by the speed of light c, otherwise physics changes into chaos. Considering the situation that the Lorentz transorm would not have been discovered before, the significant upper limit of all natural velocities by c is given then by the change of physics into chaos.

8.7 Single Particle Model of Nonlinear Force and High Internal Electric Fields Inside of Plasmas

After we have seen how the general transient nonlinear force was derived within the space-charge quasi-neutral two fluid model which had to be extended appropriately to arrive at the necessary conditions of momentum conservation of transient laser interaction with plasmas and to fulfil Lorentz and gauge invariance, we shall show now how the study of the nonlinear interaction enforced the criticism of the space charge neutrality of plasmas. As mentioned in Chapter 2, plasmas at about 1

Million degrees temperature have an electric conductivity similar to metals. As known from elementary textbooks, the electrical conductivity of the plasmas as in metals is such that if an electric field is being generated within one region, it will decay exponentially with a characteristic time of about 10^{-17} sec. From this result, it is a firm belief for plasma physicists that there are never internal electric fields in plasmas.

This is correct in metals only, if the medium is homogeneous. The same is valid only if the plasma is homogeneous. Nearly all plasmas in laboratories or space, however, are inhomogeneous, even extremely inhomogeneous in the case of laser produced plasmas. For space plasmas it was evident from other historical reasons of the Birkeland theory how the inhomogeneous plasmas do possess internal electric fields (Alfven 1981, 1988; Peratt, 1988, 1989). Their measurement is possible e.g. by long fiber cables where linearly polarized light is transmitted and the turning of the polarization by the Kerr effect could be measured, when moved with a space craft through cosmic plasmas.

We shall show now how the study of laser produced plasmas arrived at the internal electric fields. One model which preceded the first publication of the nonlinear force due to the inhomogeneity of a plasma [158] was the following single particle theory (Hora, 1971). We consider the single particle motion of an electron in the laser field within a plasma and use the WKB approximation (see Chapter 7) for the electric field, Eq. (7.9)

$$\mathbf{E} = \mathbf{i}_2 \frac{E_v}{n^{1/2}} \cos F \tag{8.89}$$

where the Maxwellian equations result in a magnetic field

$$\mathbf{H} = \mathbf{i}_3 \frac{c}{2\omega} \frac{E_v}{n^{3/2}} \frac{dn}{dx} \sin F - \mathbf{i}_3 E_v n^{1/2} \cos F \tag{8.90}$$

using

$$F = -\int^{x} \frac{\omega n(x)}{c} dx + \omega t \tag{8.91}$$

in the refractive index n in the collisionless plasma is spatially varying in the x-direction by a varying electron density $n_e(x)$ by

$$n^2(x) = 1 - \frac{\omega_p^2(x)}{\omega^2} \; ; \; \omega_p^2 = \frac{4\pi e^2 n_e(x)}{m_e} \tag{8.92}$$

and if the incident electromagnetic wave is linearly polarized with the E-vector in the y-direction and is propagating into the x-direction.

The quiver motion of an electron in this field is given by velocity components in the y and x direction only as

$$\dot{v}_{ey} = (e/m)E_y \tag{8.93}$$

and

$$\dot{v}_{ex} = (e/mc)\, v_{ey}H_z \tag{8.94}$$

where temporal integration of Eq. (8.93) according to Eqs. (8.89) and (8.91) results in

$$v_{ex} = \frac{e}{m\omega}\frac{E_v}{n^{3/2}}\sin F \tag{8.95}$$

Substitution into Eq. (8.94) and using the magnetic field from Eq. (8.90) and (8.91) arrives at the force density of the electron gas of an electron density n_e of

$$\begin{aligned} f_{e,NL} &= n_e m\, v_x\, \dot{v}_x\, i_x \\ &= i_x.\ n_e \frac{e}{c}\frac{e}{m\omega}\frac{E_v}{n^{1/2}}\sin F \left(\frac{c}{2\omega}\frac{E_v}{n^{3/2}}\frac{dn}{dx}\sin F - E_v n^{1/2}\cos F\right) \end{aligned} \tag{8.96}$$

where the time averaging of the sinFcosF-term is zero and of the $\sin^2 F$ term is 1/2, leaving

$$f_{e,NL} = i_x \frac{1}{4}\frac{e^2}{m\omega^2}\frac{E_v^2}{n^2}\frac{\partial n}{\partial x} = -i_x \frac{\omega_p^2}{16\pi\omega^2}\frac{\partial}{\partial x}\frac{E_v^2}{n} \tag{8.97}$$

This is not the nonlinear force to the plasma but to the electron cloud only (acting by the laser fields in a kind of space charge neutral way inside the plasma to all electrons). This electron gas is then pushed or pulled and a strong electric field will be established between the electrons and ions until the ions will follow. The ions are then determining the inertia such that the force to the electrons will result in a force density in the plasma of

$$f_{NL} = i_x\, n_i\, m_i\, v_x = -i_x \frac{(1-n^2)}{16\pi}\frac{\partial}{\partial x}E^2 \tag{8.98}$$

This is identical to the derivation of the nonlinear force, Eq. (8.37) for a collisionless plasma for a plane wave perpendicularly incident on a stratified plasma.

What one realized is the action of the phase between the electric and the magnetic field, (Eqs. 8.89 and 8.90) which is given by the spatial derivative

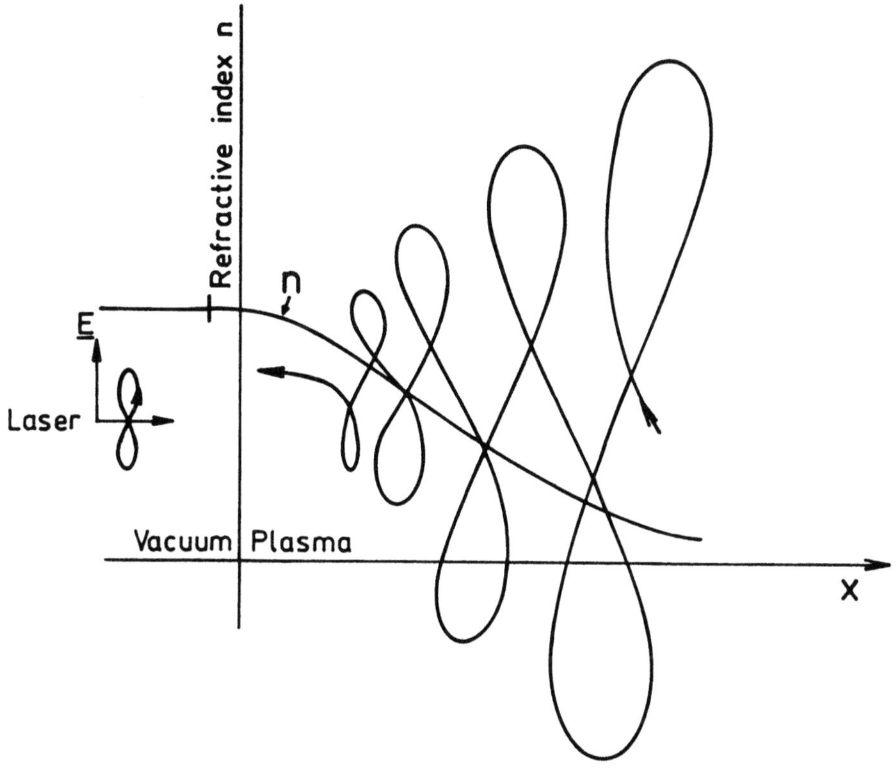

Fig. 8.1. Refractive index n in vacuum and in an inhomogeneous plasma corona depending on the depth x with monotonously increasing electron density. If laser light is perpendicularly incident with E polarized in the plane of drawing, an electron performs a closed eight-like motion in vacuum. Inside the plasma the eight is swelled up (see Eq. (8.89) due to the decreasing denominator n to values 0.1 or much less) and the phase shift causes that the electron is not a standing eight but drifting towards vacuum (towards lower electron density).

of the refractive index. Only this little phase shift if then the reason that the nonlinear force occurs: no surprize that it was not realized earlier or after 1967 [158] it was played down as a very insignificant effect until the experimental facts confirmed its dominant importance at high laser intensity interaction in plasmas. Only due to this phase shift, the force appears.

In vacuum or in a homogeneous plasma, this phase shift is zero and the quiver motion of an electron in a laser field is simply an eight-like motion, see Fig. 8.1 (left hand side). In the plasma, the eight is swelled up due to the decreasing denominator of the refractive index n in Eq. (8.89) increasing the amplitude. The phase shift causes that the eight of the motion is not fixed but is drifting towards the vacuum (towards lower electron density). It turns out that the excess of quivering energy is converted into translative energy of the electrons moving against the laser light. The inertia of this motion is determined by the ion mass since the ions are attracted electrostatically (by the very high fields we shall show later from computations) to the electrons.

One difficulty in this consideration is the fact that microscopic electron motion is combined with the phase between E and H given by the macroscopic solution of the Maxwellian equation for the laser light. A similar treatment was put forward by Ljamov (1967) who treated the force in a homogeneous metal produced by the irradiation of an electromagnetic field. The exponential decay due to collisions caused in the same way as in Eq. (8.98), a gradient of the field square and therefore the force of the radiation pressure in a metal surface. The advantage was that then the locally resolved radiation pressure was achieved for the first time and not only the integrated radiation pressure with which the integral was indeed identical. The publication of this article caused a lot of headache to the editors as seen from the two year delay of the publication. It is known that the authority of V.L. Ginzburg opposed the combination of microscopic motion with the macroscopic property of the phase shift (in this case due to absorption and not due to the inhomogeneity as in our case above.)

The correctness of the mentioned combination of single electron motion with the macroscopic phase result was simply given in our case from the identical agreement of the nonlinear force formula with that of the macroscopic theory, first derived from Schlüter's plasma model [158] and derived also from the momentum flux density description as expressed by the stress tensor formulation [138,158]. The question of the strong electric fields between the electron and ion fluid from the just mentioned single electron motion picture was the starting point to look for a more general macroscopic model as described in the following subsection.

8.8 Genuine Two Fluid Plasma Model with Full Description of Internal Electric Fields

Many of the properties of laser interaction with plasmas can be studied macroscopically with rather simplified models. We shall see in the following chapters from a number of most interesting examples that single fluid models reproduce reproduce observations very well. Also using the space charge neutral Schlüter type two fluid model including the before reported improvements and generalizations show good agreement with observations. The knowledge of the strong electric fields in laser produced plasma when the electron cloud is driven by the nonlinear forces and the ion cloud has to follow electrostatically, however, could not be ignored until a systematic study was performed. One has to go back to a type of hydrodyamic equation of the kind (6.1) and (6.2) and include the electrostatic coupling by Maxwell's equations, or in the simplified case by a Poisson equation.

This type of general formulation is not new and can be found in good plasma physics textbooks (e.g. Krall and Trivellpiece, 1973) but obviously nowhere was a systematic treatment of the solution of this general hydrodynamic problem. Moreover the mentioned formulation did not include properties of the nonlinear forces and the correct description of the electromagnetic fields with detailed solution of the incident and reflected waves.

With these hydrodynamic equations including Maxwell's equations and further assumptions of particle collisions, only numerical solutions can be tried. The best possible general formulation is being used and most advanced computer capacities are needed. In the following treatments and cases, nevertheless several simplifications are involved, as the emission of line radiation and or of continuum radiation of the Kramers type for transparent plasmas or of the blackbody Planck type. This radiation emission will be neglected apart from the fusion gain calculations in the last Chapter where the radiation losses by bremsstrahlung or their self absorption in fusion pellets had to be included. The question how Planck radiation can drive plasma in the way of radiation hydrodynamics is a topic of very recent research (Yabe, 1984; Yabe et al, 1989; Kaiser et al 1989; Meyer-ter-Vehn, 1990) where the experiments of converting laser radiation in a hohlraum into black body radiation up to 200 eV temperature (Sigel et al 1990; Tsakiri's et al 1991) are the leading motivation.

We are presenting here a numerical model of genuine two fluids of electrons (index e) and ions (index i) for a fully ionized plasma mostly for a hydrogen isotope with one spatial dimension only. The physics involved, however, is most general and the time steps are very short in order to cover all details of the plasma oscillations. The computation has then to be for a very large number of time steps such that phenomena in real time of laser interaction with plasmas can be studied. This code was first developed by Lalousis (1983) (Lalousis et al, 1983; 1984) and continued by Szichman (1988), Gu Min et al (1989), and Aydin (1990), see H. Hora et al 1990. For this one dimensional case we have to solve the following seven differential equations for the seven quantities depending on the variables x and time t: the densities n_e and n_i, the velocities v_e and v_i, the temperatures T_e and T_i, and for the longitudinal electric field E which corresponds to the plasma oscillations but are more general than "electrostatic" oscillations.

Continuity equations for electrons (index e) and ions (index i)

$$\frac{\partial n_e m_e}{\partial t} + \frac{\partial m_e n_e v_e}{\partial x} = 0 \tag{8.99}$$

$$\frac{\partial n_i m_i}{\partial t} + \frac{\partial m_i n_i v_i}{\partial x} = 0 \tag{8.100}$$

Conservation of momentum

$$\frac{\partial(n_e m_e v_e)}{\partial t} = -\frac{\partial(n_e m_e v_e^2)}{\partial x} - \frac{\partial P_e}{\partial x} - n_e eE - n_e m_e \nu(v_e - v_i) + f_{NL} \tag{(8.101)}$$

$$\frac{\partial(n_i m_i v_i)}{\partial t} = -\frac{\partial(n_i m_i v_i^2)}{\partial x} - \frac{\partial P_i}{\partial x} - n_i ZeE + n_e m_e \nu(v_e - v_i) + \frac{m_e}{m_i} f_{NL} \tag{8.102}$$

Conservation of energy

$$\frac{\partial(m_e n_e \varepsilon_e)}{\partial t} \tag{8.103}$$

$$= -\frac{\partial(m_e n_e \varepsilon_e v_e)}{\partial x} - P_e \frac{\partial v_e}{\partial x} - \frac{3}{2}\tau^{-1} K n_e (T_e - T_i) + \frac{\partial}{\partial x}(\kappa_e \frac{\partial T_e}{\partial x}) + W_L$$

$$\frac{\partial(m_i n_i \varepsilon_i)}{\partial t} \tag{8.104}$$

$$= -\frac{\partial(m_i n_i \varepsilon_i v_i)}{\partial x} - P_i \frac{\partial v_i}{\partial x} + \frac{3}{2}\tau^{-1} K n_e (T_e - T_i) + \frac{\partial}{\partial x}(\kappa_i \frac{\partial T_i}{\partial x})$$

and Poisson's equation

$$\frac{\partial}{\partial t}E = 4\pi e\,(n_e\, Zn_i) \tag{8.105}$$

In these equations (8) to (15), the wave equation for the external electric laser field E_L of radian frequency ω is solved using the temporally and spatially variable refractive index n which depends on the plasma dynamics which determines the values of the electron density n_e and electron temperature T_e in the plasma frequency ω_p and collision frequency ν

$$\nabla^2 E_L + \tilde{n}^2(\omega/c)^2 E_L = 0 \tag{8.106}$$

$$\tilde{n} = 1 - \omega_p^2/[\omega^2(1 + i\nu/\omega)] \tag{8.107}$$

here ν represents the Coulomb collision frequency, Eq. (2.37) [70] including the nonlinear dependence on the laser intensity I, see Eq. (6.58a)

$$\nu = \left(\frac{m\pi}{2}\right)^{1/2} \frac{\omega_p^2 e^2 Z \ln\Lambda}{16(KT_e + I/n_c c)} \tag{8.108}$$

with the Coulomb logarithm

$$\ln\Lambda = \ln\{(\tfrac{3}{2}kT)^{3/2} / [(\pi n_e)^{1/2} Z e^3]\} \tag{8.109}$$

which contains the effective temperature, Eq. (6.57), in the form of

$$T = T_{th} + I/(n_{ec}\, cK) \tag{8.110}$$

n_{ec} is the critical density for electrons. The optical frequency of the laser field is ω and ω_p is the plasma frequency (Eqs. 2.6 and following)

When the laser irradiates the plasma, the electron heating power density W_L produced by the laser optical absorption in Eq. (8.103) is given by

$$W_L = (I\,\omega/c)\mathrm{Im}(\tilde{n}) \tag{8.111}$$

while the time resolved nonlinear force in Eqs. (8.101) and (8.102) can be derived as, see Eq. (8.37)

$$f_{NL} = \frac{\omega_p^2}{8\pi\omega^2}\sin^2(\omega t)\,\frac{\partial E_{Lx}^2}{\partial x} \tag{8.112}$$

for mostly driving the electron fluid.

This system including initial and boundary conditions determines the seven quantities n_e, v_e, T_e (the electron number density, velocity and temperature), n_i, v_i and T_i the same quantities for ions, and the internal longitudinal electric field E. The expressions ε_e, ε_i, P_e and P_i are internal energy densities and scalarly assumed pressures for electrons and ions viz.,

$$\varepsilon_e = 3KT_e/2m_e \tag{8.113}$$

$$\varepsilon_i = 3KT_i/2m_i \tag{8.114}$$

$$P_e = n_e kT_e \tag{8.115}$$

$$P_i = n_i kT_i \tag{8.116}$$

The following assumptions in the equations (8.99) to (8.105) have to be noted:

1) the thermal conductivities of the electrons κ_e (Eq. 8.103) and ions κ_i (Eq. 8.104) are used in the classical form of Spitzer [107]. This is contrary to the knowledge that the thermal conductivity of the plasma may be decreased by a factor 1/f of 50 to 100, as seen from fitting the calculations of the thermal flow from the hot laser irradiated corona to the plasma interior in fusion pellets to experimental determinations. This has been discussed in numerous papers and was simply explained by a double layer (DL) mechanism (Cicchitelli et al 1984) showing that the electrons are mainly reflected between the hot and cold plasma and the ions only contribute to the thermal conductivity. Indeed we could have concluded similar mechanisms of spread double layers (Hora, 1982) in the following calculations of the density ripple and would have had to include the modified electronic thermal conductivity.

The complications of the thermal conduction, however, are not of importance in the short time steps where this model is mostly used. However, earlier numerical experiments with the numerical code (Szichman, 1988) however have shown that, during the fast period of few picoseconds, a change of the thermal conductivity from the Spitzer value by a factor 70 does not greatly change the results within the corona. The processes between the hot corona and the cold plasma interior are not discussed here since this is a range of depth of plasma which goes far beyond the values of the following initial density profiles. A discussion of possible very fine variations would then be quite academic since they would be beyond experimental proof.

2) In Eq. (8.99) to (8.107), the electron density and the ion density for the refractive index ñ were assumed for each time step of the computations to solve the differential equation which depends only on space. This neglects that the time dependence of ñ apart from its spatial dependence does not exactly permit the separation of the spatial and temporal wave equation in order to arrive at Eq. (8.106). Thus the approximation implies the approximation the following first order condition

$$(\partial \tilde{n}/\partial x) = (1/c)\,(\partial \tilde{n}/\partial t) \qquad (8.117)$$

This is fulfilled even in extreme cases as seen from the following numerical results which justify the use of Eq. (8.106). Indeed this includes the very important slow time dependence of the refractive index. The stepwise solution of Eq. (8.106) uses the correct boundary conditions towards high plasma densities (decaying field) based on a specially economic matrix procedure as explained in before (Lalousis 1983; Hora et al 1984).

3) In the following computations the density and other plasma parameters are not fixed on the boundaries, but are allowed to vary. So that changes brought about by the laser may be initiated within the plasma and then develop in the direction towards the oncoming incident laser light. Since we do not use herein an initially zero density with our code for the first simulation step, our result of 1974 (see e.g. following Fig. 10.10a) with a much more simplified one fluid computation with a density beginning at the value zero at laser irradiation 0 cannot be reproduced directly. In the latter case it is possible to calculate how the laser intensity varies with depth into the plasma, initially ranging from the incident value at the edge to a peak at the critical density where reflection occurs.

4) The equipartition time τ for the thermal exchange between the electron and ion fluids is used in the standard formulation [107] where the electron ion collision frequency ν is based on the classical value and allowance for the anomalous collision frequency due to the quantum deviation (see subsection 2.6) has been ignored at this stage.

The solutions of the wave equation (8.106) are expressed in the following form

$$E_L = E_{Lx}(x,t)\sin\omega t$$

where the amplitude is a complex value of a non-elementary function. This result takes into account the presence of a propagating and a partially standing wave. The time dependences of plasma density and temperature are slow compared to the wave period, satisfying the approximate condition of equation 8.117. After achieving the numerical solution with the boundary condition of outgoing waves only at high (supercritical) density (see Chapter 7) separation into the real part and imaginary part permits the identification of the incident laser intensity I by the relation

$$E^2_{Lx} = (8\pi/c)I \tag{8.118}$$

which is used in Eq. (21).

Eqs.(8.99)to(8.118) describe the successive motion of the plasma under the given initial and boundary conditions. An Eulerian difference scheme has been developed by a basically new numerical technique (Lalousis 1983; Lalousis et al 1984). The Two-Step Lax-Wendroff Method was chosen to solve the continuity and momentum equations. The energy and Poisson's equations were solved using an implicit scheme. An explicit difference equation was used to compute the wave equation. In order to resolve the oscillations of the electric field E evolving from Poisson's equation (6.106) and the laser standing wave caused by the nonlinear force, one needs to adopt a time step and a space step as follows

$$\Delta t = 0.1/\omega_{pm} \tag{8.119}$$

$$\Delta x = 0.05\lambda \tag{8.120}$$

where ω_{pm} is the maximum plasma frequency and λ the vacuum wavelength of the laser field. With such time and space steps, the spurious growth of short wavelength oscillations from nonlinear interaction will emerge because the laser pulse length is usually longer than a few picoseconds. By using the smoothing method (Shapiro, 1970) a satisfactory hydrodynamic code has been obtained.

With the aim to demonstrate the strenght of the genuine two fluid model, this hydrocode was used first for a plasma without laser irradiation. The initial conditions were chosen in the following way in order to have conditions close to the later cases of plasmas irradiated by neodymium glass laser light with a critical density of 10^{21} cm^{-3}: a fully ionized hydrogen plasma slab of 10µm thickness with a linear increase of the electron and ion density from 5 x 10^{20} cm^{-3} at x =0 to 10^{21} cm^{-3} at x = 10µm. The initial temperatures were taken to be $T_c = T_i = 10^3$ eV. The initial velocities were $v_e = v_i = 0$ and the electric field E = 0 at t = 0. Working with time steps of 100 attoseconds for covering the detail of the longitudinal osciallation of the electrode, the expanding plasma showed very strong oscillation of the electric field characterized by electrons moving down the ramp and being returned (see fig. 8.2). At later times the oscillations were damped (Fig. 8.3 until the oscillations were nearly

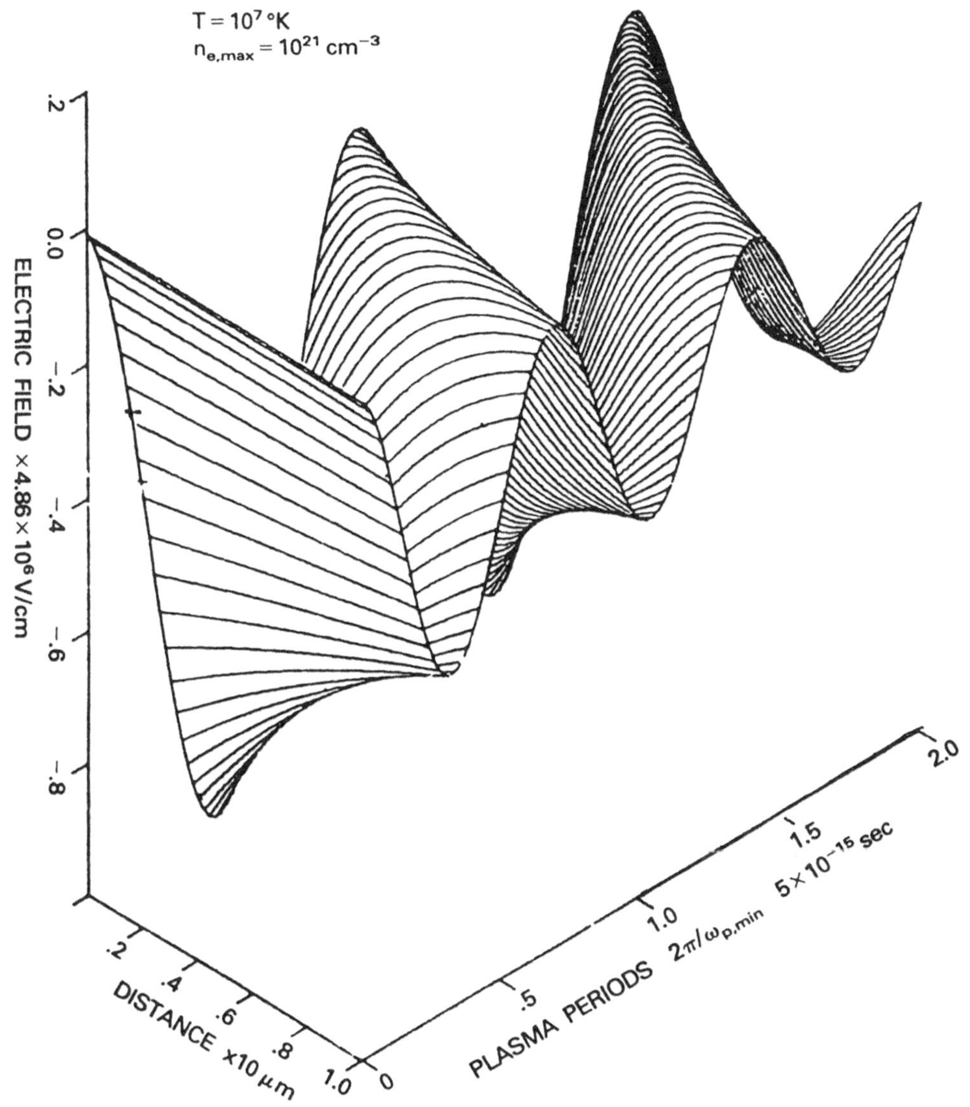

Fig. 8.2. Time dependent development of the longitudinal dynamic electric field E_s along the density with an initial resting ramp of linear plasma of initial temperature 10^7 K and 5×10^{20} cm^{-3} at x = 0 and 10^{21} cm^{-3} at x = 10 μm (Lalousis et al. 1983).

damped) out and a bent profile of the electric field resulted, nearly unchanged along the whole expanding plasma profile (see Fig. 8.4). The field had the highest negative value at x = 0 of 2.2×10^6 V/cm. This value can easily be understood: the plasma has a temperature of 1 keV and with respect to its length of 10^{-3} cm this produces a field of 10^6 V/cm due to the electrons driving out of the plasma thermally.

This numerical solution can be approximated analytically in the following way: taking the time derivative of Poisson's equations (8.105), using the equations of continuity (8.99) and (8.10), and integrating over the spatial coordinate (assuming a zero integration constant) one gets

$$\partial E/\partial t = 4\pi e(n_e v_e - Z n_i v_i). \qquad (8.121)$$

Further time differentiation (of eq. 8.121), using the equations of motion (8.101) and (8.102) and

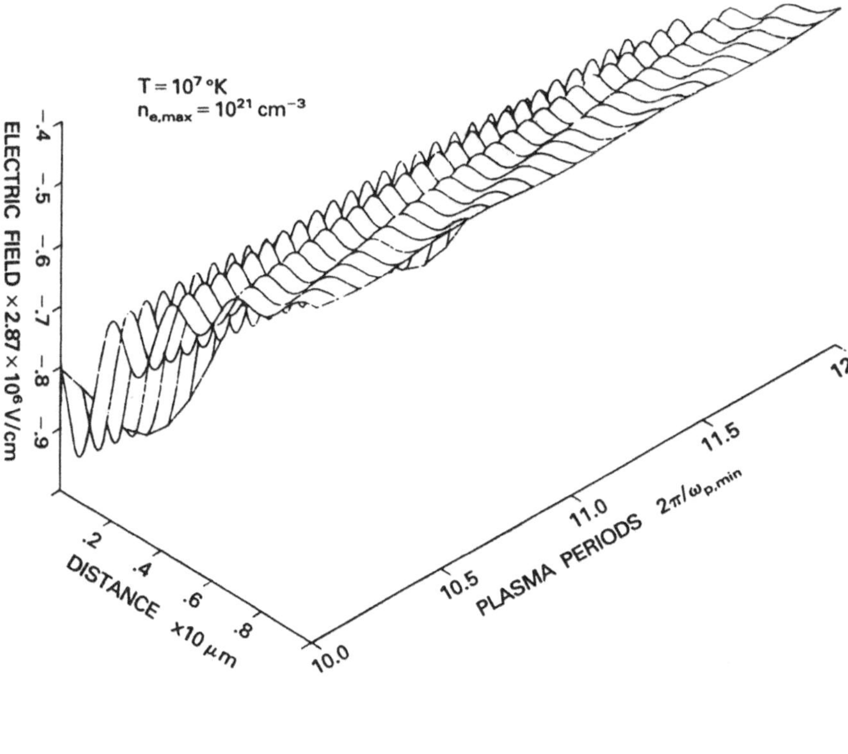

Fig. 8.3. Same as in Fig. 8.2 for times 10 to 12 plasma oscillation periods (Lalousis et al 1983).

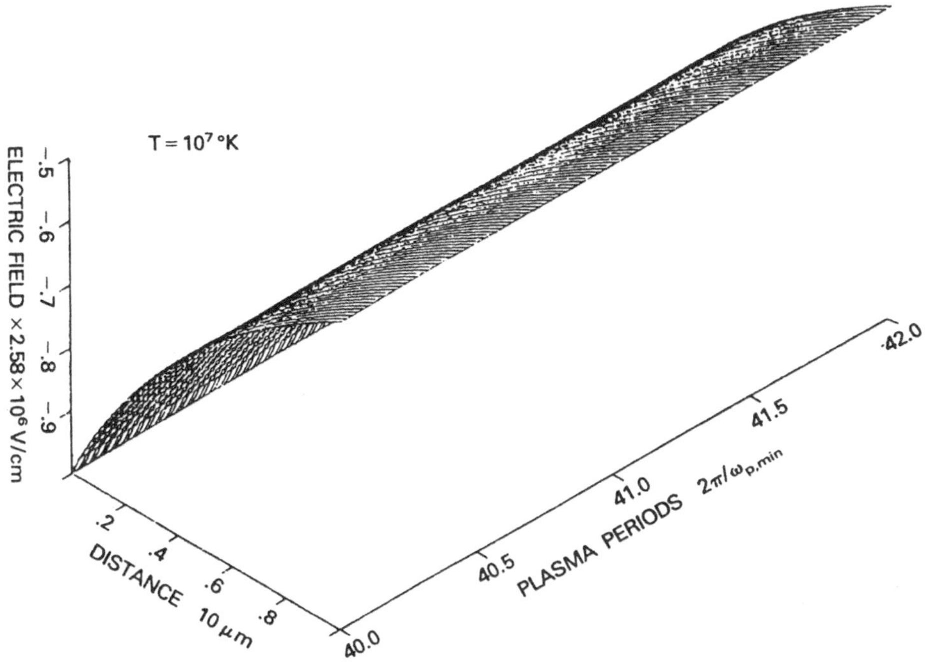

Fig. 8.4. Same as Fig. 8.2 for times of 40 to 42 periods (Lalousis et al. 1983).

rearrangements of the terms result in (Lalousis et al. 1983)

$$\frac{\partial^2 E}{\partial t^2} + \nu \frac{\partial E}{\partial t} + \omega_{p0}^2 E = \tag{8.122}$$

$$E_{s0}\omega_{p0}^2 + \frac{4\pi e}{m_e}\frac{\partial}{\partial x}(E_L^2 + H_L^2)/8\pi + 4\pi e\nu(n_i v_i - Zn_e v_e)$$

where

$$E_{s0} = \frac{4\pi e}{\omega_{p0}^2}\left(\frac{\partial}{\partial x}\left(\frac{3n_i kT_i}{m_i} + Zn_i v_i^2\right) - \frac{\partial}{\partial x}\left(\frac{3n_e kT_c}{m_c} + n_e v_e^2\right)\right)$$

and

$$\omega_{p0}^2 = 4\pi e^2\left(\frac{n_e}{m_c} + \frac{Z^2 n_i}{m_i}\right) \tag{8.124}$$

Neglecting the last term in eq. 8.122 for $\nu \ll \omega_{p0}$ and assuming a vanishing laser field ($E_L = H_L = 0$), the local solution of 8.122 results in an electric field.

$$E = E_{s0}\{(1 - \exp\left(\frac{-\nu}{2}t\right)[\cos(\sqrt{\omega^2_{p0} - \nu^2 t})$$
$$+ \frac{\nu}{2\sqrt{\omega^2_{p0} - \nu^2}}\sin(\sqrt{\omega^2_{p0} - \nu^2 t}]\} \tag{8.125}$$

which oscillates with a frequency close to the plasma frequency. These oscillations, however, are damped (expnentially decaying) by the collision frequency such that after a time $t \gg 1/\nu$ a nearly constant electric field remains, as seen numerically (Figs. 8.3 and 8.4) equivalent to to the stationary cases of the preceding sections. This field is determined by the spatial gradients of the enthalpy of the ions and electrons given in the brackets within the square brackets of eq. (8.123), divided by the particle masses. Neglecting the electron and ion velocities and assuming $T_e \gg T_i$, one gets approximately from (8.123)

$$eE_{s0} \cong \frac{1}{n_e}\frac{d}{dx}(3n_e kT_e) \tag{8.126}$$

We can conclude from this equation that the electric field E_{s0} is simply caused by the gradients of the electron density and temperatures. At the critical surface these gradients might cancel each other; therefore in the case of high laser irradiance the extra contribution from the nonlinear force will play the crucial role in determining E_{s0} (very strong double layers, see [Hershkowitz 1985].

Using similar simplifying approximations as in (8.123) but including now the oscillating laser field, the longitudinal electric field E is given by (Hora et al. 1983), 1985, Goldsworthy et al. 1986).

$$E_s = \frac{4\pi e}{\omega_p^2}\left[\frac{\partial}{\partial x}\left(\frac{3n_i kT_i}{m_i} + Zn_i v_i^2\right) - \frac{\partial}{\partial x}\left(\frac{3n_e kT_e}{m_e} + n_e v_e^2\right)\right. \tag{8.127}$$

$$+\frac{1}{m_c}\frac{\partial}{\partial x}(E_L^2 + H_L^2)/8\pi](1 - \exp(-\frac{\nu}{2}t)\cos\omega_c t)$$

$$+\frac{\omega_p^2 - 4\omega^2}{(\omega_p^2 - 4\omega^2)^2 + \nu^2\omega^2}\frac{4\pi e}{m_e}\frac{\partial}{\partial x}\overline{(E_L^2 + H_L^2)}\cos 2\omega t$$

$$+\frac{2\nu\omega}{(\omega_p^2 - 4\omega^2)^2}\frac{4\pi e}{m_c}\frac{\partial}{\partial x}\overline{(E_L^2 + H_L^2)}\sin 2\omega)$$

where the first term represents the former quasi-static ambipolar field E_{s0} (8.123) with its damped fast oscillations but modified by the amplitude of the fast time averaged laser field density $E_L{}^2 + H_L{}^2$ which is dominant before the gas dynamic pressure n_ekT_e acts.

It is to be mentioned that the last term in the first square bracket of Eq. (8.127) is not a "ponderomotive potential", because this "electrodynamic pressure" can well be representing a nonconservative force in the general case as seen also in numerical calculations. It could be shown, that fast electrons moving thought the dynamic fields of the kind of (8.127) gain or lose energy due to the non-conservative property of the last term in the first square bracket. The single electron pressure gradient term, is well known from generating the ambipolar field, as can be seen also as a result of Schlüter's generalized Ohm's law (or the more general form (C.21) if currents and velocities are zero. Here, however, in Eq. (8.127) we have an electric field due to all static and dynamic pressures of electrons and ions and of the part derived from the electromagnetic (non-conservative) field interaction given by the laser field amplitudes (index L in first square bracket). The basic generalization, however, apart from having now all kinds of pressures on top of the ambipolar field, consists in the fact that all is oscillating with the local plasma frequency and is being damped by the collision frequency.

The last two terms in Eq. (8.127) are essential terms with a second harmonics oscillation of the laser frequency (ω). We shall come back to these terms later since these led to the discovery of a new type of plasma resonance (Hora et al 1983, 1985 Goldsworth et al 1986) and of a new type of second harmonics emission from a laser irradiated plasma corona in a wide spread way and with typical spatial oscillation.

The usefulness of this general formulation of the genuine two fluid model will be demonstrated in further examples discussed in the following Chapters, though it is only a one dimensional formulation. The three dimensional formulation (see Goldsworthy, 1988) consists of 16 equations for the electron and the ion fluid (two continuity equations, six equations of motion for all components of the vector equations, two equations of energy and the six equations for all components of Maxwell's equations) for determining the 16 quantities, n_e, n_i, T_e, T_i, the three each components of the electron and ion velocity, and the three components of the electric field and the three ones for the magnetic field.

This all has to be solved for the four dimensions x,y,z, and t for given initial conditions and for boundary conditions mainly determined by the external fields as the laser fields (as in the cases discussed before) or the fields of magnetic coils (e.g. in tokamaks). The irradiated laser radiation consists in solutions which fields itself have to be determined at each time step in the whole space as Maxwell's solutions including the response by the plasma. This treatment automatically should reproduce not only the internal electric fields of plasmas and their general nonconservative oscillations, it should automatically also reproduce the self generated magnetic fields (discovered and first measured by J. Stamper et al [304] in most general form.

Finally, this general hydrodynamic treatment is not complete compared with microscopic theory since it needs equilibrium distributions of the thermal energy locally at each time and it cannot describe the interpenetration of plasma. The general treatment along the lines of microscopic theory however, was shown to have enormous other problems.

8.9 Double Layers and Surface Tension of Plasmas

The result of Fig. 8.4 clearly indicated the wide spread internal electric fields inside of a plasam where not at all any outside electric field is acting. This field is simply speaking an ambipolar field and it corresponds to a surface potential. Taking its strength of about 1 Million V/cm integrated over its lengths of 0.001 cm, one arrives, as mentioned before at the potential of about keV just what corresponds to the plasma temperature. Electrons within this plasma have a work function of about this value for being emitted outside of the plasma. For a generalized formulation of the work function with respect to Richardson equations, the reader is referred to Eq. (1.26) in the monography of Eliezer et al (1986).

Where such a potential is present, this is related also to a separation of charges and an electric **double layer** (DL). In the case of the internal electric field of the inhomogeneous plasma from Fig. 8.4 (one may remember the much more complicated dynamic field than the ambipolar component only of Eq. (8.127) what was followed up numerically in the two preceding Figures), the charge separation and the double layer is indeed very wide spread, much wider than in the case of the surface potential of the plasma of Fig. 2.1 in which case the spread is of the value of the Debye length only.

It is very important to realize the wide spread double layer. In the historical plasma theory, any charge separation or electric field within the dimension of a Debye length was considered as a matter of microscopic fluctuations which intentionally are neglected by macroscopic hydrodynamics such that the picture of a space charge quasi neutral plasma could have been saved. The result of Fig. 8.4, however, makes clear that the internal electric fields, just that described by Alfvén which "intuitively are not clear" (Kulsrud, 1983), do exist in dimensions very much larger than the Debye length without external fields.

How these fields are responding with external magnetic fields has been seen in experiments with tokamaks (Sigmar et al, 1974) where side-on measurements of the H-alpha line showed a Doppler shift corresponding to a rotation of the plasma with a velocity of several 10^6 cm/sec around the torus axis. But the authors could not believe in such a rotation and assumed that "this is no rotation of the plasma but there are suprathermal ions due to banana instabilities". The rotation velocity could be directly calculated from the radialelectric field in the torus given by the value of Eq. (8.127) combined by the external magnetic field .

It was the merit of M. Bell (1979) that he was measuring the rotation of the plasma in the tokamak precisely resulting in exactly the same numbers we could evaluate from the ExB drift (Hora et al 1983, Eliezer et al 1989). Bell measured a rotation velocity of $2x10^5$ cm/sec in a small tokamak with a radius of 2 cm only, a magnetic field of B=0.5 Tesla and an electron temperature of T=50 eV.

Taking the stationary electrostatic field without external electric fields from Eq. (8.127) one arrives at the radial electric fields in the tokamak of

$$E_S = \frac{3}{en_e}\frac{d}{dr} n_e kT_e \tag{8.128}$$

resulting in a plasma rotation velocity in the torus in the presence of the external magnetic field B of

$$v_{rot} = 3T/rB \tag{8.129}$$

one arrives at $v_{rot} = 2.4x10^5$cm/sec. The radial electric field in the tokamak was just at this time measured by Razumova (1983) to a rather surprize of the tokamak experts. But these facts of plasma rotation and internal electric fields in tokamaks are still considered as marginal properties only through theoretical work (Zehrfeld, and Green 1972) elaborated these conditions.

With respect to laser produced plasmas, the result of the electric fields inside of plasma or especially in the outermost parts of the plasmas motivated then the discovery of **surface tension** in plasmas.

In plasma there are no intermolecular forces, and attractive and repulsive forces between the electrons and ions are fully compensated. Therefore, surface tension was never expected or considered in the literature for high temperature plasmas. This situation has changed since generalization of the plasma theory to the non-space-charge neutral case was possible, and the internal electric fields and the DLs at the surface of laser-produced plasmas are a basic property. We report now on the derivation of surface tension (Eliezer et al 1989) in plasmas and the following stabilization of surface waves against the Rayleigh-Taylor instability.

The existence of electric fields and DLs of the thickness of the order of a Debey length or more in the surface of a laser-produced plasma expanding into vacuum shows that this produces surface tension. In particular, if the DLs are very strong, as in normal and inverted DLs, due to the nonlinear force, the effect of surface tension may be very important.

Considering the electric field **E** as from Fig. 2.1 or from an internal electric field E_s as in Fig. 8.4, this causes an electric field energy density $\mathbf{E}^2/8\pi$. Integrating this over the surface layer of a thickness l (or in spherical geometry from a radius R to R+l), and a surface $S = 4\pi R^2$, the surface tension α is given by the definition of "Energy per surface" as

$$\alpha = \frac{4\pi R^2 \int (\mathbf{E}^2/8\pi) dR}{4\pi R^2} \tag{8.130}$$

Since the electric field E does not change its sign within the DL, one can approximate the average values of E and its square values in the following way:

$$\alpha = \overline{E}^2 l/8\pi \cong |\overline{E}|^2 l/8\pi \tag{8.131}$$

Using the fact that the DL produces the potential of the thermionic work function $e|E|el = gkT$ and that its thickness l is that of a Debye length Eq. (2.7) results in

$$\alpha = \frac{(gkT)^2}{e^2 8\pi l} = 4.75 \times 10^{16} g^2 n^{1/2} T^{3/2} \mathrm{erg/cm^2}; \quad [T] = K. \tag{8.132}$$

For example, a plasma of density 10^{22} cm^{-3} having a temperature $kT \cong 100$ eV implies $\alpha \cong 5.3 \times 10^5$ erg/cm^2 assuming g=3. For a spherical pellet, the surface pressure P_α due to the surface tension of the charged layer is given by (Landau et al 1959)

$$P_\alpha = \frac{2\alpha}{R} \tag{8.133}$$

If the DL is produced in the density minima cavitons at laser/plasma interaction where the nonlinear force is predominant, the temperature determining the DL is given by an effect T_{eff} - see Eq. (6.57),

$$kT_{eff} = mc^2\, I/I_{eff} \tag{8.134}$$

where the relativistic threshold, (Eq. 6.74) for Nd glass lasers is $I_{rel} = 3 \times 10^{18}/\lambda^2$, with the laser wavelength λ in micrometres.

There results of plasma surface tension can be applied to the discussion of surface waves and their stabilization, depending on the wavelength of the surface waves and the strength of the DL producing the surface tension. The results coincide with recent observations of the stabilization of plasma surfaces against the Rayleigh-Taylor instability. The dispersion relation for electric DLs is taken in analogy to the treatment of fluids (Landau et al 1966). The DL of a plasma is produced by an electric field E defined by the surface charge σ of each of the positive or negative layer

$$E = E_z = 4\pi\sigma, \tag{8.134}$$

which should be in the z direction perpendicular to the surface. If the distance between the charged layers, the thickness of the DL, is l, the surface charge density is given by the potential V_{DL} of the DL

$$\sigma = \frac{Q}{S} = \frac{V_{DL}}{4\pi l}, \tag{8.135}$$

as given before in Eq. (8.130) by the temperature T of the plasma and the factor g. The effective temperature T is given either by the chaotic motion or by the nonlinear force (ponderomotive) potential if the laser interaction is strong enough. The electrostatic potential is given by

$$\phi = -4\pi\sigma\xi + \phi(0) \quad , \tag{8.136}$$

where, as usual, $V_{DL} = \phi(l) - \phi(0)$. We assume that the surface wave propagating in the x direction perpendicular to the surface results in a displacement in the z direction given by

$$\xi = B\exp[i(kx-\omega t)] \tag{8.137}$$

In this case, the potential field above the oscillating surface can be written as

$$\phi(x,z,t) = -4\pi\sigma z + \psi(x,z,t) \tag{8.138}$$

and

$$\psi = B \exp\,[i(kx - \omega t)]\exp(-kz); \tag{8.139}$$

ψ is a small perturbation that satisfies the Laplace equation

$$\nabla^2\psi = 0\,,\ \psi(z \to \infty) = 0\ . \tag{8.140}$$

Choosing the boundary equation $\phi(z=0) = 0$, one gets from Eq. (8.139)

$$\psi(\xi) \ = \ 4\pi\sigma\xi \tag{8.141}$$

On the charged surface of the DL, the electric pressure can be approximated to the first order in ψ by

$$\frac{E^2}{8\pi} \cong \frac{E_z^2}{8\pi} \cong 2\pi\sigma^2 + (k\sigma\psi)_{z=0} = 2\pi^2 + 4\pi^2 k\xi \tag{8.142}$$

Assuming that the DL has an acceleration a, one can write the equation of motion for the surface of the liquid as follows:

$$P = -\alpha\frac{\partial^2\xi}{\partial x^2}\ , \tag{8.143}$$

where α is the surface tension coefficient and P is the pressure approximated by

$$P = \rho a\xi - \rho\left(\frac{\partial\psi}{\partial t}\right)_{z=0} - \frac{E^2}{8\pi}\ ; \tag{8.144}$$

ρ is the plasma density ($\rho = m_i n_i$) and ψ is the velocity potential

$$\frac{\partial\xi}{\partial t} = \left(\frac{\partial\psi}{\partial t}\right)_{z=0} \tag{8.145}$$

Using Eqs. 8.142 to 8.144, one gets

$$\sigma a\xi + \rho\left(\frac{\partial\psi}{\partial t}\right)_{z=0} - \alpha\frac{\partial^2\xi}{\partial x^2} - 4\pi\sigma^2 k\xi = 0\ . \tag{8.146}$$

Substituting

$$\xi = B \exp[i(kx - \omega t)] \tag{8.147}$$

and

$$\psi = A \exp[i(kx - \omega t)] \tag{8.148}$$

into Eq. 8.146, one gets two equations (Re = 0 and Im = 0) with two parameters A and B. A nontrivial solution requires a vanishing determinant that implies the following dispersion relation:

$$\omega^2 = \frac{k}{\rho}\,(a\rho - 4\pi\sigma^2 k + \alpha k^2)\ . \tag{8.149}$$

The surface wave dispersion relation for a nonaccelerating charge layer is given by

$$\omega^2 = \frac{k^2}{\rho}\,(\alpha k - 4\pi\sigma^2) \quad . \tag{8.150}$$

The stability criterion in this case ($\omega^2>0$) implies

$$\alpha > \frac{4\pi\sigma^2}{k} \tag{8.151}$$

or equivalently

$$\lambda_{sw} < \frac{\alpha}{2\sigma^2} = \left(\frac{8\pi^2}{g}\right)\ \lambda_D, \tag{8.152}$$

where λ_{sw} is the surface wavelength. For a weak DL, $g \cong 1$, yielding $\lambda_{sw} < 8\pi^2\lambda_D$, while for a strong DL, $g \cong 10$, one gets $\lambda_{sw} < 8\lambda_D$ for stabilization.

The resulting surface tension stabilizes against the Rayleigh-Taylor instability (Nishihara 1988) in spherical plasma surfaces in the same way as in water droplets. If there would not be surface tension, the droplets would break into small parts by the instability, but the molecular surface tension of the water counteracts and results in a fixed droplet diameter. The same happens to spherical plasmas at laser irradiation. For large diameter, the Rayleigh-Taylor instability is bending the surface, but any surface ripple or surface waves can be unstable only for a wavelength larger than the value in Eq. (8.152), for smaller surface wave length there is stabilization. This is effective especially if the Debye length in Eq. (8.152) is large, i.e. for large pellet temperatures or for even much higher effective "temperatures" given by the nonlinear force. Unexpected strong stabilization in laser fusion experiments were observed (Lerche et al. 1987, Key et al 1990) which may be explained by the double layer surface tension mechanisms.

It should be mentioned that the plasma model for surface tension has an important consequence also for solid state physics, especially for metals.

In the same way as in plasmas, see Fig. 2.1, the degenerate electron gas within the rigid ion lattice likes to move out "thermally" as in a plasma. This occurs as long until the electrons outside produce such a space charge and double layer with an electric field (and the subsequent surface potential) that the electrostatic field energy density per surface produces a surface tension. The energy with which the electrons leave the plasma, however, is not the thermal one but the quantum energy (Fermi energy) of about 5 to 10 eV.

One can then calculate the surface tension which agrees very convincingly with measured values (Hora et al 1989). This model is basically different from the jellium model (Laud et al 1970), where the surface tension of aluminum arrives at negative values contrary to the measured positive values which are reproduced with our plasma model immediately (Hora, 1989). This plasma model even can be extended to nuclei, however there the positive charges have the nuclear forces as counterpart what needs then the inclusion of a factor two in calculating the thickness of the layer of a kind of a degenerate Debey length for the charge decay. This decay length agrees immediately with Hofstadter's (1987) measured charge decay in nuclei. This plasma surface tension model for nuclei has the advantage that it provides an understandable mechanism contrary to preceding models. The resulting surface energies in the range of 40 MJ agree very well with the measured values (Hora 1989).

NINE

Momentum and Instabilities by the Nonlinear Forces

This section considers some general properties of plasmas due to the nonlinear forces at incident laser radiation—or equivalent microwave interaction. One long-known property is the usual radiation pressure which turns out to be as a trivial case from the following general treatment. Another well-known property of low-density plasma within a standing electromagnetic wave is the pushing of the plasma toward the nodes of the standing wave just by the gradients of $\mathbf{E}^2$ of the wave field [see Eq. (8.86)] [159 to 161]. What was discussed at laser plasma interaction first [158] was the reaction to high-density plasma due to dielectric effects. If a propagating electromagnetic wave runs into a high-density plasma, the spatial derivative of the refractive index is essential. The dielectric increase of the $\mathbf{E}^2+\mathbf{H}^2$ value for perpendicular incidence of the electromagnetic wave due to the dielectric swelling is the driving process. Dielectric swelling means the increase of

$$\mathbf{E}^2+\mathbf{H}^2=E_v^2\left(\frac{1}{|n|}+|n|\right)\exp(-\bar{k}x) \tag{9.1}$$

above its vacuum value, due to the decreasing refractive index n; this causes forces in the plasma. The corona is driven toward lower density (as long as the integral absorption constant $\bar{k}$ is small enough) causing an appropriately strong recoil to the plasma interior as an increased radiation pressure due to the swelling. The swelling factor is

$$S=\frac{1}{|n|} \tag{9.2}$$

Figure 9.1 schematically describes the explosion process of the plasma corona due to the nonlinear forces because of dielectric swelling.

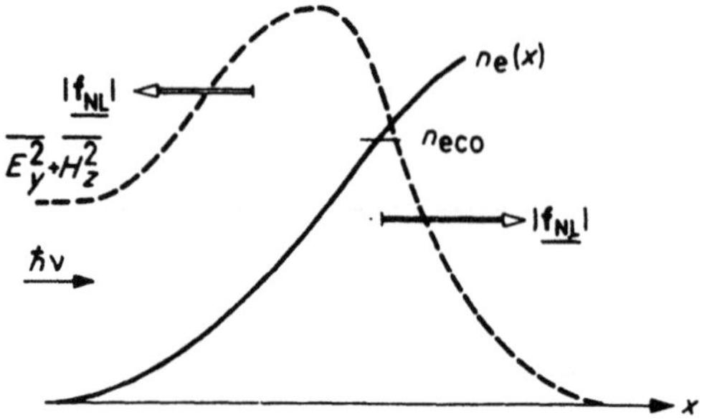

Figure 9.1 Schematic description of the nonlinear force in a plasma corona of an electron density profile $n_e(x)$ where the swelling of $\mathbf{E}^2+\mathbf{H}^2$ gives rise to the nonlinear force $\mathbf{f}_{NL}$.

This section first evaluates under what conditions the nonlinear forces are larger than the thermokinetic forces. The transfer of the momentum to the plasma corona will then be demonstrated generally, as well as the transfer of kinetic energy to the plasma. From that result, a conclusion about the momentum of photons or of the electromagnetic energy in plasma can be drawn. This question touches the 70-year-old Abraham–Minkowski controversy. From the aspect of energy density in blackbody radiation, a short interpretation of the energy density in a plasma will be given, though a general solution of the Abraham–Minkowski controversy—at least for nonabsorbing or little absorbing media—has been evaluated now by M. M. Novak [179]. This general result supports indirectly the validity of the special formulations of the nonlinear force, as described in the preceding subsection as appropriate formulation for plane waves. A further general property of the nonlinear forces at perpendicular incidence on plasmas is the generation of parametric instabilities about which the last subsection will report. The details of the numerical evaluation of the nonlinear force for perpendicular incidence and a comparison with various experiments will be presented in the following section.

9.1 Range of Predominance of the Nonlinear Force

The nonlinear force $\mathbf{f}_{\mathrm{NL}}$ will be observed only when the thermokinetic force $\mathbf{f}_{\mathrm{th}}$, contained in the total force $\mathbf{f}$ [Eqs. (8.4) and (8.5)], is comparable to or less than the nonlinear force. The formula (8.84) for plane waves at perpendicular incidence is rewritten

$$\mathbf{f} = -\nabla p + \nabla(\mathbf{E}^2+\mathbf{H}^2)/8\pi \tag{9.3}$$

The comparison of the nonlinear force with the thermokinetic force would be simply the comparison of the potentials $n_e kT$ with that value of the nonlinear force before spatial differentiation in Eq. (9.3) where, however, an appropriate integration constant has to be added. This integration constant is determined by the fact that for a constant dielectric constant there has to be no net nonlinear force if the electromagnetic waves have a temporally constant amplitude. The value of $n_e KT$ must be compared, therefore, with the excess of $\mathbf{E}^2+\mathbf{H}^2$ over its vacuum value. E^* is defined as the electric field amplitude in vacuum, where the nonlinear force equals the thermokinetic force.

$$n_e\left(1+\frac{1}{Z}\right)KT_{\text{th}}=\frac{E_v^{*2}}{16\pi}\left[\left(\frac{T^{3/4}}{a}+\frac{a}{T^{3/4}}\right)\exp_0-1\right] \tag{9.4}$$

The minimum absolute value of the refractive index n has been used from Eq. (6.49) with Eq. (6.50). The exponential function in Eq. (9.4) is abbreviated by $\exp_0$. The temperature T for the nonlinear refractive index consists of the temperature T_{th} of random motion and the energy of coherent motion, thus T^* in Eq. (6.57). The electron density n_e in Eq. (9.4) is then the cutoff density. Equation (9.4) is then a high-order equation between the plasma temperature T_{th} and the critical laser field E_v^* for the predominance of the nonlinear force. The solution is found by an iteration, where the first-order approximation is $T=T_{\text{th}}^{(1)}$. This is used for the calculation of the second-order solution $T_{\text{th}}^{(2)}$ including Eqs. (6.55) to (6.58)

$$\begin{aligned} n_e\left(1+\frac{1}{Z}\right)KT_{\text{th}}^{(2)}=\frac{E_v^{*2}}{16\pi}\Bigg\{\Bigg[\left(T_{\text{th}}^{(1)}-\frac{E_v^{*2}T_{\text{th}}^{(1)3/4}}{16\pi a n_{ec}K}\right)^{3/4}\frac{1}{a} \\ +a\Bigg/\left(T_{\text{th}}^{(1)}-\frac{E_v^{*2}T_{\text{th}}^{(1)3/4}}{16 a n_{ec}K}\right)^{3/4}\Bigg]\exp_0-1\Bigg\}\end{aligned} \tag{9.5}$$

The iteration proceeds until the difference between the last two iterations is less than 1%. The threshold electric field strength depends on the plasma temperature, on $\exp_0$, and on laser radiation; see Fig. 9.2, in which the electric field strength is expressed in laser intensities, Eq. (6.56). The threshold for the predominance of the nonlinear force over the thermokinetic force is a little above 10^{14} W/cm^2. For CO_2 lasers this threshold is more than 100 times less.

Apart from the mentioned numerical calculation of the nonlinear threshold intensities I^* [146], a more direct estimation can be used by an extension of the calculation of Steinhauer and Ahlstrom [180]. The pressures compared were those generated by the thermalization of the radiation and those governed by the nonlinear effects. The ratio

$$\frac{|\mathbf{f}_{\text{th}}|}{|\mathbf{f}_{\text{NL}}|}=R\geqslant\frac{1.14\times10^5}{T^{54}}\left(\frac{Z}{\lambda_0}\right)^{1/2} \tag{9.6}$$

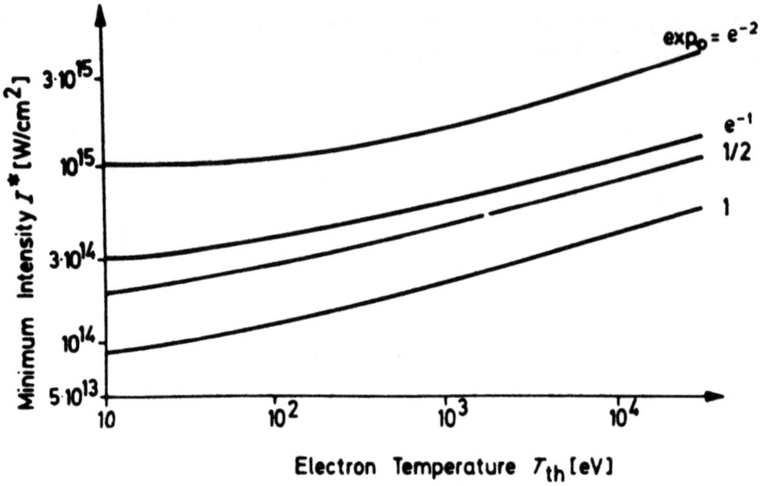

Figure 9.2 Minimum intensity I^* for neodymium glass laser radiation to create larger nonlinear forces $\mathbf{f}_{NL}$ than thermokinetic forces with an undefined exponential function for the collision induced attenuation [146].

was a realistic upper bound, because the maximum values of temperatures T were used. An increase due to thermal conductivity was neglected. The laser wavelength λ_0 is given in micrometers and T in electron volts. The result is that even at intensities less than the above-given threshold I^*, the nonlinear force should predominate, if the kinetic temperature T of the plasma is larger than 10 keV for neodymium glass laser wavelengths. The connection with the above-mentioned result for I^* can be seen in the following way. Because Steinhauer and Ahlstrom's temperature determines the optical constants, Eq. (6.57), we can use [146]

$$T = T_{\text{th}} + \frac{\varepsilon_{\text{osc}}}{2K} = T_{\text{th}} + \frac{\omega_p^2 E_v^2}{16\pi\omega^2 n_e K|n|} \approx \frac{E_v^8}{(16\pi a n_{ec} K^{3/4})^4} \tag{9.7}$$

The predominance of the nonlinear force claims from Eq. (9.6), that R is less than unity, therefore

$$1 > (7.2 \times 10^8/E_v)^{10} \tag{9.8}$$

with E_v in V/cm and for the neodymium glass laser wavelength of 1.06 μm. This immediately shows that there is a strong increase of the nonlinear force, as soon as I_v exceeds 7.2×10^8 V/cm, which corresponds to a laser intensity of about 10^{14} W/cm^2. The nonlinear intensity dependence of the refractive index was not used by Steinhauer and Ahlstrom but their result could be rectified later [146]. The then following result (9.8) shows the strong, resonance like increase of the nonlinear force, as soon as the threshold is exceeded. This resonance like behavior of the nonlinear force is a basic property which could be seen from the mentioned interaction process in a more precise way. The use of the constant R in Eq. (9.6) to be one is somehow arbitrary and can give the right order of magnitude only.

9.2 Momentum Transfer to the Plasma Corona and Compression

Considering again plane electromagnetic waves perpendicularly incident on a stratified plasma with a depth given by the x-coordinate, the momentum transferred by the nonlinear force—if it is predominant under the conditions given in the preceding subsection—following the scheme of Fig. 9.1, will be evaluated. A laser pulse is incident between the times t_1 and t_2. If the time variation of the amplitude E_v of the electric laser field strength in vacuum is slow enough, the Poynting vector can be neglected. It can be shown that even a picosecond substructure of the laser pulses does not change these assumptions [181, 182]. The total energy ε_L of the laser pulse is then

$$\varepsilon_L = c \int_{\tilde{K}} dy\, dz \int_{t_1}^{t_2} dt\, \frac{E_v^2(y, z, t)}{8\pi} \tag{9.9}$$

Here the integration is performed across the entire cross section $\tilde{K}$ (coordinates y and z), where light interacts with the plasma. The cross section is assumed to be sufficiently large to use a plane wave description. The momentum of all photons in the vacuum is

$$P_0 = \frac{\varepsilon_L}{c} \tag{9.10}$$

Under the condition for applying the WKB approximation (7.10), the nonlinear force produces a total momentum P_{inh} transferred to the inhomogeneous plasma between the depths x_1 and x_2 in the direction of $\mathbf{f}$ given by

$$P_{\text{inh}} = \int_K dy\, dz \int_{x_1}^{x_2} dx \int_{t_1}^{t_2} \mathbf{f}_{\text{NL}}\, dt \tag{9.11}$$

Using Eq. (8.72) for the nearly collisionless case is found

$$\begin{aligned} P_{\text{inh}} &= -\int_K dy\, dz \int_{x_1}^{t_2} dt \int_{x_1(t)}^{x_2(t)} dx \left[\frac{\partial}{\partial x} \frac{E_v^2(y, z, t)}{16\pi} \frac{1}{|n|} - |n|\right] \\ &= -\frac{P_0}{2|n_2|}(1 - |n_2|)^2 \end{aligned} \tag{9.12}$$

n_2 equals $n(x_2)$ by definition. x_1 is assumed to be outside the plasma in the vacuum. Equation (9.12) expresses the momentum of the accelerated inhomogeneous plasma layer in terms of the momentum P_0 of the laser pulse in the vacuum. To get a high momentum one needs low values of n_2, which means a high swelling S according to Eq. (9.2). The magnitude of the momentum is limited by damping; this causes an effective collision frequency. The momentum [146] transferred to the plasma can be much more than the usual radiation pressure, which is given by the photon momentum, because the momentum at the front of the plasma can be compensated by the momentum transferred to the plasma interior. Figure 9.3 shows the ablative acceleration of the plasma corona towards negative x if the swelling is sufficient, to produce the negative momentum P_{inh}. If x_2 exceeds the value for the maximum swelling, the net momentum transferred to the plasma between x_1 and the then reached depth x_3 is then increasing again. It reaches positive values up to the value of the usual radiation pressure P_0. The conservation of momentum gives

$$P_0 = P_{int} + P_{inh} \tag{9.13}$$

When all photons are absorbed in the plasma interior, then the total momentum transferred to the plasma interior, that is, to the plasma below the critical density, is

$$P_{int} = P_0 + P_{inh} = \frac{P_0}{2}\left(\frac{1}{|n|} + |n|\right) \tag{9.14}$$

In terms of P_0 can be written

$$\frac{P_{int}}{P_0} = 1 + \frac{P_{inh}}{P_0} \tag{9.15}$$

The momentum for compressing the interior of the plasma by the nonlinear force can be multiples of the radiation pressure, if the swelling provides a sufficiently large momentum to the inhomogeneous surface.

In order to see the maximum values of the swelling and of the explosion by

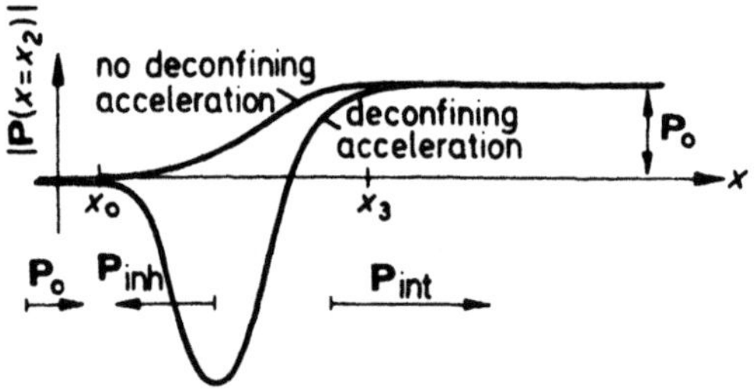

Figure 9.3 Momentum P transferred to the plasma between the vacuum and the interior of the inhomogeneous plasma. At sufficient swelling, the momentum can be negative P_{inh}, showing an ablation of the corona up to the minimum value of the refractive index [146].

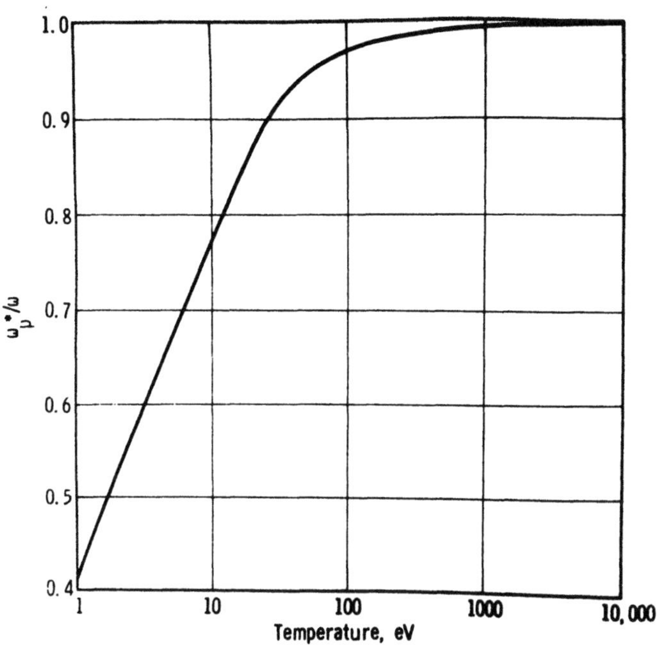

Figure 9.4 Electron densities corresponding to a plasma frequency $\omega_p{}^*$ below which the collision produced absorption (the exponential factors in Eq. (8.72) for the nonlinear force can be neglected at WKB conditions) [138].

the nonlinear forces, it has been evaluated [138] numerically up to plasma densities (expressed by a plasma frequency ε_p^*) for which the exponential damping factor in the formulation of Eq. (8.36) can be neglected. A pessimistic estimate of absorption is determined considering Coulomb collisions without nonlinear effects if the WKB approximation is used with a value of $\theta < 0.25$ (Eq. 7.10). The results are given in Fig. 9.4. Using these values, the ratio of P_{inh}/P_0 or the nonlinear increase of the radiation pressure by swelling corresponding to the nonlinear force is given as dependent on the plasma temperature in Fig. 9.5. Note that these are optimum values, while the practical cases can provide conditions of lower swelling. On the other hand, the evaluations of the Figs. 9.4 and 9.5 were done for a linear behavior of the collision frequency. The resonancelike increase for the nonlinear collision processes can then increase the values given in Figs. 9.4 and 9.5.

9.3 Energy Transfer by Integration of the Nonlinear Force

In order to discuss some experimental cases, in which the conditions for the predominance of the nonlinear force hold, an exact integration of the equa-

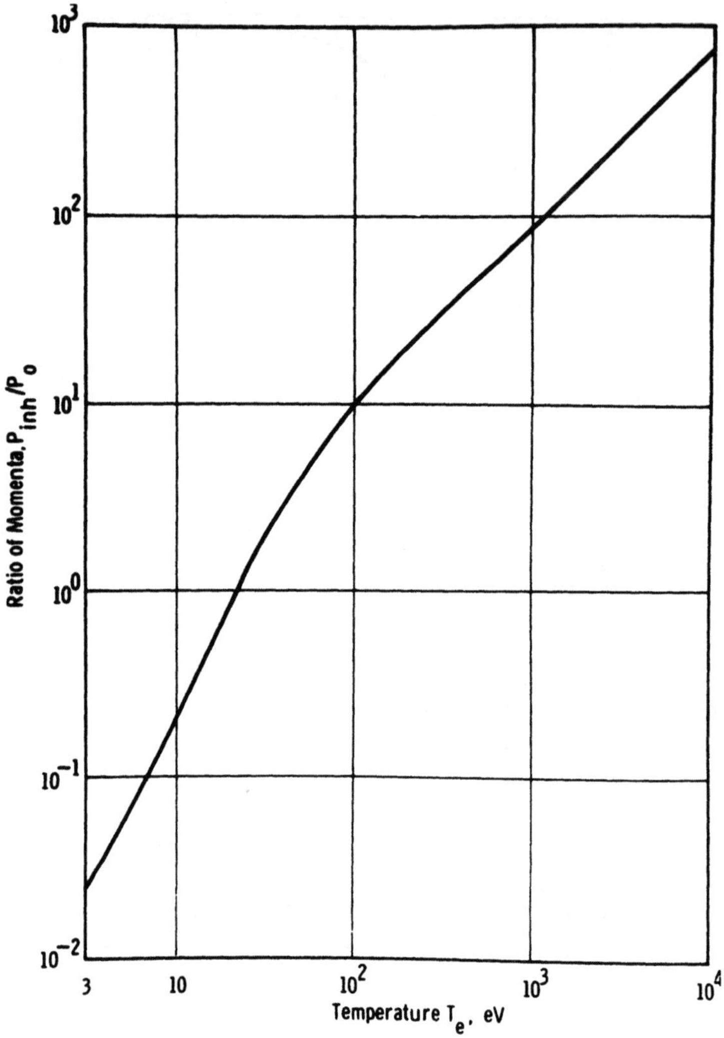

Figure 9.5 Ratio of the maximum momentum transferred to a WKB-like plasma corona by the nonlinear force depending on the temperature, if the densities of Fig. 9.4 are used and if the collision frequency is linear [138].

tion of the nonlinear force is possible if the resulting plasma profiles vary sufficiently slowly in time. The same geometry of perpendicularly incident plane wave and the WKB approximation are considered. The equation of motion is used in formulation

$$\mathbf{f}_{NL} = -\mathbf{i}_x \frac{E_v^2}{16\pi} \frac{n_0}{n_{ec}} \frac{\partial}{\partial x} \frac{1}{|n|} \tag{9.16}$$

We recall that the velocity v_N gained by a body falling along a distance x

with an acceleration dv_i/dt is

$$v_0=[2(dv_i/dt)x]^{1/2} \tag{9.17}$$

The differential gain of the square of the velocity for varying acceleration is

$$\delta v_0^2=2(dv_i/dt)\Delta x \tag{9.18}$$

The increase of the kinetic energy of the ions due to the force (9.16) is then (using $n_e=Zn_i$, with the ion charge Z, and avoiding the restriction of the collisionless case)

$$d(\tfrac{1}{2}m_i v_i^2)=m_i \frac{dv_i}{dt}dx=\left|\frac{E_v^2}{16\pi}\frac{Z}{n_{ec}}\frac{\partial}{\partial x}\frac{\exp(-\bar{k}x/2)}{|n|}\right|dx \tag{9.19}$$

The exact integral for the ion energy is then

$$\begin{aligned}\frac{m_i}{2}v_i^2&=\frac{E_v^2}{16\pi}\frac{Z}{n_{ec}}\int_{x_1}^{x_2}\frac{\partial}{\partial x}\left[\frac{\exp(-\bar{k}x/2)}{|n|}\right]dx\\ &=\frac{E_v^2}{16\pi}\frac{Z}{n_{ec}}\left[\frac{\exp(-\bar{k}x_2/2)}{|n(x_2)|}-\frac{\exp(-\bar{k}x_1/2)}{|n(x_1)|}\right]\end{aligned} \tag{9.20}$$

The integration is performed between the depth x_1, a point in vacuum ($n=1$; $\bar{k}=0$), and x_2, the point in the plasma corresponding to minimum refractive index $|n|$. $\bar{k}$ is assumed to be sufficiently small as shown in Fig. 9.4; see also Marhic [183]. The maximum energy of the ions in the nonlinear-force driven ablation region is then

$$\varepsilon_i^{\mathrm{trans}}=\frac{E_v^2}{16\pi}\frac{Z}{n_{ec}}\left(\frac{1}{|n|_{\min}}-1\right) \tag{9.21}$$

The ion energy of the plasma interior corresponding to a compression is given by

$$\varepsilon_i^{\mathrm{compr}}=\frac{E_v^2}{16\pi}\frac{Z}{n_{ec}}\frac{1}{|n|_{\min}} \tag{9.22}$$

The limits of the integration in Eq. (9.20) are now those for $|n|_{\min}$ and for the interior of the superdense plasma with $\bar{k}\gg 2/x$ a nonvanishing refractive index. The acceleration of the plasma parallel to the laser beam corresponds to a (dynamical macroscopic) nonlinear absorption process.

The results (9,21) and (9.22) can be interpreted very easily by taking into account the fact that the average kinetic energy of the oscillation, $\varepsilon_{\mathrm{osc}}^{\mathrm{kin}}$ of the electrons in the laser field, Eq. (6.56), can be expressed for a WKB case with dielectric swelling by

$$\varepsilon_{\mathrm{osc}}^{\mathrm{kin}}=\frac{E_v^2}{16\pi n_{ec}|n|}\exp(-\bar{k}x/2) \tag{9.23}$$

$\bar{k}$ is assumed to be much less than one at the maximum of the swelling in agreement with the evaluation of Fig. 9.4, but a damping mechanism is included to prevent an infinite resonancelike process. Finally, the maximum energy of the ablated ions is found to be

$$\varepsilon_i^{transl} = \frac{Z(\varepsilon_{osc,max} - \varepsilon_{osc,vac})}{2} \tag{9.24}$$

$\varepsilon_{osc.max}$ and $\varepsilon_{osc.vac}$ are the oscillation energy of the electrons in the plasma and in vacuum, respectively. In cases where the swelling of the oscillation energy

$$\varepsilon_{osc.max} = S\varepsilon_{osc.vac} = \frac{\varepsilon_{osc}}{|n|_{min}} \tag{9.25}$$

reaches a factor $S=100$, the oscillation energy of the vacuum can be neglected. This correction represents the radiation pressure in the vacuum. For strong swelling is found

$$\varepsilon_i^{transl} = Z\varepsilon_{osc.vac}/2 \tag{9.26}$$

The ion energy and the ion velocity are predominant for the motion of the plasma. This is used in the above consideration. The nonlinear force is acting on the plasma electrons, and the driven electrons are electrostatically attached to the ions. Therefore, the whole model works for plasma dimensions only.

9.4 Photon Momentum in Plasma (Abraham–Minkowski Problem)

The results of the subsection before last can be used to discuss the momentum of the electromagnetic energy and the momentum of photons in plasma. It has to be pointed out, that it is not evident from the beginning that the momentum of the electromagnetic energy is the same as that of the photons, as plasmons are involved too. Considering the momentum of the plasma material in the corona, Eq. (9.12), this is a recoil against the propagation of the laser photons. Therefore, it can be concluded that the momentum of the electromagnetic energy in the plasma has been increased by this recoil. The momentum of the electromagnetic energy is then

$$P = |P_{inh}| + P_0 \tag{9.27}$$

and from (9.10) and (9.12),

$$P = \frac{P_0(1+|n|^2)}{2|n|} \tag{9.28}$$

Transferring this to the energy per photon in vacuum

$$p_\phi = \frac{\hbar\omega}{c} \tag{9.29}$$

the photon momentum (in the sense of electromagnetic energy) in the plasma is obtained

$$p_{\phi,\mathrm{pl}}=\frac{\hbar\omega}{2|n|c}(1+|n|^2) \tag{9.30}$$

This is the result of consideration along the lines of the last but one subsections [138], which has been reproduced completely and confirmed by Lindl and Kaw [154]. Furthermore, it has been derived in a completely different way for the electromagnetic energy density of a wave packet in a homogeneous nonabsorbing plasma by Klima and Petrzilka [184]. It has been pointed out by the latter authors that this momentum (9.30) differs from the momentum

$$p_A=\frac{\hbar\omega}{nc} \tag{9.31}$$

derived by Abraham, and

$$p_M=\frac{\hbar\omega}{c}n \tag{9.32}$$

by Minkowski, respectively. Formally, the energy density (9.30) would correspond to a photon momentum which is (half that of Abraham and half that of Minkowski, the average of the two momenta) [185]

$$p_{\phi,\mathrm{pl}}=\frac{p_A+p_M}{2} \tag{9.33}$$

The fact that the momentum (9.33) is reasonable for the plasma (confirming in this way the whole derivation of the preceding section) will be seen in the following evidence [185] that a photon momentum of the kind in Eq. (9.33) in a plasma results in the Fresnel formulas for the reflection of light, when passing from vacuum discontinuously to a homogeneous plasma of refractive index n without collisions. If $\tilde{R}$ is the fraction of reflected photons and $\tilde{T}$ that of transmitted photons, the conservation of energy requires

$$1-\tilde{R}=\tilde{T} \tag{9.34}$$

The conservation of photon momentum with a correct sign for the reflected photons at the interface results in

$$p_\phi+\tilde{R}p_\phi=\tilde{T}p_{\phi,\mathrm{pe}} \tag{9.35}$$

Together with (9,34), the reflection R and transmission T can be eliminated simply

$$\tilde{R}=\left(\frac{1-n}{1+n}\right)^2; \qquad \tilde{T}=\frac{4n}{(1+n)^2} \tag{9.36}$$

resulting in the Fresnel formulas.

The conclusion is that a photon in the aforementioned sense of the radiation energy density, has a higher momentum inside the plasma according to the formula (9.30). In order to push such a photon into a plasma, a recoil is necessary to produce this increase of momentum over the vacuum value. This recoil is that of the reflected photons according to the Fresnel formulas for a discontinuous transition. For a reflectionless, WKB-like, inhomogeneous interface between vacuum and plasma, the inhomogeneous layer has to take the momentum difference as a mechanical recoil directed against the incident photons. If the photons with their increased momentum are absorbed in the plasma, their total momentum of Eq. (9.33) [identical to (9.14) per photon] is then transferred to the absorbing region in the plasma. This causes the recoil, which is increased by the nonlinear swelling of the radiation pressure. For $|n| \ll 1$, this increased radiation pressure is

$$P_{\text{int}} = \frac{P_0}{2|n|} = \frac{S}{2} P_0 \tag{9.37}$$

One should recall the factor $\frac{1}{2}$ at this point in Eq. (9.37).

As has been pointed out [185], the formal rewriting of Eq. (9.30), using (6.47) for $\nu = 0$,

$$p_{\phi,\text{pl}} = \frac{\hbar\omega}{nc}\left(1 - \frac{\omega_p^2}{2\omega^2}\right) = p_A - \frac{\omega_p^2}{2\omega^2} p_A \tag{9.38}$$

indicates that the photon propagates with an Abraham momentum. However, it is reduced by an exchange process of the radiation energy with the oscillating photons. Formally, Eq. (9.38) can be written as

$$p_{\phi,\text{pl}} = \frac{\hbar\omega}{c}\left(n + \frac{1-n^2}{2n}\right) = p_M + \frac{\omega_p^2}{2\omega^2} p_A \tag{9.39}$$

This indicates that the photon behaves like a Minkowski photon with an additional momentum due to the plasma oscillations.

Under the latter aspect of the Minkowski momentum, the photoelectric interaction of the blackbody radiation is understandable. If the blackbody radiation density $U(\omega, T)$, depending on frequency ω and temperature T is calculated from the quantum electrodynamic interaction of this radiation with particles by spontaneous emission and stimulated emission and absorption, one arrives for nonabsorbing media with refractive index n at [186]

$$U(\omega, T) = U_p(\omega, T) n^3 \left(1 + \frac{\omega}{n}\frac{\partial n}{\partial \omega}\right) \tag{9.40}$$

where

$$U_p(\omega, T) = \frac{8\pi\hbar\omega^3}{c^3[\exp(\hbar\omega/KT) - 1]} \tag{9.41}$$

is Planck's spectral distribution for vacuum. Using the refractive index n for a plasma without collisions ($\nu=0$), Eq. (6.46) in Eq. (9.40), the result is

$$U=U_p n \tag{9.42}$$

This result agrees with that derived by Bekefi [187] and Dawson [188] for plasmas. The only problem is the restriction $\omega_p \leqslant \omega$; otherwise formula (9.42) can become purely imaginary. The result (9.42) confirms the Minkowski picture, as U can be connected only to the part of the electromagnetic energy, which consists of "photons in vacuum." The additional photon part of the total electromagnetic energy is then due to the electron oscillations only, see Eq. (9.39). These plasmon oscillations do not contribute to the photoelectric excitation and de-excitation, which was the basis of Einstein's derivation of Eq. (9.41) [189].

The neglect of $\exp(-\bar{k}_x/2)$ for the energy density in a plasma is possible, as emission and absorption are in equilibrium. The total electromagnetic energy density of blackbody radiation in a plasma is

$$U_{\text{tot}}=U_p\left(|n|+\frac{\omega_p^2}{2\omega^2|n|}\right) \tag{9.43}$$

Using the absolute value of the refractive index, formula (9.43) is valid for plasmas with collisions and even for frequencies ω at or below ω_p. The photon contribution to photoelectric interaction is then

$$U(\omega, T)=U_p(\omega, T)|n| \tag{9.44}$$

This is a radiation energy density in contrast to the electromagnetic energy density of Eq. (9.43). In Figs. 9.6 and 9.7, examples of the function U for temperatures $T=10^{6\circ}$K and densities $n_e=3.14\times10^{24}$ cm^{-3} and 1.26×10^{25} cm^{-3} are given in comparison with the Planck function U_p in vacuum (Badertscher [190]).

The spectral integration of U

$$U_{\text{tot}}=\int_0^\infty U\,d\omega=\sigma_{\text{op}}T^4 \tag{9.45}$$

arrives at the Stefan Boltzmann constant in the usual magnitude for vacuum

$$\sigma_{\text{op}}=\sigma_0=5.67\times10^{-5}\ \text{erg/cm}^2\ \text{sec}\ {}^\circ\text{K}^4 \tag{9.46}$$

For plasma, however, σ_{op} has a very slight dependence on temperature and depends on plasma density. For the cases in Figs. 9.6 and 9.7, the values are [190]

$\sigma_{\text{op}}=5.393\times10^{-5}$ cgs for $T=10^{6\circ}$K; $n_e=3.14\times10^{24}$ cm^{-3}

$\sigma_{\text{op}}=5.074\times10^{-5}$ cgs for $T=10^{6\circ}$K; $n_e=1.26\times10^{25}$ cm^{-3}

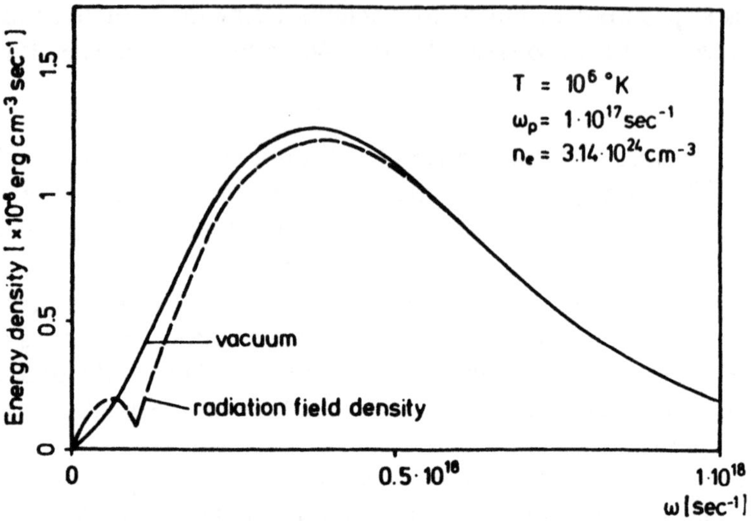

Figure 9.6 Planck's radiation formula (vacuum) for $T=10^{6\circ}$K compared with the spectral distribution of the radiation energy density for a plasma of the same temperature and an electron density of 3.14×10^{24} cm^{-3} [190].

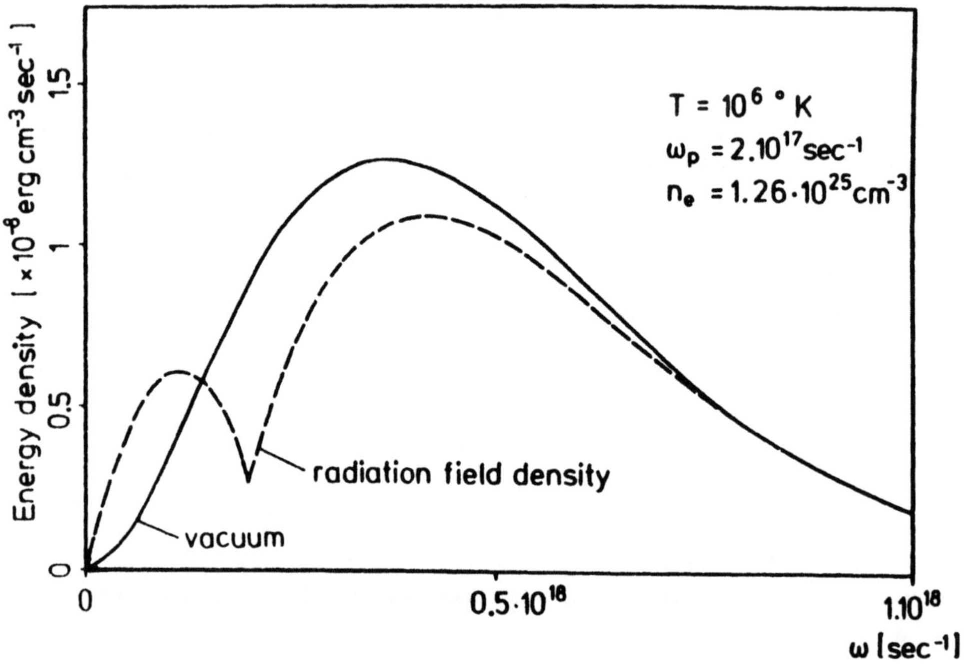

Figure 9.7 Planck's radiation formula as in Fig. 9.6 compared with the plasma of an electron density $n_e=1.26\times10^{25}$ cm^{-3} [190].

Evaluation for higher densities can formally produce σ_{op} values higher than ω_0, but this happens for densities that are above the limit of degeneracy, for which the above theory of optical constants has to be revised.

While we were considering the total energy density (determined by half of the Abraham and half of the Minkowski momentum), only the energy density of the vacuum field (bare photons = photoelectrically acting) excluding the energy exchanged with the electrons (dressed photons), led to the adequate description of the blackbody radiation in a plasma. This result was the starting point for Novak [179] to derive a basically new concept for the discussion of the Abraham–Minkowski problem.

We begin with the assumption (to be justified below) that the Minkowski momentum refers to a spineless photon while that of the Abraham describes a photon of spin unity. As is well known, the concept of spin occurs as a low-frequency phenomenon only, since its contribution at the other end of the spectrum may be neglected.

The equilibrium between the radiation and a medium is studied along the lines of Einstein [189]. Only the Minkowski form of the momentum density satisfied the equilibrium conditions, Eq. (9.40). In order to obtain this result, one imposes a condition $h\nu \gg kT$, effectively, we may say that it holds in the geometrical optics approximation.

In order to study the whole range of frequencies, we include the similarity between the dispersion relation

$$\omega^2 = \omega_p^2 + \mathbf{k}^2 c^2 \tag{9.47}$$

and the relativistic form of the energy E and momentum $\mathbf{p}$ of the particle

$$E^2 = m_0^2 c^4 + \mathbf{p}^2 c^2 \tag{9.48}$$

In view of the relations

$$E = \hbar\omega; \qquad \mathbf{p} = \hbar\mathbf{k} \tag{9.49}$$

We assert that a photon, which before entering a plasma had a zero rest mass, acquired, while inside it, an effective rest mass $\hbar\omega_p/c^2$. As a consequence, the Maxwell field in a medium can now be replaced by the Proca field, describing vector mesons. We now utilize the methods of the Proca electrodynamics and obtain the canonical energy-momentum tensor

$$T^{\mu\nu} = \frac{\partial \mathscr{L}}{\partial(\partial_\mu A^k)} \quad \partial^\nu A^k - g^{\mu\nu}\mathscr{L} \tag{9.50}$$

where the Proca Lagrangian $\mathscr{L}_p$ is

$$\mathscr{L}_p = -\frac{1}{16\pi} F_{\alpha\beta}F^{\alpha\beta} - \frac{1}{c} j_\alpha A^\alpha + \frac{\mu^2}{8\pi} A_\alpha A^\alpha \tag{9.51}$$

with $\mu = m_0 c/h$ being the effective photon's rest mass in units of the universal length. This tensor is not generally accepted. In order to satisfy the conservation laws, we require, following Belinfante [191] that

$$\Theta^{\mu\nu} = T^{\mu\nu} + T_s^{\mu\nu} \tag{9.52}$$

where $T_s^{\mu\nu}$ is interpreted as the spin energy-momentum tensor. It is then found that the total field tensor is symmetric and consists of an "orbital" part, which determines the energy and momentum, and a "spin" part, which does not contribute to either but is important in calculations involving the total angular momentum.

Hence, in accord with the promise of this work, the same conclusion must apply also to the tensor describing the electromagnetic field in a plasma. It is agreed that the Minkowski tensor readily adjusts itself to the canonical formalism. Then, in analogy to Eq. (9.52), the tensor

$$S_{sp}^{\mu\nu} = S_A^{\mu\nu} - S_M^{\mu\nu} \tag{9.53}$$

ought to be interpreted as the spin energy-momentum tensor. In the component form, this tensor has the form

$$S_{sp}^{oo} = 0; \qquad S_{sp}^{oi} = 0; \qquad S_{sp}^{io} = \frac{n-1}{4\pi c}\, \mathbf{E} \times \mathbf{H} \tag{9.54}$$

Then, it follows from the above that the force corresponding to the spin in

$$f_{sp} = \frac{n-1}{4\pi c} \frac{\partial}{\partial t} \mathbf{E} \times \mathbf{H} \tag{9.55}$$

At high frequencies, the refractive index approaches unity and the (spin-) force vanishes, while its value increases toward the low-frequency end. One should remember that the spin was derived by the relativistic generalization of quantum mechanics in the correct way (Dirac equation) of spin $\frac{1}{2}$ for particles as electrons. Here we have a derivation of the spin one of photons without quantization.

Thus, from the interchangeability between the radiation in a medium and a neutral vector meson in vacuo, we conclude that the correct form for the field energy-momentum tensor is given by Abraham. However, in the limit of high frequencies, where the spin contribution may be neglected, the asymmetric Minkowski tensor represents a valid approximation. A very one of photons qualitative picture is then understanding why at low frequencies the photo effect (with exchange of spin to the electron) dominates and at high frequencies the Compton effect with no change of the electron spin. It has been derived from general principles that the electromagnetic energy exchanged between the blackbody field and the electrons is mainly classical by quivering motion even at relativistically high temperatures. The exchange by quantum processes does not exceed $5/\pi^5 \simeq 5\%$ [192].

9.5 Parametric Instabilities

Parametric instabilities are a large number of transfer processes of the laser energy into oscillation or wave modes of the plasma, especially of electrostatic (Langmuir) waves, ion-acoustic waves, and others where an increase of these oscillations or density fluctuations is due to the laser at certain conditions of frequencies (energy) and wave vector (momentum) of the involved waves. Francis Chen gets the credit for discussing and analyzing these phenomena consequently on the bases of the nonlinear force [193, 194] after Oraevski and Sagdeev [195] had introduced this type of wave interaction into plasma physics and a more consequent treatment followed by Silin [196], Dubois and Goldman [197], and Nishikawa [198]. The following broad discussion of this topic was reviewed by Cap and co-workers [199].

The basic phenomena "parametric instability" was called "parametric resonance" when Landau and Lifshitz [200] discussed an oscillating mechanical system of which one parameter was influenced by another oscillation. One example is a mathematical pendulum of length l, mass m whose origin is oscillating in the vertical direction y by a frequency ω_0, and an amplitude A ($y=A\cos\omega_0 t$), see Fig. 9.8. Using angle ϕ as generalized coordinate, the Lagrangian (kinetic energy minus potential energy) is then

$$L=\frac{ml^2}{2}\dot{\phi}^2+mla\omega_0^2\cos\omega_0 t\cos\phi+mgl\cos\phi \tag{9.56}$$

from the Lagrange equation of the second kind

$$\frac{\partial}{\partial t}\frac{\partial}{\partial\dot{\phi}}L-\frac{\partial}{\partial\phi}L=0 \tag{9.57}$$

the following equation of motion for small amplitudes $\sin\phi\simeq\phi\ll 1$) is achieved

$$\frac{\partial^2\phi}{\partial t^2}+[a+q\cos(\omega_0 t)]\phi=0 \tag{9.58}$$

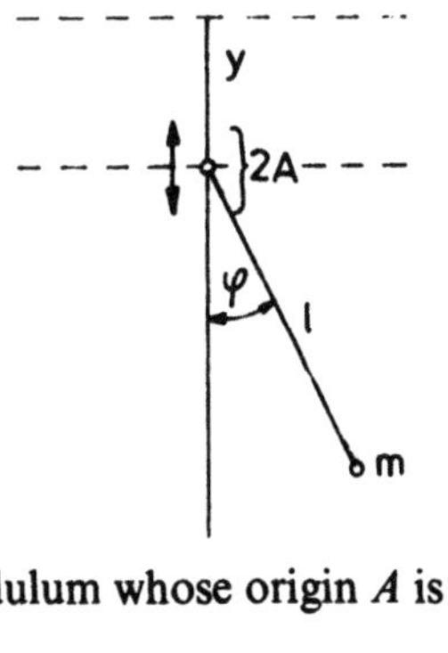

Figure 9.8 Mathematical pendulum whose origin A is oscillating in the vertical direction y.

where

$$a = \omega^2 \tag{9.59}$$

is the radian frequency of the undisturbed pendulum ($A=0$) and

$$q = \frac{4\omega^2 A}{l} \tag{9.60}$$

Equation (9.58) is Mathieu's differential equation which has the special property of quasi-periodic (stable) and nonperiodic (unstable) solution as shown in Fig. 9.9. A quasi-periodic solution is achieved if

$$\omega_0 = 2\omega + \varepsilon' \tag{9.61}$$

where $\varepsilon \ll \omega$. In this case, the solution of $\phi(t)$ is an oscillation with temporally increasing amplitude

$$\phi \sim t \sin c_1 t + \text{oscillating terms} \tag{9.62}$$

($c_1 \sim$ const), if $q \ll a$. This can be seen from (9.58) using Eqs. (9.59) to (9.61)

$$\frac{\partial^2}{\partial t^2}\phi + \omega^2[1 + \eta \cos \omega_0 t]\phi = 0 \tag{9.63}$$

by the iterative solution

$$\phi^{(m)} = \phi_{n-1} + \phi_n \qquad (n = 2, 3, \ldots) \tag{9.64}$$

where ϕ_1 is the undisturbed solution ($q=0$, $\eta=0$). The second-order correction is then from

$$\frac{\partial^2}{\partial t^2}\phi_2 + \omega^2\phi_2 = -\omega^2\eta \cos(\omega_0 t)\cos(\omega t) \tag{9.65}$$

The solutions are oscillating [202], but if Eq. (9.61) is fulfilled, the solutions are

$$\phi \sim t \sin(c_1 t). \tag{9.66}$$

as was to be shown for increasing amplitude (=parametric instability).

The meaning of the result of Eq. (9.62) in the case of a plasma is that ω (pendulum frequency) is an eigenfrequency of the plasma, for example, that of electrostatic oscillations ω_p, while the external disturbance ω_0 is that of the incident laser. An instability will then occur, see Eq. (9.61), if $\omega_p \simeq \omega/2$ corresponding to an electron density of one-quarter of the cutoff density. This has been seen immediately from single-particle simulation codes [203] or from the fact that the light reflected from a plasma contains a small component of half of the frequency of the incident laser beam.

The study of other instability ranges (other shaded areas in Fig. 9.9), the

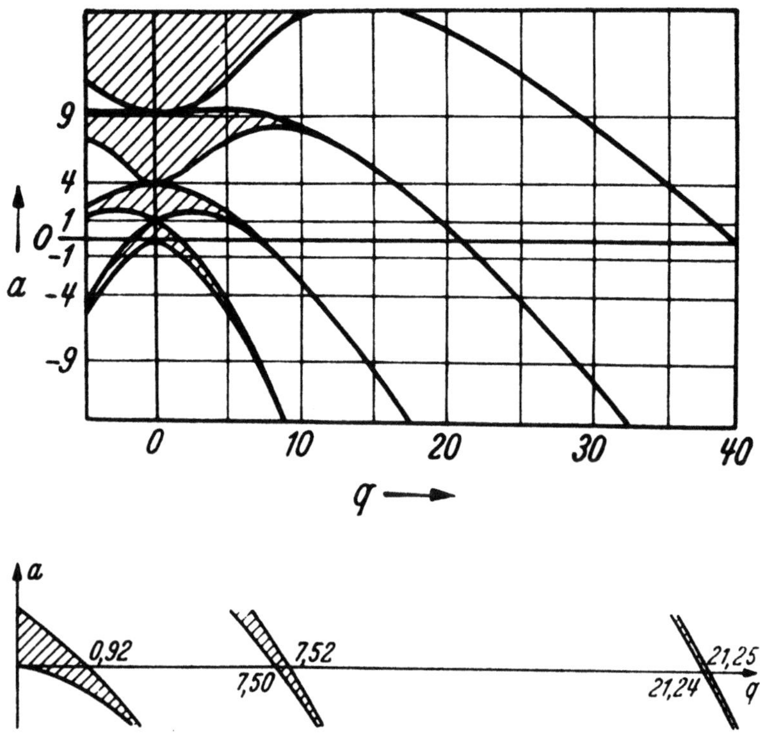

Figure 9.9 Ranges of stable (shaded) and unstable solutions of Mathieu's differential equations (9.58), after Paul and Raether [201].

possible detuning by ε', Eq. (9.61), or the complete solution [$n=\infty$ in Eq. (9.64)] are only some of the further mathematical tasks. Physics is important in following up where the energy transfer from the laser light is introduced, which needs the use of a more general Lagrangian equation (9.57), where the generalized forces depend on velocities and cannot be related to potentials.

Chen's analysis of the parametric instabilites in plasmas at laser irradiation [193] starts from the nonlinear force for perpendicular incidence of infinite plane waves on plasmas whose density profile permits a WKB approximation for the solution of the wave equation Eq. (8.86),

$$\mathbf{f}_{\mathrm{NL}} = -\mathbf{i}_x \frac{1}{16\pi} \frac{\omega_p^2}{\omega^2} \frac{\partial}{\partial x} \overline{\mathbf{E}^2} \tag{9.67}$$

or

$$\mathbf{f}_{\mathrm{NL}} = \frac{\omega_p^2}{8\pi\omega^2} \{\mathbf{E}\cdot\nabla\mathbf{E} - \mathbf{E}\times(\nabla\times\mathbf{E})\} \tag{9.68}$$

where use was made of Eq. (8.19). Chen derived this formula from the quivering motion description found in most textbooks, where the difficulties due to the phase differences between **H** and **E** and **j** [54] were not followed up (which question was cleared recently by Kentwell [169]). Chen's formulation (9.68) has the merit that the last term results in forces along the propagation of the laser light, while the first term works on any deviation from the striated structure in the direction perpendicular to that of the propagation. This permits a distinguishing between *backscattering instabilities*, due to $\mathbf{E}\times(\nabla\times\mathbf{E})$, and the *electrostatic parametric instabilities*, due to $\mathbf{E}\cdot\nabla\mathbf{E}$ which is equivalent to the $\mathbf{v}\cdot\nabla\mathbf{v}$ convection term in the equation of motion (8.1). This nonlinear force in the lateral direction of the perpendicularly incident plane wave may indeed need more analysis with respect to the Maxwellian stress tensor (see Section 12). Chen [193] was aware of some of these problems, and certain limitations of the following results may be necessary.

The *electrostatic parametric instabilities* are the result of the interaction of perpendicularly incident plane waves of laser radiation with the lateral deviations n_1 of the electron density from its equilibrium value n, due to the $\mathbf{v}\cdot\nabla\mathbf{v}$ term in Eq. (9.67), Fig. 9.10. For the electrostatic oscillations the plasma frequency ω_p, Eq. (2.6), was due to the deviation of the electrons from their equilibrium whose oscillation is attenuated by Landau damping, Eq. (3.68). Bohm and Gross [204] studied the generated electrostatic waves which have the following frequency ω_e

$$\omega_e^2=\omega_p^2+(3/2)\mathbf{k}_s^2 v_{\text{th}}^2 \tag{9.69}$$

(Bohm–Gross frequency) where the thermal electron velocity $v_{\text{th}}=2KT_e/m$ determines the transport of a signal by this wave (Langmuir wave). The wave vector $\mathbf{k}_s$ is given by the phase velocity v_ϕ of the wave, which can be very large

$$|\mathbf{k}_s|=\omega/v_\phi \tag{9.70}$$

It has to be distinguished where the laser frequency ω is a little less than

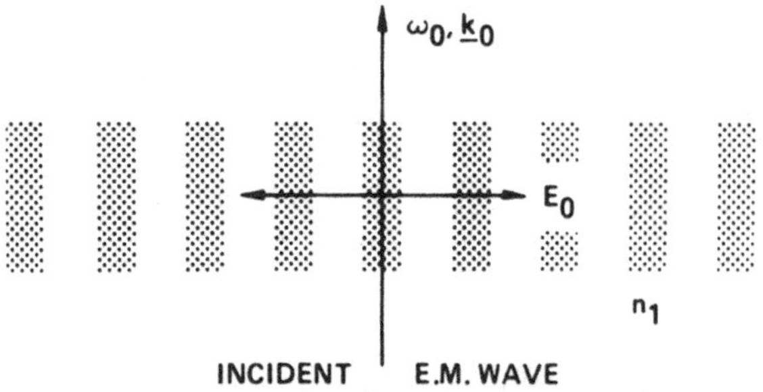

Figure 9.10 Direction of the k vector of the incident plane laser wave of frequency ω and an electric field E_0 in a plasma with some lateral (stochastic) deviation n_1 of its equilibrium electron density n resulting in electrostatic parametric instabilities. After Chen [193].

ω_e in which case we have an *oscillating two-stream instability*,

$$\omega < \omega_e \tag{9.71}$$

Density ripples in the direction of the electric laser field E_0 (Fig. 9.11) will then grow without propagation. The laterally uniform laser field E_0 will then interact with the space charge field E_1 of the density rippling by the nonlinear force $\mathbf{f}_{NL}$

$$8\pi \frac{\omega^2}{\omega_p^2} \mathbf{f}_{NL} = -2E_0 \frac{\partial E_1}{\partial x} \tag{9.72}$$

causing an increase of the ripple, Fig. 9.11.

If the laser frequency is a little larger than the Bohm–Gross frequency,

$$\omega > \omega_e \tag{9.73}$$

the *parametric decay instability* is generated. The oscillating two-stream instability does not work, and the incident wave decays into an electron wave ω_e and an ion acoustic wave ω_1. Following Fig. 9.12, the nonlinear force acts

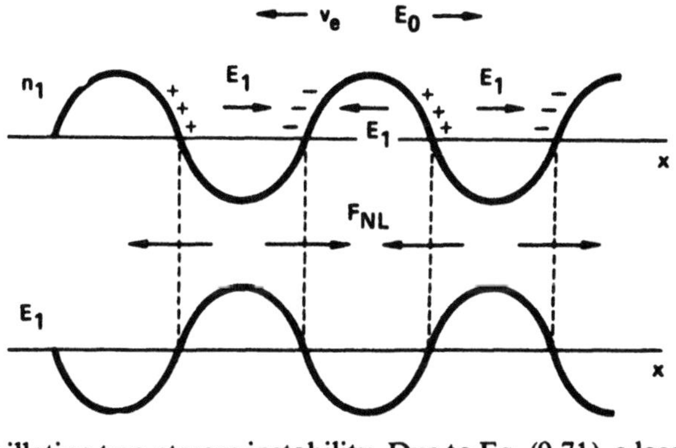

Figure 9.11 Oscillating two-stream instability. Due to Eq. (9.71), a laser frequency less than the Bohm–Gross frequency, a lateral density ripple is at rest. The electrostatic field $\mathbf{E}_1$ of the ripple interacts with the laser field $\mathbf{E}_0$ by the nonlinear force Eq. (9.72), increasing the ripple. After Chen [193].

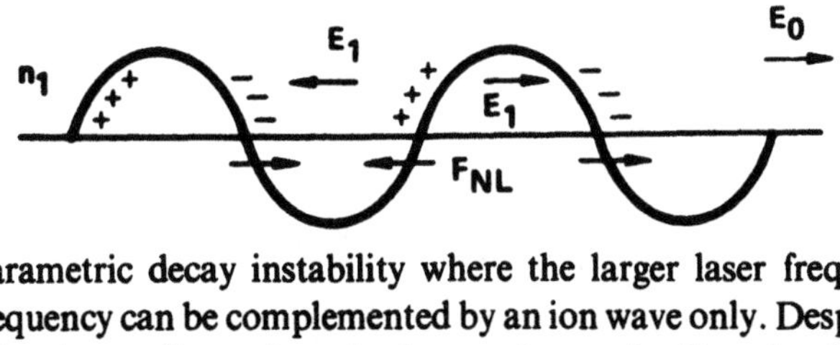

Figure 9.12 Parametric decay instability where the larger laser frequency than the Bohm–Gross frequency can be complemented by an ion wave only. Despite the decay of a density ripple by the nonlinear force in the rest frame, the Doppler shift in the frame moving with the wave results in a similar increase of the ripple in the same way as at the oscillating two-stream instability in Fig. 9.11. After Chen [193].

so as to destroy the density perturbation n_1. However, in the frame of the moving ion wave, the density perturbation would be at rest as in the oscillating two-stream case. Its mechanisms can again operate based on quasi-neutrality and a Doppler shift, causing a growing ion wave.

The lateral effects of the parametric decay instabilities will cause a *filamentation* or self-focusing of the laser beam in the plasma. This is based on the balance of the lateral nonlinear force with the gasdynamic pressure first proposed by Askaryan [205]

$$\mathbf{f}_{NL} = \nabla n K T_e \tag{9.74}$$

as shown in Fig. 9.13. Based on a Boltzmann-like density profile

$$n = n_0 \exp\left(-\frac{\omega_p^2}{\omega^2}\frac{\overline{\mathbf{E}^2}}{8\pi n_0 K T_e}\right) \tag{9.75}$$

a threshold for self-focusing at the laser power P_0 [193]

$$P_0 = 8800\left(\frac{\omega_e}{\omega_p}\right)^2 T\ \mathrm{W} \tag{9.76}$$

where $[T]=$ eV, was found [193]. This is the same value that we had derived before [206], where the mechanisms of total reflection and diffraction had been added to the balance of the forces Eq. (9.74) only (see Section 12).

The *backscattering instabilities* occur from nonlinear forces parallel to the direction of the wave vector $\mathbf{k}$ of the laser light, where, however, the details of the quiver motion and the phases of the initial and the induced electric and magnetic field have to be included. Again currents and velocities are produced perpendicular to $\mathbf{k}$, reacting then with the E-field of the laser. While the electrostatic parametric instabilities are transverting the laser energy into electrostatic waves, the backscatter instabilities result in a transversion into an electromagnetic wave of frequency ω nearly in or against the direction

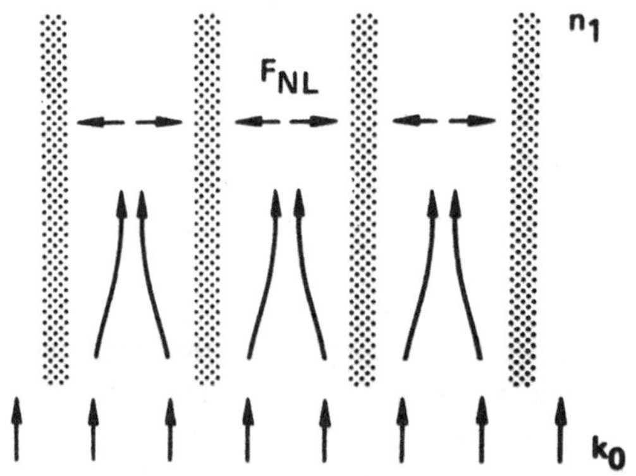

Figure 9.13 Filamentation or self-focusing due to the nonlinear force. After Chen [193].

of **k**. The lateral wave ω for this coupling can be equal to ω_e (an electron plasma wave); then we have *stimulated Raman scattering* (SRS). If ω is equal to that of the ion acoustic wave, we have *stimulated Brillouin scattering* (SBS). If ω is not perpendicular to **k**, the density perturbation can still exist if the laser field **E** is sufficiently large to maintain it against diffusion. This is then called *resistive quasi-mode scattering.* If $\omega_1 = k_1 v_e$ or $k_1 v_i$ where v_e and v_i are the thermal velocities of electrons and ions, interaction with resonant particles can cause an instability. This is the *induced Compton scattering* or *nonlinear Landau growth.*

Without looking into the detailed derivation, the following result of the threshold and growth rate in homogeneous plasma are given

	Threshold	Growth Rate	
SBS	$\frac{v_0^2}{v_e^2} = \frac{8\gamma_i \nu_{ei}}{\omega_i \omega_0}$	$\gamma_0 \simeq \frac{1}{2}\frac{v_0}{c}\left(\frac{\omega_0}{\omega_i}\right)^{1/2} \omega_{pi}$	(9.77)
SRS	$\frac{v^2}{c^2} = \frac{2\omega_p^2}{\omega_0^2}\frac{\gamma_e}{\omega_p}\frac{\nu_{ei}}{\omega_0}$	$\gamma_0 \simeq \frac{1}{2}\frac{v_0}{c}(\omega_0 \omega_p)^{1/2}$	(9.78)

where v_e are the thermal electron velocity, γ_e and γ_c the electron or ion wave damping rates, and $(\omega_p^2/\omega_0^2)(\nu_{ei}/2)$ is the damping rate of the electromagnetic waves. The following thresholds [207] are for Nd-glass and CO_2 lasers for intensities in W/cm^2

	Nd	CO_2
SBS	10^{13}	10^{10}
SRS	10^{13}	10^{9}
Oscillillation two-stream instability	10^{13}	10^{9}
Parametrametric decay instability	10^{13}	10^{10}

The action of the backscatter instabilities can be seen immediately from the electromagnetic waves reflected from the laser produced plasma. The intensity of the reflected light with half of the laser frequency or the higher harmonics is much less than the incident light, which is the most direct indication that the instabilities do not grow to infinity but are limited by saturation [208]. The use for diagnostics is very valuable. Watteau et al. [209] were able to demonstrate the change of the blue shift of the critical plasma density to a red shift, if the Nd glass laser intensity passes 10^{15} W/cm^2. Wong [210] used the fact of the small backscatter intensity as an obvious argument that the dynamics of laser-plasma interaction will not be influenced by the instabilities. This result has been confirmed theoretically by Bobin et al [211], who emphasized that the parametric instabilities can work for Nd glass lasers for intensities between 10^{14} and 10^{16} W/cm^2 only. Above these intensities, the nonlinear force is predominant for plasma dynamics. Similar conclusions were drawn by Balescu [212]. A more detailed analysis was given by Liu et al. [213]. The nonlinear force disturbs the resonance conditions for the parametric decays. At the interesting intensities near 10^{15} W/cm^2 for Nd glass lasers, only 1% of the absorbed radiation can go into decay modes [214]. Under very artificial conditions, this contribution may grow to 10%. Experiments by Ng et al. [174] confirm these conclusions.

More recent experiments elaborated these facts in more details. It is the merit of C. Labaune et al (1985) to demonstrate experimentally that SBS, while well existing with all the properties predicted theoretically, does not account strong for the energy transfer of laser radiation into the plasma. Only if very extreme and pathologically non-natural density profiles with flat plateaus of 1 mm length of laser irradiation at the neodymium glass wave length are produced artificially, a strong amount of optical energy is being transferred into the ion-acoustic waves according to SBS.

Another experimental result is that of Paul Drake (1988) which demonstrated that SRS is not contributing a large amount of energy transfer at neodymium glass laser irradiation. The mechanism of SRS well exists with most of the theoretically predicted properties, especially the generation of the easily detectable 3/2 harmonics in backscattered light which is just due to the SRS process feeding laser energy into longitudinal electron oscillations of a frequency given by the local electron density according to the plasma frequency, Eq. (2.6).

The fitting of the energy of the incident photon with the plasmon energy and the emitted photon, as well as the fitting of the corresponding momenta can best be seen from microscopic theory well including the before mentioned description by the ponderomotive term of the nonlinear force [193]. The earlier mentioned saturation [213] of parametric instabilities can be seen very drastically from the experiments of Batha et al (1990), Fig. 9.14, where the stimulation of the forward Raman scattering SRFS first follows the expectation from the convective theory but then saturates after reaching a certain gain level.

The suggestion is that this is due to the hydrodynamic driving of the longitudinal large amplitude oscillations by the laser radiation as a result of the genuine two fluid model. We note the long years experience from computations of laser plasma interaction following the genuine two fluid model, see subsection 8.7, that the transversal laser radiation causes longitudinal oscillation of large amplitude, e.g. for a laser intensity of

10^{16} W/cm^2 dielectrically swelled up by a factor four, amplitudes of 1.2 x 10^8 V/cm have been calculated (Hora et al 1984). This shows that the amplitude of the longitudinal electron wave was about 0.1 times of the amplitude of the electric field of the irradiated laser.

Numerous other examples showed always a ratio of 0.1 or less (to 0.01).

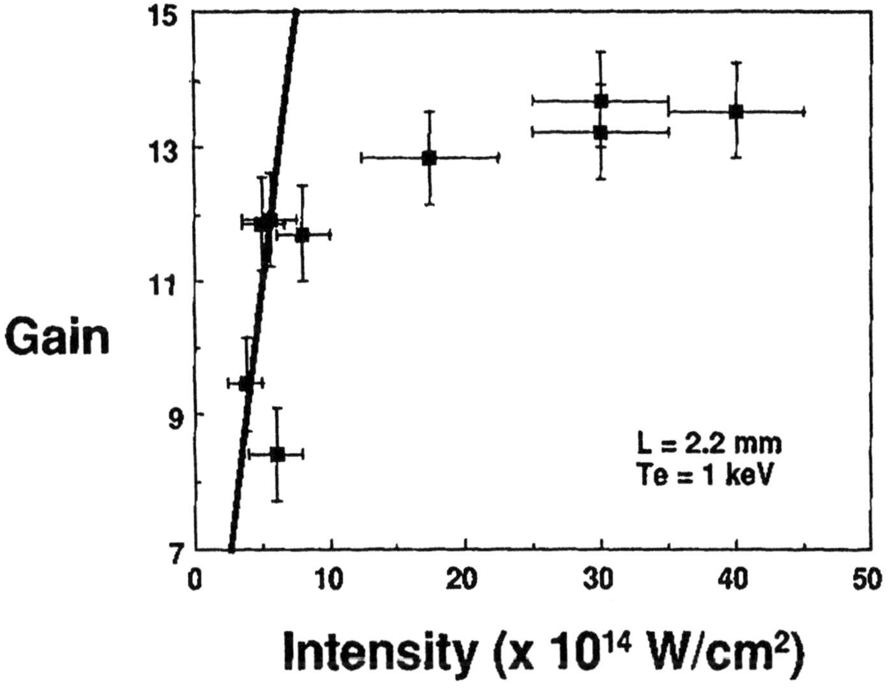

Fig. 9.14. Measured gain of stimulated Raman forward scattering SRFS with neodymium glass laser pulses of 15 kJ in 0.95 mm diameter spot on plane CH target depending on the laser intensity, where a second harmonics laser beam is being used for stimulation of the scattering. The straight line shows the expected gain from convective theory (Batha et al, 1990).

Numerical experiments with flat density plateaus of critical, half critical, or quarter critical density did not result in higher ratios than 0.1. It may be suggested that these basically different hydrodynamically driven large amplitude longitudinal electron oscillations interact with the otherwise produced SRS oscillations and from a certain amplitude on the process is determined by the hydrodynamically driven oscillation which may disturb the conditions of energy and momentum conservation for the transfer of laser energy into longitudinal oscillations of the local plasma frequency.

The backscattered signals from laser irradiated plasma are most important for understanding the mechanisms involved. While we shall come back to a very unexpected explanation of the broadening of the spectrum of the backscattered fundamental and of the second harmonics radiation from neodymium glass laser produced plasmas, one observation of the wide local spread for the emission of second harmonics from a laser irradiated plasma corona could be explained in a straight forward way from the result of Eq. (8.127). Taking the second last term,

$$E_s^1 = \frac{\omega_p^2 - 4\omega^2}{(\omega_p^2 - 4\omega^2)^2 + \nu\omega^2} \frac{4\pi e}{m} \frac{\partial}{\partial x}(E_L^2 + H_L^2) \cos 2\omega \quad (9.79)$$

one realized that a second harmonics longitudinal field oscillation occurs where local gradients (as in at least partially standing waves) of the laser field (index L of the field strengths) occurs as a **reactive response.**

According to the first factor of this oscillation, it is nearly independent of the plasma density given by the plasma frequency ω_p. This immediately gave an explanation of the very strange observation shown in Fig. 1.13, (Aleksandrova, Brunner et al 1985) that the laser irradiation of a spherical target well showed a radially strong decay of the scattered laser light according to the strong decay of the electron density, but the emission of the second harmonics was nearly of unchanged amplitude up to the outermost plasma of very low density. The just mentioned result of the second harmonics term immediately could explain this density independence. Only the observation of a certain radial periodicity was not immediately shown.

A numerical evaluation of the mentioned second laser term of Eq. (8.127) (Goldsworthy 1988; Goldsworthy et al 1988; 1990) for a linear electron density profile for a plasma corona of 150 wave length thickness using the Airy solutions for the laser field, results in Fig. 9.15 for this part of the longitudinal field. While the net amplitude varies only by about 40% along the whole corona despite its strong decay of the density, oscillations are being seen. There is a large oscillation length of about 50 μm which very well fits with the results of Aleksandrova et al (1985), Fig. 1.13, while another oscillation with a wave length of about four laser wave lengths appears. This spatial period of the widespread second harmonics emission from a laser irradiated plasma corona was just then observed by Tan Weihan et al (1978) when the result of Fig. 9.15 was published. When these authors used a wide band gap laser, the spatial modulation

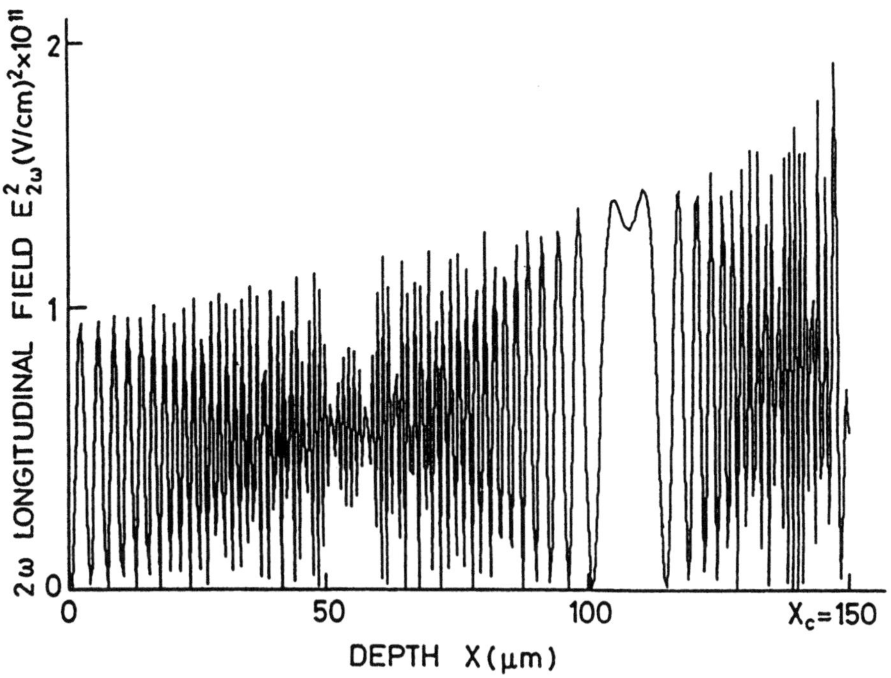

Fig. 9.15. Amplitude of the second harmonics reactive response term (second latest term in Eq. (8.127) for a linear plasma corona from x = 0 to a critical density at 150 μm for 10^{14} W/cm^2 incident Nd glass laser irradiation (Goldsworthy 1988, Goldsworthy et al 1988; 1990). The fine oscillations correspond to the observed striations (Tan et al 1987) and the long periods of about 50 μm correspond to the observations by Aleksandrova et al (1985), see Fig. 2.10.

disappeared what is immediately understandable from Fig. 9.15, if there is a superposition of several wavelengths that then the periodic oscillation is being washed out.

These examples may indicate how the parametric instabilities are well interlinked with the observed basically different processes due to the macroscopic hydrodynamic large amplitude longitudinal osciallations driven by the intense laser radiaiton. The experience with this transparent hydrodynamic model may be an access for separating the phenomena involved when interpreting N-particle simulations where both mechanicsms, the kinetic model of the parametric instability as well as the macroscopic oscillations should appear simultaneously, however, including collisions.

TEN

Numerical and Experimental Examples–Solitons

This section is devoted to plane waves perpendicularly incident on inhomogeneous plasmas. The forces in the plasma will be calculated numerically and compared with experiments. First, calculations without the nonlinear forces are considered, where the plasma dynamics are determined by the thermal pressure of the plasma after heating by the laser radiation. Subsequently, nonlinear forces are included. One of the early results was the discovery of the generation of a density minimum (caviton) by Shearer, Kidder, and Zink [171]. These minima can never be produced by thermokinetic pressures and are therefore typical for the plasma dynamics with nonlinear forces. The observation of these minima and the subsequent steepening of density profiles is the first tool for checking the action of nonlinear forces in experiments. The development of the nonlinear-force driven plasma dynamics is then shown to be typical for the generation of solitons. Macroscopic nonlinear absorption, caused by the net transfer of optical energy into kinetic energy of plasma without heating, occurs and leads to the mentioned soliton.

10.1 Thermokinetic Forces

The numerical study of laser-plasma interaction without nonlinear forces was done simultaneously by Mulser [88] and Rehm [215] for the one-dimensional case of plane electromagnetic waves perpendicularly incident on a stratified plasma. The basic hydrodynamic equations of conservation [the equations of continuity, Eq. (4.17), of motion, (4.6) (or (8.3) without the quantities **E** and **H**), and of energy conservation, Eq. (4.39)] are used to calculate the plasma density $\rho(x, t)$, given by the ion density $n_i(x, t)=\rho(x, t)/m_i$, the plasma temperature $T(x, t)$, and the plasma velocity in the x-direction $v(x, t)$. The initial conditions $\rho(x, 0)$, $T(x, 0)$, and $v(x, 0)$ are given. The boundary condition is the time dependence of the incident laser radiation. The problem is the formulation of the power generation term $W(x, t)$ in the energy equation. The question of heating the initially condensed material, its ionization, and Saha-equilibrium turns out not to be important for laser intensities above 10^9 W/cm^2 for neodymium glass lasers. For $W(x, t)$, the solution of the Maxwell equations (6.17) and (6.18) is necessary for the incident and the reflected wave (generated by the inhomogeneous plasma) for each instant. It was assumed that the transfer of absorbed energy in the plasma, according to the optical constants, is without any delay. This is correct if the time of interaction exceeds 1 nsec. It is interesting that Mulser [88] used a Lagrangian-type numerical code (localizing the intervals of computation to the moving mass density of the plasma). Rehm [215] used an Eulerian code (fixing spatial intervals to the coordinates). Both calculations arrived at the same result.

Figure 10.1 shows the result of a stepwise neodymium glass laser pulse

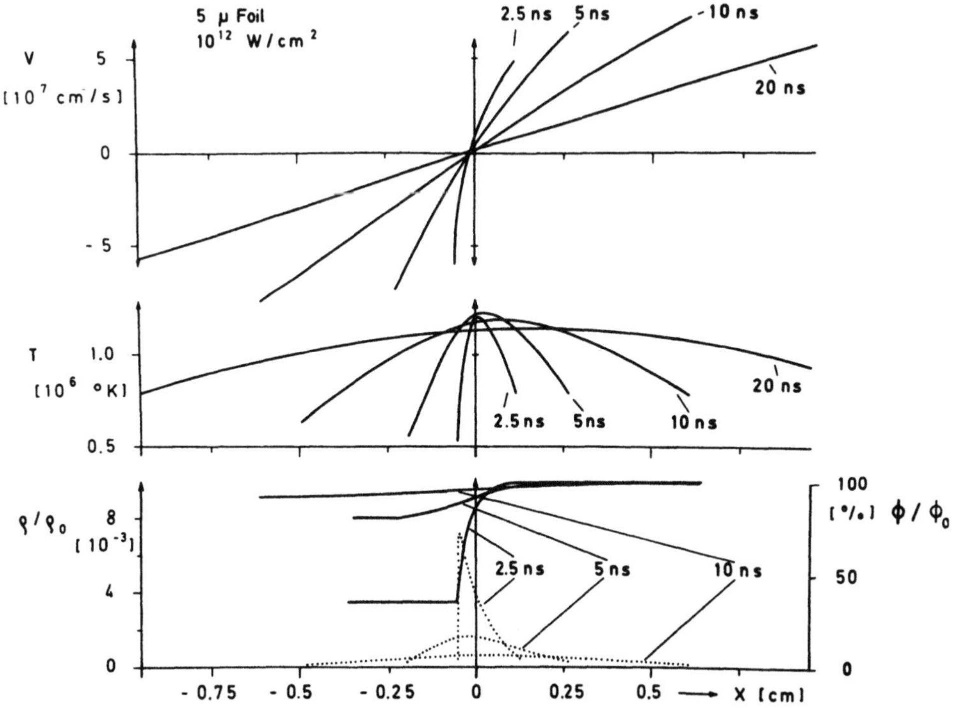

Figure 10.1 One-dimensional numerical solution of laser plasma interaction for a hydrogen foil of 5 μ thickness. A linear velocity profile and Gaussian density profile resulted at 5 nsec and later. After Mulser [88].

irradiating a solid hydrogen foil of 5 μm thickness. At and after 5 nsec, the velocity profile is nearly linear and the density profile is nearly Gaussian. The dynamics change approximately into that of the self-similarity model (Section 5).

A thick hydrogen block shows a quite different behavior (Fig. 10.2). The light is absorbed in a plasma density, which is 60 times less than the solid density. The ablation of the corona causes a compression of the plasma interior. This is shown by the negative velocity v in Fig. 10.2. In this way, a compression of the interior to multiples of the initial density results. Plasma densities of up to 250 times that of the corona are calculated.

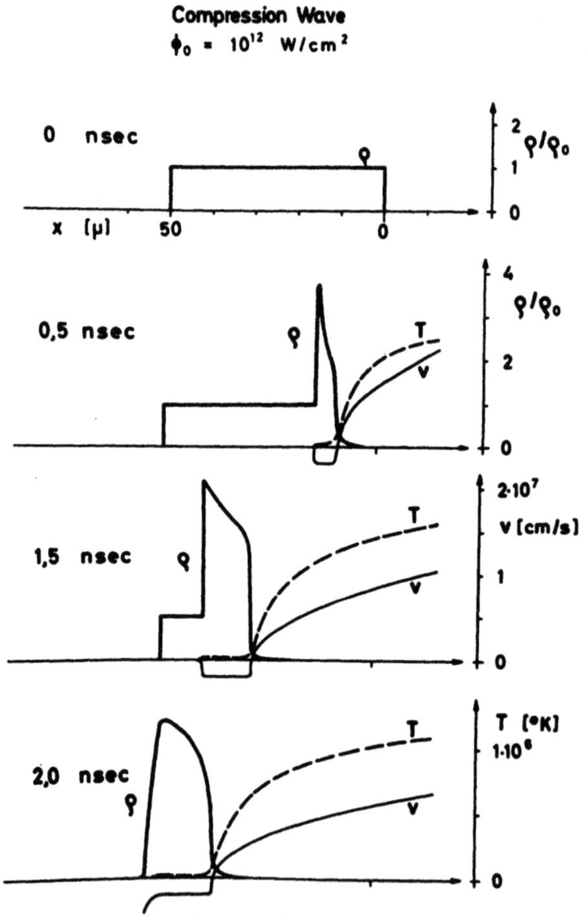

Figure 10.2 One-dimensional numerical solution of the hydrodynamic equations of conservation for laser light with a steplike intensity of 10^{12} W/cm^2 incident on a 50 μm thick slab of solid hydrogen (density ρ_0). The resulting density $\rho = n_i m_i$, velocity v, and temperature T are shown for times $t=0$, 0.5, 1.5, and 2.0 nsec. After Mulser [88].

This shock process happens only if there is no self-focusing, which could cause a more homogeneous heating of the plasma and a self-similarity expansion of thick foils also. The shock process was also shown by some analytical studies, based on hydrodynamic similarity laws by Krokhin and Afanasyev [216], Caruso and Gratton [217], or in a more general derivation, by Perth [218]. The most straightforward result, however, is that of the numerical calculations [88, 215].

A much more general hydrodynamic compression was calculated by Nuckolls [219], where use was made of a temporally increasing laser intensity. The compression process was then a growth of density similar to the concept of Guderley [220], where the addition of increasing shocks, following an appropriate sequence for meeting at one point, can produce densities of 10^4 times that of a solid [221] for a spherical geometry.

This successive compression by gasdynamic ablation requires a sufficiently short equipartition time (time to equilibrate the electron and ion temperatures). For Mulser's calculations with nanosecond pulses and no nonlinear derivations of the collision frequency, the instantaneous equipartition could be assumed. This is not the case for the calculations of Nuckolls [219]. To demonstrate the long duration of collisional equilibration, the electron collision time (6.58a)

$$\tau_{col} = 1/\nu \qquad \text{where } T = T_{th} + \varepsilon_{osc}/2K \tag{10.1}$$

is plotted in Fig. 10.3 for various lasers and varying intensity.

In order to correlate the collision time with the interacting laser pulse and its thermokinetically caused mechanical pulse, it will be assumed for simplicity that the pulse of mechanical power density I_{th}, arising from the thermalizing interaction of the radiation with the plasma, has the form

$$I_{th} = I_0 \sin^2\left(\frac{\pi t}{\tau_0}\right) \qquad \text{for } 0 \leqslant t \leqslant \tau_0 \tag{10.2}$$

τ_0 is the half-width of the pulse. A generalization to a more complicated pulse shape does not substantially change the following results. The laser pulse must then arrive earlier (see Fig. 10.4) by a precursion time, which depends on the laser intensity, and can be identified with the collision time τ_{col}. Thus, for short pulses, the laser light behaves in the plasma like a light beam in transparent glass and will produce no thermalizing coupling and no remarkable thermalizing energy transfer (except by nonthermalizing nonlinear force). The relation between the slope angles α and α' of the pulses (Fig. 10.4) can be used to find the greatest possible increase of a laser pulse. This is the instantaneous increase, corresponding to $\alpha' = \pi/2$, the highest possible increase I_{th}, which limits the gasdynamic compression models [219, 221]. Quantitatively, the maximum increase of I_{th} for a pulse of the

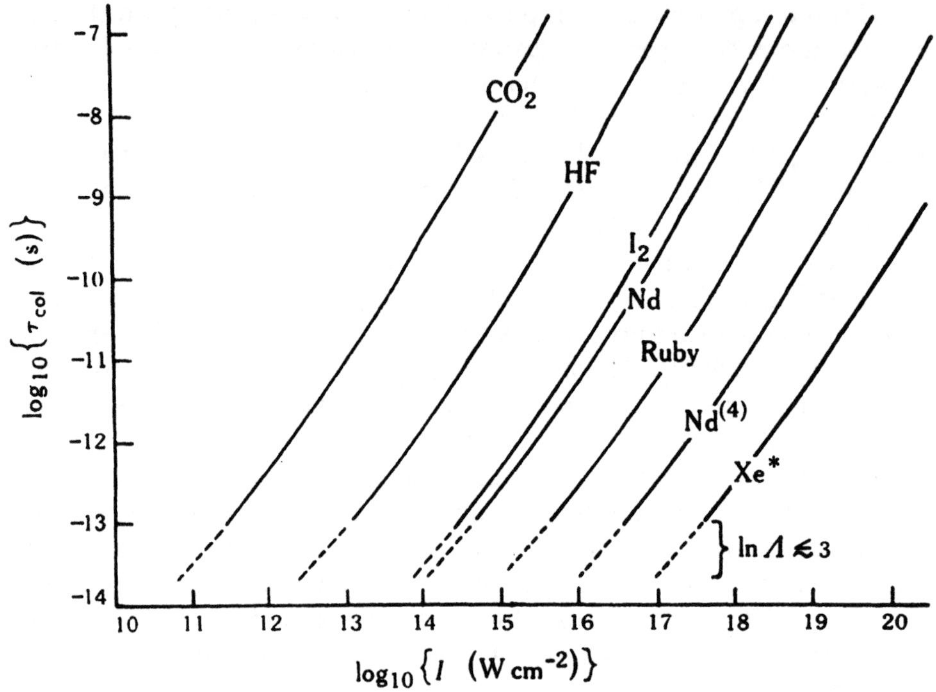

Figure 10.3 Plot of the minimum time for thermalization, as given by the collision time ω_{col} of the electrons, Eq. (10.1), as a function of intensity by the indicated lasers ($\varepsilon_{osc} \gg kT_{th}$; $|n|=1$; $n_e < 1/2n_{ec}$). The values of n_{ec} in cm^{-3} are: CO_2, 10^{19}; HF, 8.6×10^{19}; I_2, 6.6×10^{20}; Nd–glass, 10^{21}; ruby, 2.3×10^{2}; fourth harmonic Nd–glass, 1.6×10^{22}; Xe*, 3.7×10^{22}.

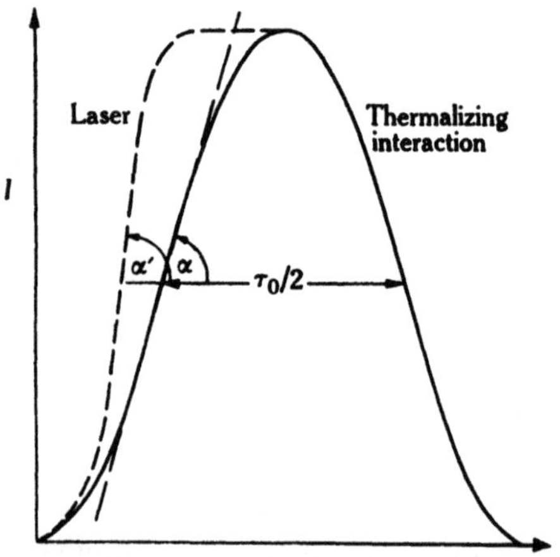

Figure 10.4 The laser pulse (dashed curve) must precede the thermalizing interaction pulse (continous curve) by an irradiance-dependent precursion time τ^* [identified with τ_{col} of Eq. (10.1)], in order to drive a gas dynamic ablation-compression process. The limitation on the thermalization is reached for $\alpha=\pi/2$ [222].

shape (10.2) is given by

$$\frac{\partial I_{th}}{\partial t} = \frac{I_0 \pi}{\tau_0} \sin\left(\frac{2\pi t}{\tau_0}\right) \qquad \text{at } t=(1/4)\tau_0 \tag{10.3}$$

or

$$\left.\frac{\partial I_{th}}{\partial t}\right|_{max} = \frac{I_0 \pi}{\tau_0} \tag{10.4}$$

Inserting $\tau^* = \tau_{col}$, leads to

$$\tau^* = 1.23 \times 10^{18} \left(\frac{1}{n_e \ln \Lambda}\right)\left(\frac{I_v}{|n| n_{ec}}\right)^{3/2} \tag{10.5}$$

$I_v = I_{0'}$ as is evident from Fig. 10.4. From the geometry of Fig. 10.4, the following condition, corresponding to $\alpha \leqslant \pi/2$, is obtained

$$\frac{\partial I_{th}}{\partial t} \leqslant \left(\frac{\partial \tau^*}{\partial I_v}\right)^{-1} \tag{10.6}$$

With Eqs. (10.4) and (10.5) and the inequality, Eq. (10.6), it is found that the laser pulse duration τ_0 has to be longer than the precursion time τ^* or

$$\frac{I_0}{\tau_0} \leqslant \frac{(|n| n_{ec})^{3/2} n_e \ln \Lambda}{(3/2)\pi 1.23 \times 10^{18} I_v^{1/2}} \tag{10.7}$$

Once more identifying I_0 with $I_{v'}$, the restriction for gasdynamic compression is

$$\frac{I_v^{3/2}}{\tau_0} \leqslant 1.72 \times 10^{-19} n_e (n_{ec} |n|)^{3/2} \ln \Lambda \tag{10.8}$$

The units of I_v are W/cm^2, of n_e, and n_{ec} are cm^{-3}, and τ_0 is in seconds.

From the inequality (10.8) and with $|n| = 1$ (requiring that $n_e = \frac{1}{2} n_{ec}$), $\ln \Lambda = 8.1$ (for $I = 10^{16}$ W/cm^2), and for neodymium glass laser radiation

$$\frac{I_v^{3/2}}{\tau_0} \leqslant 2.1 \times 10^{34} (\text{W/cm}^2)^{3/2}/\text{sec} \tag{10.9}$$

Therefore, for $\tau_0 = 100$ psec, I_v has to be less than 1.7×10^{16} W/cm^2. The calculation by Nuckolls [219] has exceeded the limit (10.9). This indicates that some revision is necessary in the application for nuclear fusion calculations.

10.2 Static Case with Nonlinear Forces

Before going to fully dynamic calculations of plasma motion at laser irradiation including the nonlinear forces, some static cases will be considered.

Assuming a certain plasma density and temperature profile at a certain time, the resulting forces in the plasma can be calculated.

The most simplified case is that of a collisionless plasma with a linear density profile, as studied by Lindl and Kaw [154]. The solutions of the Maxwell equations for the electromagnetic waves are given by the Airy case, Section 7, and the resulting force is

$$\overline{\mathbf{f}_{NL}} = \mathbf{i}_x \frac{\omega}{c} E_v^2 \exp(-2\rho s)\{(a_{iR}a'_{iR} + a_{iI}a'_{iI}) + \rho^{-2/3}[a'_{iR}(\zeta_R a_{iR} - \zeta_I a_{iI}) + a'_{iI}(\zeta_I a_{iR} + \zeta a_{iI})]\} \qquad (10.10)$$

where the real parts and the imaginary parts of the Airy functions a_i are denoted by the indices R and I. The prime means differentiation with respect to the arguments. Figure 10.5 reports on the numerical result of the evaluation of f_{NL} from Eq. (10.1) (oscillating curve). Because of the collisionless plasma, total reflection occurs. The oscillation of the nonlinear force shows the acceleration of the plasma toward the nodes of the standing wave. The

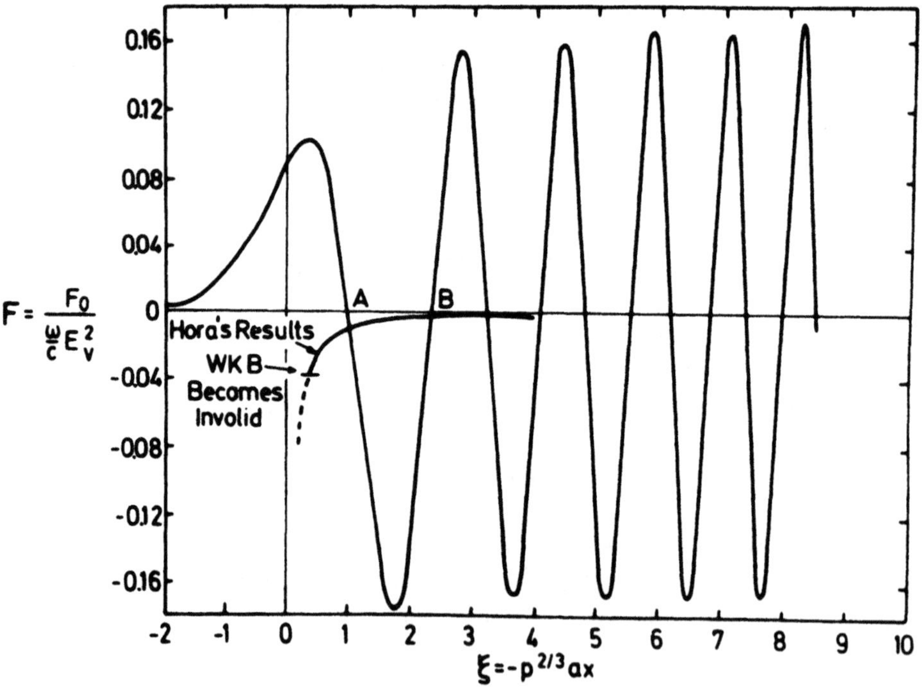

Figure 10.5 The nonlinear force for a collisionless plasma of linear density gradient is given. The light (incident from the right) creates a standing electromagnetic wave and the locally oscillating force. The net acceleration of the plasma is given by the double value of the monotonic force close to the abscissa. It is compared with the WKB approximation (Lindl and Kaw [154]).

spatially averaged net forces are drawn and show agreement of the exact results [154] with those where the WKB approximation, Eq. (8.50), is applicable.

Another static example uses a plasma with collisions given by a constant temperature $T_e = T_i$ of 1 keV and a density profile, where the WKB approximation is applicable [$\theta \leqslant 0.1$ in Eq. (7.10)]. In these calculations, n_e is derived from n_e and x, using $\Delta|n|$, Eq. (6.48), with a linear collision frequency [$T = T_{\text{th}}$, Eq. (6.58)]. The approximation $T = T_{\text{th}}$ causes a lower bound of the final forces, because the implication of the electron oscillation causes a higher temperature, hence lower values $|n|$ and then stronger force densities. The result in Fig. 10.6 shows a very slow increase of $n_e(x)$ around the cutoff density, which for neodymium glass laser radiation is around 10^{21} cm^{-3}. The second step is to calculate the intensity

$$I = \frac{E^2}{8\pi c} = \frac{E_v^2}{8\pi|n|c} \exp\left(\frac{-\bar{k}x}{2}\right) \tag{10.11}$$

The reflected wave can be neglected because of the WKB condition. The

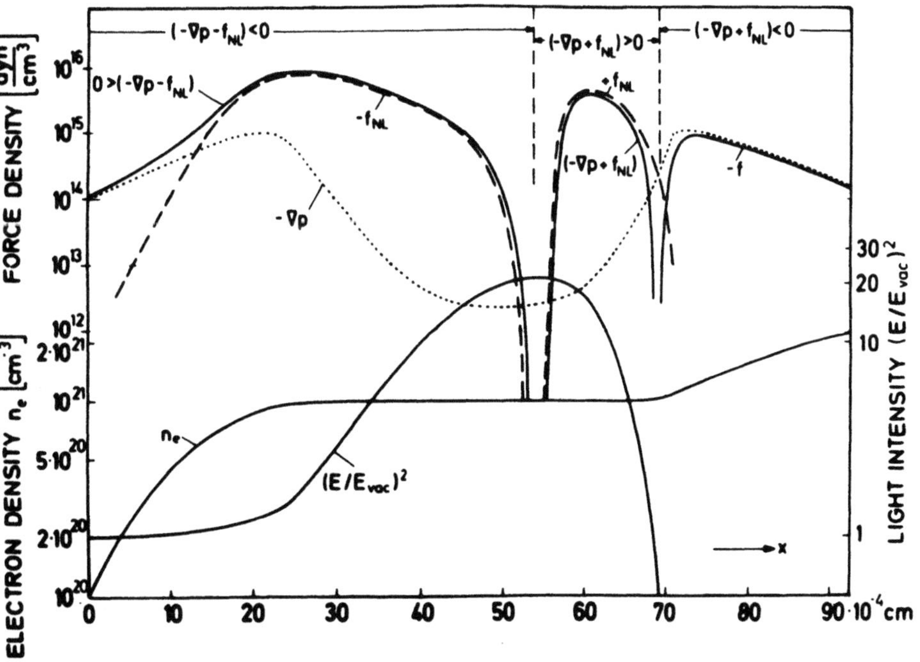

Figure 10.6 Numerical calculation of WKB-like density profile $n_e\,(x)$ of a plasma of temperature $T = 1$ keV. The radiation of 10^{15} W/cm^2, incident from the left side, produces an amplitude E of the electric field of the penetrating light wave compared with the amplitude E_v in vacuum before incidence. The resulting thermokinetic force f_{th} and the nonlinear force f_{NL} are calculated on the basis of the linear collision frequency [223].

strong increase of $(E/E_v)^2$, up to values beyond 20, proves that the decrease of $|n|$ occurs much earlier than the decrease of I, influenced by the integral absorption constant $\bar{k}(x)$. Therefore, the maximum values of $|n|^{-1}$ are higher than 20. This results in an increase of the effective wavelength λ^{eff} compared to the vacuum wavelength λ by more than 20.

$$\lambda_{\mathrm{eff}} = \lambda S = \frac{\lambda}{|n|} \tag{10.12}$$

From this point of view, the nearly pathologically flat density profile from 25 to 65 μm (Fig. 10.6) is only over about four effective wavelengths.

The next step is to calculate the thermokinetic force density $\mathbf{f}_{\mathrm{th}} = -\nabla p = -kT\nabla n_e$, which is given by the pointed curve in Fig. 10.6. The nonlinear force $\mathbf{f}_{\mathrm{NL}}$ is negative up to a depth of 52 μm. It is larger than the thermokinetic force for a depth larger than 13 μm and can be nine times as large as the maximum of the thermokinetic force. The total force $\mathbf{f} = \mathbf{f}_{\mathrm{NL}} + \mathbf{f}_{\mathrm{th}}$ and its components have a negative sign, which means an acceleration of the plasma in the negative x-direction. For depths larger than 53 μm the nonlinear force f_{NL} is positive (directed in the $+x$-direction), and, because its absolute value is very high, the total force $\mathbf{f}$ is positive up to a depth from a little more than 53 to 68 μm which is nearly the skin depth, which is so large because the effective wavelength is at some depth larger than 20 vacuum wavelengths. For greater depth, n_e is increased monotonically in an arbitrary manner. A detailed dynamic calculation may also result in a decreasing n_e within the highly shocked material below the radiation zone, but this question is irrelevant at this point.

10.3 Approximative Dynamic Cases

The examples of the preceding subsection are of static nature, as these show the magnitude of the nonlinear force at a certain plasma configuration only. In real cases, the whole time-dependent dynamics should be calculated. The first calculations of this kind were performed by Shearer, Kidder, and Zink [171], using the Wazer code, which treats electrons and ions separately with respect to temperature. The only critical point of the model is that the laser radiation is approximated by an incident and a reflected wave only without a general solution of the wave field in the inhomogeneous corona. This results in too optimistic properties. The standing wave process (pushing plasma toward nodes) causes a fast growing density rippling, which leads to macroscopic Brillouin instability, which cannot be shown. Nevertheless, this approximation is useful as it shows a long time interaction process with a significant effect. The nonlinear force produces a density minimum in the initially monotonic density profile. Beyond the minimum, a very sharp increase of the density is produced, a shock wave compressing the plasma to high densities. The density minimum is called a caviton and is a characteristic of the nonlinear force. In a plasma with a uniform temperature distribution, any thermal pressure can produce monotonic density profiles only. This can be seen from the fact that the temperature of the laser irradiated plasma is proportional to the collision frequency.

$$T \sim \nu \sim c_1 \frac{n_e}{T^{3/2}}$$

For the second proportionality see Eq. (2.35). Therefore, for the monotonic thermokinetic force $\partial(n_e kT)/\partial x > 0$ is found a positive value for which

$$T\left(1+\frac{c_1}{T^{5/2}}\right)\frac{\partial}{\partial x} n_e - \frac{3}{2}\frac{n_e c_1}{T^{5/2}}\frac{\partial T}{\partial x} > 0$$

because T, n_e, and c_1 are positive (if there is no stimulated emission). The thermokinetic force could only be nonmonotonic if the laser light produced a higher temperature at a larger depth than at a lower depth, but this is not possible in chemically uniform plasmas.

The results of Shearer, Kidder, and Zink are shown in Figs. 10.7 to 10.9. The time dependence of the laser pulse is given in the upper part of Fig. 10.7 with a maximum of 10^{16} W/cm^2. The electron temperature is much higher, Fig. 10.9, than the ion temperature because of the very long equipartition time. The strong ion temperature maximum in Fig. 10.9 is due to the adiabatic heating of the compressed plasma. One should note the first derived density minimum (caviton) in Fig. 10.7. The calculation was a first step of approximation with respect to the Maxwellian equations. This was improved in the next following case where the nonlinear intensity dependence of the optical constants was included.

The complete solution of the Maxwell equation for the whole plasma corona is used with nonlinear optical constants (see Section 6). The critical point, however, is the assumption of a stationary solution of the Maxwell equations, for which the following examples of subpicosecond interaction will not be correct. However, the basic properties of a very general nonlinear behavior can be seen [225, 226].

The equation of motion for the plasma in a one-dimensional geometry is given by the force density, distinguishing between electrons and ions. Besides the equation of continuity [Eq. (4.17) for both components], the energy equation (4.39) contains internal and external thermodynamic energies, thermal conductivity, and the incident laser radiation absorbed by collisions or by nonlinear electrodynamic motion as a source term.

The calculation begins at a time $t=0$ with a distribution of plasma density,

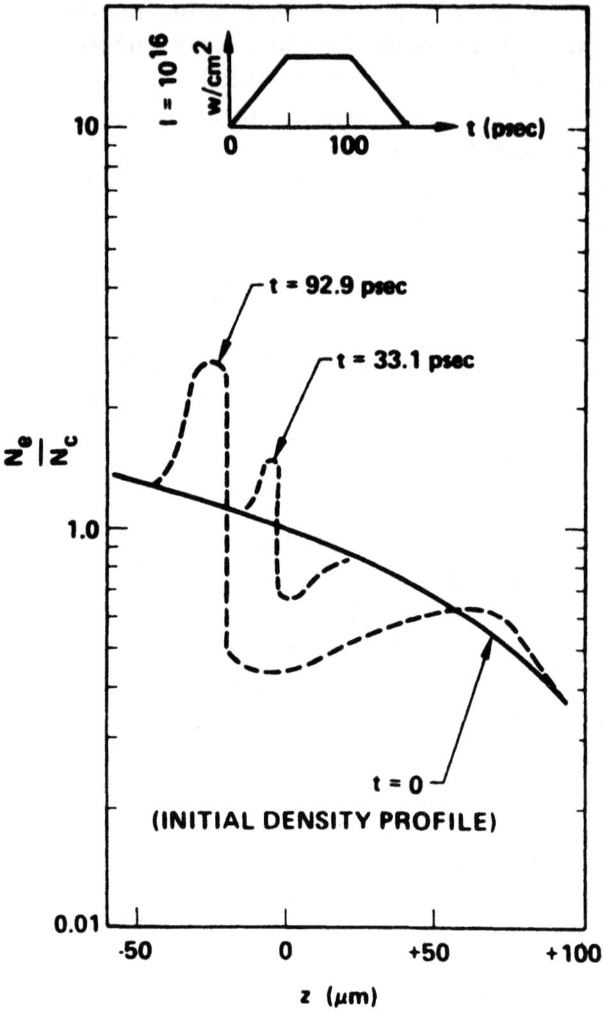

Figure 10.7 Density profile at different times calculated with the dynamic computer programme WAZER by Shearer [171, 224]. The low-density maximum is caused by the nonlinear force. The assumed laser pulse is given in the upper part. This was the first numerically discovered density minimum (caviton) and profile steepening due to nonlinear forces. After Shearer, Kidder, and Zink [171].

starting from solid-state density and a temperature profile. Without any laser field, there would be gasdynamic expansion and conservation of the total energy. At the successive time steps, a time-dependent incident laser intensity is prescribed, for which at each time step an exact stationary solution of the Maxwell equations is calculated including the actual plasma density and its refractive index; the retardation of the waves, the switching-on mechanism, and the development of the reflected wave are neglected.

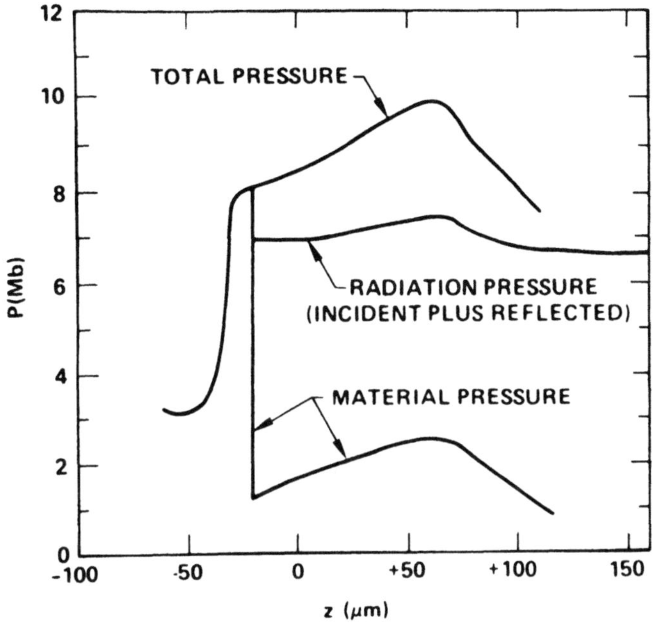

Figure 10.8 Pressure versus distance at $t=90.9$ psec for the same WAZER computer run [224], see Fig. 10.7. After Shearer, Kidder, and Zink [171].

The motion of the plasma within the next time step (with appropriately varying step size) is described with the gasdynamic force and the nonlinear electrodynamic force. The following two examples from extensive series of computer runs illustrate characteristic results. In both cases, the initial temperature of the electrons and ions (100 eV) is constant for the whole plasma (this could be produced by a prepulse from the laser). The laser pulse increases within 10^{-13} sec up to an intensity of 2×10^{16} W/cm^2. The increase is linear for 5×10^{-14} sec and then follows smoothly a Gaussian profile. After reaching 2×10^{16} W/cm^2, the laser intensity remains constant.

The initial plasma density increases (for $x<0$) quadratically above the cutoff density up to the solid-state density of LiD. It then changes smoothly in Fig. 10.10*a* into a linear decrease up to the length $x=50$ μm. The resulting electromagnetic momentum flux density $(E^2+H^2)/8\pi$ is given in Fig. 10.10*a*. Curve A is taken at an early time, when the laser intensity is 2×10^{14} W/cm^2. In a very thin plasma ($x\simeq50$ μm), the value of E^2+H^2 is constant. At about 20 μm an oscillation of E^2+H^2 is found, which increases in amplitude and wavelength between 0 and 8 μm. This behavior is well known from the analytical work for the same linear density profile [154] (dealing, however, with temperature $T=0$ and collisionless plasma). Near $x=0$, the $|\mathbf{E}^2+\mathbf{H}^2|$

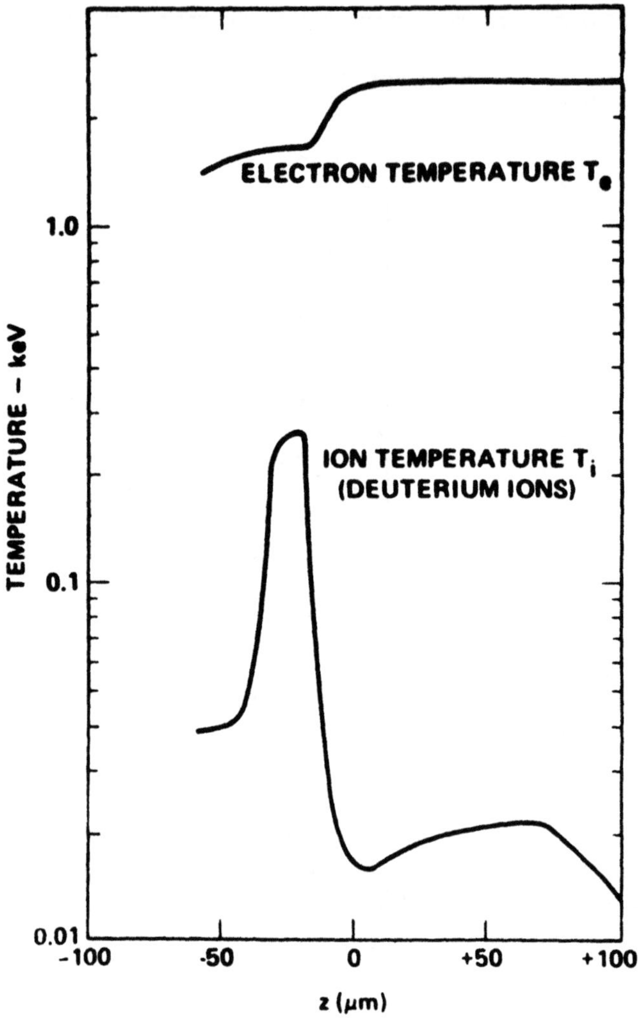

Figure 10.9 Electron and ion temperature vs distance at t=92.9 psec for WAZER computer run [224] of Fig. 10.7 [171].

value has increased due to the dielectric properties of the plasma.

The increase of the laser field **E** and of the wavelength (10.12) over its vacuum values E_v and λ_0 is given by a swelling factor $S=|n|^{-1}>1$

$$|\mathbf{E}|=E_v S=\frac{E_0}{|n|} \tag{10.13}$$

n is the optical (complex) refractive index of the plasma, provided the electromagnetic field can be described by the WKB approximation. In the case of curve A in Fig. 10.10a, the plasma fits the WKB conditions at 1.3 μm quite

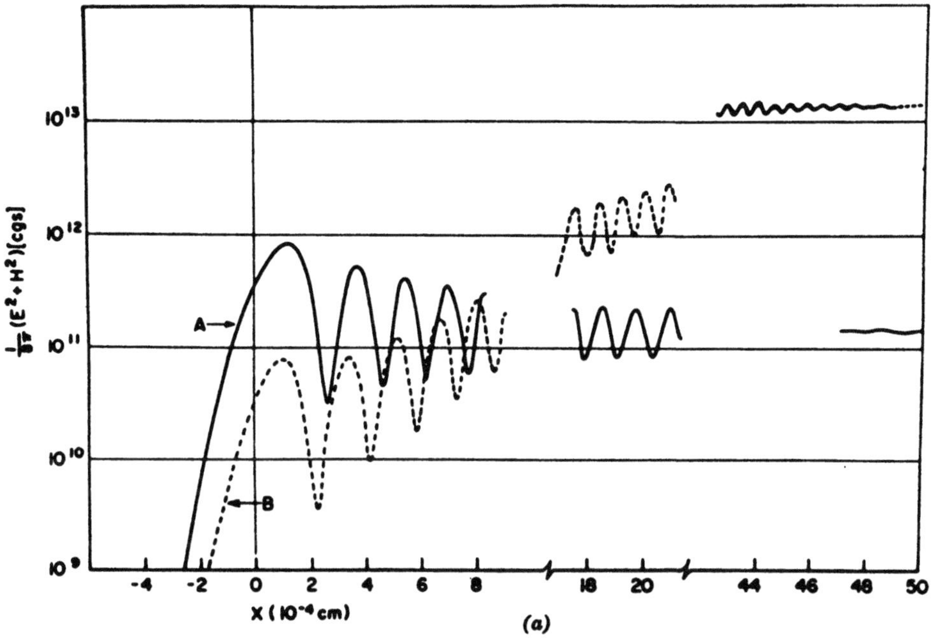

Figure 10.10a A laser beam is incident from the right side on a plasma of initial temperature of 100 eV and linear density increasing from zero at $x=50$ um to the cutoff density at $x=0$ and then increasing more rapidly. The exact stationary (time-dependent) solution without retardation of the Maxwell equations with a nonlinear refractive index n, based on a collision frequency [Eq. (6.59)], results in an oscillation due to the standing wave and dielectric swelling of the amplitude (curve A). At a later time, (2×10^{-13} sec), the laser intensity in 2×10^{16} W/cm^2 (curve B), where the relative swelling remains, but the intensity at $x=0$ is attenuated by dynamic absorption [1].

well, and the swelling factor $|n|^{-1}=6$ (taken from Fig. 10.10*a*) agrees with the value calculated from the actual refractive index.

For later times, curve A moves in nearly parallel fashion to higher values. However, its upward shift is reduced near the cutoff density ($x=0$). This decrease becomes so marked at curve B, that the intensity in the thin plasma ($x \simeq 50$ μm) decreases by a factor of 10 up to $x \simeq 20$ and to a thousandth part of the initial intensity at $x \simeq 0$. The reason is very simple. The standing wave pushes the plasma toward its nodes with an ion velocity as high as 10^7 cm/sec even at $x \simeq 48$ μm. The gasdynamic velocities reached at that time are 10^3 cm/sec or less. The result is shown in Fig. 10.10*b*.

At $x=46$ to 50 μm the initial linear density (upper part of Fig. 10.10*b*) acquires a ripple. The slight total increase is due to net motion driven by the nonlinear force. The ion energy increases, oscillating up to 200 eV (from the initial value of 100 eV), out of phase with the ripple. There is a curious effect seen in the electrons, in which the temperature increase is only a fifth of that

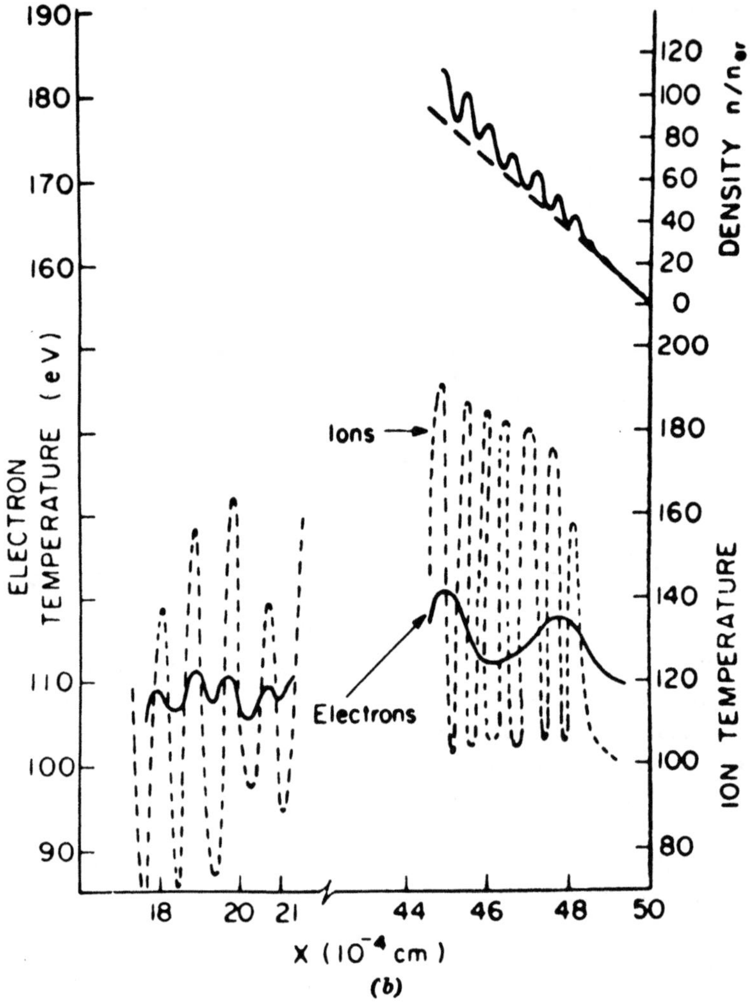

Figure 10.10b The initial density (dashed line) and the density along curve *B* of Fig. 10.10*a*, where a ripple is created by the nonlinear force, pushing plasma towards the nodes of the standing wave. The electron and ion temperatures are increased following the ripple by dynamic compression at conditions identical to curve *B* [1].

of the ions, as it should be for a collisionless shock; however, the periodicity is clearly different from that of the ions. At $x \simeq 20$ μm the periodicity of the temperature is again less than the ion temperature.

The density ripple explains the strong decrease of laser intensity with depth at later time. It causes strong reflection of light and a transfer of the optical energy into the ions by collisionless ripple shocks. Because it is not connected with decay of photons into microscopic acoustic modes, this process of dynamic ion absorption is called a "collisionless ion heating by nonlinear-force-induced macroscopic dynamic ion decay (MDID)." The depression of the laser intensity in interior regions becomes marked only after the density acquires pronounced maxima and minima (ripple).

An obvious idea is to start with a modified density profile, such that a change of the laser intensity I near cutoff results in a change of the actual nonlinear refractive index n, by virtue of the energy of the electron oscillation $\varepsilon_{osc}(I)$ determining the electron temperature (Eq. (6.57)). In Fig. 10.10*a* at 1.3 μm the absolute value of the refractive index is determined only by its real part (given by the actual density), while the imaginary part (determined by the temperature) is too small. Therefore, a computation is made for a density profile which is the same in the overdense region and which, in the underdense region varies linearly first from $x=0$ to 10 μm from cutoff to $1-10^{-6}$ times cutoff, then from there to $x=20$ μm to 0.99 times cutoff, and last linearly to zero at $x=50$ μm. Figure 10.11*a* shows the $(\mathbf{E}^2+\mathbf{H}^2)/8\pi$ values for an initial time (0.5×10^{-14} sec) at a laser intensity of 10^{15} W/cm^2 and at sub-

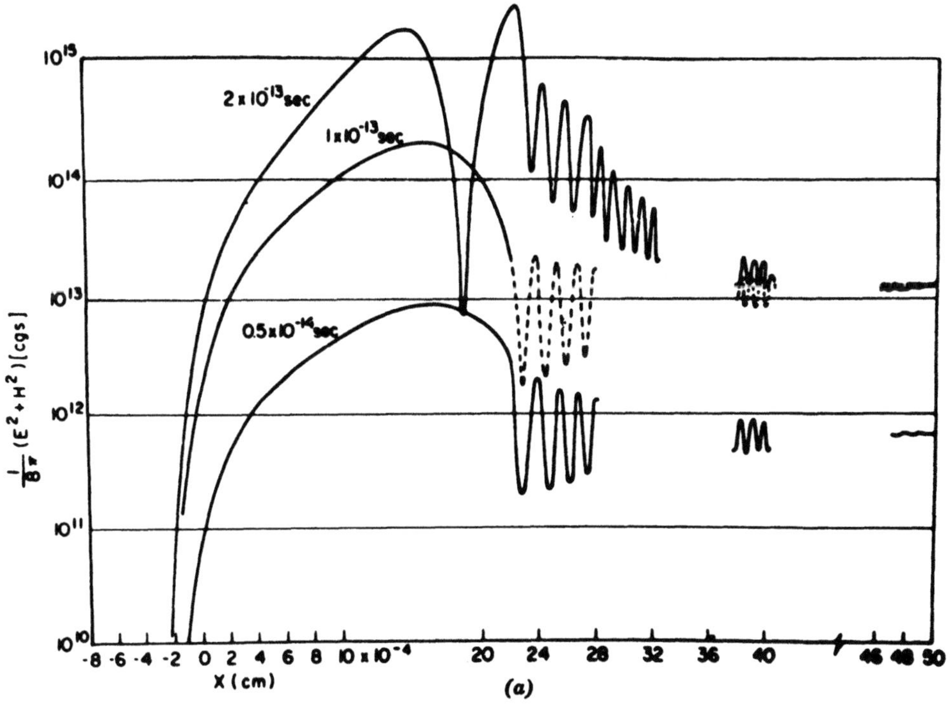

Figure 10.11a Similar to Fig. 10.10*a*, but with an initial density decreasing linearly from n_{ec} at $x=0$ to $n_{ec}\times(1-10^{-6})$ at $x=10$ μm, and to $n_{ec}\times(1-10^{-2})$ at $x=20$ μm and then to $n_e=0$ at $x=50$ μm. The $(\mathbf{E}^2+\mathbf{H}^2)$ values show an increase by a factor of 31.2 at time 5×10^{-15} sec. At 10^{-13} sec, the again-constant laser intensity of 2×10^{16} W/cm^2 is reached, with much less attenuation due to dynamic absorption at 2×10^{-13} sec. A swelling of $S\lesssim400$ occurs [1].

sequent times of 10^{-13} sec and 2×10^{-13} sec. The swelling factor at the beginning is

$$S=|n|^{-1}=31.2 \tag{10.14}$$

This is caused by the dielectric properties of this density profile. At 0.1 psec, the little depression of the curve at $x=20$ μm indicates some tendency to decrease due to density rippling. The maxima and minima of $\mathbf{E}^2+\mathbf{H}^2$, however, change locally very quickly, and at 0.2 psec, a swelling of $S=|n|^{-1}=400$ is reached. The dynamic change of the density near $x=20$ μm is shown in Fig. 10.11*b*. A process resembling tunnelling of the electromagnetic field through the overdense plasma near $x=20$ μm at 0.2 psec is very strong, by virtue of the depth of the overdense plasma (~1 μm) being small compared with the actual effective wavelength of more than 20 μm.

The resulting velocity profile at 0.2 psec is shown in Fig. 10.11*c*. The compression of plasma between $x=2$ μm and $x=14$ μm with velocities of 4×10^7 cm/sec does not change much within subsequent time intervals, while the velocities at $x=50$ μm, for example, are still changing. The increased ion

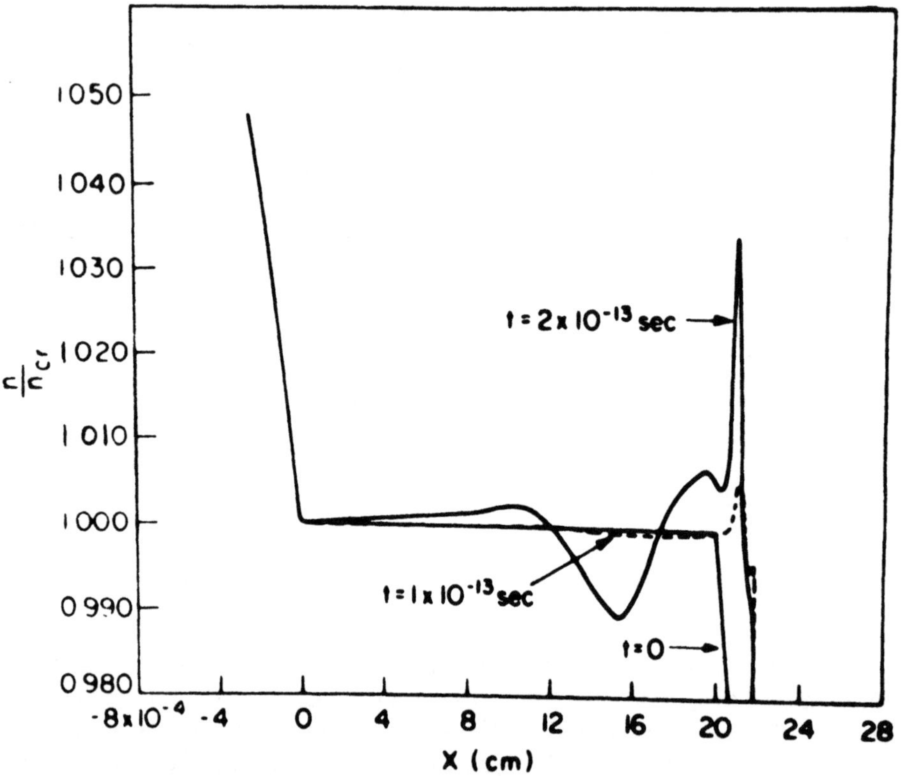

Figure 10.11b Initial and subsequent density profiles indicate motion of plasma toward the interior for x less than 14 μ [74].

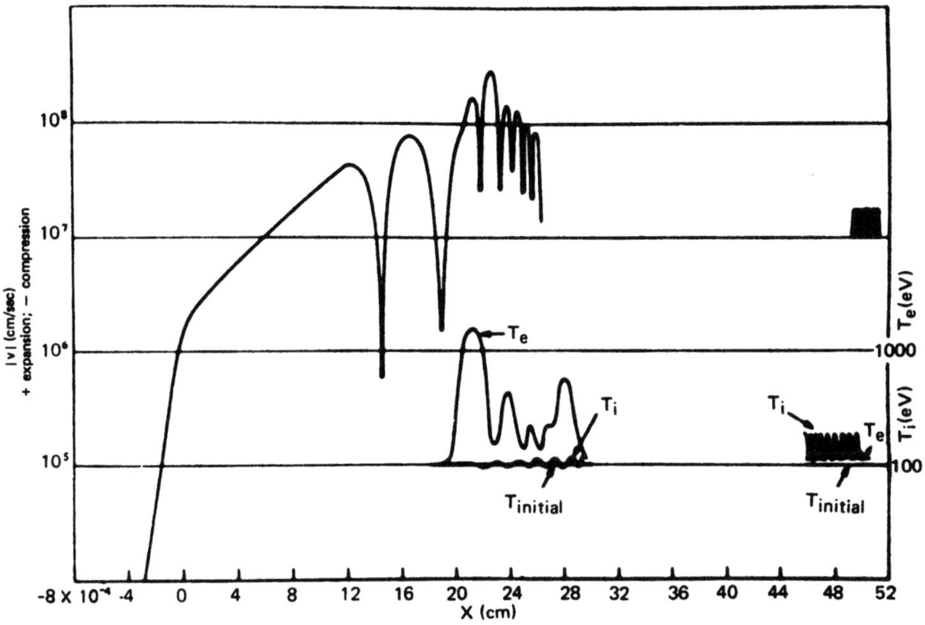

(c)

Figure 10.11c Dynamic heating of ions (at $x=48$ μm) and of electrons ($x=20$ μm) occurring at time $t=2\times10^{-13}$ sec. the resulting plasma velocity v shows a compressing motion of the whole plasma between -4 and 14 μm with speeds up to 5×10^7 cm/sec, followed by expansion and alternating compression and expansion due to the standing-wave field [1].

temperature from collisionless processes at $x=50$ μm is again much larger than the electron temperature, as could be expected from macroscopic dynamic ion absorption. A curious effect occurs in the interval from 20 to 28 μm, where the electrons are much hotter than the ions and display an irregular oscillation. It can be assumed that a longitudinal acceleration of the electrons occurs, due to the nonlinear force when the electrons have less interaction with the ions, comparable to hot electrons in solids at high electric fields. This phenomenon of dynamic electron absorption may be called "collisionless electron heating by nonlinear-force-induced macroscopic dynamic electron decay (MDED)."

The effects observed are highly sensitive to changes in the intensity I and the density profile. For an optimum case with $I=4\times10^{16}$ W/cm^2 and the identical time and density profiles, the incident laser energy per cm^2 is 4.1 kJ during up to 0.2 psec. The compression front between $x=-2$ μm and $x=14$ μm absorbs 0.96 kJ in kinetic energy, the ablating plasma take 0.68 kJ in kinetic energy of net motion and 2.2 kJ for dynamic heating. The remainder of 0.61 kJ goes into reflection and collisional heating of electrons.

The dynamic description of plasma by this numerical model for neodymium-glass laser intensities exceeding 10^{16} W/cm^2 indicates, that

1 A strong dielectric increase of the laser field and the effective wavelength occurs, described by a swelling factor $S = |n|^{-1}$ of up to 400 (n is a parameter resulting from rigorous solution of the Maxwell equations and is identical with the complex refractive index in the WKB approximation).
2 A compression of plasma driven by the nonlinear force occurs at a thickness of 15 μm with velocities exceeding 6×10^7 cm/sec. The mechanical energy in the compression front contains 43% of the incident laser radiation.
3 A rippling of the density occurs in the whole plasma at densities below cutoff with nonlinear-force generated velocities of 10^7 cm/sec or more. This may explain generation of laser produced fusion neutrons from peripheral plasma coronas. The ripple decreases laser light penetration of a plasma down to the cutoff density.
4 Rippling can be suppressed by selective profiling of the plasma density and the laser intensity with time.
5 A preferential transfer of laser energy to ions at a density gradient by collisionless shock occurs in the ripples (MDID).
6 A preferential transfer of laser energy to electrons can occur at nearly constant densities near cutoff (MDED).

10.4 Experimental Examples

The experimental study of the nonlinear force aims to find a direct proof. It is not very helpful, if indirect examples are seen, as, for example, Fig. 1.4. A nonlinear behavior of the keV ions in the fast group of plasma is evident, but a generation of the necessary intensities above the nonlinear force threshold, Section 9, is reached only indirectly including self-focusing or otherwise.

The direct action of the nonlinear force was shown first by the generation of cavitons by microwaves. Directed fast ions and a swelling of the electric field (by a factor of 700!) were measured. Wong and Stenzel [227] and Kim, Wong, and Stenzel [228] succeeded in these brilliant measurements, Fig. 10.12. The irradiation (40 kW at 1 GHz) into a 4 m long plasma with almost linearly increasing density resulted in fast ions moving against the irradiated field. Their energy was higher than the ion temperature. The latter was in the order of eV, because the interaction time was too short for a reasonable heating of the ions by collisions.

The first observation of the action of the nonlinear force in a laser irradi-

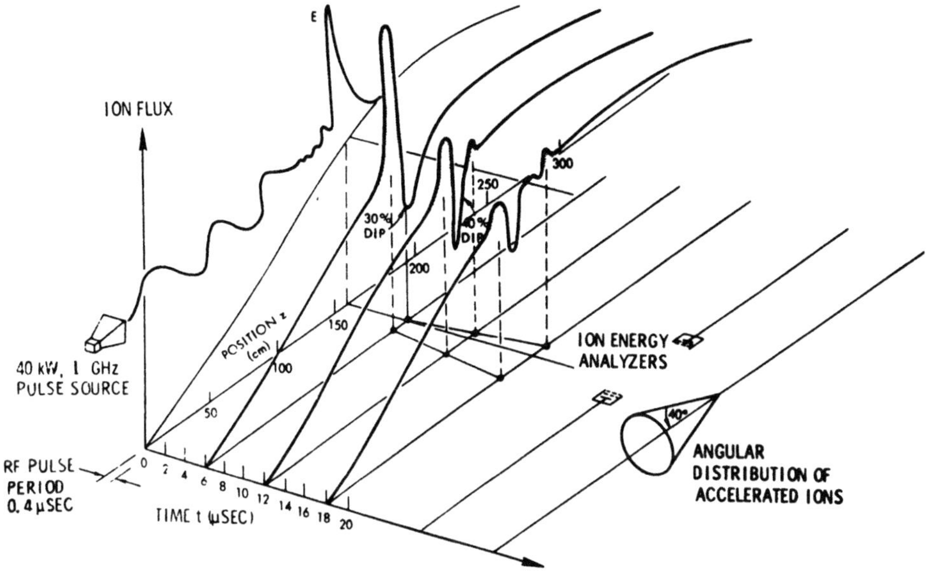

Figure 10.12 Space-time representation of ion bursts (shaded) driven by the nonlinear force. Density cavities are created as a result of ion expulsion. After Wong and Stenzel [227].

ated plasma was that by Marhic [229], Fig. 10.13. A consequent evaluation, based on the above derived theory, was in full agreement with the measurements.

A side-on measurement of the caviton produced by the nonlinear force (similar to the result of Wong et al. [227]) was seen first by Zakharenkov et al.

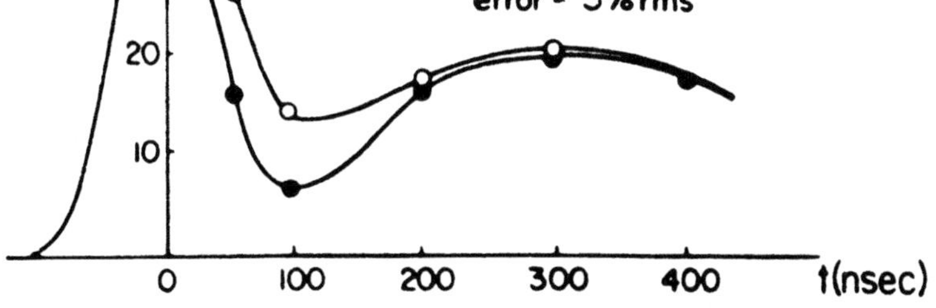

Figure 10.13 Relative change in light intensity at the focus indicating the action of the nonlinear force. $p = 0.4$ Torr, $B = 5$ kG, $J = 100$ A [229]. After Marhic [156].

[230], Fig. 10.14, with neodymium glass lasers and by Fedosejevs et al. [231], Fig. 10.15, with CO_2 lasers. Similar results including the radial caviton, Fig. 2.10, were seen by Azechi et al. [232].

The stability of these cavitons were studied theoretically [233] and experimentally [234]. The nonlinear force in connection with the generation of fast ions was discussed by several authors [235 to 240] for cases where self-focusing could not have occurred, and where the ion energies were in the order of the values expected from the nonlinear force theory.

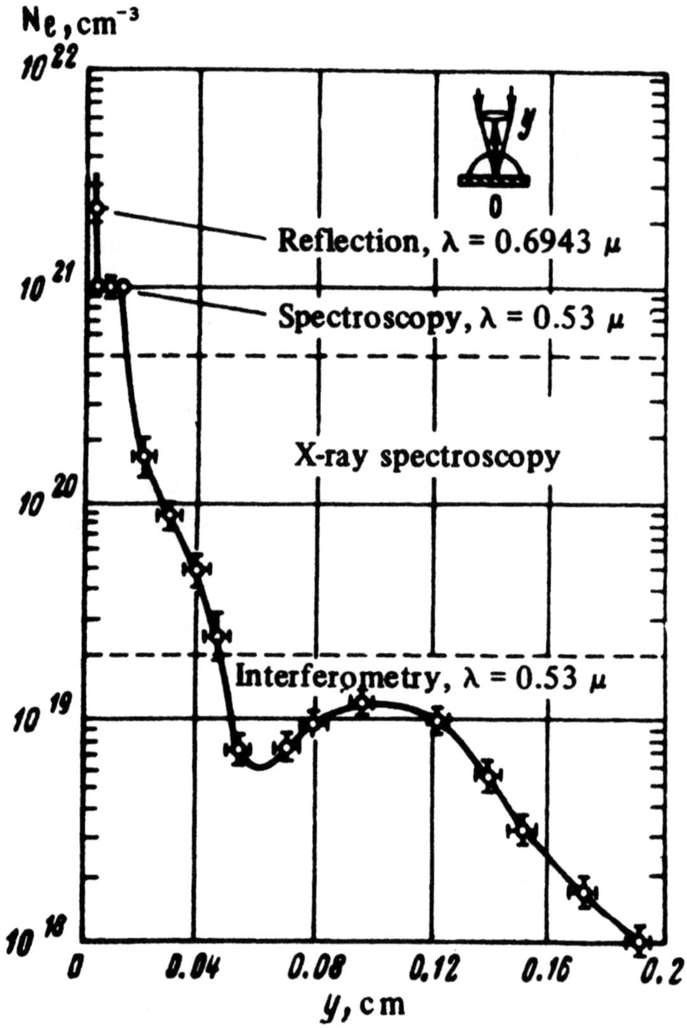

Figure 10.14 Electron density profile for laser plasma after 2 nsec, obtained for given flare by different methods. Target A, flux density 3×10^{14} W/cm^2. After Zakharenkov et al. [230].

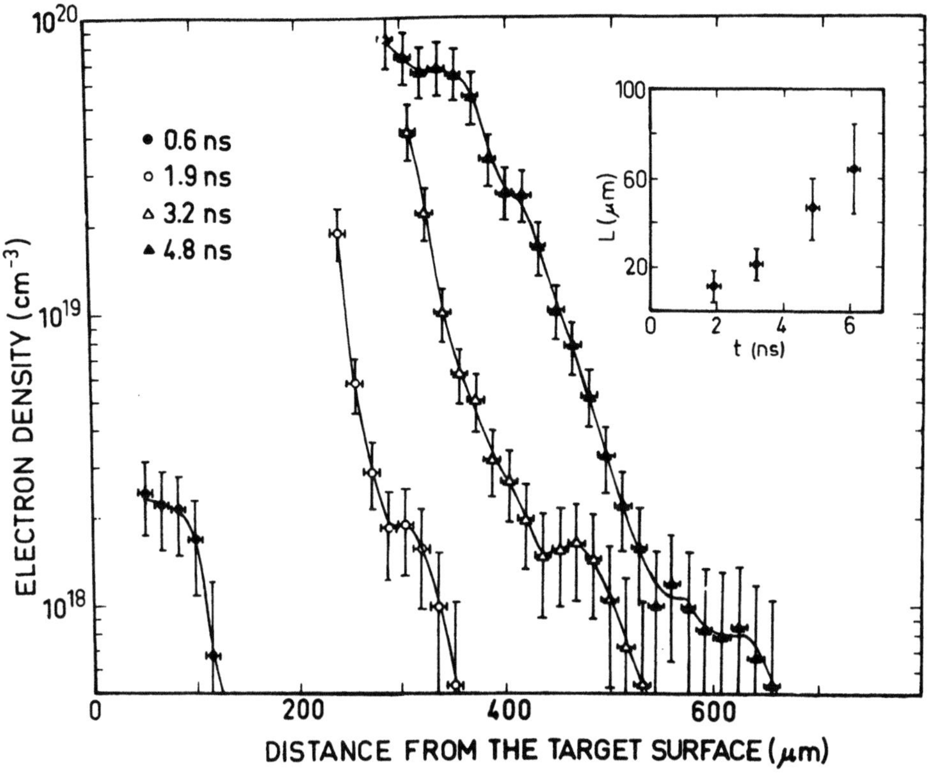

Figure 10.15 Plot of axial electron density versus distance from the original target surface at various times. The insert shows the scale length $L = n_e(dn_e/dx)^{-1}$ of the density plateau at various times relative to the leading edge of the CO_2-laser pulse. After Fedosejevs et al. [231].

10.5 Acceleration of Thick Blocks

This subsection reports on an extension of the fully gasdynamic calculations for plane electromagnetic waves at perpendicular incidence on stratified plasmas, including nonlinear forces as well as a nonlinear complex refractive index. The examples described in Figs. 10.10 and 10.11 were for short pulses of subpicosecond duration, which seems to be unrealistic for high intensities at present, although pulses of 170 femtoseconds (0.17 psec) have been generated [17, 241] and laser pulse powers of gigawatts have been achieved.

The following calculations are based on similar assumptions as used in that for Figs. 10.10 and 10.11; some numerical refinements were introduced and longer interaction times have been reached, before the computation became unstable. A parameter for this was the appearance of negative densities. The consistency of the computations was given by the conservation of energy, by the behavior of nonlinear momentum and energy transfer in agreement with global calculations, and last but not least by a numerical exact reproduction of a solitonlike behavior. The intensities of the laser pulses in all the cases were $\sin^2(mt)$-like, increasing to the maximum at 1 psec and where m is a constant. In all figures I corresponds to this constant intensity [242 to 244].

In order to avoid reflection, standing waves and the subsequent Brillouin-type density rippling and high-reflectivity—at least for the earlier stages—an initial density profile was chosen, which was expected to be very close to a case of very low reflection, namely the Rayleigh case, as discussed in Section 7. There a collisionless plasma was considered only. The fact that collisions are included can cause a superdense behavior near $x=0$ if the plasma temperature is 100 eV or less and the intensity is below the threshold for the predominance of the nonlinear force. The exponential decay of the intensity from 30 μm and below is seen in Fig. 10.16.

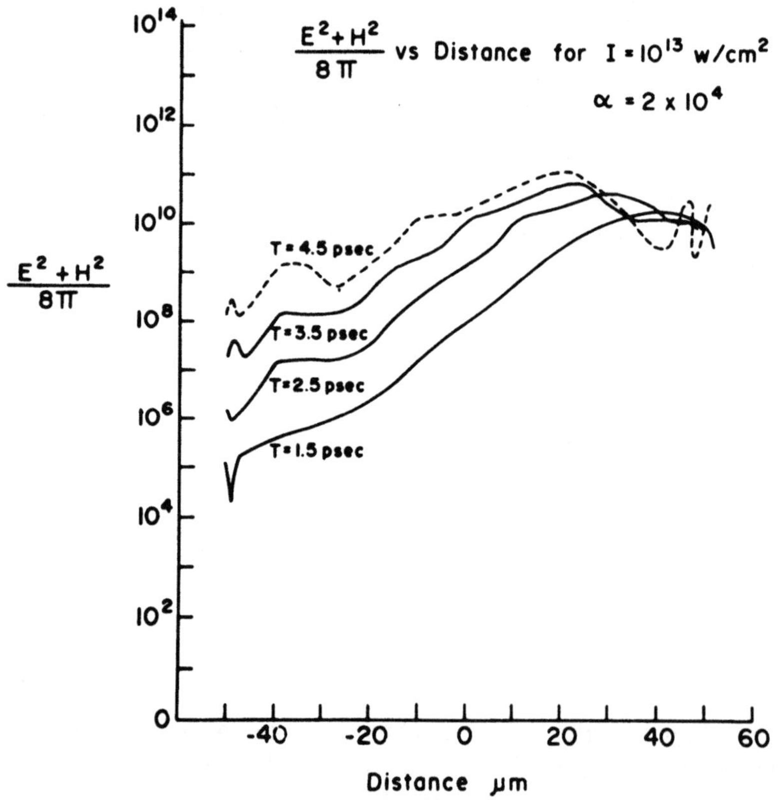

Figure 10.16 Time dependent solutions of $(\mathbf{E}^2+\mathbf{H}^2)/8\pi$ for an initial temperature of 100 eV and an initial profile as in Fig. 10.17 [243].

The fully dynamic calculation includes thermokinetic and nonlinear force motion. It causes an increasing transparency and a subsequent higher swelling in time. As shown in Fig. 10.6, at 4.5 psec the swelling is given by twice the ratio of E^2 at 50 μm at the earlier times, to that of the broad maximum at 4.5 psec. Its value is $S=82.2$. Note that $(\mathbf{E}^2+\mathbf{H}^2)/8\pi$ is always given in cgs units.

For the same initial density profile as in Fig. 10.17 and an initial temperature of 100 eV, Fig. 10.18*a* shows the energy densities at 1.5 psec for various intensities, and Fig. 10.18*b* shows the corresponding profiles of the plasma velocities. One recognizes the generation of fast moving thick blocks of plasma with velocities beyond 10^8 cm/sec, while the ion and electron temperatures have been increased to less than 500 eV. Figure 10.18*b* contains numerical extension to intensities of 10^{19} W/cm^2. This is not valid, as this is in the range of relativistic effects for the optical constants that are not yet included into the code.

The following rough estimation confirms that velocities above 10^8 cm/sec for intensities of 10^{18} W/cm^2 are achieved. Figure 10.18*a* leads to the

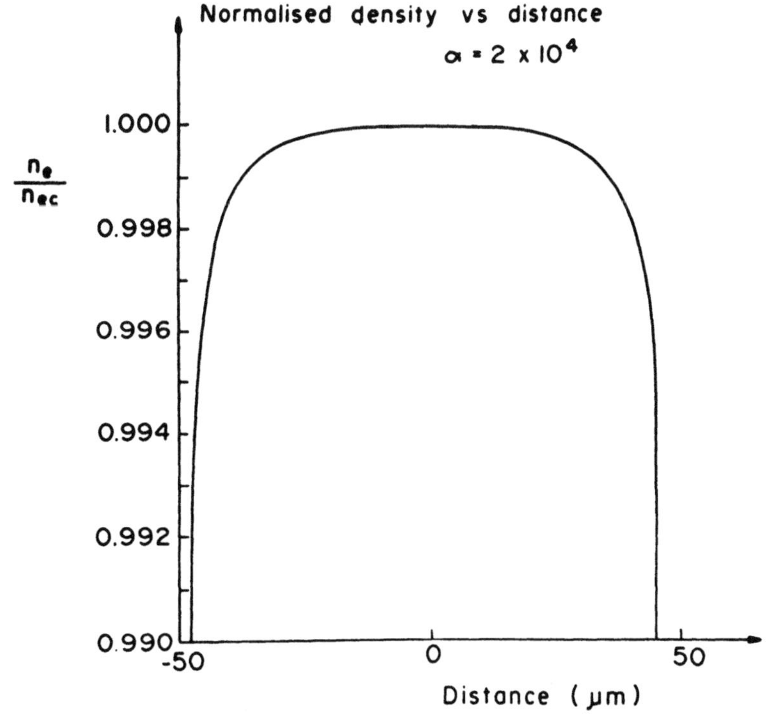

Figure 10.17 An inhomogeneous bi-Rayleigh density profile with $\alpha=2\times10^4$ cm^{-1} for neodymium glass lasers or CO_2 lasers, corresponding to Eq. (7.39). In all cases, initial temperatures are assumed to be uniform throughout the plasma [243].

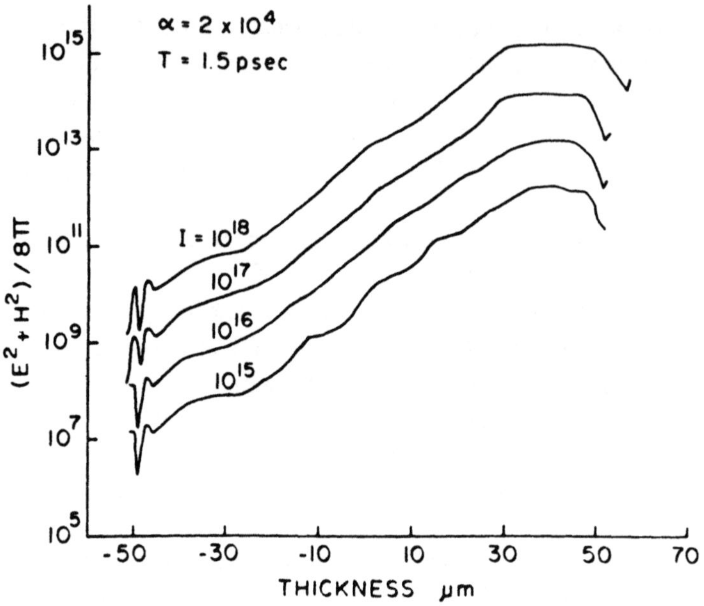

(a)

Figure 10.18a For an initial profile of Fig. 10.17 and an initial temperature of 100 eV, single broad maxima of the energy density $(\mathbf{E}^2+\mathbf{H}^2)/8\ \pi$ are generated at 1.5 psec, which vary directly with intensity of the incident laser light [244].

acceleration a from

$$\mathbf{f}_{\mathrm{NL}}=m_i n_i \mathbf{a}=\mathbf{i}_x \frac{\partial}{\partial x}(\mathbf{E}^2+\mathbf{H}^2)8\pi \tag{10.15}$$

$$\mathbf{a}=\mathbf{i}_x \frac{1}{m_i n_i} \frac{\Delta[(\mathbf{E}^2+\mathbf{H}^2)/8\pi]}{\Delta x} \tag{10.16}$$

For $m_i=2.5\times 1.67\times 10^{-24}$ g (DT-plasma) and the Nd glass cutoff density n_i with $\Delta[\mathbf{E}^2+\mathbf{H}^2]/8\pi=10^{15}$ cgs along 5 μm, $|\mathbf{a}|=4.79\times 10^{20}$ cm/sec^2. The velocity achieved during a time $t=1.5$ psec at constantly approximated acceleration is 7.1×10^8 cm/sec. Considering the temporally increasing acceleration, the velocities in Fig. 10.18b are then understood.

The total kinetic energy transfered to the plasma between the thicknesses x_1 and x_2 is given by

$$E=\frac{1}{x_2-x_1}\int_{x_1}^{x_2} \frac{m_i n_i(x)}{2} v_i^2(x)\, dx \tag{10.17}$$

It is evaluated numerically for the case of $\alpha=3000$ cm^{-1} $x_1=-0.05$ mm, and $x_2=+0.05$ mm. The result shows a superlinear increase (Fig. 10.19)

$$E_{\mathrm{kin}}=I^m; \qquad m=1.8 \tag{10.18}$$

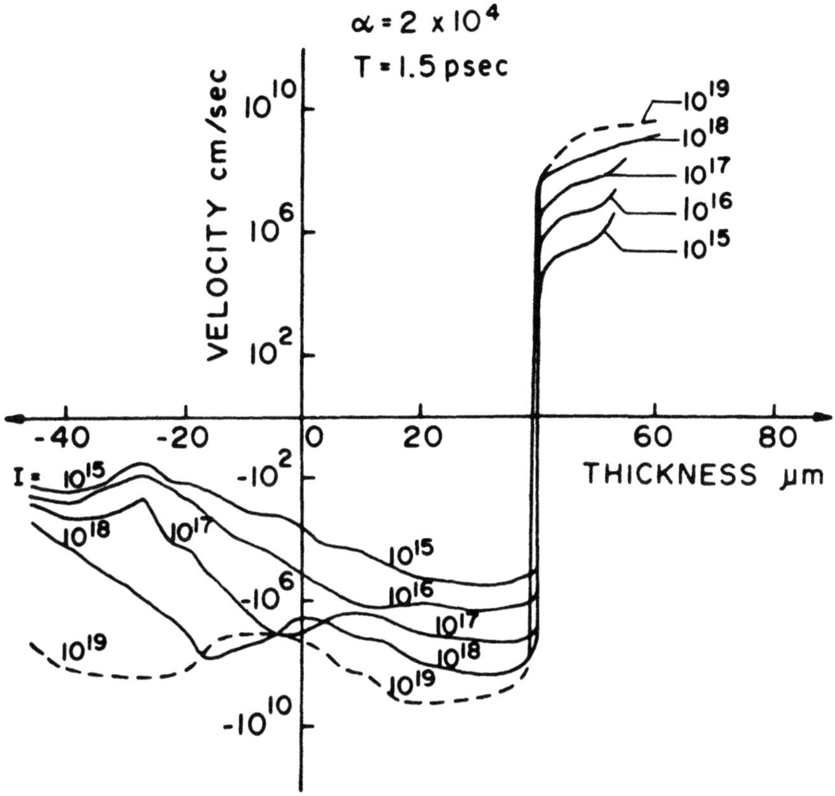

(b)

Figure 10.18b The velocity of plasma corresponds to the electromagnetic energy density of Fig. 10.18*a*. These blocks of plasma move toward the interior of the plasma due to the nonlinear force [224].

according to the nonlinear macroscopic absorption process of energy transfer by the nonlinear force.

For a longer interaction, the initially smooth profiles for the electromagnetic energy density and plasma density are split up (see Figs. 10.20*a* to *c*). For a Nd glass laser, the same behavior is observed in the case of a CO_2 laser (see Figs. 10.21*a* to *c*). It is remarkable that the maxima of the electromagnetic energy densities $(\mathbf{E}^2+\mathbf{H}^2)/8\pi$ are—as it must be—different by about 100 times (for CO_2 10^{16} and Nd 10^{18} W/cm^2), but the resulting maximum velocities of the blocks, Fig. 10.21*b*, are nearly the same in both cases. This confirms that the squares of the maximum plasma velocities, due to the nonlinear force acceleration are proportional to $I\lambda^2$.

$$v^2 \propto I\lambda^2 \tag{10.19}$$

This result had been confirmed experimentally for numerous quantities:

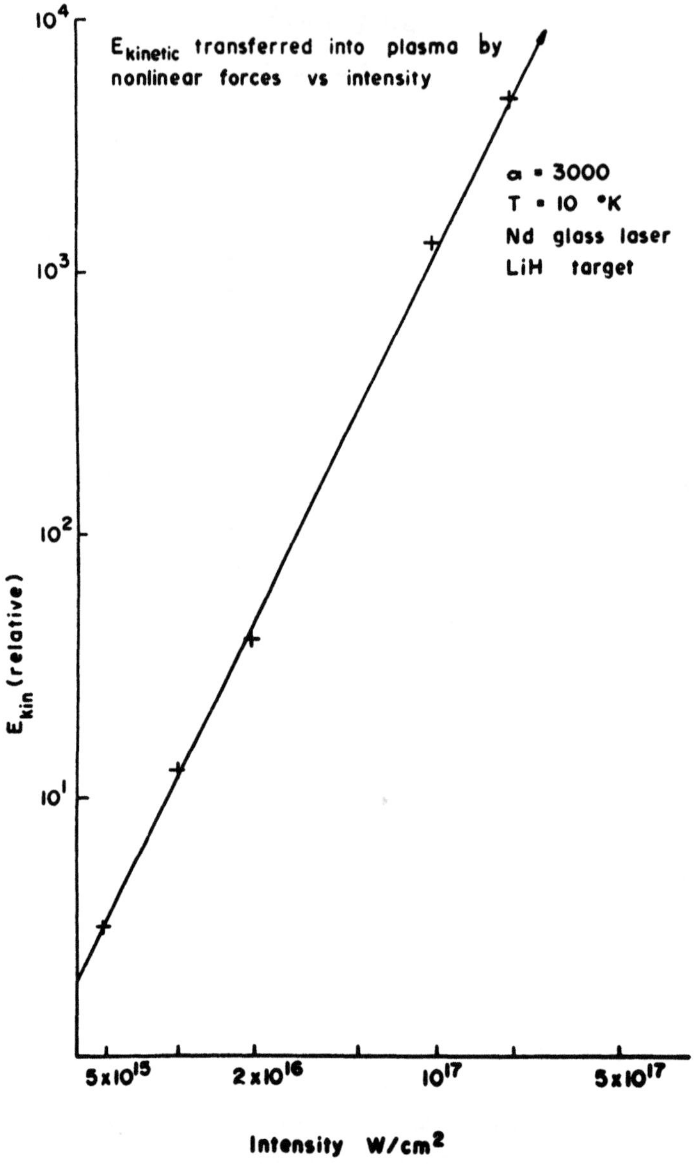

Figure 10.19 Kinetic energy transferred to the plasma depends on laser intensity and justifying the nonlinear nature of the interaction [243].

for the fast ion energy of laser produced plasmas, for the similar nonlinear "temperature," and for similar nonlinear quantities [245]. The explanation is very easy: these quantities, as well as the nonlinear force [see Eq. (9.23)] are proportional to the (nonrelativistic) oscillation energy [Eq. (6.56)].

$$\varepsilon_{osc} \sim I\lambda^2 \qquad (10.20)$$

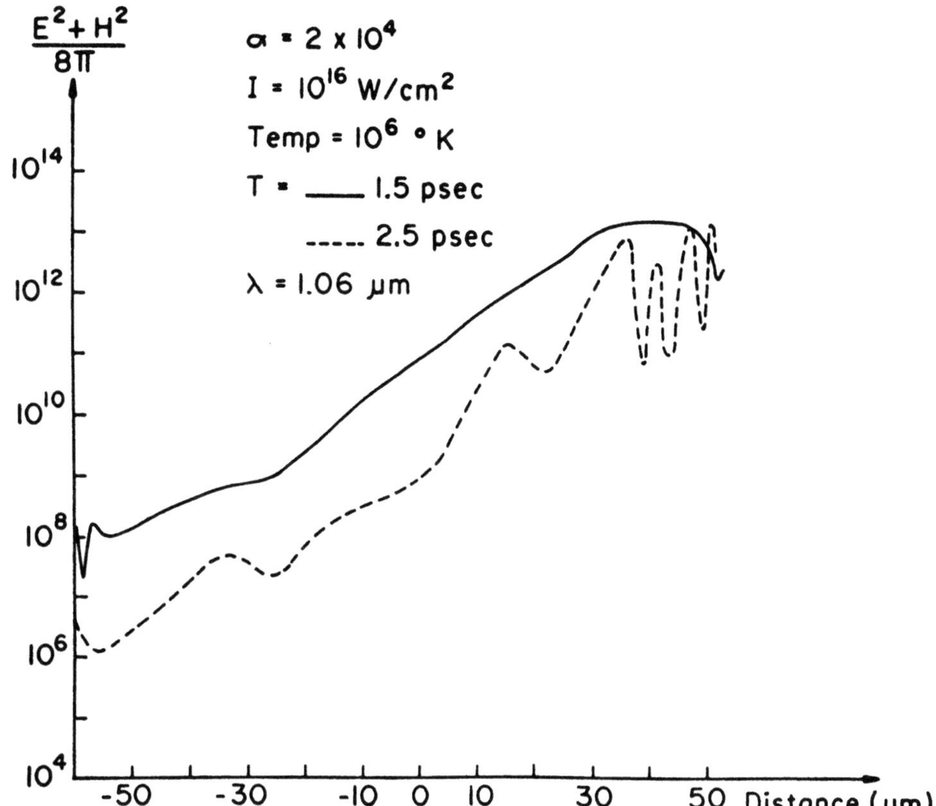

(a)

Figure 10.20a Electromagnetic energy density $(\mathbf{E}^2+\mathbf{H}^2)/8\pi$ for same initial conditions of a bi-Rayleigh profile $\alpha = 2 \times 10^4$ cm^{-1} as in Fig. 10.18 but for time $t = 1.5$ and 2.5 psec, at Nd glass laser intensity 10^{16} W/cm^2 [243].

This explains the results [245], which were gained by very expensive experiments.

10.6 Solitons

The previously described change of the smooth behavior into the oscillating one at later times (Fig. 10.20*a* to *c*) suggests the possibility of developing solitons at the plasma dynamics. The absorption process seems to follow a solitonlike behavior according to the Korteweg–de Vries equation with all the well-known consequences of solitons. This is shown by an evaluation of the computer results of the type in Fig. 10.20*a* to *c*.

There are various processes that have a solitonlike behavior and fulfill a Korteweg–de Vries (KdV) equation. If a system is within this regime, it

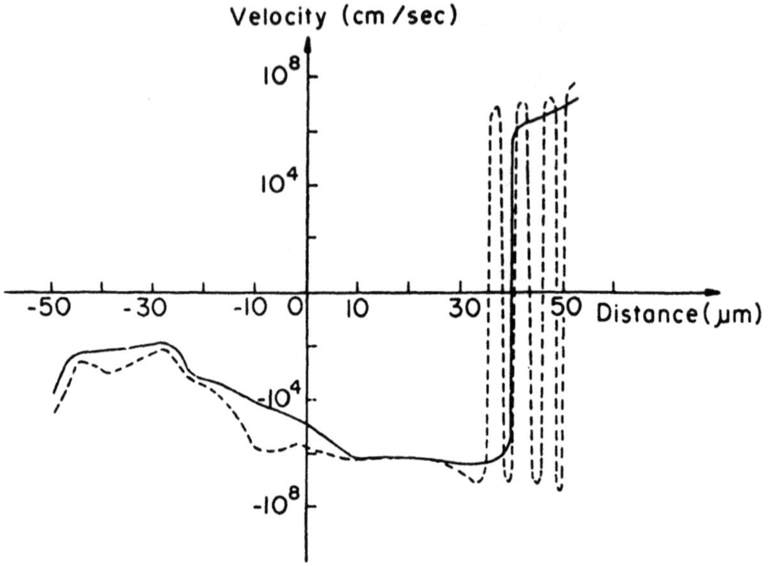

(b)

Figure 10.20b The velocity profiles for the cases of Fig. 10.20*a* [243].

behaves with some stable properties; but to arrive at these conditions or to deviate, dissipative processes (energy absorption or transfer) are necessary. In this case, the nonlinear force causes this dissipation.

One well-known example of solitons in plasma are acoustic waves. The KdV equation for ion acoustic waves in a plasma describes solitons in full agreement with experiments [246].

$$\frac{\partial}{\partial t} n_i + \frac{\partial}{\partial t} x \frac{\partial}{\partial x} n_i = -\mu' \frac{\partial^3}{\partial x^3} n_i \tag{10.21}$$

where n_i is the ion density. A dispersion function μ' is used, or a term representing the nonlinear force must be added. In the case considered here, numerical results of nonlinear laser plasma interaction of the type prescribed before are evaluated. The plasma velocity $\mathbf{v}$ is considered, following a KdV equation

$$\frac{\partial}{\partial t} \mathbf{v} + \mathbf{v} \frac{\partial}{\partial x} \mathbf{v} = -\mu' \frac{\partial^3}{\partial x^3} \mathbf{v} \tag{10.22}$$

These are the "velocity-type" solitons and not ion acoustic solitons as in Eq. (10.21). The left-hand side of Eq. (10.22) can be identified with the acceleration, given by the force density **f**.

$$\frac{1}{m_i n_i} \mathbf{f} = \frac{\partial}{\partial t} \mathbf{v} + \mathbf{v} \frac{\partial}{\partial x} \mathbf{v} \tag{10.23}$$

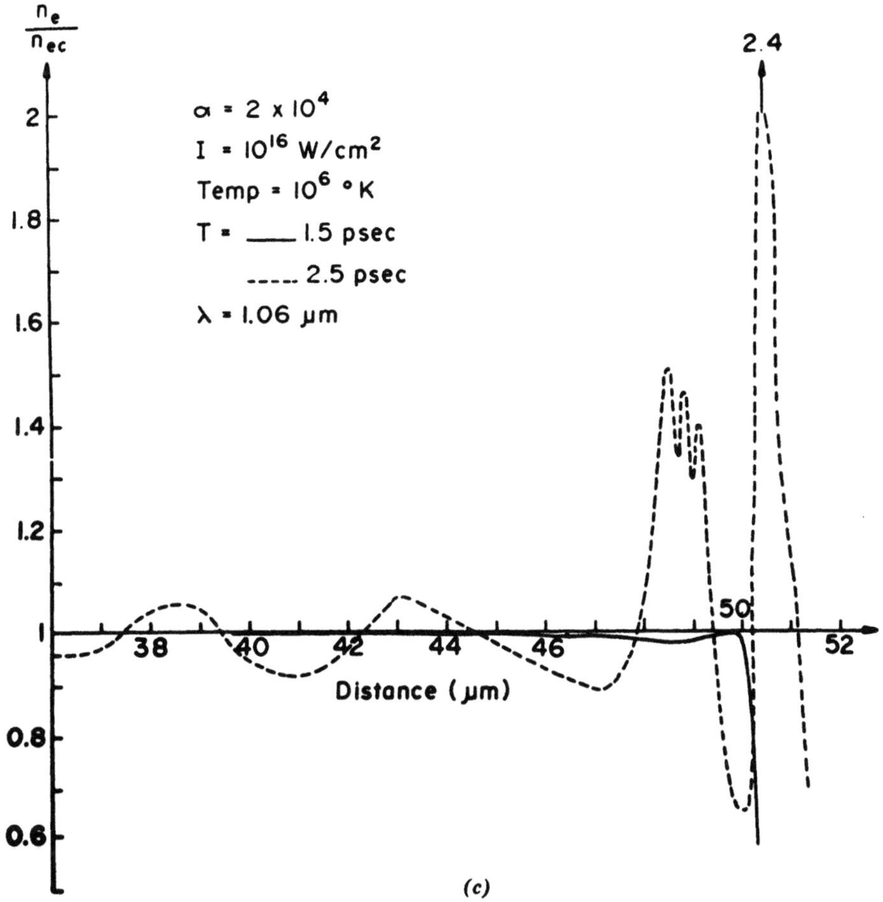

Figure 10.20c The density rippling due to the oscillations of the velocities corresponding to Figs. 10.20*a* and 10.20*b* [243].

For an evaluation of numerical results of the type in Fig. 10.20*a* to *c*, **f** has to be compared in Eq. (10.23)

$$\mathbf{f} = -\nabla p + \nabla(\mathbf{E}^2 + \mathbf{H}^2)/8\pi \tag{10.24}$$

with the third spatial derivation of **v**, Eq. (10.22). It will turn out that an agreement is possible only if the gasdynamic pressure *p* is neglected and only if the nonlinear force with the strengths **E** and **H** are included. The energy deposition by the radiation, based on the nonlinear intensity dependence of the optical constants, has to be considered [242, 243]. This results in examples such as in Figs. 10.20*a* to *c*.

The numerical evaluation of these results shows a soliton process of Eq. (10.22). Examining whether a similarity or a relation exists in the sense of

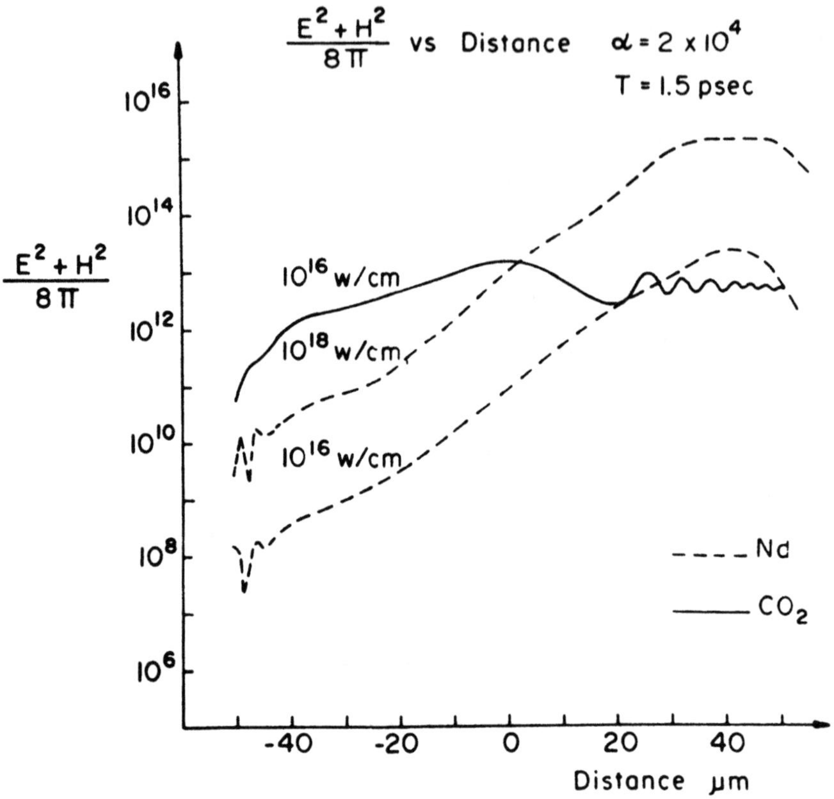

(a)

Figure 10.21a Electromagnetic energy density at 1.5 psec for various intensities for Nd glass and CO_2 lasers [243].

a Korteweg–de Vries equation, from Eqs. (10.22) to (10.24), we should have

$$-\mu' \frac{\partial^3}{\partial x^3} v = \frac{1}{8\pi m_i n_i} \frac{\partial}{\partial x} (\mathbf{E}^2 + \mathbf{H}^2) = \frac{|\mathbf{f}_{\mathrm{NL}}|}{m_i n_i} \tag{10.25}$$

The thermokinetic force has to be dropped as shown from the following.

For the evaluation, several cases of the type in Fig. 10.20 are used, where there results a very transparent decay of the $(\mathbf{E}^2 + \mathbf{H}^2)/8\pi$ field or the density into ripples, or the velocities into oscillations after an earlier very smooth behavior [247]. The evaluation of $\partial^3 \mathbf{v}/\partial x^3$ and of the right-hand side of Eq. (10.25) by numerical differentiation is shown in Fig. 10.22. There are polelike maxima of which every second is coincident.

After this result, several dispersion functions μ were tested and a complete satisfaction of Eq. (20.25) was found.

$$\mu' = \frac{\partial/\partial x(1 - n_e/n_{ec})}{1 - n_e/n_{ec}} \tag{10.26}$$

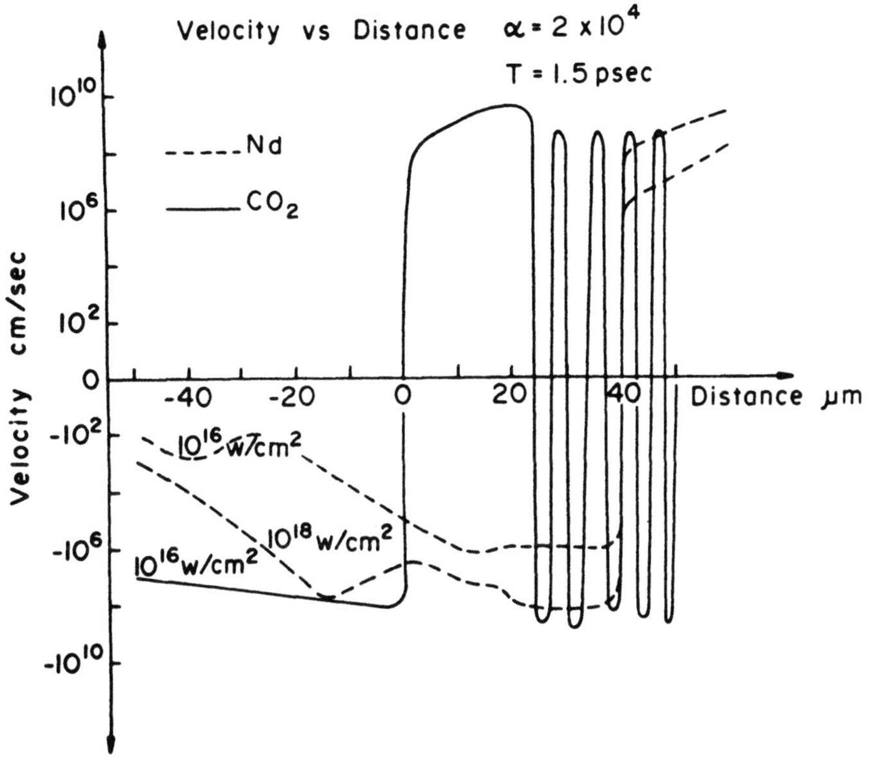

(b)

Figure 10.21b The velocity profiles at 1.5 psec for the cases of Fig. 10.20*a* [243].

where n_e is the electron density, and n_{ec} is the cutoff density. This can be seen from Fig. 10.21 when comparing the part of n_e/n_{ec} for the time of the plots of the quantities in the upper diagram. Going from *A* to *B* by multiplying $(\partial/\partial x)(\mathbf{E}^2+\mathbf{H}^2)/8\pi$ (dashed line) by $-1/\mu$, keeps the sign but drops the lines to zero at *B* to be coincident with the $\partial^3 v/\partial x^3$ values. Due to the pole of $-1/\mu$ in Eq. (10.26), one arrives at the same pole for the dashed curve multiplied by $-1/\mu$ as the pole of $\partial^3 v/\partial x^3$ shows at *C*. The same procedure can be followed from *C* to *E*.

Within the numerical accuracy, a complete identity of

$$\frac{\partial}{\partial t}v+v\frac{\partial}{\partial x}v=\frac{1}{8\pi n_i m_i}\frac{\partial}{\partial x}(\mathbf{E}^2+\mathbf{H}^2) \tag{10.27}$$

with a KdV equation (10.22) is found [limited to positive brackets, Eq. (10.26)], otherwise the more general expression [Eq. (10.26)] has to be used:

$$\frac{\partial}{\partial t}v+v\frac{\partial}{\partial x}v=\frac{\partial \ln \varepsilon'}{\partial x}\frac{\partial^3}{\partial x^3}v \tag{10.28}$$

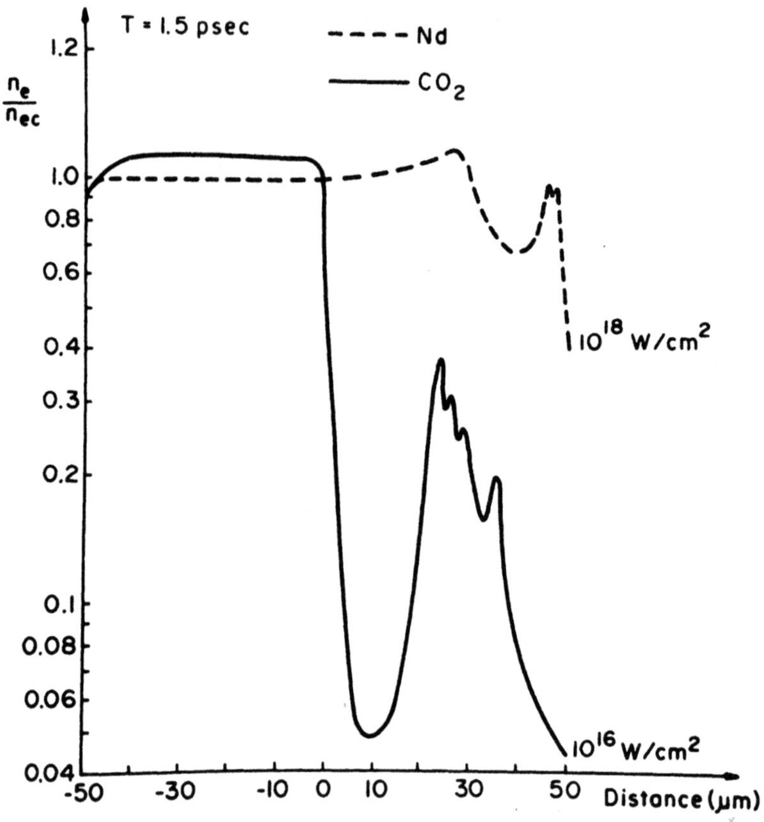

(c)

Figure 10.21c The density profiles for Ng glass lasers at 10^{18} W/cm² and for CO_2 lasers at 10^{16} W/cm² at 1.5 psec. Both density formations show the existence of a caviton [243].

The real part of the dielectric constant ε'

$$\varepsilon = \varepsilon' + i\varepsilon''; \qquad \varepsilon' = 1 - \omega_p^2/\omega^2 = 1 - n_e/n_{ec} \tag{10.29}$$

(ω_p=plasma frequency, ω=laser frequency) has to be used. The numerical accuracy is very low in the region of the function poles, the coincidence of which, however, is very sharp. The accuracy near the points *B* and *D* is very high and confirms that no imaginary part has to be used from ε.

Any correlation of Eqs. (10.27) and (10.28) for the special time is a proof of the numerical stability of the calculations described in the preceding subsection. Furthermore, as expected, the gasdynamic pressure *p* for the motion in Eq. (10.24) is negligible compared with the nonlinear force. The only strangeness of the result is that the dispersion factor is not the dielectric constant but a logarithmic derivative of the real part of the dielectric constant

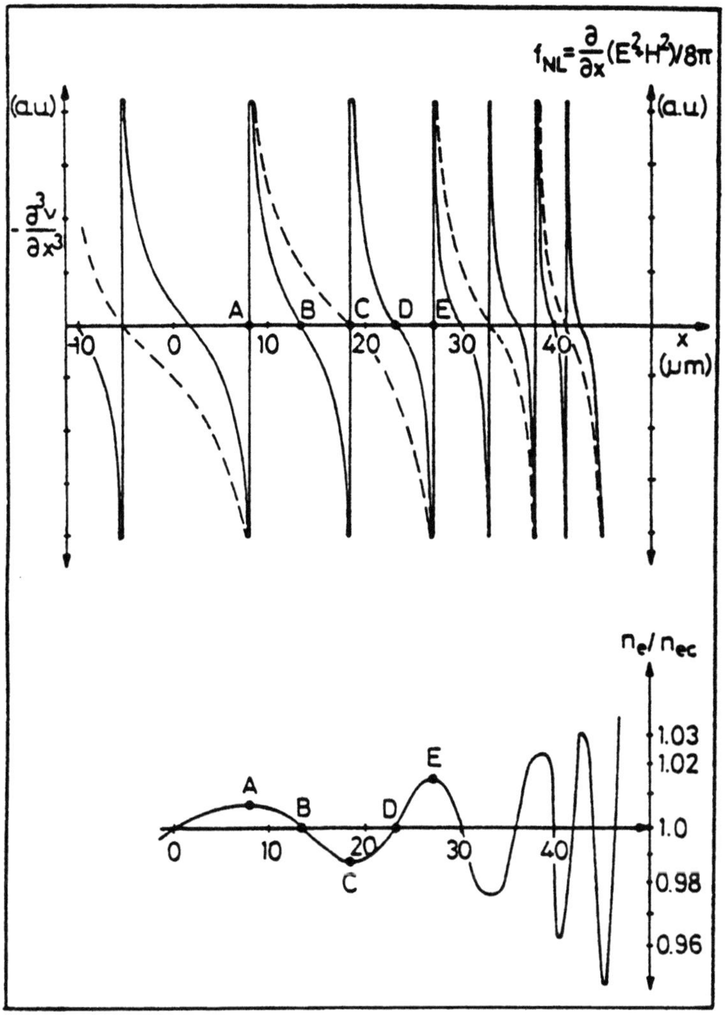

Figure 10.22 $\partial^3 v/\partial x^3$ (——) and f_{NL} (- - -) are evaluated from numerical examples of nonlinear dynamic calculations of laser plasma interaction [247] and compared with the generated density ripple. Introducing the special dispersion function, Eq. (10.26), a behavior according to the Korteweg de Vries equation is established [266].

only. However, the similarity to the expression $\partial \ln \varepsilon/\partial x$ in the theory of the resonance absorption [248] is not surprising. It has an importance similar to the attempts to explain the density rippling [242, 249] as a structure resonance [250, 251].

It has to be noted that at earlier times of the interaction, correlation to a KdV equation is not possible. The change from this case to the soliton case is due to nonlinear-force produced macroscopic and nonthermalizing absorption [252].

In order to gain further numerical facts about the solitonlike process, Lalousis [253] calculated cases with the same parameters as in Fig. 10.20*a* but with varying parameter α of the bi-Rayleigh initial density profile. For relatively low α, for example, $\alpha = 10^{-3}\ \mathrm{cm}^{-1}$, the cutoff density is not reached up to 3 psec along the whole plasma of about 100 μm thickness. Attention is then due to usual and nonlinear absorption only, and a wave field with a high degree of reflection (standing wave) is created. It is remarkable that at 2 psec, the profiles of the nonlinear force and of the derivatives of the velocity are following roughly the Korteweg–de Vries equation (see Fig. 10.23) but the lower range from 40 down to -50 μm follows the Banjamin–Ono equation [254]

$$\frac{\partial}{\partial t} v + v \frac{\partial}{\partial x} v = -H \frac{\partial^2}{\partial x^2} v = \frac{\partial}{\partial x} \frac{\mathbf{E}^2 + \mathbf{H}^2}{8\pi} \tag{10.30}$$

where the Hilbert transform H has to be unity. For an analysis of these processes, more numerical examples are necessary for understanding the change into the solitonlike behavior by the dissipation processes.

The result of this subsection permits the following conclusion. It is shown numerically and experimentally that the nonlinear force can transfer optical energy into thick fast-moving blocks of plasma without heating. This highly efficient transfer of optical energy into kinetic energy of compressed plasma is the basis of a high-efficiency concept of laser fusion [255]. It avoids the difficulties of the interaction (absorption) and transport processes for laser

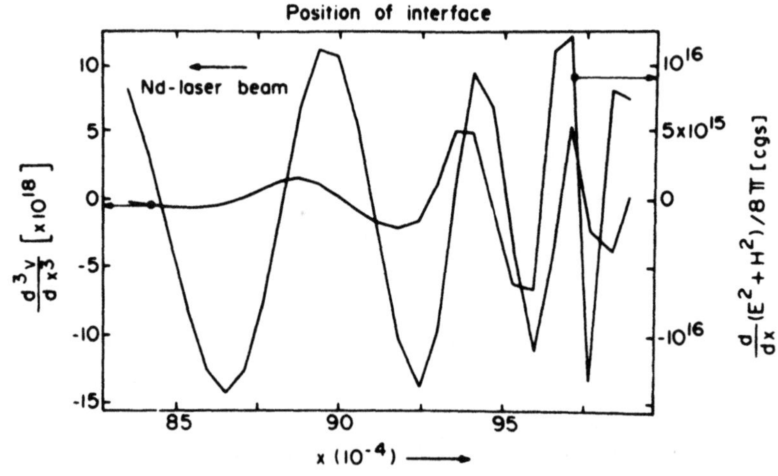

Figure 10.23 Output for same initial conditions as in Fig. 10.20 for $I = 10^{16}\ \mathrm{W/cm^2}$ but for $\alpha = 10^3\ \mathrm{cm}^{-3}$. The peripheric part (43 to 49 μm) follows nearly the Korteweg–deVries equation [253].

fusion [255], but it needs better laser technology (shorter and more precise pulses). The result of the soliton decay, however, provides much faster thermalization of laser radiation to ions in the corona than by Coulomb collisions (see Eq. (10.8)). Any such mechanism for faster thermallization is of interest for extending the gasdynamic ablation-compression scheme of Nuckolls (219) to higher laser intensities.

10.7 Numerical Results from the Genuine Two Fluid Model and Electric Double Layers

A straight forward extension of the numerical calculations of the plane wave perpendicular incidence one dimensional laser-plasma interaction following Fig. 10.21b was possible by application of the genuine two fluid model described in subsection 8.7. The only difference is that then the internal electric fields in the plasma are generated which correspond to electric double layers. Furthermore we immediately calculate the longitudinal electric fields in the plasma driven at the one hand by the gas dynamics of the inhomogeneous plasma as shown in Fig. 8.2 to 8.4, and at the other hand much stronger by the laser radiation.

We return to the computations in subsection 8.8 and discuss now cases where laser radiation is incident. This laser radiation is not only producing the nonlinear force to the electron equation of motion, it also determines the collisional absorption of laser radiation in the equation of energy conservation of electrons heating the electron fluid. This heat is spread by thermal conduction and transferred (with appropriate time delay) to ions by equipartition. Further viscosity between the electron and ion fluid is included. Heating by adiabatic compression, especially of ions, happens due to the nonlinear force action since thermal conduction of the ions and heat exchange to electrons by equipartition is rather slow.

The computations are with time steps of about 0.1 fsec in order to have at least ten time steps for the fastest plasma oscillation for typical electron densities of 10^{21} cm^{-3}, the critical density of the mostly considered neodymium glass irradiation. A very extensive computation for each time step during the complex temporal and spatial development of the plasma was the Maxwellian exact solution of the electromagnetic field in the plasma using the optical constants with collisions on the basis of the nonlinear intensity dependence as explained in Chapter 6. Only with these changing laser fields the nonlinear force and the collisional absorption could be calculated.

Initial conditions for the fluid densities, velocities and temperatures had to be chosen and the boundary conditions were determined by the time dependence of the irradiated laser intensity. The hydrodynamic boundary conditions were those of free variation of the boundary values as connected to vacuum or lower density plasma or to higher, e.g., supercritical density. The initial-boundary value problem could not be solved along the usual textbook techniques but a new method had to be discovered (Lalousis, 1983, Lalousis et al 1984).

For the case of 10^{16} W/cm^2 Nd glass radiation, the computer output of the following cases is discussed (Hora et al 1983). A 25 μm-thick plasma slab of initial 10^3-eV temperature and zero velocity with an ion and electron density of symmetric parabolic shape very close to the value in

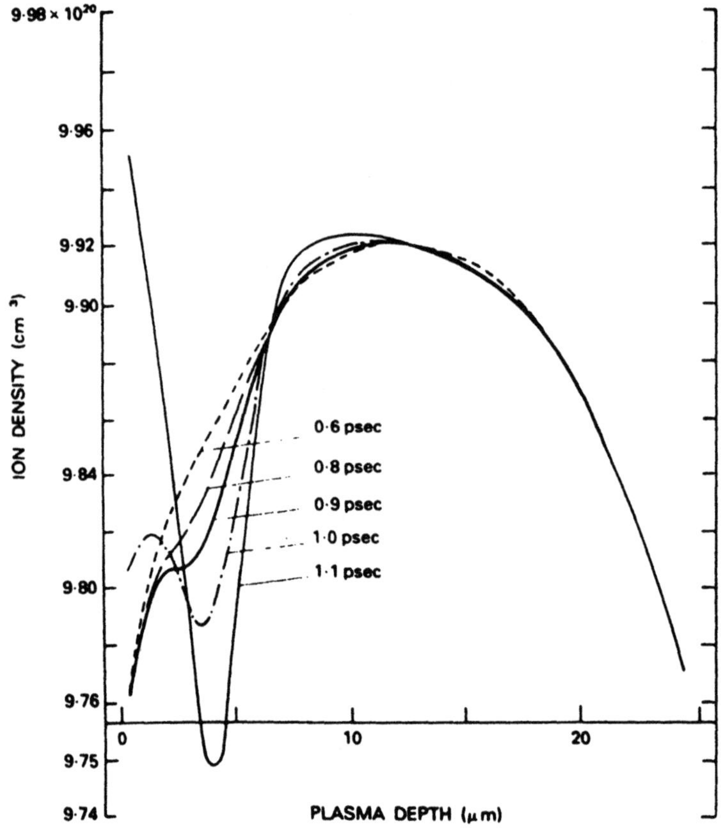

Fig. 10.24. Ion density of a 25 μm-thick hydrogen plasma slab initially at rest and 1-keV temperature irradiated from the left side by a 10^{16} W/cm^2 Nd:glass laser. At t=0.6 ps the density is very similar to its initial value. The energy maximum near x =4 produced a cavitation by nonlinear forces.

Fig. 10.24 for t= 0.5 ps is given. No laser interaction occurs during the first 0.5 ps, so that the minor thermal expansion does not change much of the initial density profiles while this time is long enough to damp down the fast electric oscillations. At t = 0.5 ps, a Nd:glass laser field incident from the left side is switched on with a vacuum amplitude of 10^{16} W/cm^2. The resulting electric field density $E_L{}^2/8\pi$ averaged over a laser period is given in Fig. 10.25, showing an exponential decay for x > 8 μm because of

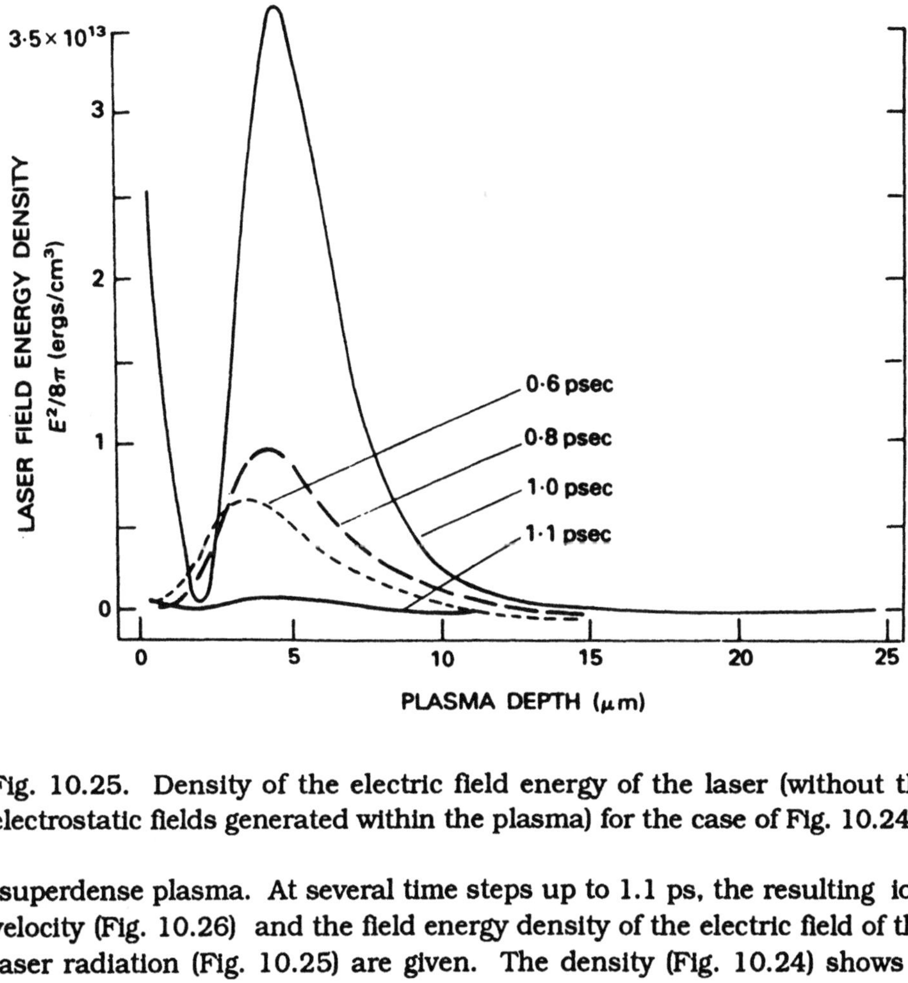

Fig. 10.25. Density of the electric field energy of the laser (without the electrostatic fields generated within the plasma) for the case of Fig. 10.24.

superdense plasma. At several time steps up to 1.1 ps, the resulting ion velocity (Fig. 10.26) and the field energy density of the electric field of the laser radiation (Fig. 10.25) are given. The density (Fig. 10.24) shows a strong minimum caviton at x = 5μm after 1 psec interaction, indicating the predominance of the nonlinear-force-driven ponderomotion. Plasma blocks with ion velocities up to 10^7 cm/s are created in agreement with simplified estimates of the strong acceleration densities.

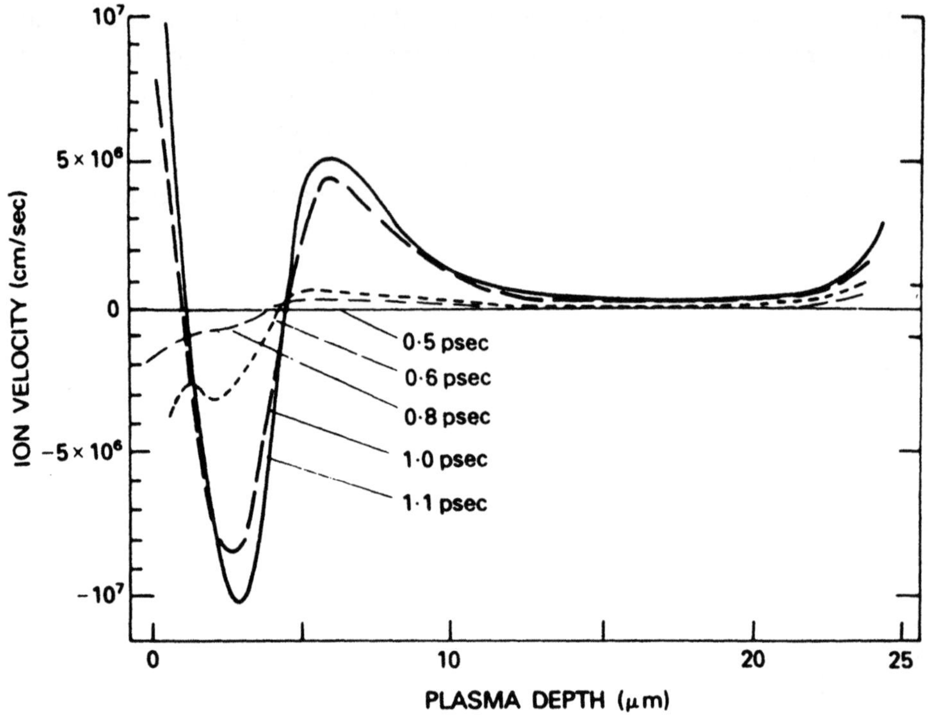

Fig. 11.26. Ion velocity v_o at several time steps for a plasma, as in Fig. 10.24. A block of plasma is generated with a velocity up to 10^7 cm/s.

The resulting differences in the ion and electron densities are given in Fig. 18.27. They cause fast-changing longitudinal electric fields E_s (Fig. 10.28), reaching values beyond 10^8 V/cm. This value corresponds to the expected numbers: The dielectrically swollen laser field E_L in the plasma can be up to 10^{11} V/cm, decaying to zero within 10^{-3} cm.

The numerical result of Fig. 10.28 can explain the observed (Eliezer et al 1983) inverted DLs in laser-produced plasmas if cavitons are produced by the nonlinear forces. The existence of the electric fields in plasma surfaces was shown directly by electron beam probes (Mendel et al 1978) and from electrostatic acceleration of a small number of nonlinear-force-accelerated ions. A more systematic experiment was performed by Eliezer and Ludmirsky (1983) where the temporal dependence of charge of the expanding plasma and the temporal change of the target potential were measured. A very unexpected observation was that the plasma leaving the target was first positively charged and then negatively charged. This was in contrast to the general expectation (see Fig. 2.2) that an electron cloud should first leave the plasma. The picture changes,

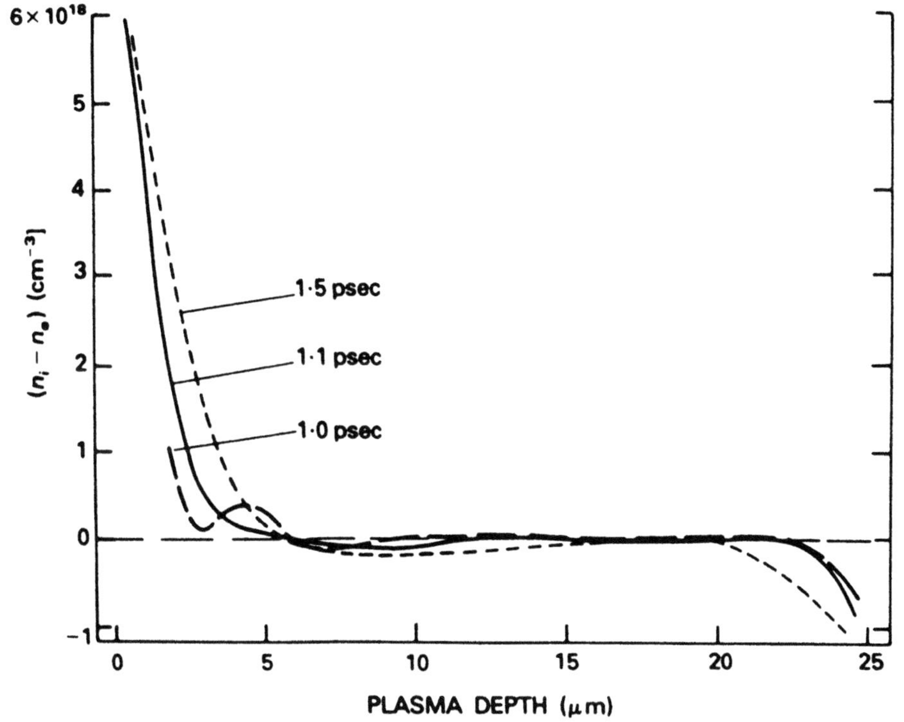

Fig. 10.27. The genuine two-fluid model shows the difference between ion density n_i and electron density n_e with the surprising result of a positive difference (space charge) before the caviton and a negative region behind the caviton (inverted DL as observed by Eliezer and Ludmirsky (1983)). Near x=25μm the laserless plasma expands normally with a negative periphery.

however, if we look at all fields at the surface and in the interior of the plasma in the genuine two-fluid model if a nonlinear-force-driven caviton is generated. Figure 10.27 shows, near x = 25μm where no laser light acts, that a negatively charged plasma expands before the positively charged plasma follows. Near x = 0, a strong positively charged plasma is emitted first, then a negatively charged plasma, and finally a nearly neutral plasma. This is the result of the caviton generation. Though the experiment (Eliezer et al 1983) was on the nanosecond time scale, the comparison with the picosecond processes should be justified not only by the correct polarity of the plasma charges but also from the other experiments that showed the picosecond buildup of the cavitons (Hora,

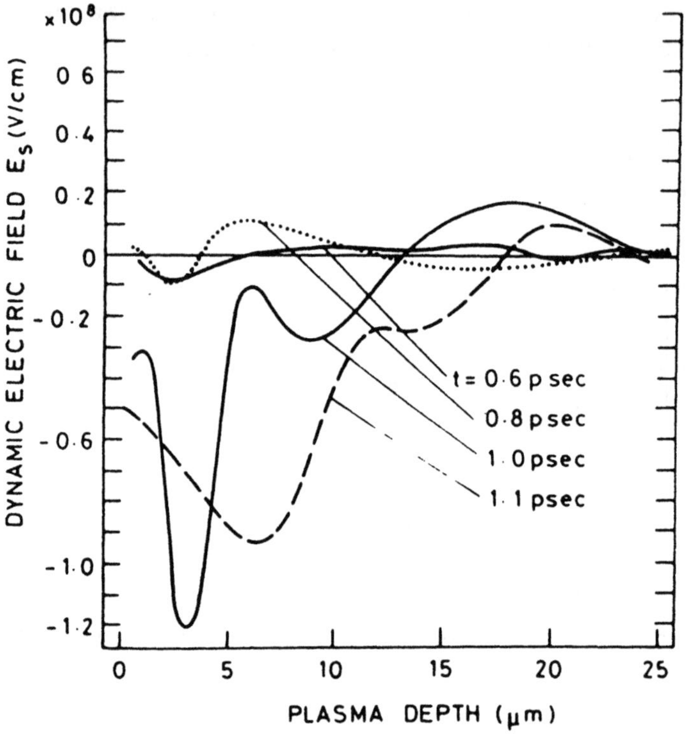

Fig. 10.28. Electric field E_S inside the plasma of Fig. 3, dynamically evolving with absolute values beyond 10^8 V/cm near the caviton produced by the nonlinear laser forces.

1983; Briand et al, 1986). Eliezer and Ludmirsky's (1983) experiment is an additional proof that they also generated cavitons.

Another experiment that can be explained is the energy upshift of alpha particles from laser fusion pellets. It was observed (Gazit et al 1979) that the deuterium-tritium (D-T) alpha particles from laser fusion pellets did not have the expected maximum energy of 3.56 MeV but showed an upshift of $\Delta\varepsilon$ of up to 0.5 MeV. The exact description of the interaction of the alpha particles with the spatially and temporally varying electric field E(x,t) in the (one-dimensional) plasma corona is very complicated as the field is nonconservative. The velocity of the alpha particle v, with an initial velocity v_0 and mass m_α, is given by the complex integral equation:

$$v(x) = v_0 + \frac{2e}{m_\alpha} \int_{t_L}^{t_2} E[e(t)t],dt \; ; \; x = \int v(t)dt \tag{10.31}$$

For a very simplified estimate, we use

$$d\left(\frac{m_\alpha}{2}v^2\right) = 2eE[x,t(x)]dx \ , \qquad (10.32)$$

with an average value $\overline{E}$ of E to give the increase of the alpha energy,

$$\Delta\varepsilon = 2e\,\overline{E}\,\Delta x \, , \qquad (10.33)$$

after acceleration along a length ΔX of the plasma corona. To reach $\Delta\varepsilon$ = 0.5 MeV for Δx = 10μm, we find $\overline{E}$ = 2.7 x 10^8 V/cm. Such fields for Nd:glass laser pulses of 10^{16} W/cm^2 are possible only if the nonlinear-force-produced cavitons (Fig.10.28) are present, since lengths >> 10μm are not realistic. Thermally produced fields of up to 10^6 V/cm could not produce the measured upshifts of 0.5 MeV. Our results, therefore, are not only a rough explanation of the alpha upshift by the large electric fields in the cavitons but also a clear indication that no thermal electric fields can cause the measured upshifts.

We have preliminary results on the exact numerical solution of Eq (10.31) from E values derived from laser-plasma dynamics. It was discovered that broad E maxima move within 0.3 to 0.9 times the speed of light. The correct phasing of the charged particles in the field leads to an acceleration by multiples of the estimate of Eq. (10.33). It can now be shown how the formerly available CO_2 laser (Antares) with 80-TW short laser pulses and a sequence of several pulses can shift electron clouds of giga-electron-volt energy to tera-electron-volt electron energy. The 10^{11} V/cm caviton (nonlinear-force) fields of the type in Fig. 10.28 act like the (nonconservative) pump fields in the microwave cavities of an accelerator. The phasing of the nonlinear-force-field electron accelerator is an extension of the concept proposed by Willis (1977) based on many years of work on the nonlinear force and the results on high electric fields in plasmas.

10.8 Smoothing of Laser-Plasma Interaction

The preceding subsections of this Chapter reported about confirmed properties of laser interaction with plasmas where the experimental results are rather completely understood theoretically. These properties refer to the thermal heating, ablating and compressing of the plasma irradiated at medium laser intensities. The ablation and compression has been established in numerous experiments and follow the description first evaluated numerically from hydrodynamics by Mulser [88] and Rehm

[215]. At higher intensities, the nonlinear force action [138] causes the generation of density minima as first predicted from numerical studies by Shearer, Kidder and Zink [171] and confirmed experimentally by numerous laboratories (see Fig. 10.12 to 10.15 and others).

The more recent numerical studies with the genuine two fluid model (see preceding subsection) confirmed both, the gasdynamic, or thermokinetic ablation and compression process at lower intensities and furthermore the nonlinear force produced caviton generation at higher intensities apart from the understanding of normal and inverted double layers (Hora et al 1984). Furthermore, the double layer processes confirmed the widespread nearly uniform emission of the second harmonics from a laser irradiated plasma corona including the large and the small wave length local periodicity as shown in last subsection of the preceding Chapter.

Apart from these rather settled properties of laser interaction with plasma, there were, however, a very confusing and most complex number of observations which could not be understood so clearly. There was the broad spectrum of the backscattered radiation of the fundamental or the higher harmonics from the plasma. The amount of absorption at increasing intensity and its dependence on the laser wave length (after frequency doubling, tripling, etc) indicated a rather similar behaviour but there was a wide difference of the results from different laboratories and there was agreement that little differences of the laser pulse quality may additionally vary this all. About the desperate alternatives of "indirect drive" for laser fusion to avoid these difficulties we shall report in the last Chapter.

Here we shall report on the direct way out of this problem which was found by smoothing of the laser interaction of the plasma and how this may be understood in view of theoretical explanations. Initially it was assumed that the mentioned anomalies of the interaction are due to parametric instabilities, especially SRS and SBS. After the experiments of Joshi et al (1983), Labaune et al (1985) and Drake (1988) confirmed that the instabilities are well present but not dominating and determining the transfer of the laser energy, the question was what else is then the reason for the complicated interaction.

The understanding of the complexity may start from the numerical result of 1974 as given in Figures 10.10a and 10.10b. When laser light irradiates (from the right hand side) a ramp of plasma with linearly increasing electron density, the light moves - as expected - up to the critical density and is reflected there strongly. The inclusion of collisions and the method of computing Maxwellian exact solutions of the laser field in the plasma corona results then in a partially standing wave field. The early computation of Lindl and Kaw [154], Fig. 10.5, had no absorption and the completely modulated standing wave appeared. The computation of Shearer, Kidder and Zink [171], Fig. 10.7, had no reflection at all and - fortunately - could then follow up the calculation to rather long irradiation times and clearly led to the discovery of the cavitons. In the case of Fig. 10.10, however, the dynamically changing details of the real collisional nonlinear optical constants were included to calculate the Maxwellian solution with the reflected components.

The partially standing waves at the sufficiently high laser intensities (where the thermokinetic forces are not important), cause a pushing of plasma into the nodes of the standing wave. This axial density ripple of the plasma is then a self-produced ideal Bragg grating which causes a high reflection of laser light even at very low plasma densities as seen in Fig. 10.10a at later times. Then, the laser field energy density is decreasing from its value outside in the corona to one hundredth or less long before it reaches the critical electron density. The self-generated Bragg density ripple is seen in the upper part of Fig. 10.10b compared with its initial linear profile. This all happens within two psec.

This was the reason by M. Lubin measured the psec resolution of the reflectivity from a plasma where the irradiated laser power was proved to be rather smooth. The result was sketched in Fig. 1.12 (see also Jackel et al 1976). This result was indeed downcasting for the hope to achieve laser fusion: the plasma does everything to prevent the smooth penetration and absorption of laser radiation by this pulsation in the 10 psec range. A more profound experimental clarification was due to the work of Luther-Davies et al (1987) together with Perry (1986) and Maddever (1988) (Maddever et al 1990). They observed that the backscattered light (fundamental frequency or second harmonics) from a laser irradiated plane target detected within a very small aperture of the end of a bundle of fiber optics, had a broad spectrum which had a modulation of about 4 Angstroms. This modulation was washed out when collecting the wide angle back scattered light as done usually. The spectral modulation was present even at nearly 90° scattering, was not dependent on the ion mass and was stochastic with respect to the periodicity.

The clarification was rather unexpected: it turned out that the laser light is accelerating the plasma during few psec and stops then the interaction such that a plasma cloud of a velocity of about 10^7 cm/sec is reached. After about 10 psec the interaction begins again and pushed the plasma by a further additional velocity of about the same number. The broad spectrum of the back scattered light is then simply a sequence of Doppler shifted light with a rather irregular pulsation in the range of 10 to 30 psec. Such a pulsation was seen also in the reflected light changing during this time up and down from few to nearly 100% similar to Lubin's early result, Fig. 1.12. Recent experiments with the K_α radiation showed a similar

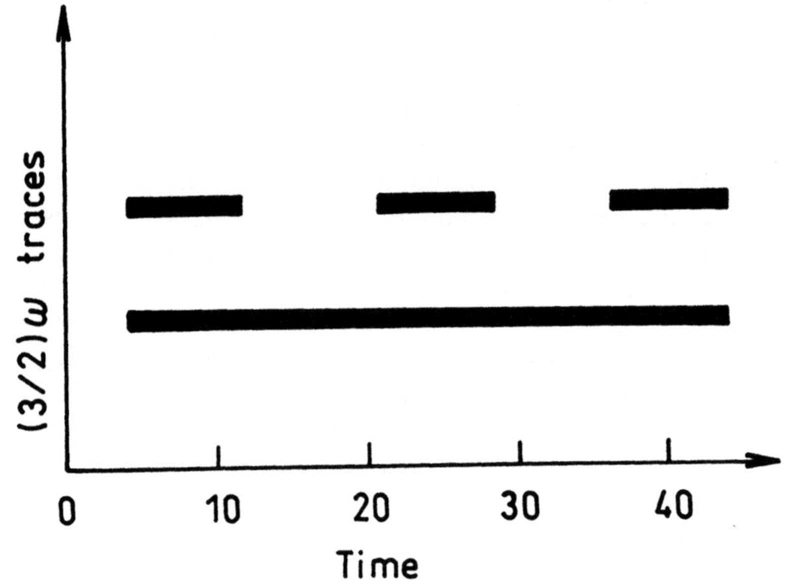

Fig. 12.29. Time dependence of the 3/2 harmonics emission from a laser irradiated plasma according to Giulietti et al (1989), upper trace, if no random phase plate (Kato et al 1984) is put into the laser beam. If such a plate is used, the smooth lower trace has been observed.

pulsation (Rode et al 1990), and it was observed by Giulietti et al (1988), Fig. 10.29, that the SRS produced 3/2 harmonics emitted from the laser irradiated target has an irregular pulsation of about 20 psec duration.

It was possible to clarify experimentally (Maddever et al 1990) that the laser light is alternately reflected at the critical density showing low reflectivity, while the other reflection after about 10 psec each occurs in the outermost low electron density periphery of the plasma corona when the reflectivity is high. An understanding of this was possible by applying the hydrodynamic computation using the genuine two fluid model (Gu Min et al 1989; Aydin et al 1990). The question was what happens if the computations of the kind of Fig. 10.10 would be continued to longer times

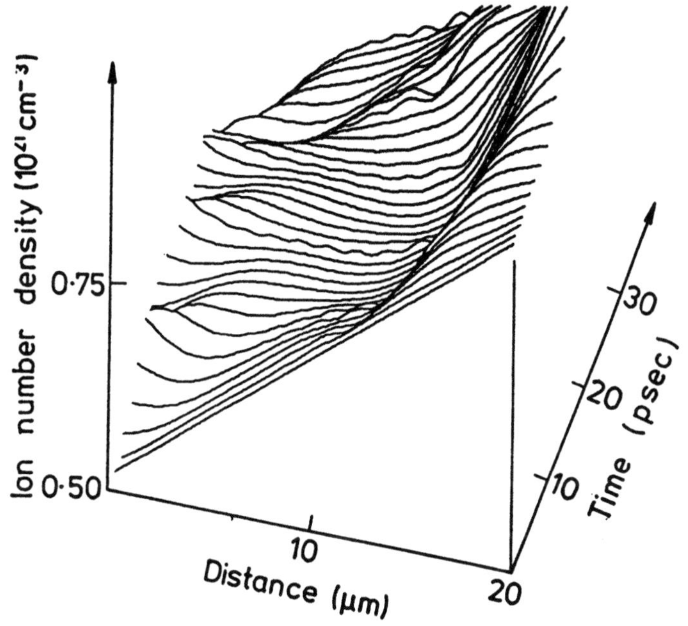

Fig. 10.30. Time dependence of the ion density profile of an initial ramp when irradiated by neodymium glass laser radiation of 10^{15} W/cm^2 from the left hand side with a pulsating generation and relaxation of density ripples. At about 5, then at 25 and 40 psec (Aydin et al (1990).

of laser interaction. If the laser light does not longer penetrate the corona after generating the Bragg grating of density ripple, the untouched part of the corona should then have a decay of the ripple by hydrodynamic motion.

Exactly this happens as shown in Fig. 10.30 to 10.32. Neodymium glass laser light is incident from the left hand side onto a 20 μm deep deuterium plasma of initially linear profile of electron and ion density growing from 0.5 to 1.2 of the critical density, initial temperatures 30 eV and zero initial velocity. Using the genuine two fluid model, the time evolution of the ion density at irradiation of 10^{15} W/cm^2 laser intensity is shown in Fig. 10.30, the ion velocity in Fig. 10.31 and the electromagnetic energy density in Fig. 10.32. It is clearly seen that the laser light penetrates to the critical density during the first 7 psec (strong maxima in Fig. 10.32) until the first density ripples (see Fig. 10.30 at 7 psec) have been produced. The laser field is then decaying and not penetrating the plasma corona until at about 18 psec. Light penetrated to the critical density again, causes ripples until 25 psec and a reduced laser intensity is in the corona. At 34 psec, light penetrates again to the critical density.

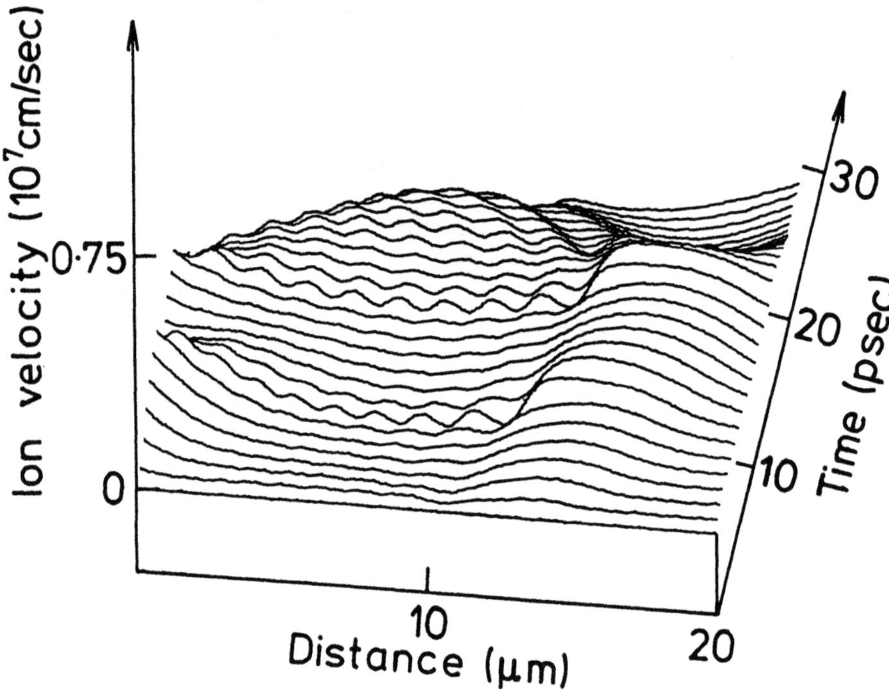

Fig. 10.31. Time dependence of the ion velocity for the case as in Fig. 10.31 with rippling and relaxation.

The ion velocity shows similar behaviour of rippling and relaxation, Fig. 10.31, for the corresponding time steps. It is remarkable that the bulk velocity goes to the value of about 10^7 cm/sec after the first driving of the plasma corona during the first 6 psec, and remains then nearly constant until the second push by the laser light occurs adding similar velocity to the ions etc.

We see that the pulsating acceleration with a sequence of Doppler shift of the reflected light occurs as observed (Maddever et al 1990) and that the reflectivity from low plasma density is reproduced as measured. A more detailed evaluation of the reflectivity with a rather irregular pulsating behaviour was shown (Gu Min et al 1989).

Independently from all these experimental and theoretical developments, methods for a smooth interaction of laser radiation with plasmas have been tried by appropriate suggestion. One of these is the

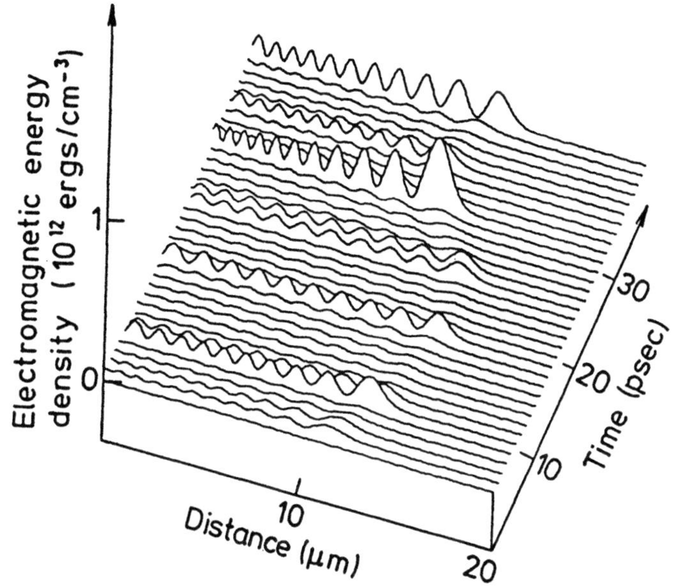

Fig. 10.32. Time dependence of the electromagnetic energy density of the laser field as for the case of Fig. 10.30 showing times of strong penetration of the light to the critical density (strong maxima) and those with low penetration.

method of the induced spatial incoherence, ISI, (Lehmberg et al 1983, Obenschain et al 1986) where the coherence time of the laser light is about 2 psec. It is evident immediately from our hydrodynamic computation, (Figures 10.30 to 10.32, that any standing wave profile is then changing after 2 psec and no density ripple can be produced. Laser irradiation is smooth with low reflectivity and high energy deposition in the plasma. Instabilities (SRS or SBS) are reduced by a factor 100(!) (Mistovich et al 1987; Obenschain et al 1989).

The action of smoothing by the random phase plate RPP, introduced by Kato et al (1984) consists in the generation of beamlets within the laser beam of which the phase is shifted forward and backward randomly. If a standing wave pattern is produced in one beamlet, the next beamlet has the standing wave pattern shifted. Any generation of a density ripple in one beamlet would then be laterally washed out hydrodynamically by the neighbour beamlet structure following our hydrodynamic result (Figures

10.30 to 10.32) or due to the superposition of the diffraction spread fields of the beamlets. Again a good smoothing occurs and the achieved high deposition of laser energy into the irradiated plasma was one of the reasons for the famous production of high gain fusion neutron yields at direct laser drive (Yamanaka and Nakai 1986).

Alternatively, smoothing was intended to be achieved by broad band laser irradiation or the "fly eye" optics (Deng Ximing et al 1984). A combination of this led to a very successful smoothing by spectral dispersion (SSD) (McCrory et al, 1990). The action of a broad spectrum of laser irradiation is immediately understandable from our model as this broad spectrum may not permit the standing wave patterns and the generation of density ripples as discussed before.

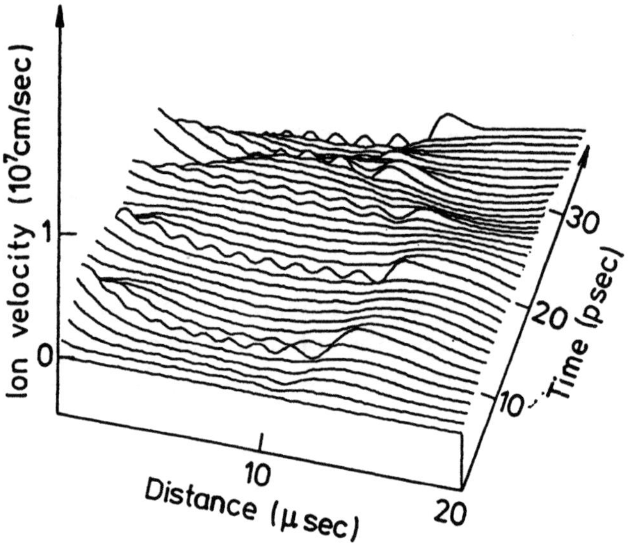

Fig. 10.33. Ion velocity profiles similar to the case of Fig. 10.30, only for 0.6 times of the laser intensity when irradiated by one single laser frequency with a pulsating generation of ripples.

In order to prove this (Aydin et al 1990), a computation of the model similar to the cases of Fig. 10.30 was performed for a neodymium glass laser intensity of 6×10^{14} W/cm^2 and the same parameters as in the case

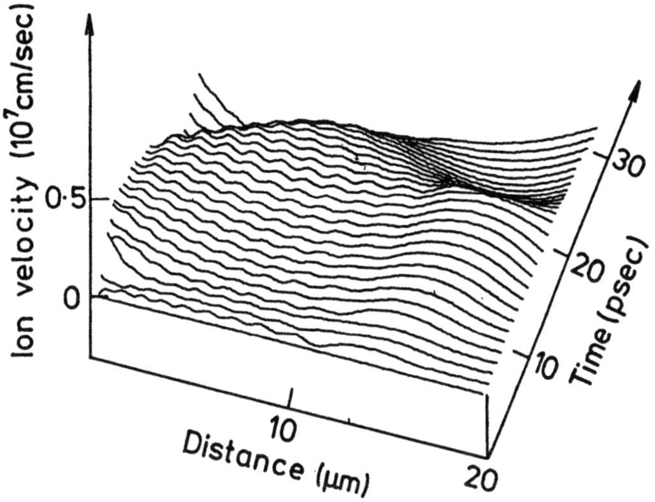

Fig. 10.34. Same as Fig. 10.33 but with irradiation of a spectrum of 1% spectral width represented by three laser frequencies 0.5% spectrally apart.

of Fig. 10.30. If a single laser frequency was used, the velocity profile had a time dependence as shown in Fig. 10.33 with ripples at 8 and 15 psec (synchroneous with ripples and pulsation in the density or the field properties). If a set of three laser frequencies each 0.5% different were incident, the velocity profiles showed the result of Fig. 10.34, where the pulsation was suppressed nearly completely.

These calculations were all done for a one dimensional plane wave geometry. In reality the laser beam is of finite size, and divergent, and the plasma is in three dimensions. One extension of the plane wave geometry to beams may be justified by considering Sigel's question mark experiment (Eidmann et al 1973). Putting a question mark into the laser beam and looking for the picture produced by the reflected light after irradiating a plane target, it was found that the question mark was not turned around as expected from the reflection from an ideal mirror, but it appeared unchanged. This corresponds just to the reflection of the kind of a Bragg grating where the beamlets are reflected into themselves in full agreement with our (Aydin et al 1990) described density ripple result.

One modification may be the following suggestion: The Bragg reflection with high reflectivity from the very low plasma density occurred only for some time. It may be asked whether the weak reflection for the short times occurring from the critical density are not having a turning of the question mark. Therefore an experiment with a time resolved question mark may have a pulsating up and down, or in time integrated experiments there may be a further weak picture of the question mark the other way round than reported (Eidmann et al 1973).

ELEVEN

Striated Motion and Resonance Absorption

Sections 9 and 10 discuss the nonlinear forces, the photon momentum, the created ion energies, and several fully dynamic, numerical treatments of the plasma expansion following laser interaction, including the nonlinear forces and optical properties. Plane waves from a laser are assumed to be perpendicularly incident onto a stratified (inhomogeneous) plasma. This section discusses the case of plane waves obliquely incident onto a stratified plasma.

There are two questions of interest. While the net motion of the plasma corona under laser irradiation is found to be directed toward the decreasing density (parallel to the negative density gradient) and no net motion can happen in the plane of the plasma surface because of conservation of momentum, the standing wave field in the plasma corona can produce a straited motion. Parts of the plasma can move toward the one direction and other parts to the contrary direction with vanishing net motion. The general formulas for these forces from the general equation of motion are valid for all kind of wave fields, even for the evanescent wave field beyond the turning point of the wave in the corona. But the evaluation of the wave fields is restricted to the WKB approximation only for an angle of wave propagation of less than 50° [256].

The other question is the evaluation of the classical and linear solution of the wave field for the densities below the turning point of the wave. It has been found by Denisov [248], see also Ginzburg [257], that for *p*-polarization the component of the electric field perpendicular to the plane of the plasma has an extraordinarily high maximum at the cutoff density [258]. This is far below the turning point at oblique incidence, inside the evanescent region of the wave field. The reason is the resonance with the electrostatic waves. This process has been called resonance absorption [257]. Thus the process of aborption has to be calculated separately.

11.1 Striated Motion

Using the x-direction along the depth of the stratified plasma and using linear polarized radiation, the case of the p-polarization of an obliquely incident wave is then given by field components E_x, E_y, and H_z only. From the equation of motion in the formulation of Eqs. (8.81) or (8.82); the y-component of the nonlinear force is then [256]

$$f_{\mathrm{NL}py}=\frac{1}{8\pi}\frac{\partial}{\partial x}[E_xE_y-(1-n^2)E_xE_y]+\frac{1}{16\pi}\frac{\partial}{\partial y}[-E_x^2+(2n^2+1)E_y^2-H_z^2) \tag{11.1}$$

For the case of the s-polarization (**E** perpendicular to the plane of incidence) the wave field has the components E_z, H_x, and H_y only. The resulting component of the nonlinear force in the y-direction is then

$$f_{\mathrm{NL}sy}=\frac{1}{8\pi}\frac{\partial}{\partial x}H_xH_y+\frac{1}{16\pi}\frac{\partial}{\partial y}(-E_z^2-H_x^2+H_y^2) \tag{11.2}$$

There will be no time-averaged forces in the z-direction. The x-component of the general forces for oblique incidence are discussed in Section 8.

The solution for **E** and **H** for stationary solutions with a frequency according to Eq. (6.19) is given from Eq. (6.30)

$$\Delta\mathbf{E}+\frac{\omega^2}{c^2}n^2\mathbf{E}-\nabla\frac{2}{n}\mathbf{E}\cdot\nabla n=0 \tag{11.3}$$

The magnetic field can be derived from the Maxwellian equation

$$\dot{\mathbf{H}}=c\nabla\times\mathbf{E} \tag{11.4}$$

It is evident, for the case of s-polarization, that the last term in Eq. (11.3) vanishes, as there is a z-component of **E** only and the refractive index n is dependent on the x-coordinate only. In this case the solutions of the WKB case, Eqs. (7.28) and (7.30), can be used in Eq. (11.3). A vanishing time-averaged y-component of the nonlinear force follows, according to Eq. (11.2).

Any force can result for the p-polarization only $\mathbf{f}_{\mathrm{NL}}=0$. For this case, the last term of Eq. (11.3) has to be taken into account. This term couples the transversal components of the electromagnetic wave with the longitudinal components generated in the plasma when turning the direction of the obliquely incident wave. As this coupling is not necessary for the discussion of the net force component in the x-direction (to lower plasma density), this part is neglected in Sections 7 and 8. Now this coupling is included, and the WKB approximation of the wave field in the corona of the plasma is derived for p-polarization. The geometry is described in Fig. 11.1, which shows the resulting nodes of the standing wave and the finally resulting forces for the striated motion.

From Eq. (11.3) for p-polarization—the index p will be neglected within this section—the following equations are found

$$\left(\frac{\partial^2}{\partial x^2}+\frac{\partial^2}{\partial y^2}+\frac{\omega^2}{c^2}n^2(x)\right)E_x+\frac{\partial}{\partial x}\frac{2E_x}{n}\frac{\partial n}{\partial x}=0 \tag{11.5}$$

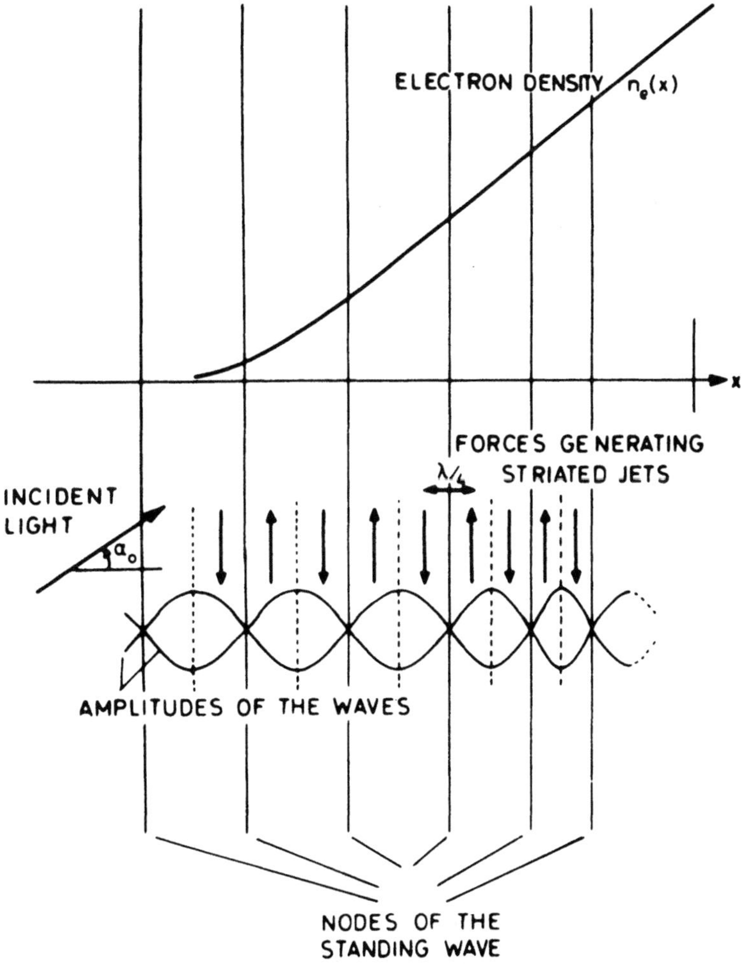

Figure 11.1 A plane p-polarized wave incident at an angle α_0 on a stratified plasma with an electron density $n_e(x)$ together with the corresponding reflected wave generates spatially averaged forces and a net motion along the fixed nodes of the standing wave (striated jets) [256].

$$\left(\frac{\partial^2}{\partial x^2}+\frac{\partial^2}{\partial y^2}+\frac{\omega^2}{c^2}n(x)\right)E_y+\frac{\partial}{\partial y}\frac{2E_x}{n}\frac{\partial n}{\partial x}=0 \tag{11.6}$$

These equations can be separated by the ansatz

$$E_x=e_{xx}(x)e_{xy}(y); \qquad E_y=e_{yx}(x)e_{yy}(y) \tag{11.7}$$

The appropriate boundary conditions for plane waves incident at an angle α_0 in vacuum (Fig. 11.1) leads to Snell's law

$$\sin \alpha_0=n(x) \sin \alpha(x) \tag{11.8}$$

From Eq. (11.5) is obtained

$$\frac{\partial^2}{\partial y^2}e_{xy}+\frac{\omega^2}{c^2}(\sin^2 \alpha_0)e_{xy}=0 \tag{11.9}$$

$$\frac{\partial^2}{\partial x^2}e_{xx}+\frac{\omega^2}{c^2}n^2(x)\cos^2\alpha(x)e_{xx}+2\frac{\partial}{\partial x}\frac{d\ln n}{dx}e_{xx}=0 \tag{11.10}$$

For the separation of Eq. (11.6), the relation

$$e_{xx}=e_{yx}\frac{\sin\alpha(x)}{\cos\alpha(x)} \tag{11.11}$$

has to be, which is used in agreement with the case of homogeneous plasma, and which leads to

$$\frac{\partial^2}{\partial x^2}e_{yx}+\frac{\omega^2}{c^2}n^2[\cos^2\alpha(x)]e_{yx}=0 \tag{11.12}$$

$$\frac{\partial^2}{\partial y^2}e_{yy}+\frac{\omega^2}{c^2}(\sin^2\alpha_0)e_{yy}+2\tan\alpha(x)\frac{d\ln n}{dx}\frac{\partial}{\partial y}e_{xy}=0 \tag{11.13}$$

A necessary condition for this separation is

$$\tan\alpha(x)\frac{\partial \ln n(x)}{\partial x}=\text{const} \tag{11.14}$$

which is immediately given by Snell's law, Eq. (11.8).

Using the integration constants $C_n(n=1, 2, \ldots)$, the exact solution of Eq. (11.9) is

$$e_{xy}=C_1\cos[y(\omega/c)\sin\alpha_0]+C_2\sin[y(\omega/c)\sin\alpha_0] \tag{11.15}$$

The solution of Eq. (11.10) is given for the assumptions of the WKB approximation and by neglecting the last term of Eq. (11.10), which is justified at this point because it does not produce forces in the y-direction, but it must not be neglected at the other points of this derivation.

$$e_{xx}=[n(x)\cos\alpha(x)]^{-1/2}[C_3\cos(G_x)+C_4\cos(G_x)] \tag{11.16}$$

where

$$G_x = -\frac{\omega}{c}\int^x n(\xi)\cos\alpha(\xi)\,d\xi \tag{11.17}$$

The WKB solution of Eq. (11.12) is

$$e_{yx} = [n(x)\cos\alpha(x)]^{-1/2}[C_5\cos(G_x) + C_6\sin(G_x)] \tag{11.18}$$

In Eq. (11.13) the solution (11.15) has to be substituted. The exact solution is obtained [202, 256]:

$$\begin{aligned} e_{yy} = {} & C_7 \sin[y(\omega/c)\sin\alpha_0] + C_8\cos[y(\omega/c)\sin\alpha_0] \\ & + 2\tan\alpha(x)\frac{c\ln n(x)}{dx}(\omega/c)(\sin\alpha_0)(y - y_0) \\ & \times\left[C_1\cos\left(y\frac{\omega}{c}\sin\alpha_0\right) - C_2\sin\left(y\frac{\omega}{c}\sin\alpha_0\right)\right] \end{aligned} \tag{11.19}$$

The constant y_0 is coupled with the integration constants.

We try to fit the solution in the inhomogeneous plasma with the propagating wave in the vacuum at an angle of incidence α_0 (Fig. 11.1) and an amplitude E_v and a magnetic vector

$$\mathbf{H} = \mathbf{i}_3 E_v \cos[G_{yt} - (\omega/c)y\sin\alpha_0] \tag{11.20}$$

where

$$G_{yt} = -(\omega/c)y\sin\alpha_0 + \omega t \tag{11.21}$$

One finds the conditions for the integration constants with a propagating wave at an angle of incidence α_0 in Eqs. (11.13) to (11.16). With Eq. (11.5) is obtained

$$E_x = \frac{E_v\sin\alpha_0\cos^{1/2}\alpha_0}{(n\cos)^{1/2}}(\cos G_x\cos G_{yt} - \sin G_x\sin G_{yt}) \tag{11.22}$$

$$\begin{aligned} E_y = {} & \frac{E_v\cos\alpha_0\cos^{1/2}\alpha_0}{(n\cos)^{1/2}} \\ & \left[\cos(G_x + G_{yt}) + 2\tan(x)\frac{d\ln n}{dx}\frac{\omega}{c}(y - y_0)(\sin\alpha_0)(\cos G_x\cos G_{yt}\right. \\ & \left. + \sin G_x\cos G_{Nt})\right] \end{aligned} \tag{11.23}$$

The difference in the solutions of E_x and E_y from the case in vacuum is the spatial variation of the amplitude depending on x according to the WKB condition, and further the term with the factor $d\ln n/dx$, which determines the longitudinal component of the electric vector with respect to the actual direction of propagation of the wave. This expression, called the "longitudinal term," increases linearly on y, where y_0 is constant. Although an infinite plane wave is assumed, the real conditions are determined by a wave bundle of finite width. To describe exactly this complex problem with a wave bundle in the optical case (in contrast to the wave mechanical case [257]), the plane wave must be cut laterally to form a bundle; the side of lower y can be identified by y_0, so the longitudinal component increases monotonically from one side. To facilitate the discussion one can restrict this treatment to the condition where the longitudinal term is much smaller than the remaining term in the bracket of Eq. (11.23): that is,

$$\tan(x)\frac{d\ln n}{dx}\frac{\omega}{c}(y-y_0)\sin\alpha_0=p_0\leqslant 0.30 \tag{11.24}$$

With this restriction, the magnetic field in the plasma can be approximated by

$$\mathbf{H}=\mathbf{i}_3\frac{E_v(n\cos\alpha_0)^{1/2}}{\cos^{1/2}\alpha}(G_x+G_{yt})-\mathbf{i}_3\frac{c}{\omega}\frac{E_v\cos^{1/2}\alpha_0(0.5-\sin^2\alpha)}{n^{3/2}\cos^{3/2}\alpha}\frac{\partial n}{\partial x}\sin(G_x+G_{yt}) \tag{11.25}$$

With this result for a plane wave propagating at an angle of incidence α_0 (Fig. 11.1) we can derive the field for a standing wave by adding a wave with an angle of propagation resulting from total reflection of the first wave. The complications due to the Goos–Haenchen effect are excluded from the discussion here and may not be of significant influence [257].

The reflected wave results in expressions similar to those given in Eqs. (11.23), and (11.25), where the sign of G_x and of $\mathbf{H}$ have to be changed. Adding the solutions for the incident and reflected wave, the standing wave for the polarization of $\mathbf{E}$ parallel to the plane of incidence (p-polarization) results in

$$\begin{aligned}\mathbf{E}=&\frac{2E_v\cos^{1/2}\alpha_0}{(n\cos\alpha)^{1/2}}\cos\left(-\frac{\omega}{c}\int_0^x n(\xi)\cos\alpha(\xi)\,d\xi\right)\\ &\times\cos\left(-\frac{\omega}{c}y\sin\alpha_0+\omega t\right)\Bigg[\mathbf{i}_7\sin\alpha_0+\mathbf{i}_2\cos\alpha_0\\ &\times\left(1+2\tan\alpha(x)\frac{d\ln n}{dx}\frac{\omega}{c}(y-y_0)(\sin\alpha_0)\right)\Bigg]\end{aligned} \tag{11.26}$$

$$\begin{aligned}\mathbf{H}=\mathbf{i}_3&\frac{2E_v(n\cos\alpha_0)^{1/2}}{\cos^{1/2}\alpha}\sin\left(-\frac{\omega}{c}\int^x n(\xi)\cos\alpha(\xi)\,d\xi\right)\\ &\times\cos\left(-\frac{\omega}{c}y\sin\alpha_0 t+\omega t\right)\left(1-\frac{dn}{dx}\frac{\omega}{c}\frac{E_v\cos^{1/2}\alpha_0(0.5-\sin^2\alpha)}{n^{3/2}\cos^{3/2}\alpha}\right)\end{aligned} \tag{11.27}$$

The evaluation of the nonlinear force in the y-direction follows Eq. (11.1), where terms averaged by the y-coordinate are considered. This averaged value of the second term on the right-hand side of Eq. (11.1) vanishes, because the y-dependence of $\mathbf{H}^2$, E_x^2, and E_y^2 is given by terms containing only $\sin^2[y(\omega/c)\sin\alpha_0]$, $\cos^2[y(\omega/c)\sin\alpha_0]$, or $(y-y_0)\sin[y(2\omega/c)\sin\alpha_0]$ as factors. The derivation by y results in spatially averaged vanishing values. This vanishing does not hold if condition (11.24) is not fulfilled. From the first term on the right-hand side of Eq. (11.1) and Eqs. (11.22) and (11.23) is found, after time averaging

$$f_{NLpy} = \frac{1}{8\pi}\frac{\partial}{\partial x}\frac{E_v^2\omega\sin^3\alpha_0\cos^2\alpha_0}{2cn^2\cos^2}\sin\left(2\frac{\omega}{c}\int^x n(\xi)\cos\alpha(\xi)\right)\frac{\partial\ln n}{\partial x}(y-y_0) \tag{11.28}$$

Differentiation and neglect of second and higher order terms in $\partial n/\partial x$ leads to

$$f_{NLsy} = \frac{1}{8\pi}\frac{E_v^2\sin^3\alpha_0\cos^2\alpha_0\omega^2}{n\cos\alpha}\frac{\omega^2}{c^2}(y-y_0)\frac{\partial\ln n}{\partial x}$$
$$\times\cos\left(2\frac{\omega}{c}\int^x n(\xi)\cos(\xi)\,d\xi\right) \tag{11.29}$$

At a constant x-value, the force increases along the y-coordinate and is proportional to the gradient of the refractive index n. An increase in the force occurs especially at densities close to the cutoff density n_{ec}. The validity of this condition can be assumed within the depth given by a large number of vacuum wavelengths $\lambda_0 = 2\pi c/\omega$ as a special calculation demonstrates [223]. Also at very small dn/dx, the force can be large owing to sufficiently high y. The nonlinear force component f_{NLpy} oscillates in time and is comparable to the field of a standing wave (Fig. 11.1). The force is zero at the nodes and antinodes of the standing wave, which have a distance of a quarter of the actual wavelength,

$$\lambda^* = \lambda_0\cos(x)/n \tag{11.30}$$

Between the nodes, the force is directed toward the positive or negative y-direction for $y > 0$ and $y_0 < 0$. The maximum (or minimum) value of the force is, using $p_0 = 0.3$ from Eq. (11.24) in cgs units,

$$f_{NLpy} = f_0 = 0.075(E_v^2/\lambda)|n|\sin\alpha_0\cos^2\alpha_0 \tag{11.31}$$

The resulting force density can reach remarkably high values. The restriction to $\alpha < 40°$ was only necessary [138] in decoupling the waves of linear polarization. In the following only linear p-polarization is considered. For a reasonable example of an intensity $I_v = 10^{14}$ W/cm^2 of neodymium laser irradiation, with $\alpha = 25°$ and a value of the refractive index $|n| = 0.9$ is found

$$f_0 = 2.25\times10^{14}\ \text{dyn/cm}^3 \tag{11.32}$$

The thermokinetic force of a deuterium plasma with a temperature

$T_e=100$ eV and a density $n_e=1.9\times10^{20}$ cm^{-3}, dropping linearly to vacuum within 10^{-2} cm, results in

$$f_{th}=3.2\times10^{12}\ \text{dyn/cm}^3 \tag{11.33}$$

The essential term in the nonlinear force in the y-direction of a propagating wave in an inhomogeneous plasma is the nonlinear term. Without it, the spatially averaged force vanishes. The case of p-polarization is determined by Eqs. (11.3) and (11.4), where the last term in Eq. (11.3) vanishes. Therefore, no longitudinal component is created for the conditions used here in the case of s-polarization. The y-component of the nonlinear force density, Eq. (11.2), results, as mentioned, in a vanishing spatially averaged value similar to that for p-polarization without the longitudinal term.

Several conditions have to be discussed in connection with striated motion in the plasma due to the force given by Eq. (11.1). Besides the questions of how the standing wave is created and why the gasdynamic motion of the plasma through the standing wave pattern is slow enough to build up a striated motion, the *mean free path* of the particles with respect to the thickness of the layers must be considered, and, furthermore, the limitation between a *laminar motion of the layers and a turbulent motion* must be determined.

The thickness of one layer is one-quarter of the vacuum wavelength multiplied by $n^{-1}(x)\cos\alpha(x)$. Because we consider conditions of $S=1/n<10$ and $\alpha_0<45°$, we can begin with neodymium glass laser radiation from a thickness $d>2.3\times10^{-5}$ cm. The mean free path for ions in fully ionized deuterium is

$$\bar{l}=v_i/\nu_{ii} \tag{11.34}$$

In a partially ionized plasma, the mean free path is less because of collisions with neutrals. For our purposes, l of Eq. (11.34) is an upper bound, where v is the thermal velocity of the ions $v_i=(2kT_i/m_i)^{1/2}$, while T_i is the ion temperature. The ion collison frequency $\nu_{ii}=\nu_{ee}(m_e/m_i)^{1/2}$ (with the electron mass m_e), which is used from Spitzer's derivation, Eq. (2.37), results in

$$\nu_{ii}=9\times10^{-9}n_i\ln\Lambda/T_i^{3/2} \tag{11.35}$$

where n_i is in cm^3 and T_i is in eV. The Coulomb logarithm ln Λ is simply set equal to 10. To get a mean free path l smaller than d, the following condition is obtained

$$n_i\geqslant n_{ic}=4.4\times10^{17}T_i^2 \tag{11.36}$$

Figure (11.2) contains a plot of the critical ion density n_{ic} as function of temperature T_i. Densities lower than n_c are excluded in the following considerations.

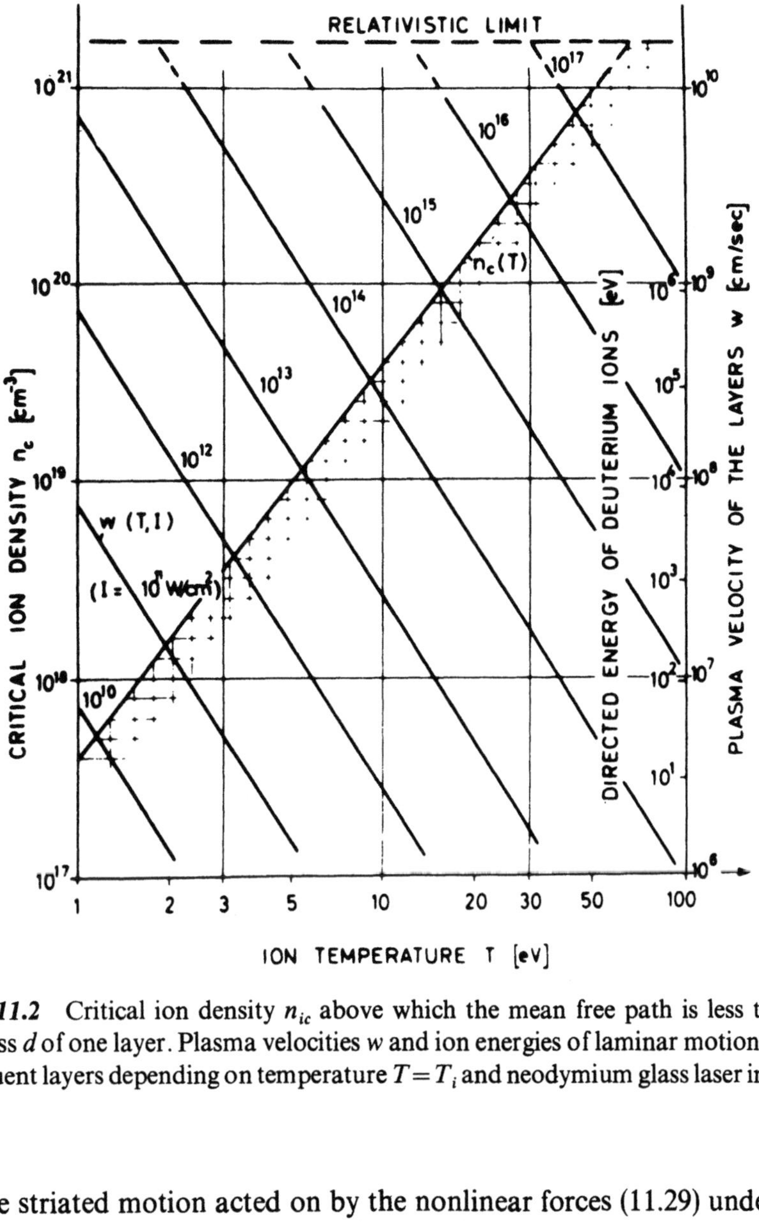

Figure 11.2 Critical ion density n_{ic} above which the mean free path is less than the thickness d of one layer. Plasma velocities w and ion energies of laminar motion for two subsequent layers depending on temperature $T=T_i$ and neodymium glass laser intenities I [256].

The striated motion acted on by the nonlinear forces (11.29) under laminar conditions is calculated. At a distance d,

$$d=|n|(\lambda_0/4)\cos\alpha_0 \tag{11.37}$$

the stationary velocity $\bar{w}$ due to a force F at a viscosity η is

$$\bar{w}=2Fd/\eta \tag{11.38}$$

The factor 2 takes into account that at both sides of d, forces F act in opposite directions. F is the force per cm^2 of the layer of laminar motion, and it is given from the force density f by

$$F = fd \tag{11.39}$$

Using the viscosity η of the liquid of the ions with an ion temperature T_i

$$\eta = \frac{n_i K T_i}{\nu_{ii}} \tag{11.40}$$

From Eqs. (11.37) to (11.40) and Eq. (11.25) is found

$$w = \frac{\lambda_0}{107|\eta|} E_v^2 \sin\alpha_0 \cos^4\alpha_0 \frac{\nu_{ii}}{nkT} \tag{11.41}$$

E_v is in V/cm, T_i is in eV, and the wavelength λ_0 is in cm. The velocity of the laminar motion does not depend on the density, as is well known from the kinetic theory of gases. Figure 11.2 contains the plots of the difference of the velocity w of deuterium between one maximum and one minimum of the forces at the maximum value $u_0 = 25.5°$, for a reasonable $|n| = 0.25$ and $\ln\Lambda = 10$ for various intensities of neodymium glass laser radiation. The corresponding ion energies are also given by the ordinate on the right-hand side of the diagram.

The striated motion of the plasma is calculated for laminar conditions. This is limited by the critical Reynolds number Re_{kr}, above which the motion is turbulent. In the case of turbulence, the final velocities of the layers will be much higher under stationary conditions than in the laminar case. The turbulent state is of very complex nature, and the straited motion will then be disturbed by Helmholtz–Kelvin instabilities. Anyhow, the larger velocities of the layers and the stronger, subsequent thermalization has to be taken into account for the processes in the corona with maximum effects around 25° incidence for p-polarization.

The following evaluation, for the restricted case of laminar motion, should give some knowledge of what processes of straited motion can occur in the corona. The necessary restriction to low plasma temperatures is only a limitation for this case. Under real conditions, the striated motion will then happen at higher temperatures with higher velocities in the regime of turbulent motion. The critical Reynolds number has a value

$$\mathrm{Re}_{kr} = \frac{\bar{w} d n_i m_i}{\eta} \tag{11.42}$$

between 10^3 and 10^6. The higher limit may be more realistic because of the fact that the motion of the layers is not disturbed by surfaces or corners of solid boundaries. Using Eqs. (11.37) to (11.41) for deuterium is found

$$\mathrm{Re}_{kr}=2.5\times 10^{-3}\frac{\lambda_0^2}{kT^5}(\ln\Lambda)^2 E_v^2 n_k m_k$$
$$=8.7\times 10^{-30}\frac{E_v^2 n_i}{T^5} \tag{11.43}$$

E_v is in V/cm, T in eV, and the ion density n_i is in cm^3. Using a reasonable Reynolds number of 3×10^5 (about 100 times its minimum value), one arrives at maximum ion densities n_i in Fig. 11.3 at maximum laser intensities at given temperatures where laminar conditions are fulfilled.

From the results of Figs. (11.2.) and (11.3), the conditions for the case are selected, where an ion velocity in the various layers of the laminar striated motion representing 1 keV ion energies is reached. Figure 11.4 then gives a synopsis; the necessary laser intensity I_1 is a well-defined function of the plasma temperature. The density n_i of the ions first must be larger than the

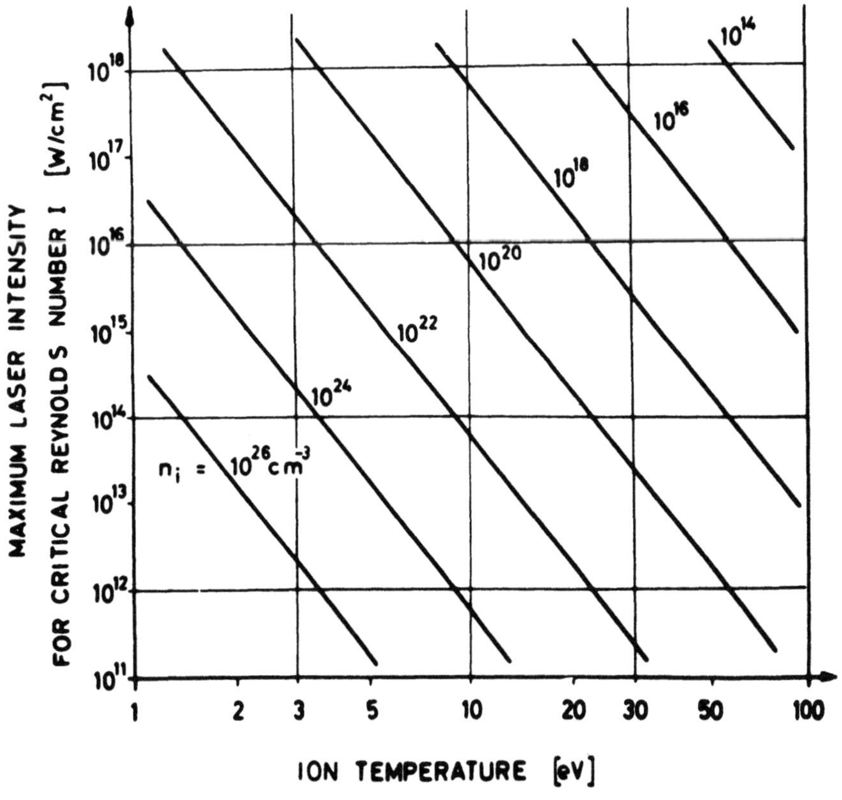

Figure 11.3 Maximum neodymium glass laser intensities at miximum ion densities n_i for laminar motion of the striated jets [256].

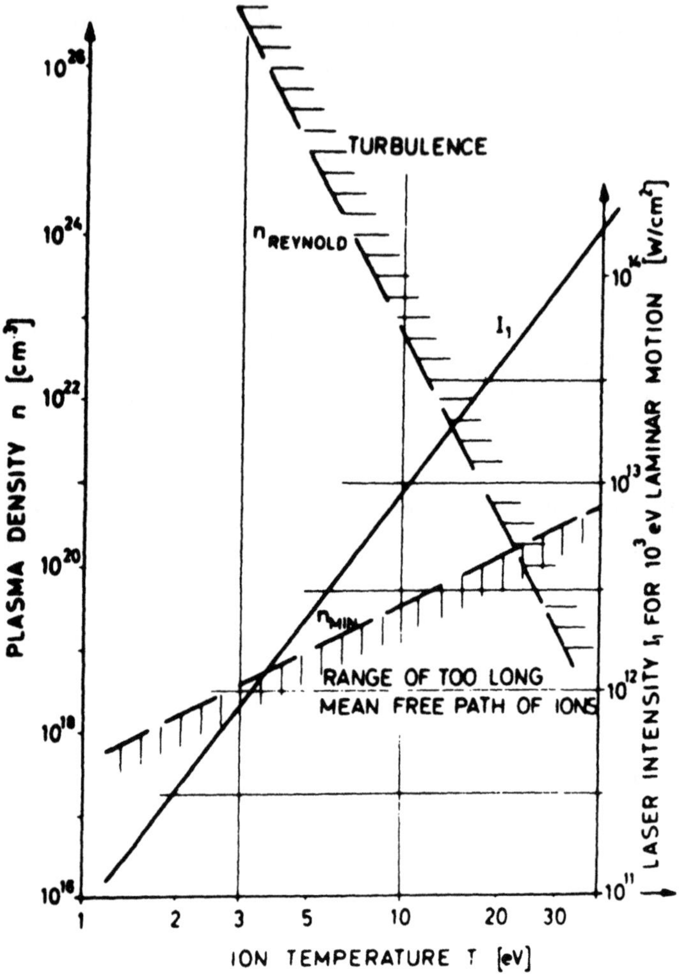

Figure 11.4 Laser intensities for laminar motion of 1 keV ion energy depending on the plasma temperature. Limitation of the ion densities by the mean free path and by Reynolds number [256].

minimum value n_{min} defined by the mean free path $l \leqslant d$. A further limitation of the density is given by n_{Re}, which distinguishes between the cases of laminar and turbulent motion. Reasonable conditions of keV energy of the deuterium ions for obtaining laminar motion can be reached for densities up to 100 times lower than that of the cutoff density of 10^{21} cm^{-3} (neodymium glass laser) at reasonable laser intensities between 10^{12} and 10^{14} W/cm^2 and for temperatures below 20 eV. This is not unrealistic for the conditions of a laser produced plasma. Otherwise, turbulent faster motions occur.

Now is calculated the time within which the striated motion is built up. The force density f accelerated the plasma, if viscosity is neglected, up to the velocity w.

$$\bar{w}=at_a; \qquad a=\frac{f}{n_i m_i} \tag{11.44}$$

where a is the acceleration. The acceleration time t_a is given by

$$t_a=\frac{wn_i m_i}{f}=1.11\times 10^{-29}\frac{n_i}{T^{5/2}|n|^3}\ \text{sec} \tag{11.45}$$

This acceleration time does not depend on the laser intensity I, because each of w and f are linear with I according to Eqs. (11.30) and (11.41). In the numerical factor of Eq. (11.41), the angle of incidence is $\alpha_0=25°$. The reasonable temperature $T=20$ eV results in

$$t_a=6.2\times 10^{-33} n_i\ \text{sec} \tag{11.46}$$

It shows a buildup time for the striated motion of less than a picosecond for densities $n_i<1.6\times 10^{20}$ cm^{-3}. At these low temperatures and high densities, the Coulomb collision time is high enough to conserve the conditions of the equation of motion, Eq. (8.77), for the space charge neutrality of the plasma. The time t_a for establishing the striated motion is shorter than the time the plasma takes to penetrate the length between the thickness of one layer by its thermokinetic expansion, if the nodes of the stationary wave are assumed to simply rest in space. The thermokinetic expansion results in plasma velocities not seriously exceeding 10^7 cm/sec at the laser intensities in question. To penetrate the length of a quarter of the wavelength, a time of less than 2 psec is then necessary. Therefore, if the propagating wave process has built up a striated motion, it can persist for a time until being destroyed by thermokinetic processes, if some more favorable stable state is not created to suppress the thermokinetic properties. Rethermalization of the striated motion with keV ions should then de detectable by the production of fusion neutrons from the plasma irradiated, even at such low intensities as 10^{13} W/cm^2, if oblique incidence and p-polarization occur. In this connection, it should be remarked that fusion neutrons have been produced even at intensities of 100 times less [258].

A further evaluation of the net nonlinear absoportion by the striated motion will now be done. The radiation energy which is gained under stationary conditions, by the layers in the laminar striated motion due to its viscosity can be calculated. A layer of a thickness d and a cross section of 1 cm^2 is considered. The consumed power density of the radiation is then equal to

$$I_c=F\bar{w}=\frac{2f^2 d^3 \nu_{ii}}{n_i k T_i} \tag{11.47}$$

For the optimum angle of incidence of 25.4° [with the maximum forces from Eq. (11.29)] is found

$$I_c = 7.2 \times 10^{-11} \frac{\lambda_0 I^2 \ln \Lambda}{T_i^{5/2} |n|^5} \tag{11.48}$$

The wavelength λ_0 is in cm, I is in W/cm^2, and $T$is in Ev. Using a Coulomb logarithm of 10, $T=10$ eV, and an ion temperature of 10 eV,

$$I_c = 2.2 \times 10^{-17} I^2 \frac{I^2}{|n|^5} \; [\mathrm{W/cm^2}] \tag{11.48a}$$

From this, the amount of laser power absorbed within a layer of d can be calculated from I_c/I. The absorption within a depth of one vacuum wavelength is about six times this value (absorbed energy)/(cm^3 sec along one wavelength)$=1.3 \times 10^{-15}$ I/n^5. This is a reasonable value, for example, 1.3% at $I=10^{13}$ W/cm^2 and 13% at $I=10^{14}$ W/cm^2. The *nonlinear nature of the absorption* is seen in Eq. (11.48) by the factor I^2. A comparison with the exponential laws of linear absorption is not readily possible, but the reasonable absorption within the depths of a few wavelengths at the usual intensities confirms the possibility of reality for this process. This process would also confirm a very fast thermalization of the laser energy for oblique incidence and p-polarization.

We here summarize the results of this subsection. For p-polarization of obliquely incident laser radiation, a striated motion is generated in the standing wave field of the plasma corona. This process is established within times of picoseconds and creates relative ion velocities, in a laminar motion, which correspond to ion energies of more than 1 keV for neodymium glass laser intensities of 10^{13} W/cm^2.

The mechanism is a typical nonlinear absorption process. For conditions of turbulent motion, the ion energies may be larger, but the details of this process will include the generation of Helmholtz–Kelvin instabilities very soon, so that a detailed analysis will depend on the individual conditions of intensities, angles of incidence, and so on, for which numerical studies are necessary.

11.2 Resonance Absorption

This subsection is devoted to the process of the so-called resonance absorption, which was treated first by Denisov [248], and which was studied subsequently by numerous authors. The process happens when a plane p-polarized electromagnetic wave is obliquely incident on a stratified plasma. It should be noted from the beginning that resonance aborption does not occur at perpendicular incidence, and it does not occur for s-polarization. It must be considered in two steps. First there is the linear solution of the Maxwell equations for the p-polarization, which leads to a strong, resonancelike increase of the longitudinal component of the **E**-field at the cutoff density. This resonance field is a simple consequence of the temporally harmonic solution of the wave equations (6.26) for $\mathbf{H}=\mathbf{i}_z\ H_z$ and Eq. (6.35) for $\mathbf{E}=\mathbf{i}_x E_x+\mathbf{i}_y E_y$ corresponding to p-polarisation (x is the direction of the gradient of the stratified plasma).

Denisov [248] calculated the **H**-vector for his treatment and discussed the interesting longitudinal component E_x subsequently, using complex coordinates for plasmas with absorption. A more straightforward way is to treat **E** from the beginning. The difficulty is the coupling of E_x and E_y, see Eq. (11.6). However, if E_x alone is discussed, Eq. (11.5) permits an independent solution. This approach was used by White and Chen [259]. Under the assumption of a stationary solution with a harmonic ansatz (6.19), after separation (11.7) of Eq. (11.6), the x-dependent factor of the E_x-component follows Eq. (11.10).

$$\frac{\partial^2}{\partial x^2} E_{xx}+\frac{\omega^2}{c^2} n^2(x)\cos^2 \alpha(x)E_{xx}+2\frac{\partial \ln n}{\partial x}\frac{\partial E_{xx}}{\partial x}+2\frac{\partial^2 \ln n}{\partial x^2} E_{xx}=0 \qquad (11.49)$$

The solution of the y-dependent factor, E_{xy}, is given in Eq. (11.15). In contrast to the case of the plasma corona [where the last two terms in Eq. (11.49) can be neglected], the discussion of a depth

$$x>x_t; \quad 1-n^2(x_t)\sin^2 \alpha_0=0 \qquad (11.50)$$

has to include the last terms of Eq. (11.49). x_t is the turning point, up to which the wave field is the superposition of a standing and a propagating wave. For $x>x_t$, below the turning point x_t, the evanescent wave field is dominant. x_t is uniquely defined by Snell's law, if n is monotonically increasing with x.

While White and Chen [259] immediately gave the transformation of Eq. (11.49) into such without a linear term, it is instructive to follow the general steps of this "reduction of a linear differential equation" of second order for $y(x)$

$$y''+g(x)y'+h(x)y=0$$

It is transformed by [260]

$$u(x)=y(x)\exp\frac{1}{2}\int -g(x)\,dx$$

into

$$u''(x)+Iu=0$$

where

$$I=h-\tfrac{1}{4}g^2-\tfrac{1}{2}g'$$

This is immediately evident by resubstitution. In the case of Eq. (11.49) the functions

$$g = 2\frac{\partial \ln n}{\partial x}$$

and

$$h = (n^2 - \sin^2 \alpha_0)\frac{\omega^2}{c^2} + 2\frac{\partial^2 \ln n}{\partial x^2}$$

are used. The resulting function $u(x)$

$$u = E_{xx} \exp \frac{1}{2} \int 2\frac{\partial \ln n}{\partial x}\, dx = E_{xx} n$$

turns out to be of a very special formula. After calculations of the function $I = \tilde{K}^2$, the reduction of Eq. (11.49) arrives at

$$\frac{\partial^2}{\partial x^2}(E_{xx}n) + \tilde{K}^2(E_{xx}n) = 0 \qquad (11.51)$$

The wave vector $\tilde{K}$ defines an "effective dielectric constant" $\tilde{\varepsilon}_{\text{eff}} = \mathbf{N}^2$

$$\tilde{K}^2 = \frac{\omega^2}{c^2}\tilde{\varepsilon}_{\text{eff}} = \frac{\omega^2}{c^2}\left[n^2 - \sin^2 \alpha_0 + \frac{c^2}{\omega^2}\frac{\partial^2}{\partial x^2}\ln n - \frac{2}{\omega^2}\frac{c^2}{n^2}\left(\frac{\partial n}{\partial x}\right)^2\right] \qquad (11.52)$$

This is formally identical with the expression of White and Chen [259], using $n^2 = \varepsilon$

$$\tilde{\varepsilon}_{\text{eff}} = \tilde{\varepsilon} - \sin^2 \alpha_0 + \left(\frac{1}{2}\frac{\varepsilon''}{\tilde{\varepsilon}} - \frac{3}{4}\frac{\varepsilon'^2}{\tilde{\varepsilon}^2}\right)\frac{c^2}{\omega^2} \qquad (11.53)$$

however, taking a dielectric constant for a plasma with collisions. This generalization causes a basically different behavior.

The fact that $|n|$ is going to very low values at $n_e = n_{ec}$ causes a very high maximum of E_{xx} at this resonance density. This has been pointed out first by Denisov [248] and is shown by a special example with a linear decrease of the plasma density in Fig. 11.5 [261]. It has been noted that the numerical solution of the wave equation for Fig. 11.5 was a new way to fit the complex amplitude of the reflected wave: a variation of phase and modulus has been done for arriving at a minimum "reflected wave" in the evanescent region. This is a special solution of the Osterberg problem (see Section 7 [152]). The result of ε_{eff} for a collisionless plasma, $\nu = 0$, was discussed by White and Chen [259]. Near x_0, the value of ε_{eff} passes a negative pole, see Fig. 11.6.

The discussion for $\nu \neq 0$ will be done now for $\tilde{\varepsilon}_{\text{eff}} = \varepsilon_{\text{eff}}(n = |n|)$ where the complex refractive index is substituted by its absolute value. This procedure is not too strange, as the nonlinear force for perpendicular incidence is

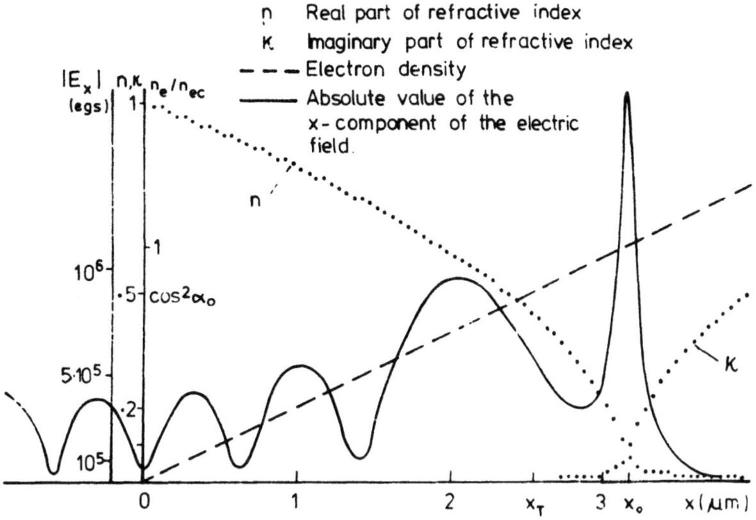

Figure 11.5 Exact solution of E_{xx} [Eq. (11.51)], for Nd glass laser radiation for a linear density profile, $T_e = 100$ eV and an angle of incidence $\alpha_0 = 26°$. The swelled maximum is reached before the turning depth x_t. The evanescent field for $x > x_t$ shows the resonance maximum at $x = x_0$, where $n_e = n_{ec}$. The figure shows an exact solution with respect to reflectivity determined by only penetrating waves at $x \gg x_0$, derived by Ladrach [261].

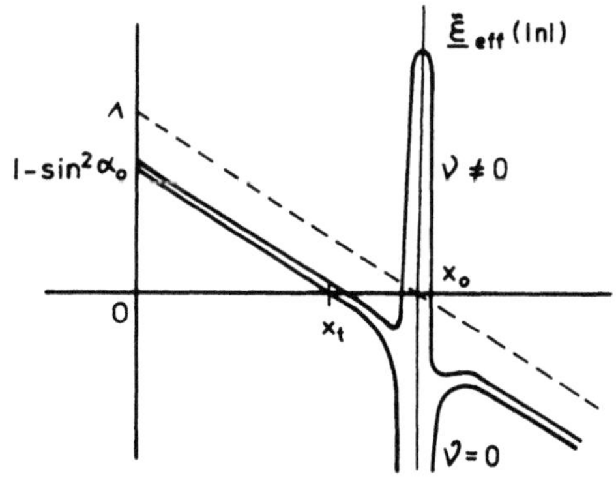

Figure 11.6 Schematic slope of $\tilde{\varepsilon}_{\rm eff}$ near the turning point x_t and the cutoff density x_0 for $\nu = 0$, following White and Chen [259]. With collisions $\gamma \neq 0$, a positive $\tilde{\varepsilon}_{\rm eff}$ ($|n|$) results from the discussion in this section, Hora, [4].

achieved correctly by using the absolute values of *n*, see [138, Eq. (25a)]. It has the advantage, and is justified that we shall be able to derive the Denisov length from our generalized equation (11.53) for *any* density profile of a plasma with collisions. The agreement of our generally considered profiles with those of a linear decay of the electron density will be shown after the general derivation.

Figure 11.6 shows $\tilde{\varepsilon}_{\rm eff}$ schematically compared with the case of $\nu = 0$. At the critical density $x=x_0$, $\tilde{\varepsilon}_{\text{eff}}$ jumps from negative infinity to a maximum which can have high positive values, even larger than unity, reminiscent of a dielectric medium with dipoles, corresponding to the electrostatic oscillation.

Following Eq. (11.52), a positive $\tilde{K}^2$ is possible only for the evanescent region if

$$\frac{1}{|n|}\frac{\partial^2}{\partial x^2}|n| > \frac{2}{|n|^2}\left(\frac{\partial|n|}{\partial x}\right)^2 - \frac{\omega^2}{c^2}\left[|n|^2(x) - \sin^2\alpha_0\right] \tag{11.54}$$

As mentioned, for complex numbers n the absolute values were used. The wave field $E_{xx}n$ will then have a penetrating wave in a narrow zone near x_0, as in a wave guide [262].

This is possible only if there is some absorption ($\nu \neq 0$) or any kind of damping. As Landau damping is unavoidable in principle, this condition is fulfilled. In Fig. 11.7, $|n|$ is given for the collisionless case and for that with collisions ($\nu \neq 0$). The sharp minimum near x_0 causes a very steep positive slope of the first derivative and therefore a very small but sharp and high positive peak of the second derivative. As

$$\frac{\partial|n|}{\partial x} \sim \frac{\partial n_e}{\partial x}$$

a small damping and a strongly steepened profile of the electron density near x_0 causes the wave guide effect there [262].

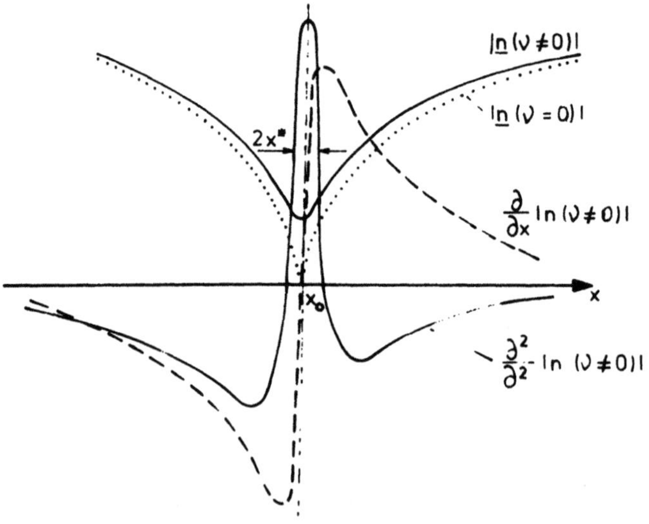

Figure 11.7 $|n|$ and its spatial derivatives near $x=x_0$ without collisions (dotted line) and with collisions (other lines). Obviously, the first and second derivatives for $\nu=0$ have poles at $x=x_0$ [4].

A further property of Eqs. (11.51) and (11.52) is the proportionality of $E_{xx}n$ and the square root of the electromagnetic energy density in the plasma. To preserve the constant energy flow into the y-direction (as is well known from the evanescent wave and the Goos–Haenchen effect at total reflection [263]), either the transport velocity of the energy has to decrease in this wave guide, or the value of E_{xx} has to be very large. This would immediately confirm the conclusion of Denisov [248] that E_{xx} has a maximum near x_0 and a very steep slope to small values for larger or smaller values of x.

Following Fig. 11.7, the thickness $x^* = L/2$ is calculated. x^* is half the range, where the second-order term in Eq. (11.52) can be positive starting from the relation of Eq. (6.48)

$$|n|\bigg|_{\min} = \left(\frac{\nu}{\omega}\right)^{1/2} \qquad (n_e = n_{ec}) \tag{11.55}$$

for $\nu \neq 0$, and from the fact that $|n(\nu=0;\, x = x_0 + x^*)| \approx |n|\big|_{\min}$ is found

$$\frac{1 - n_e(x + x^*)}{n_{ec}} = \frac{\nu}{\omega} \tag{11.56}$$

and by expanding $n_e(x + x^*) = n_e(x_0) + (\partial n_e/\partial x)|_{x_0} \cdot x^*$

$$x^* = \frac{\nu}{\omega} \frac{1}{(\partial \ln n_e/\partial x)_{x=x_0}} \tag{11.57}$$

In order to find the maximum value of the second derivative $\partial^2 |n|/\partial x^2$, the values of the differences for the limiting case of small x^* is used

$$\frac{\partial^2}{\partial x^2} |n(\nu \neq 0)| = \frac{1}{x^*} \left[\frac{\partial}{\partial x} |n(\nu \neq 0)|_{x = x^* + x_0} - \frac{\partial}{\partial x} |n(\nu \neq 0)|\bigg|_{x = x_0} \right] \tag{11.58}$$

The second term in the bracket on the right-hand side has to be zero by definition (Fig. 11.7). Approximating the other term with sufficient accuracy by its value without collisions ($\nu = 0$), from Eq. (11.49) is obtained

$$\frac{\partial}{\partial x} |n|_{x = x_0 + x^*} = \frac{\partial}{\partial n_e} |n| \frac{\partial n_e}{\partial x} = \frac{\partial \ln n_e}{\partial x} \frac{1}{2(\nu/\omega)^{1/2}} \tag{11.59}$$

The discriminating value for the second-order term in Eq. (11.52) is then

$$\frac{1}{|n|} \frac{\partial^2}{\partial x^2} |n| = \frac{1}{2} \frac{\omega^2}{\nu^2} \left(\frac{\partial \ln n_e}{\partial x} \right)^2 \tag{11.60}$$

which is always positive. Using this result, Eq. (11.52) is rewritten for $x = x_0$, where $|n| = \nu/\omega \ll 1$, and $\partial |n| \partial x = 0$

$$\tilde{K}_{\max} = \left[\frac{\omega^2}{\nu^2} \frac{1}{|n|^2} \left(\frac{\partial \ln n_e}{\partial x} \right)^2 - \frac{\omega^2}{c^2} \sin^2 \alpha_0 \right]^{1/2}$$

or using the width x^* [Eq. (11.57)]

$$\tilde{K}_{max}=\left(\frac{1}{x^{*2}}-4\frac{\pi^2}{\lambda^2}\sin^2\alpha_0\right)^{1/2}$$

(where $\omega^2/c^2=2\pi/\lambda$) and finally

$$\tilde{K}_{max}=\frac{2\pi}{\lambda}\left(\frac{\lambda^2}{4\pi^2x^{*2}}-\sin^2\alpha_0\right)^{1/2}\equiv\frac{\omega}{c}\mathrm{N} \tag{11.61}$$

Positive radicals in Eq. (11.61) are realized if

$$\frac{\lambda}{2\pi x^*}>|\sin\alpha_0| \tag{11.62}$$

As soon as x^* is sufficiently small (one wavelength or less, depending on θ_0), the wave equation (11.51) results in a wave guide type layer of $2x^*$ thickness around x_0 with propagating waves. The effective thickness of the wave guide is extended by the Goos–Haenchen effect [263] in a similar way to that discussed extensively for dielectric wave guides [262]. Thus, the solution discussed was $E_{xx}n$ and not E_{xx} alone. Finally, it is very important to remark that the length $2x^*$ is identical with the length L, which Eliezer and Schuss [264] have derived for the density profile

$$n_e=n_{ec}\exp\left(\frac{x-x_0}{L}\right)$$

if ν/ω is interpreted as the same coefficient as in the analysis at the critical density near x_0. This result was derived from the stochastic interaction of the nonlinear force (Section 8) for the conditions of resonance absorption, tacitly including the result of the exact value of the nonlinear force (Section 8) for these conditions.

Eliezer and Schuss evaluated the special cases for L (and therefore for x^*) for 10^{15} W/cm^2 intensity for CO_2 lasers and neodymium glass lasers to 1.5 μm and to 1.5 nm, respectively. Neglecting $\sin^2\alpha_0$ would result in "effective refractive indices" $N=112$ for both lasers. The case of a refractive index of 112, or similar values near the conditions assumed, leads to a very new concept for the discussion of resonance absorption. The very high energy density accumulated in the wave guide drastically influences the conditions of the nonlinear forces. The detailed study, under these extreme refractive indices, needs clear understanding of the Abraham–Minkowski problem with absorption (Section 9) at the interface to the plasma.

We have derived the Denisov length $L'=2x^*$ in Eq. (11.62) without specifying the slope of the electron density. Our derivation was based only on a schematic slope and the calculations were done on the distances of maxima and turning points only. It would not need a further specification, because proof by comparison with the Denisov length is absolute. Only for further illustration, we show here the results Läderach and Balmer [265] achieved continuing their work by elaborating the real and imaginary part of $\tilde{\varepsilon}_{\text{eff}}$, Eq. (11.53), Fig. 11.8, for a plasma with linearly increasing electron density with collisions. The behavior of our quantity $\tilde{\varepsilon}_{\text{eff}}$ ($n=|n|$) is similar to Re(ε_{eff}). The maximum can exceed 1, as was shown for another case in [265, Fig. 1*a*]. For illustrating the schematic (but for our derivation of L sufficient) curves of Fig. 11.7 with a special case of linear density, Fig. 11.9 is shown to confirm full agreement.

After the phenomenon of the resonance field with a strong maximum of

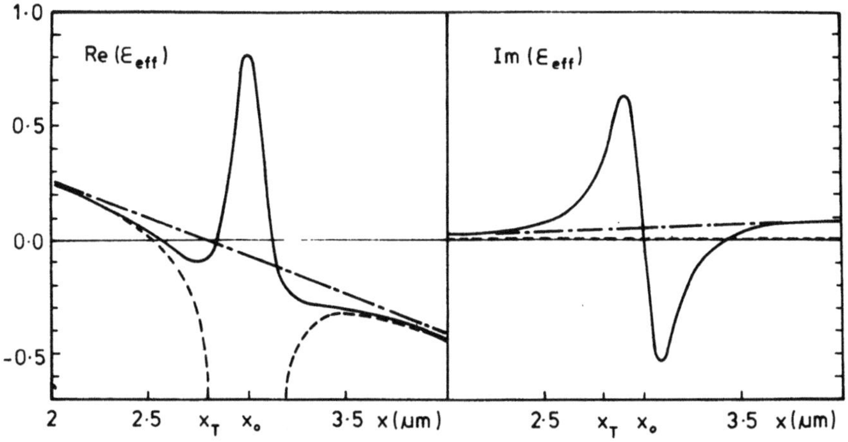

Figure 11.8 Numerical result of ε_{eff}, (Eq. 11.53) for a plasma with collisions and with a linear density profile. This is an extension of the work of Lädrach and Balmer [265] which was kindly provided for this discussion.

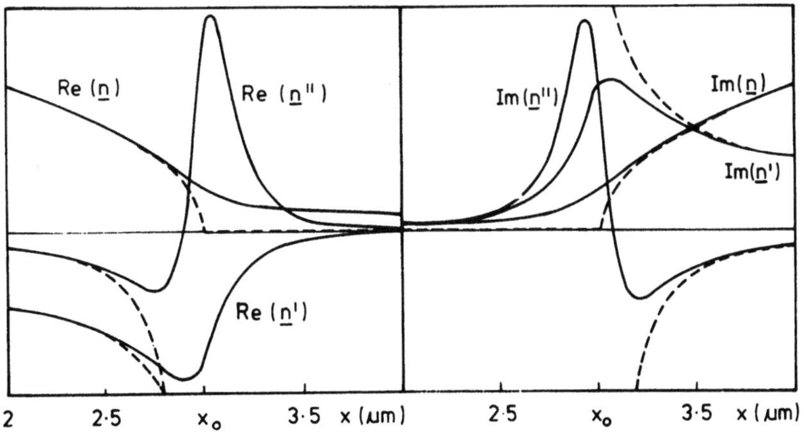

Figure 11.9 Numerical result of n for the same case of Fig. 11.8 [265].

the (longitudinal) E_x-component near x_0 has been shown, it should be mentioned in what sense this phenomenon is a "resonance *absorption*." The resonance field can contribute to an absorption due to the coupling with plasma waves at $x=x_0$. Several models have been discussed to describe this absorption. A very transparent calculation was the evaluation of the collisional absorption in the region between x_t and x_0 [267]. The absorption rate calculated by this way is shown in Fig. 11.10.

Another process is the single-particle simulation of the penetration of an electron through the E_x-resonance maximum [268]. It turns out that an electron moving at a certain initial velocity v_{in} into the maximum is accelerated. Maki [269] has an analytical solution showing that the superfast electrons of a Maxwellian distribution can gain energy. Ten keV electrons can gain energies up to 200 keV, directed perpendicular to the resonance layer. This is in agreement with single-particle simulations [268]. The same process was described by a wave braking mechanism [270] or by a soliton process [271].

Another description is the acceleration of an initially resting or slowly moving electron (anyhow without the need of an initial velocity v_{in}) near the E_x-resonance maximum. The resulting quivering motion leads to a drift and acceleration of the electrons similar to the quivering process [272] resulting in the nonlinear force for an inhomogeneous high-density plasma.

The high electromagnetic energy density $E_x^2+E_y^2+H_z^2$ and its strong gradient obviously act as a potential generating a nonlinear force. For a more careful analysis, the immediate application of an equation of motion of the type of Eq. (8.82) must be taken into account, as the conditions of space charge neutrality and restriction by the Debye length may not be fulfilled. Similar difficulties and the possibility of a simple $\mathbf{E}^2+\mathbf{H}^2$ description arise from the theory, when the nonlinear forces in a plasma lateral to a laser beam are considered [273]. To be on the safe side, the quivering model is

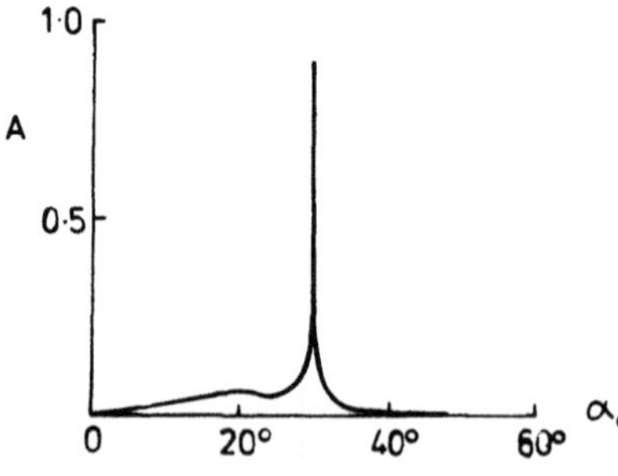

Figure 11.10 Coulomb collisional absorption in the evanescent region between the turning point and the critical density following Maki and Niu [267], at x_0 for varying angle of incidence.

followed for the description of the dynamics due to nonlinear forces in the region of the resonance absorption.

Physically, the E_x does not produce an "electrostatic field" immediately, as E_x oscillates at high frequency. Only the fact that the quivering electron is moving into areas of a smaller electric field and back into a larger field results in a net drift motion. This is an essentially different process from the drift due to the quivering motion of the electron in a plane wave, which is perpendicularly incident onto a stratified inhomogeneous plasma. Then the phase between **E** and **H** is the reason for the drift [138]. These facts show that the nonlinear forces and typical nonlinear properties of laser-plasma physics are not just simple extensions of the theory of force densities, however sophisiticated the refinements in Eq. (8.3) by additional nonlinear terms may have been. The nonlinear forces are a more general phenomenon, and the model of quivering motion is one of the guiding tools for analysis only.

The electric field maximum in the area of the resonance absorption x_0 has a very large value of E_x with a very steep decay for larger and smaller x. A drifting motion of quivering electrons is there not so much considered, but is determined by any phase difference between E_x and H_z, which together cause an eight-like motion. Gradients of time-averaged H_z values along y are relatively small and therefore neglected here. The net drift-motion of electrons is determined by the fact that the electric field E_x at a point $x=x_1$ near x_0 strongly depends on x by [266]

$$E_x(x)=E_x(x_1)[1+c_1(x-x_1)+c_2(x-x_1)^2+\cdots]\cos\omega t \qquad (11.63)$$

The equation of motion for a single electron is then

$$\ddot{x}=\frac{e}{m}E_x(x_1)[1+c_1(x-x_1)+c_2(x-x_1)^2+\cdots]\cos\omega t \qquad (11.64)$$

The first approximation ($c_i=0$, $i=1, 2, 3\cdots$) is

$$x^{(1)}=-\frac{e}{m}\frac{E_x(x_1)}{\omega^2}\cos\omega t \qquad (11.65)$$

which determines the second approximation

$$\ddot{x}^{(2)}=\frac{e}{m}E_x(x_1)\left[1-c_1\frac{eE_x(x_1)}{m\omega^2}\cos\omega t\right]\cos\omega t \qquad (11.66)$$

resulting, by integration, in

$$\dot{x}^{(2)}=\frac{eE_x(x_1)}{m\omega}\sin\omega t-c_1\left[\frac{eE_x(x_1)}{m\omega}\right]^2\left(\frac{t}{2}+\frac{1}{4\omega}\sin 2\omega t\right) \qquad (11.67)$$

The time-averaged (nonquivering) part of the next integration is taken only, giving

$$\overline{x^{(2)}} = -c_1 \left(\frac{eE_0}{2m\omega}\right)^2 t^2 \tag{11.68}$$

E_0 is the maximum of the resonance field E_x at $x=x_0$. Any symmetric part given by c_2, c_4, does not contribute to the drift.

Now the special case of a linear decrease of E_x with x from x_0 (to both sides) is considered.

$$E_x = \begin{cases} E_x(x_0)(1-2|x-x_0|/d_0) & \text{if } |x-x_0| \leqslant d_0/2 \\ 0 & \text{if } |x-x_0| \geqslant d_0/2 \end{cases} \tag{11.69}$$

At the end of the steep profile of E_x, at x^* with $|x^*-x_0|=d/2$.

The kinetic energy of the drifting electron is then given by

$$\frac{m}{2}\dot{x}^2 = \frac{1}{2}\frac{e^2E_x^2(x_0)}{m\omega^2} = \frac{E_x^2(x_0)}{8\pi n_{ec}} \tag{11.70}$$

This is equal to the maximum oscillation energy of the quivering motion of the electrons due to the field E_x at $x=x_0$. Though a very restrictive specialization for the E_x-profile, Eq. (11.69) is assumed. Any other general profile, however, results in the same relation (11.70), due to the energy conservation law for an electron passing around a closed circle, accelerating from x_0 to translational motion and back to x_0 using two different $E_x(x)$-profiles. The similar result for general profile for ions was derived for the completely different case of a net nonlinear force with perpendicular incidence, Eqs. (9.21 and (9.26)). Note that in Eq. (11.70), the kinetic energy is larger by a factor of 2.

In this sense of electrons drifting due to the quivering motion at the localized maximum of the high-frequency E_x-field near x_0, one can see the expansion of the plasma at the resonance absorption as if there were an explosion due to a "quasi-electrostatic" potential. One then has an "absorption," which is the same kind of absorption as in a macroscopic nonlinear motion without collisions, as when the net acceleration of the stratified inhomogeneous plasma by perpendicularly incident plane waves resulted numerically in a dynamic absorption (Fig. 10.19).

If the Debye length is less than half the thickness x^* of the resonance field, the accelerated electrons are coupled electrostatically to the ions, and the plasma is accelerated. The ion energy of translation is then

$$\varepsilon_i^{\text{transl}} = Z\varepsilon_{\text{osc}} \tag{11.71}$$

From Eq. (11.70), the total oscillation energy, in contrast to the half in Eq. (9.26) for the deconfining nonlinear force acceleration, is transferred into ion energy per ion charge. In Eq. (11.71), ε_{osc} is the maximum oscillation energy of the electrons due to the E_x-field.

The ion energies generated by the striated motion of the plasma corona at p-polarization for 10^{14} W/cm^2 neodymium glass laser radiation can reach 20 keV, especially if turbulent motion is generated. The process of resonance absorption, based on the quivering (nonlinear force) motion in the E_x-maximum, can easily result in similar energies. One has to take into account that the maximum of E_x can be 10 to 100 times the E_v-value, corresponding to a "swelling" of 100 to 10,000. The range of ε_{osc} in (11.71) will be near 20 keV at 10^{14} W/cm^2 neodymium glass laser intensities.

It should be mentioned that a further competitive anomaly was discovered by Goldman and Nicholson [274]. Whenever a caviton is generated during the nonlinear force interaction, this density minimum self-focuses electrostatic (Langmuir) waves in a stimulated way. The amplitudes of the electrostatic waves in the caviton can then reach high values and cause collapses. If the density in the center of the caviton is very low or even zero, as in the case of nonlinear-force-driven self-focusing, the number of electrons involved is small. For laser-produced plasmas, the effect might be not very strong.

11.3 A New Resonance at Supercritical Density

The last mentioned longitudinal wave focusing in nonlinear force produced cavitons according to Goldman and Nicholson [274], is one of the numerous processes which may cause anomalies at laser-plasma interaction. Similar processes were the striated motion at oblique incidence of laser radiation, or the generation of electron streams when electrons move through the longitudinal field maximum at resonance absorption as first convincingly shown by Maki and Niu [267], or simply by the quiverdrift of electrons in these high field gradients by quiverdrift as shown in Eq. (11.70).

Following our results about the genuine two fluid model with respect to second harmonics terms, we are now showing another resonance mechanisms as first preliminarily noted (Hora et al 1983) and then with the full acceptance of the referees of Phys. Rev. elaborated as a "New Resonance" (Hora et al 1985). Though this was mainly derived for perpendicular incidence of the laser radiation on a stratified plasma, it has far reaching consequences for oblique incidence why this mechanism is mentioned within this Chapter.

From the beginning we should mention that all these and other anomalous processes at laser interaction with plasmas were considered whether they play a role or not in generating the energetic (or "hot") electrons as first identified experimentally by Eidmann [75]. The large number of possible explanations, however, request cautiousness when any decision should be given to determine which one of the numerous mechanisms will be dominant. This undecided situation is the reason why we would not dare to prefer the one or other mechanisms as the solution of this problem and establish this result as final in a monograph as the present one.

There would indeed be a temptation to assume that the nonlinear force produced swelling of the oscillation energy of electrons especially by means of the following described relativistic self focusing (subsection 12.2) is the convincing explanation of the energetic electrons. It was clearly shown by the very transparent and precise measurements of Ehler [76] with the helios carbon dioxide laser irradiating aluminum, that the energetic electrons had an x-ray defined energy of 300 keV. The ions emitted with single charge had an energy in the time of flight signals of 150 keV, that of doubly ionized of 300 keV and so on linearly increasing to the ion energy of 1.65 MeV for eleven times ionized aluminum ions. This observation followed exactly the nonlinear force acceleration which is linear on the charge number, and corresponds to the result of Eq. (9.26) that the x-ray detected 300 keV energy would then simply be the electron oscillation energy in the relativistic focus.

This opinion that the energetic electrons are simply given by quiver energy was favoured by Brueckner (1975) but - without any reasonable argument - rejected by dominating theoreticians in the field. Another argument for the quiver energy model was the very precise and most convincing observation by Nakai (1980) that these hundred keV energetic electrons have a spectrum with a highly pronounced cut at high energy. This would just correspond to the quiver motion contrary to any thermal "hot" state of these electrons.

Nevertheless, though this picture of the quiver motion seems to be experimentally and theoretically convincing, in view of the hefty discussion of alternative models it should not be dared to try a final statement. At least we may outline with one further model of the mentioned kind in a qualitative way, how one has to consider or to exclude complicated mechanisms before reaching a final decision.

Our consideration starts from the general result of laser and hydrodynamically driven longitudinal oscillations in a plasma as derived from the genuine two fluid model. Equation (8.127) has the two last terms with the longitudinal oscillations by the frequency of the second harmonics of the laser light. The second last term has been identified as explaining the wide spread nearly density independent second harmonics emission from a plasma corona including the observed large and small spatial wave length modulation (see subsection 8.7).

The last term in Eq. (8.127)

$$E_{r1} = \frac{2\nu\omega}{(\omega_p{}^2-4\omega^2)^2+\nu^2} \frac{4\pi\varepsilon}{\omega^2 m} \frac{\partial}{\partial x} \overline{(E_2{}^2+H_L{}^2)} \sin 2\omega t \qquad (11.72)$$

obviously is a resonance term. It causes indeed longitudinal fields E_s where laser field energy is present in form of spatial gradients, but its amplitude factor - contrary to the preceding term in (8.127) - has a resonance dominator seen from a zero for the bracket in the denominator where the plasma frequency ω_p is equal to twice the laser frequency ω. This means that the resonance will happen at supercritical density, at four times the critical density, if sufficient laser light yet is present there. We shall see from the following evaluations that this expectation in special cases is not unrealistic.

What is essential is that the collision frequency has to different from zero, otherwise the resonance factor is zero or an undefined singularity will appear at exactly the resonance. This dependence on the collision frequency may be an essential difference to the observation from collisionless N-particle simulations by Kruer and Estabrook (1985) where a resonance at four times the critical electron density was reported, or if an analogy to our result (Hora et al 1983, 1985) would be the case, that then an implicit damping mechanism is being produced automatically in the N-particle simulation as a very remarkable phenomenon in order to understand the role of the collision frequency ν in formula (11.72).

In order to evaluate the importance of term (11.72) we calculated the laser field for a refractive index linearly increasing on the depth x beyond the critical density for neodymium glass laser radiation. The solution of the field equations are exactly given by the Airy functions and there has to be only the exponentially decaying mode (the other is zero). Using the parameter α

$$\alpha = (3/x)^{1/2} \;\; ; \;\; n_e = n_{ec}\,(1 + \alpha^2 x) \qquad (11.73)$$

as an expression of the steepness of the increasing electron density below the critical value, we find in Fig. 11.11 and Fig. 11.12 the results of the resonance amplitude in E_{e1} for plasmas of a temperature 100 eV and 1 keV respectively for the value of depending on the depth of the plasma at incident laser radiation of an intensity of 10^{14} and 10^{16} W/cm^2 respectively.

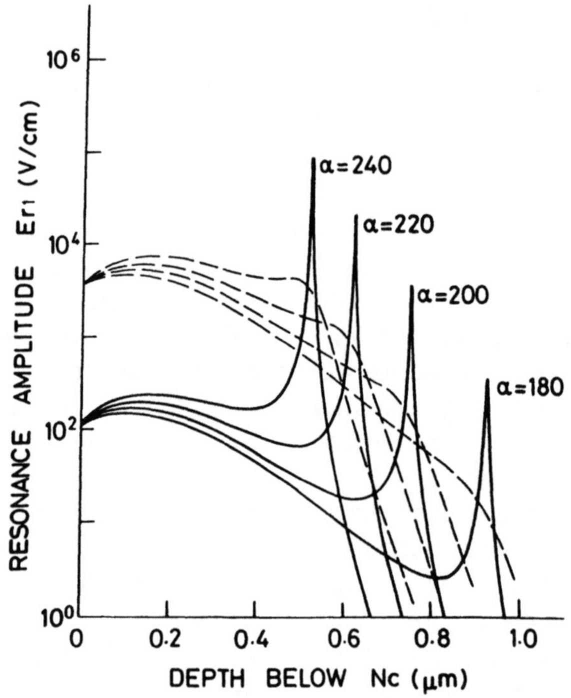

Fig. 11.11. Amplitude of the longitudinal electric field second harmonics resonance oscillation according to Eq. (11.72) calculated for neodymium glass laser fields of 10^{14} W/cm^2 for linear density profile of varying steepness (parameter α) depending on the depths below the critical density. Dashed curves are for the optical constants of a plasma of temperature 100 eV. Fully drawn curves are for the same plasma temperature however with the nonlinear optical constant including the quiver motion, Eq. (6.57) of the electrons by the laser field (Goldsworthy et al 1988).

The dashed lines are for optical constants where simply the mentioned plasma temperature was used, while the correct use of this plasma temperature plus the energy of the coherent motion as given in Eq. (6.57) arrives at the result of the fully drawn curves in Figg. 11.11 and 11.12. One realized that the nonlinear optical constants produce a very different result with a much more pronounced resonance. The peaks of the curves are just at the depths of four times the critical density. Comparing the calculated ratio of the resonance amplitude of the longitudinal oscillation with the amplitude of the laser light before reaching the supercritical

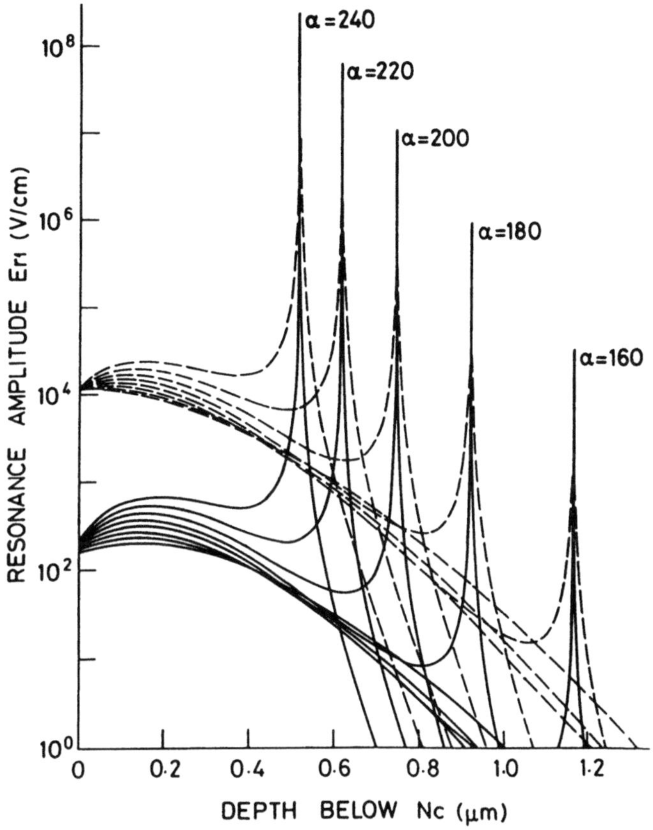

Fig. 11.12. Same result as Fig. 11.11 for a plasma of 1 keV temperature and for laser intensity of 10^{16} W/cm^2.

density, the result is given Fig. 11.13. This high resonance however is possible only for extremely steep density profiles below the critical density. This may be realistic as shown from the theory of profile steepening due to the nonlinear force produced caviton generation (Montes and Willi, 1982).

Though the strong action of the mentioned new resonance fields for laser radiation perpendicularly incident on a plasma [contrary to the Försterling-Denisov resonance absorption (Försterling, 1950) of subsection 11.2 with necessary obliqueness of incidence, otherwise resulting in no resonance at all] may have been shown here in principle, we may not draw any defined conclusion on a predominance of this

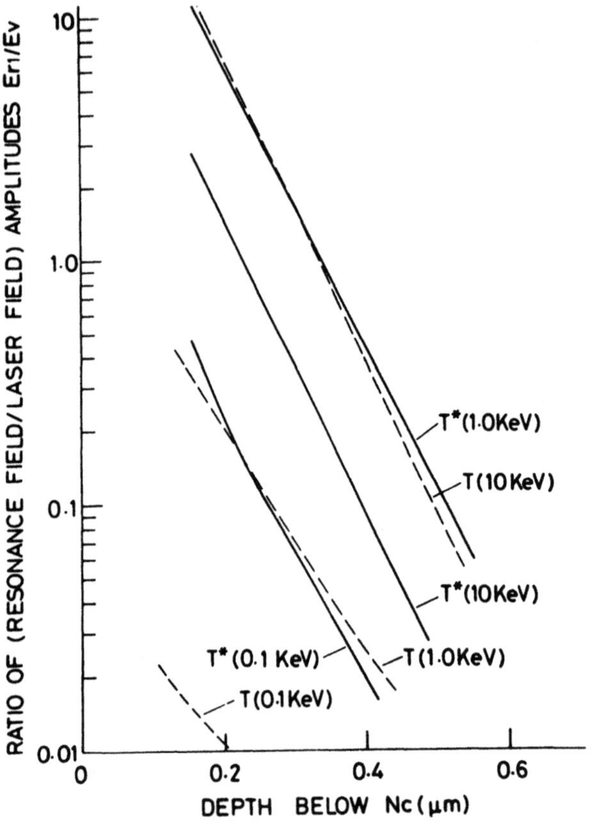

Fig. 11.13. Evaluation of the resonance field from a series of figures like 11.11 and 11.12 divided by the transversal electric field amplitude of the incident laser field for varying temperatures and for a laser intensity of 10^{16} W/cm^2 (Goldsworthy et al 1988).

resonance as long as not realistic situations of the plasma dynamics are calculated including this process. Obviously, the generation of very energetic electrons in these resonances both by quiver motion as well as by quiver drift (see Eq. 11.70) is evident if the whole plasma dynamic would provide this needed strong profile steepening.

While the mentioned cases in this subsection were exclusively elaborated for perpendicular incidence of the laser radiation, Goldsworthy (1988) formulated the resonance problems for oblique incidence receiving and mixing then both the Försterling-Denisov resonance absorption as well as the new resonance. Though a closed evaluation of all the generated terms was not possible, only the approximative study of dominant terms arrived at very strong further resonance mechanism against which the knowledge of the earlier exclusive Försterling-Denisov resonance absorption (subsection 11.2) or the just evaluated new resonance are weak components only. This all shows that these resonance process still need much more clarification before a more definite decision may be possible about the real nature of the generation of energetic electrons at very high laser intensities.

TWELVE

Laser Beams in Plasma

The discussion of laser plasma interaction was based on plane electromagnetic waves in Sections 9 to 11, exclusively. Some minor exemptions were parametric instabilities, where any deviations from plane wave or stratified geometries were marginal fluctuations only. This section discusses the behavior of laser beams in plasmas. This is the domain of self-focusing of the laser beams in plasmas of high or moderate density dominated again by the nonlinear force or by relativistic effects. At low densities, the nonlinear forces will expel the plasma out of the beam which results in interesting measurements. The finite diameter of the laser causes the spontaneous generation of magnetic fields, the interaction of which (including Alfvén waves) will then be considered.

While ideal plane (or spherical) wave fronts might be necessary for the application of lasers for nuclear fusion (see next section) and any self-focusing would have to be avoided, many other applications, such as treatment of material makes use of self-focusing. Especially at very high intensities the generation of MeV ions or the expected electron-position pair production and similar effects are of interest. A discussion of these questions is included at the end of this section.

There is an essential and basic problem we must mention at the beginning. This is the difficulty of an exact description of the laser beam. We shall see that several contradictions will arise if a solution for the complex nonlinear problems is not based on the fully exact description.

There is a long history for the derivation of the exact light beam. Debye [275] tried to describe an optical wave bundle by superposition of plane waves of various directions with a certain spectrum of intensities. To arrive at a very transparent description of an exact beam, the Schrödinger equation for electron waves permits more possibilities [276] than the optical case, as used in the description of the Goos–Haenchen effect for matter waves [263]. This better situation for matter waves is due to a basic difference of second order between optical and matter waves [277]. If a special approximation is used for optical wave beams with a Gaussian radial intensity profile as in the following, one must be aware from the beginning, that this is only an approximation of the exact Maxwellian case and the possibility of deviations from the basic nature with exact beams must be watched carefully.

Historically, the first quantitative theory of self-focusing of laser radiation, resulting in a threshold for the laser power was successful for dielectric materials (nonionized solids, liquids, and gases) [278]. The essential mechanism was the nonlinearity of the dielectric constant. Another mechanism of self-focusing in organic tissues with a much lower threshold was observed and can be related to the breakdown field strength [279]. For self-focusing in plasmas, there was no similar nonlinearity of the dielectric constant which could be used. The first successful way of deriving the self-focusing threshold was an application of nonlinear forces [205]. To arrive at a first qualitative threshold for the laser power it has to be mentioned that the first ideas on self-focusing of lasers in plasma was published by Askaryan [205]. He considered the energy momentum flux density of the laser beam $(\mathbf{E}^2+\mathbf{H}^2)/8\pi$ by which the whole plasma has been expelled and where the pressure is then balanced by the plasma pressure profile acting against the center of the laser beam. The balance should be given in this way in the form of an equation

$$\frac{\mathbf{E}^2+\mathbf{H}^2}{8\pi}=n_e K T_e\left(1+\frac{1}{Z}\right) \tag{12.1}$$

Askaryan was able to compare the necessary optical intensities to balance or compensate the gasdynamic pressure. The correct justification of this formula will be derived in Section 12.3.

12.1 Nonlinear Force (Ponderomotive) Self-Focusing

In order to derive the threshold for self-focusing of a laser beam in a plasma, three physical mechanisms (Fig. 12.1) have to be combined. Assuming that the laser beam has a Gaussian intensity profile along the y-axis while propagating in x-direction, the generated nonlinear force $\mathbf{f}_{\mathrm{NL}}$ in the y-direction has to be compensated by the thermokinetic force $\mathbf{f}_{\mathrm{th}}$ [206]

$$f_{\mathrm{th}}=f_{\mathrm{NL}}; \quad \nabla\cdot\left(\mathbf{T}-\frac{n^2-1}{4}\mathbf{EE}\right)=\nabla n_e K T(1+Z) \tag{12.2}$$

where use is made of Eq. (8.82). The second physical mechanism is the total

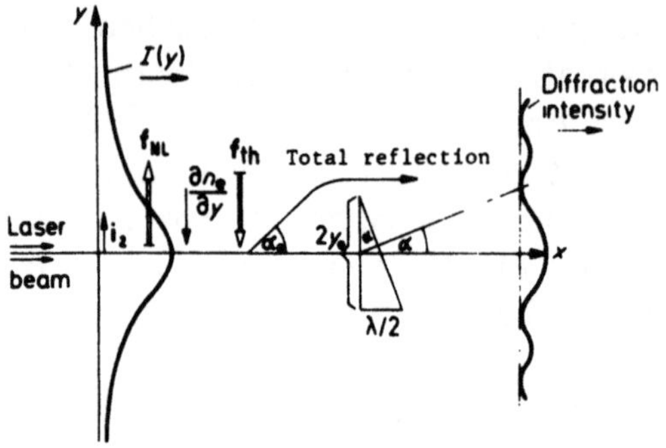

Figure 12.1 Scheme of a laser beam in plasma of a lateral intensity decrease $I(y)$, producing nonlinear forces $\mathbf{f}_{NL}$ in the plasma, rarifying the axial regions until being compensated by the thermokinetic force $\mathbf{f}_{th}$ of the gasdynamic pressure. The density gradiant causes a total reflection of partial beams. The diffraction condition for permitting partial beams of an angle of propagation less than total reflection for achieving the first diffraction minimum is the final condition for deriving the self-focusing threshold [206].

reflection of the laser beam components starting under an angle α_0 from the center of the beam and being bent into a parallel direction to the axis due to the density gradient of the plasma. The third condition is the diffraction requirement that the main part (e.g., as defined by the first diffraction minimum) of the beam has to have an angle of propagation α, which is less than the angle of total reflection. These three conditions are sufficient to calculate the threshold.

A Gaussian density profile including the refractive index n is described by the formula.

$$\overline{E_y^2} = \frac{E_v^2}{2|n|} \exp\left(-\frac{y^2}{y_0^2}\right); \qquad \overline{H_z^2} = |n|^2 \overline{E_y^2} \tag{12.3}$$

y_0 can be interpreted as the radius of the laser beam. This is only an approximation of the exact Maxwellian formulation. In analogy to the suggestions of Askaryan [205] and Schlüter [280], the nonlinear force in the direction of y is

$$\mathbf{f}_{\mathrm{NL}} = -\frac{1}{8\pi} \nabla(\overline{E_y^2} + \overline{H_z^2}) \tag{12.4}$$

It is important to note that this formulation is approximately valid only for conditions of strong swelling, that is, for refractive index differing from unity, which is the case for plasma exceeding the cutoff density in the region around the laser beam. The knowledge of difficulties for low-density plasma will be discussed in Section 12.3. The calculations [206] are repeated by several authors, on the basis of different assumptions, with always the same result achieved.

Using Eq. (12.3) in Eq. (12.4), the maximum nonlinear force in the y-direction is

$$\overline{\mathbf{f}_{\mathrm{NL}}}=\mathbf{i}_y \frac{1+n^2}{16\pi n} \frac{E_y^2}{y_0} \sqrt{2} \exp\left(-\frac{1}{2}\right) \tag{12.5}$$

If this has to be compensated by a thermokinetic force under the assumption of a spatially constant plasma temperature (the general treatment for varying temperature was studied extensively by Sodha and co-workers [281]. The formulation is found

$$\mathbf{f}_{\mathrm{th}}=-\mathbf{i}_y k T_{\mathrm{th}}\left(1+\frac{1}{Z}\right)\frac{dn_e}{dy} \tag{12.6}$$

Equating this force and the nonlinear force of Eq. (12.5) provides an expression for the electron density gradient of the plasma at the laser beam.

$$\frac{\partial n_e}{\partial y}=\frac{\sqrt{2/\exp(1)}}{16\pi K T_{\mathrm{th}}}(1+|n|^2)\frac{E_v^2}{y_o|n|(1+1/Z)} \tag{12.7}$$

The second physical condition of total reflection is given by the refractive index in the center of the beam, n and its value at y_0, for which with Eq. (12.7) is formulated [206]

$$\sin\left(\frac{\pi}{2}-\alpha_0\right)=\frac{|n|}{|n_{y_0}|} \tag{12.8}$$

Using the following Taylor expansion, for the case of a negligibly small collision frequency

$$n_{y_0}=n+\frac{\partial n}{\partial y}y_0; \qquad n^2=1-\omega_p^2/\omega^2 \tag{12.9}$$

gives from Eq. (12.8)

$$\sin\alpha_0=\left(\frac{2}{n}\frac{\partial n}{\partial n_e}\frac{\partial n_e}{\partial y}y_0\right)^{1/2} \tag{12.10}$$

If—as a third physical condition—a particular wave with an angle α for the first minimum of diffraction has to be reflected totally, the condition is found

$$\sin\alpha=\frac{\pi c}{2\omega y_0}\leqslant\sin\alpha_0 \tag{12.11}$$

Expressing the right-hand side by Eq. (12.10) and using there Eq. (12.7) and the relation for the electrical laser field amplitude $E_{v0}=c_1 P^{1/2}/y_0$ (where P is the averaged laser power and c_1 is a constant of 1.63×10^{-5} cgs) one arrives at [206]

$$P \geqslant \frac{(\pi c)^2 n^3 m_e}{e^2[2/\exp(+1)]^{1/2} c_1^2(1+n^2)} \tag{12.12}$$

It is remarkable that this threshold for self-focusing is a laser power and not an intensity. This is surprising, but not strange, as the threshold for the self-focusing of a laser beam in a dielectric nonionized medium is also a power and not an intensity [278], although both processes are basically different.

For an evaluation of Eq. (12.12), one can use the value of n given by Eq. (6.32), valid for temperatures above 10 eV. Expressing the plasma temperature T in eV and the laser power P in watts, one arrives at

$$2P \geqslant \begin{cases} 1.46 \times 10^6\ T^{-5/4} & \text{for } \omega_p \lesssim \omega \\ 1.15 \times 10^4\ T & \text{for } \omega_p \ll \omega \end{cases} \tag{12.13}$$

In the derivation given, only a laser slab has been used and no cylindrical beam. The use of a cylindrical beam results in a modification of Eq. (12.11) by the use of the Rayleigh diffraction factor 1.22 for the diffraction condition. Instead of Eq. (12.11) we then have for the beam the diffraction condition

$$\sin\alpha = \frac{1.22\pi c}{2\omega r_0} \leqslant \sin\alpha_0 \tag{12.14}$$

All the preceding derivations with the coordinates y and y_0 can be substituted by the radial coordinate r and the beam radius r_0 of a cylindrical coordinate system. The power threshold for the beam instead of that of the light slab is then

$$P \geqslant \frac{(1.22\pi c)^2 n^3 m_e}{e^2[2/\exp(1)]^{1/2} c_1^2(1+n^2)} \tag{12.15}$$

and in numbers with P in watts

$$P \geqslant \begin{cases} 1 \times 10^6 T^{-5/4} & \text{for } \omega_p \lesssim \omega \\ 8 \times 10^3\ T & \text{for } \omega_p \ll \omega \end{cases} \tag{12.15a}$$

This was the first quantitative theory of the nonlinear force self-focusing or, as it was initially called [206], the ponderomotive self-focusing. This result was rederived by several authors [282] and fully reproduced. The same thresholds are also achieved by Chen's nonlinear force treatment of the filamentation instability; see Section 9.5 [193].

A further surprising result [206] is the fact that the power threshold for self-focusing in plasma is very low, in the range of megawatts or less. This is in agreement with measurements first published by Korobkin and Alcock [283] and other authors [282, 284]. The measurements of Richardson et al.

[283] especially demonstrated in detail that the beam center shows a depletion of plasma. Another success of the theory is the agreement of the measured beam diameters [282] of a few microns for a laser power of 3 MW. If one can assume that the stationary conditions for self-focusing are reached when all plasma is moved out of the center of the laser beam, the electromagnetic energy density is then equal to $n_e(1+1/Z)KT$. This is the case for densities close to the cutoff density, where the laser intensity is equal to the threshold intensity I^*, as given in Fig. (9.2) for neodymium glass laser radiation. It is evident that the beam has then to shrink down to such a diameter to reach the necessary 10^{14} W/cm^2 from a laser power of 3 MW. The resulting beam diameter is then a few micrometers, in full agreement with the measurements.

12.2 Relativistic Self-Focusing

Another type of self-focusing happens [16] if the relativistic effects are considered. The relativistic change of the electron mass, due to oscillation energies close to or above m_ec^2, causes a modification of the optical constants, as shown in Eqs. (6.77) to (6.81). For the optical constants, Eqs. (6.33) and (6.34), or for the absolute value of the refractive index, Eq. (6.47), have to be used. With this relativistic intensity dependence, the effective wavelength of propagating laser radiation in a plasma is then given by

$$\lambda = \frac{\lambda_0}{|n(I)|} \tag{12.16}$$

where λ_0 is the vacuum wavelength. In Fig. 12.2, a Gaussian-like intensity profile of a laser beam moving through a homogeneous plasma is considered. The relativistic refractive index results in the condition

$$|n(I_{max})| > |n(I_{max}/2)| \tag{12.17}$$

showing, that the effective wavelength (12.16) is shorter for the higher laser intensity in the center of the beam than at the lower intensity of the half maximum intensity value. As shown in Fig. 12.2, an initially plane wave front is then bent into a concave front, which tends to shrink down to a diffraction limited beam diameter of about one wavelength. From the geometry of Fig. 12.2 this shrinking can be approximated by an arc resulting in a self-focusing length l_{SF}

$$l_{SF} = \left[d_0 \left(\rho_0 + \frac{d_0}{4} \right) \right]^{1/2} \tag{12.18}$$

d_0 is the initial beam diameter, and the radius of the arc with ρ_0 is given by

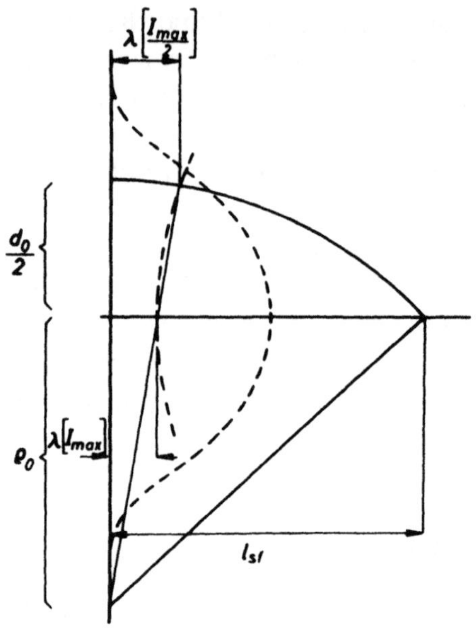

Figure 12.2 Evaluation of the relativistic self focusing length from the initial beam diameter d_0 and from the effective wavelengths. The relativistic effects cause a shorter wavelength at the maximum laser intensity I_{max} than at the half maximum intensity [16].

the effective wavelengths of the various intensities. From the geometry of Fig. 12.2, the following relation is derived:

$$\frac{|n(I_{max}/2)|^{-1}}{(d_0/2+\rho_0)}=\frac{|n(I_{max})|^{-1}}{\rho_0} \tag{12.19}$$

In combination with Eq. (12.18), this results in the ratio of the self-focusing length related to the beam diameter [16]

$$\frac{l_{SF}}{d_0}=0.5\left(\frac{|n(I_{max})|+|n(I_{max}/2)|}{|n(I_{max})|-|n(I_{max}/2)|}\right)^{1/2} \tag{12.20}$$

Using the exact absolute value of the refractive index n, as given by Eq. (6.47), with the intensity dependent relativistic values of the plasma frequency and the collision frequency [Eqs. (6.77) to (6.81)], a numerical evaluation of Eq. (12.20) is given in Fig. 12.3 for neodymium glass laser radiation for plasma densities of 10, 1, and 0.1% of the nonrelativistical cutoff density value. It is remarkable that the self-focusing length is as low as seven times the beam diameter for 10% of the cutoff density if the laser intensity is 3×10^{18} W/cm^2. This intensity is the relativistic threshold corresponding to an electron oscillation energy of m_ec^2. It is further interesting to note that the process of the relativistic self-focusing also occurs for laser intensities that are much less than the relativistic threshold, even 1000 times less. This

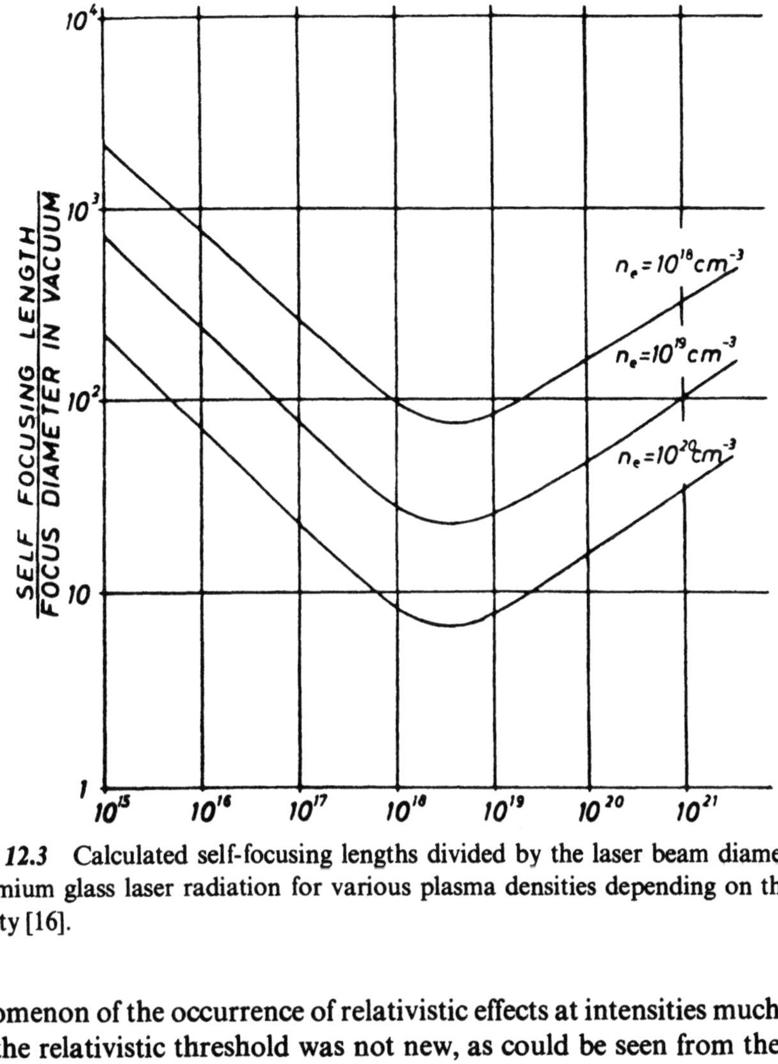

Figure 12.3 Calculated self-focusing lengths divided by the laser beam diameter for neodymium glass laser radiation for various plasma densities depending on the laser intensity [16].

phenomenon of the occurrence of relativistic effects at intensities much lower than the relativistic threshold was not new, as could be seen from the work of Tsindsatse et al. [285] for relativistic instabilities in plasmas.

The relativistic self-focusing has its maximum effect at the relativistic threshold. Its effect is lower for higher intensities. This can be easily understood from the fact that, at these higher intensities, there is an intensity dependent increase of the cutoff density, Eq. (6.78), so to speak, the plasma becomes transparent for propagating waves at densities, where the non-relativistic conditions would require evanescent waves.

The extension of the calculation of Eq. (12.20) to higher densities, shown in Fig. 12.3, is possible numerically. Simultaneously, the dependence on the plasma temperature and on the degree of ionizations is included. It is very surprising that self-focusing lengths of the same value as the initial beam diameter result, see Fig. 12.4 [286]. For lower intensities, a numerical cutoff

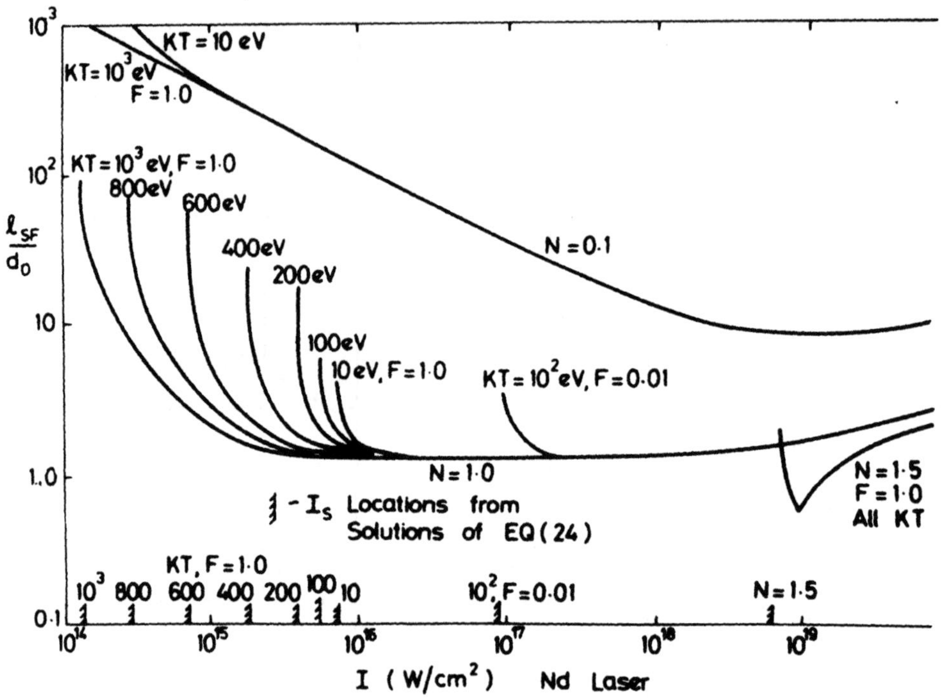

Figure 12.4 Ratio of the self-focusing length l_{SF} over the initial laser beam diameter d_0 for laser intensties near the relativistic threshold of 3×10^{18} W/cm^2 for neodymium glass laser radiation for varying plasma temperatures. The plasma density is equal to the nonrelativistic cutoff value ($N = n_e/n_{ec} = 1$) and 10% of this value ($N = 0.1$), respectively. The factor F is given by an effective collision frequency $\nu_{eff} = \nu/F^{2/3}$ to understand an eventual increase by anomalous effects [286].

of the plots is observed, where nevertheless the action of relativistic self-focusing is still working for intensities of less than 1% of the relativistic threshold. This is a remarkable result. While the well-known difficulties of designing an optical lens systems for focusing a laser beam in vacuum limit the minimum beam diameters to about 10 wavelengths, the plasma at the cutoff density realizes the very fast shrinking of a laser beam down to one wavelength diameter automatically by relativistic plasma effects.

It is worth noting that a later theory of relativistic self-focusing based on a completely different model which, however, is restricted to intensities below the relativistic threshold [287], results in nearly the same self-focusing length, as in the calculation based on the generally valid Eq. (2.20) [16], see Fig. 12.5.

It must be noted that both models mentioned, for nonlinear force self-focusing and for relativistic self-focusing, do not describe the complete process. The first model describes the threshold condition for a stationary

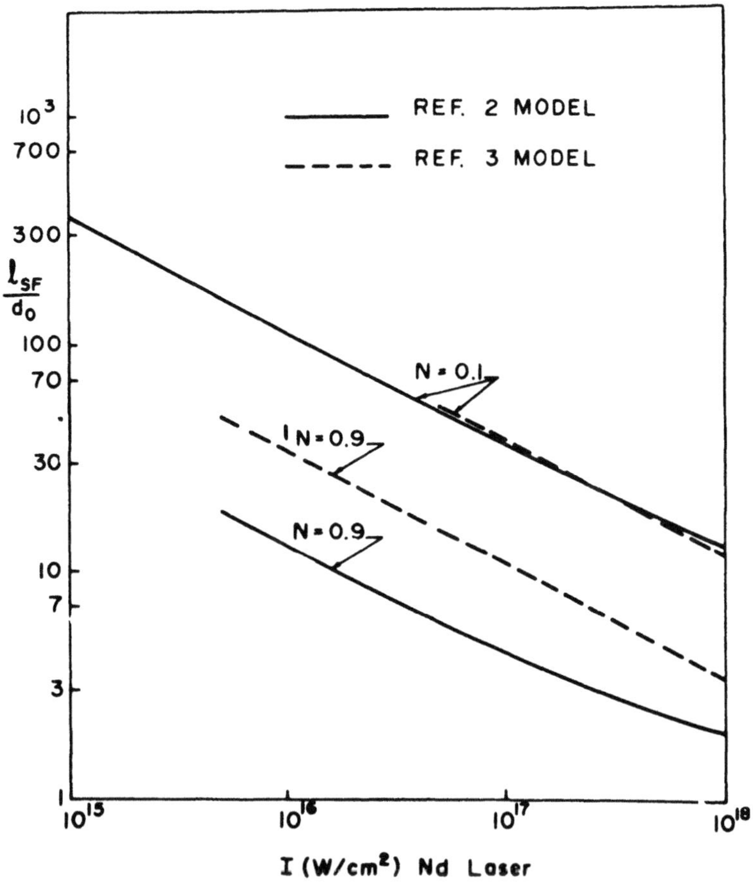

Figure 12.5 Comparison [288] of the relativistic self-focusing length l_{SF} per initial beam diameter d_0 depending on the initial neodymium glass laser intensity I following the general model (2) [16] and the subrelativistic model of Spatschek (3) [287].

case after sufficiently long interaction. The transient mechanism, the generation of the self-focusing tunnel and the resulting self-focusing length is not covered. The second model describes the relativistic self-focusing process in a homogeneous plasma, where the nonlinear forces disturb the homogeneity very quickly. The combination of all these mechanisms can be studied numerically. A very instructive example was derived by Siegrist [289], see Fig. 12.6. A very general numerical study including the nonlinear force, the relativistic self-focusing, and the transient behavior was performed by Kane [290], of which Figs. 12.7 and 12.8 are examples. The results describe the stationary solution of the axial intensity of a beam depending on the propagation length Z of a neodymium glass laser wavelength in a plasma, where the initial density and temperature are given and where the initial beam diameter of a Gaussian intensity profile is 30 μm [206].

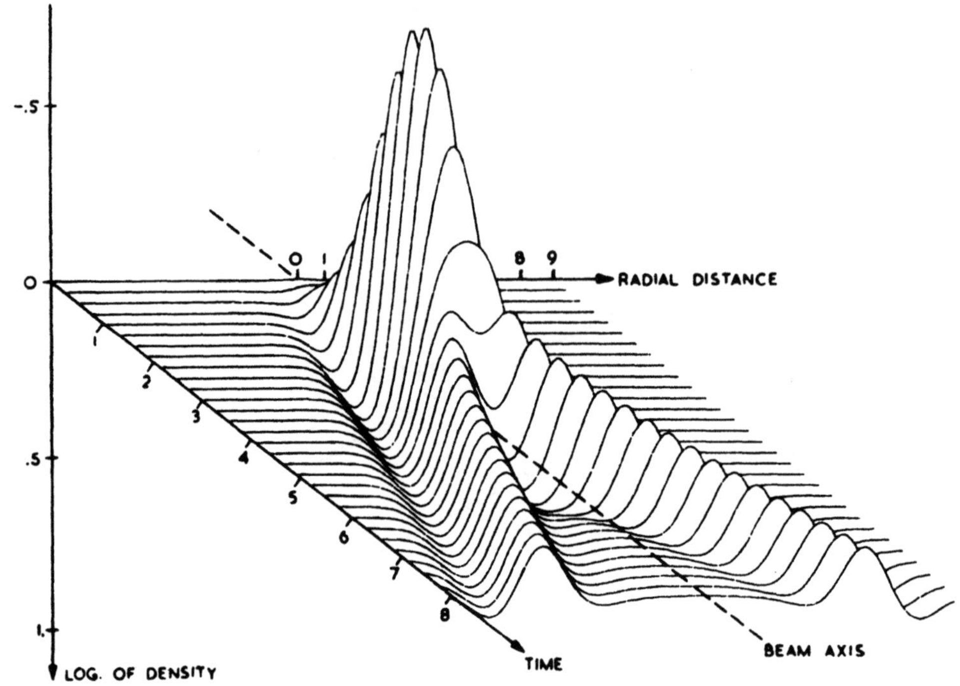

Figure 12.6 Density profile in a self-focusing channel in a plasma calculated by Siegrist [289].

There is nearly no effect for MW laser powers. The self-focusing can be seen at the next higher case of 5 MW and reaches the saturation beam diameter at a faster rate, the higher the laser power is. The saturation laser intensity at the beam center is the same for all powers: nearly 10^{15} W/cm^2, with a slight increase due to thermal effects, as for the threshold of the nonlinear force interaction, see Fig. 9.2, in full agreement with the result of nonlinear-force self-focussing [206]. see Eq. (12.15a).

The complete numerical calculations of the relativistic and nonlinear force self-focusing was done by a very general numerical treatment [290]. A Nd glass laser beam of nearly Gaussian radial intensity profile with 10 μm diameter of 30 psec duration is incident on an initially homogeneous hydrogen plasma of 80% of the cutoff density. The relativistic effect causes an immediate shrinking of the beam, while the nonlinear forces expel plasma from the interior of the beam. The radial velocities exceed 10^7 cm/sec. Because of the small radial field gradients in the center of the beam, a residual plasma remains and causes a hollow beam after 9.2 psec operation, see Fig. 12.9. This central plasma disappears, for example, at 18.4 psec due to fast axial motion, see Fig. 12.10. Axial ion energies of 5 MeV are achieved at 9.2 psec [290]. An example with a 10^{13} W-5 psec neodymium glass laser

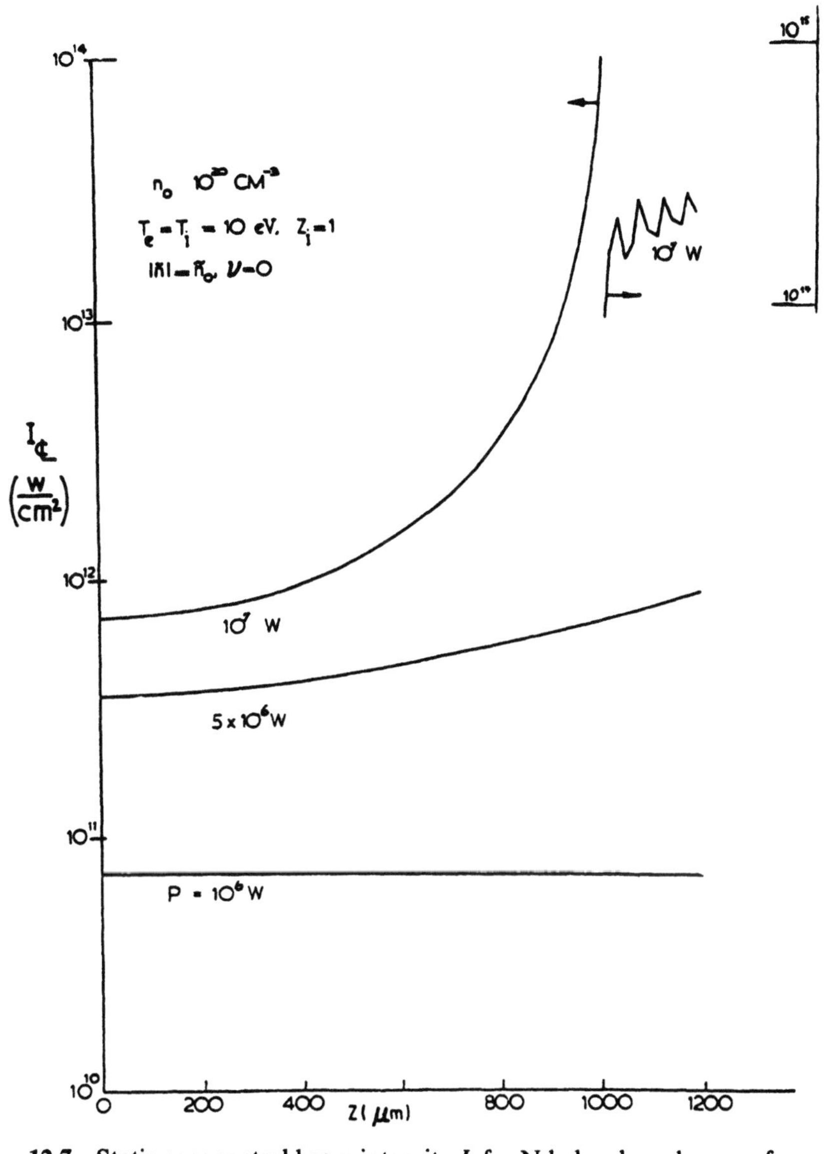

Figure 12.7 Stationary central beam intensity I_c for Nd glass laser beams of powers P in plasma of given density and temperature [290].

pulse of 20 μm diameter incident on a deuterium target of cutoff density showed the fast shrinking along an axial distance of 18 μm to a cross section of about one wavelength diameter. Ion energies exceeding 100 MeV were the result [291]. Applications in nuclear reactions and in the safe breeding of fission material by nuclear photoeffect are very promising.

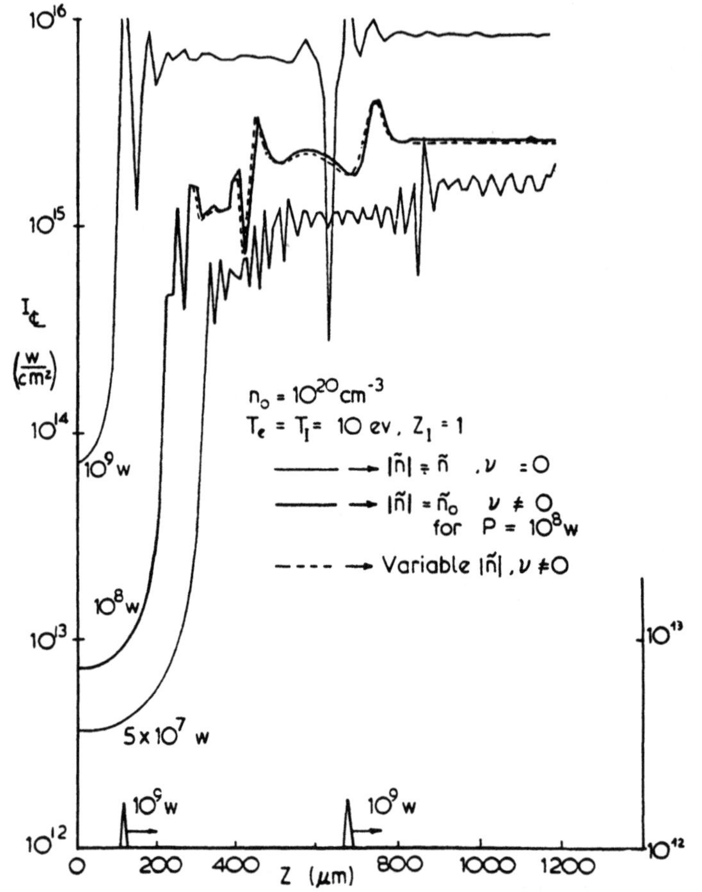

Figure 12.8 The same as Fig. 12.7 with cases of higher laser powers P [290].

12.3 Tenuous Plasmas, Exact Beams, and Free Electron Lasers

While laser beams in plasmas at or below the cutoff density will cause a dynamic reaction and result in self-focusing, a very low-density gas or the plasma (after ionizing of the gas atoms) will not modify the beam severely but will expel electrons by the nonlinear force. If the plasma dimension (diameter of laser beam) is less than the Debye length, the nonlinear force acts at the electrons only, leaving the ions behind untouched. The macroscopic plasma theory is then no longer valid, but the model of the quivering electron in the laser field and its drift results in formally the same nonlinear forces [274]. These results were used to demonstrate the action of the non-

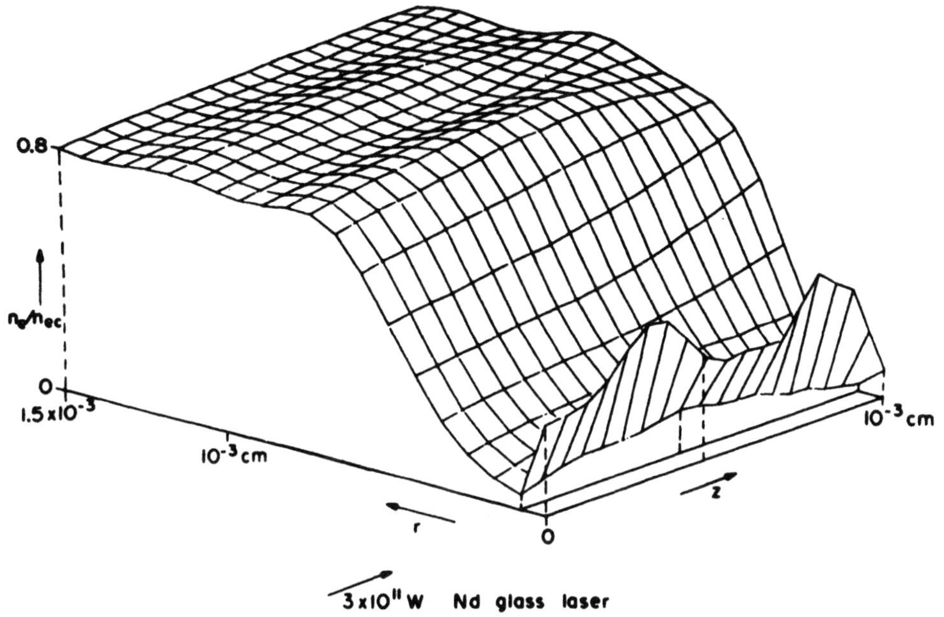

Figure 12.9 Density profile in an initially homogeneous H-plasma of 80% critical density at interaction with a beam of 10 μm half-radial width and 30 psec duration after 9.2 psec. A hollow beam has been created with a residual plasma in the beam center [291].

linear force immediately. Hollis [292] and Boreham [14, 293] focused neodymium glass laser beams into helium and other gases at about 10^{-4} Torr pressures. The laser beams had a focal intensity of about 10^{15} W/cm^2. The electrons emitted were measured along the direction of the laser light **E**-vector. Later, nearly no difference was found for other directions. It was found that the maximum electron energy

$$\varepsilon_e^{\text{transl}} = \varepsilon_{\text{osc}}^{\text{kin}} = \frac{\varepsilon_{\text{osc}}}{2} = \frac{I}{2cn_{ec}} \tag{12.21}$$

was equal to the average kinetic energy of the oscillation of the electrons according to Eq. (6.56), which was about 100 eV. This process can be explained in the same way as the energy gained by the drift of a quivering electron in a high-frequency field with a spatial gradient in field strength along the field direction, as was successful in the case of the resonance absorption for calculating the generated electron energy (Z times the ion energy), see Eq. (11.71) and the preceding derivation. For electrons in the low-density laser focus, we get the same results for electrons only if the Debye length is larger than the focus diameter.

The influence of the ions can be seen when the density produced by the

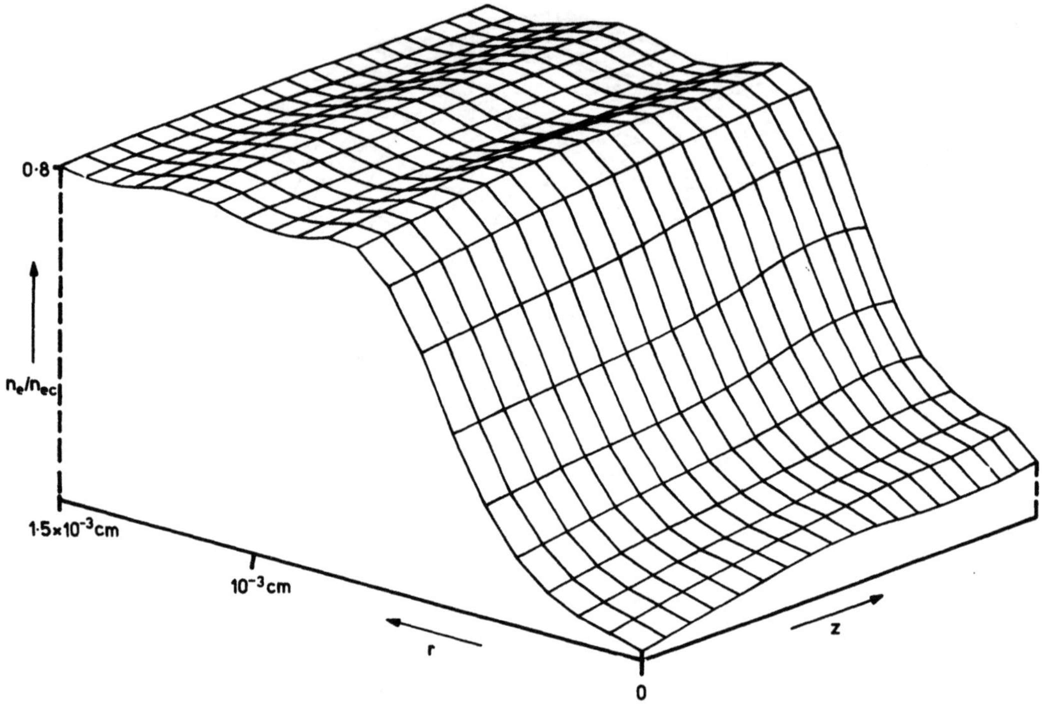

Figure 12.10 Same as Fig. 12.9 at time 18.4 psec [291].

gas pressure p is increased. Calculating a "Debye length" given by a temperature corresponding to the maximum electron energy of emission in terms of a pressure $p=p^*$, follows where the focus diameter d is equal to the Debye length, that

$$p^* = 1.37 \times 10^{-24} \frac{Z\bar{P}}{d^4} \tag{12.22}$$

where p is in torrs, $\bar{P}$ in watts, and d in centimeters. The laser intensity, given by a laser power $\bar{P}$, determines the maximum electron energy. For the conditions of the experiment by Boreham [293], the critical pressure p^* from Eq. (12.22) was 2.1×10^{-4} Torr. Figure 12.11 shows the result of the measurements. At these pressures, the emission of ions begins with the electrostatic attraction of the emitted electrons.

The experiment could be used to measure the delay of the ionization processes of the electrons bound to the helium atom at different energy levels. From this time delay it was possible for the first time [294] to measure the tunnel-type ionization process [295] according to the Keldysh theory [296] where the conditions of avalange ionization or the multiphoton ionization [297] are not given.

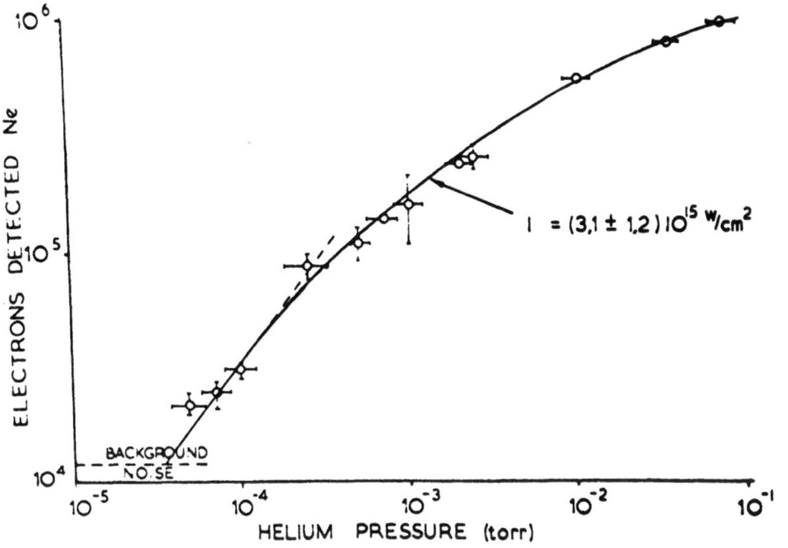

Figure 12.11 Number of electrons emitted from the focus of a laser beam in helium gas of varying pressure if the laser intensity is about 10^{15} W/cm^2. The linear increase on the number of electrons is saturated at about 2×10^{-4} Torr, due to the Debye length becoming equal to the focus diameter [14].

We come back to the question of whether the radial force of a laser beam interacting with plasma is correctly given by Eq. (12.4). There were no doubts for Askaryan [205] and all the following authors when using Eq. (12.1) for self-focusing. If one would have used the general expression of the Maxwellian stress tensor, the force in the y-direction (of the geometry of Section 12.1) is the gradient of $(E_y^2 - H_z^2)/8\pi$, if no x-component would be taken into account. It is necessary to emphasize that the field given by Eq. (12.3) is an approximation which does not fulfill the Maxwellian equations exactly.

We shall see now how a Maxwellian exact formulation of a laser beam (or slab) will result in a lateral nonlinear force given by Eq. (12.4). While plane electromagnetic waves are transversal only, as known from optics, the exact beam of finite diameter in vacuum needs a longitudinal component. Though it is well known that electromagnetic waves in media can have a longitudinal component, see Eq. (11.23) for p-polarization at obliquely incident plane waves, or in Dirac's quantum electrodynamics [298], we find the longitudinal component in vacuum as a simple result of the Maxwellian equations; only then will the lateral nonlinear force in the beam be found in agreement with Boreham's experiment [14].

To reproduce Eq. (12.21) according to the experiment, it was possible to use the drift of the quiver motion in the laser field along the E-direction even with the approximation (12.3)[273]; only for the perpendicular polarization, the inclusion of the longitudinal component of the field was necessary.

Using a slab of laser radiation with a decay in the x-direction (parallel to the polarization direction of **E**) we reproduce the earlier result [273]. The E_y field is then

$$E_y = E_0(1 + \alpha_1 |y| + \alpha_2 y^2 + \cdots) \cos\left(\frac{\omega}{c}x - \omega t\right) \tag{12.23}$$

with y being the displacement in the y-direction if the center of the beam is at $y=0$. The acceleration of electrons therefore has the form

$$\frac{d^2y}{dt^2} = -\frac{e}{m} E_0(1 + \alpha_1 y + \alpha_2 y^2 + \cdots) \cos\left(\frac{\omega}{c}x - \omega t\right) \tag{12.24}$$

Neglecting second and higher orders, a solution by small perturbation results in

$$\frac{d^2y}{dt^2} = -\frac{e}{m} E_0 \cos\left(\frac{\omega}{c}x - \omega t\right) + \alpha_1 \frac{E_0^2 e^2}{m^2\omega^2} \cos^2\left(\frac{\omega}{c}x - \omega t\right) \tag{12.25}$$

Using $\alpha_1 = (1/E_0)(\partial E_y/\partial y)$ from a Taylor expansion at $x = x_0$ and averaging over a period we have [273]:

$$m\overline{\frac{d^2y}{dt^2}} = -\frac{e^2}{4m\pi\omega^2}\frac{\partial}{\partial y}\overline{E_y^2} = -\frac{\partial}{\partial y}\overline{(\mathbf{E}^2 + \mathbf{H}^2)}/8\pi \tag{12.26}$$

The generality of this solution for higher orders has been proved by reasons of energy conservation for the net acceleration of the electron to large values of x [273].

For the case of perpendicular polarization, we use a laser slab where the electric field polarization is again in y-direction, but the field decay is in the perpendicular direction z. Similarly we can write:

$$E_y = E_0(1 + \beta_1 z + \beta_2 z^{-2} + \cdots) \cos\left(\frac{\omega}{c}x - \omega t\right) \tag{12.27}$$

with z being the displacement from the beam center at $z=0$. The Maxwell equations arrive not only at the usual transversal component

$$H_z = E_0(1 + \beta_1 z + \beta_2 z^2 + \cdots) \cos\left(\frac{\omega}{c}x - \omega t\right) \tag{12.28}$$

but also in a longitudinal component. The longitudinal z-component of the magnetic field is given by

$$H_x = \frac{E_0}{\omega}(\beta_1 + \beta_2 z + \cdots) \sin\left(\frac{\omega}{c}x - \omega t\right) \tag{12.29}$$

which usually has been neglected. The acceleration of electrons, therefore, has the form

$$\mathbf{i}_y \frac{d^2y}{dt^2}+\mathbf{i}_z \frac{d^2z}{dt^2}=-\mathbf{i}_y \frac{e}{m} E_0(1+\beta_1 z+\beta_2 z^2+\cdots)\cos\left(\frac{\omega}{c}x-\omega t\right)$$
$$+\mathbf{i}_z \frac{e}{m}\frac{dz}{dt}\frac{E_0}{\omega}(\beta_1+2\beta_2 z+\cdots)\sin\left(\frac{\omega}{c}x-\omega t\right) \quad (12.30)$$

Considering again only the first-order effect and substituting the zero-order solution of (12.30) for y, z, and dz/dt, we obtain

$$\mathbf{i}_y \frac{d^2y}{dt^2}+\mathbf{i}_z \frac{d^2z}{dt^2}=-\mathbf{i}_y \frac{e}{m} E_0 \cos\left(\frac{\omega}{c}x-\omega t\right)$$
$$+\mathbf{i}_z \frac{e}{m}\frac{E_0\beta_1}{\omega}\left(-\frac{e}{m}E_0\right)\sin^2\left(\frac{\omega}{c}\,x-\omega t\right) \quad (12.31)$$

with $\beta_1=(1/E_0)(\partial E_y/\partial z)$ at $y=0$ and averaging over a time period results in a vanishing y-component but an acceleration in the z-direction

$$m\,\overline{\frac{d^2y}{dt^2}}=-\frac{e^2}{4m\omega^2}\frac{\partial}{\partial z}\overline{E_y^2}=-\frac{\partial}{\partial z}\frac{\overline{(\mathbf{E}^2+\mathbf{H}^2)}}{8\pi} \quad (12.32)$$

This confirms the correctness of the use of Eq. (12.4) for the polarization-independent lateral nonlinear forces at electrons or in a plasma due to a laser beam. It is remarkable that the very tiny longitudinal component changes the action from no to yes. This teaches how carefully one has to proceed with nonlinear phenomena. One might have assumed that nonlinear extension should be less sensible against neglections. The inverse is the case: the basic assumptions have to be correct or nonapproximative to a greater degree than in linear theory.

The general formulation of an optical beam in vacuum which satisfies the Maxwellian equations exactly arrives always at a longitudinal component. The general formulation was given by Legendre functions [299].

We add here the result that the longitudinal component represents the property of the internal diffraction of the beam. Again a slab in vacuum is discussed where the symmetric terms are used only

$$E_y=E_0(1+\alpha_2 z^2+\alpha_4 z^4+\cdots)\cos\left(\frac{\omega}{c}x-\omega t\right) \quad (12.33)$$

where $\alpha_\nu=0$ for $\nu=4$. The Maxwellian equations then result in the usual component

$$H_z=E_0(1+\alpha_2 z^2+\cdots)\cos\left(\frac{\omega}{c}x-\omega t\right) \quad (12.34)$$

and in the longitudinal component

$$H_x = 2\frac{c}{\omega} E_0 \alpha_2 z \sin\left(\frac{\omega}{c} x - \omega t\right) \tag{12.35}$$

Obviously, E_y and H_x are out of phase, resulting in a zero lateral Poynting vector for this completely parallel beam $\bar{S}_z = 0$.

The definition of E_y in Eq. (12.33) is only for

$$z \leqslant z^* = \frac{1}{\alpha_2^{1/2}} \tag{12.36}$$

resulting in a cross section given in Fig. 12.12. The longitudinal component of H, Fig. 12.13 results in an angle α' of a converging wave field. We find from Eqs. (12.35) and (12.36)

$$\tan \alpha' = \frac{H_x(z=z^*)}{H_z} = \frac{2c\alpha_2 z^*}{\omega} = \frac{\lambda}{\pi} \alpha_2^{1/2} \tag{12.37}$$

For the diffraction limitation the far-field first minimum of the slab results in the angle α_d of the deviation from the axis

$$\sin \alpha_d = \frac{\lambda}{2z^*} = \frac{\lambda}{2} \alpha_2^{1/2} \tag{12.38}$$

For small angle α' and α_d and neglecting a factor $\pi/2$, the converging wave front of the longitudinal field of H is the same as the diffraction limitation needs for convergence of the beam. It should be possible by phase shifts to

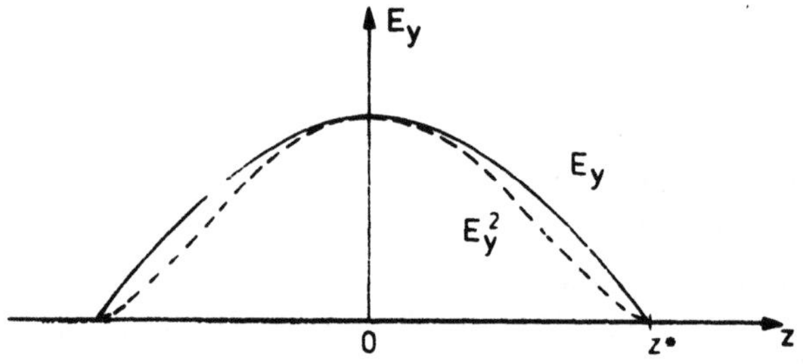

Figure 12.12 Lateral decay of E_y and E_y^2 of a slab beam according to Eq. (12.33).

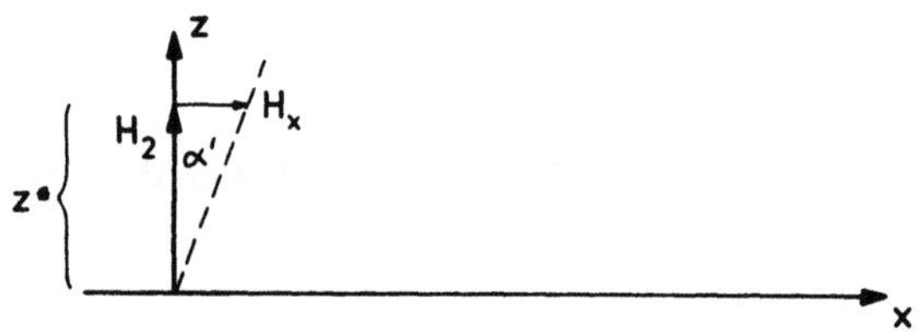

Figure 12.13 Longitudinal component H_x of the field for E_y of Fig. 12.12.

produce such absolutely parallel beams as described. This agrees with Einstein's needle radiation [189] including the finite width due to quantization.

The main result is that the radial nonlinear force due to a laser beam in a plasma is given by the gradient of $(\mathbf{E}^2+\mathbf{H}^2)/8\pi$ and all preceding calculations of self-focusing are unchanged. The confirmation of the polarization independency by Boreham's experiment is, however, to 30% accuracy only. An unexpected longitudinal component appeared in the exact description of laser beams in vacuum. We further experienced that only exact solutions lead to reasonable results at nonlinear effects.

This result is an example of how the nonlinear forces of laser radiation acting on electrons can be extended to low-density plasmas, where the macroscopic magnetohydrodynamic theory is no longer valid (the same model was successful for plane electromagnetic waves perpendicularly incident on dense plasmas [300]). It is also an example of the concept for a new type of free electron laser [301].

The presently known successful free electron laser is based on a synchrotron radiation process [302], where an electron beam has to move along the rippled magnetic field of a superconducting solenoid. The free-electron laser of the nonlinear force type, according to this new concept, can be explained by the experiment of Boreham [14]. The emission of electrons from the focus of the beam requires energy and results in an absorption of laser radiation of a nonlinear dynamic type. If—by inverting this process—an electron of energy equal to the maximum oscillation energy in the center of the laser beam is fired perpendicularly into the laser beam, the kinetic energy will be changed totally into oscillation energy. If the laser beam is switched off, when the electron is in the center of the laser beam, its oscillation energy will be transferred into optical energy of the laser beam. The different behavior of the electron in an axial direction compared to the radial direction of the laser beam follows from the general theory [138], especially from the results of Klima and Petrzilka [184].

In contrast to the synchrotron free-electron laser, the nonlinear force free-electron laser is an amplifier only of an otherwise produced laser beam. The interacting laser pulse has the same duration t_L as the crossing electron beam, which has to be incident in the direction of polarization. The focal radius is given by the energy spread ΔE of the electron beam, as $r=t_L(\Delta E/m)^{1/2}$. Using ΔE in cgs units, and the electron charge e in the same units as defines the electron beam density j (e is 1.602×10^{-19} Cb if j is in A/cm^2), the amplification factor is then

$$A=\frac{j}{n_{ec}\sqrt{\Delta E}}\,1.19\times10^5 \tag{12.39}$$

The amplification increases with the square of the wavelength, as can be seen from the cutoff density in the denominator. This is similar to the synchrotron free-electron laser. The amplification is possible with presently available CO_2 lasers, if the amplification is repeated in a totally reflecting cavity a large number of times. The free-electron laser concept has the advantage of working at intensities, where any solid-state or molecular amplifier would break up or would be ionized.

This method could amplify 10^5 J/nsec CO_2 laser beam to 10^7 J by a successive pumping during a period of seconds or more [303].

12.4 Spontaneous Magnetic Fields—Alfvén Waves

The model of electron acceleration along the **E**-vector of the laser beam in low-density plasma could be used to explain self-generated magnetic fields, at least with respect to the presently known experiments. Self-generated dc magnetic fields in laser produced plasmas have been discovered by Stamper et al. [304]. The observation of megagauss fields [305] has been established and several differing models, partly on the basis of the nonlinear force, have been developed [306]. The recent measurements by Key et al. [307] and by Yamanaka et al. [308] show that the magnetic field in the megagauss range is directed perpendicularly to the plane of the irradiated target and at least a dipole field has been observed.

Using these facts, the following model seems to describe the phenomenon, if the observation of a dipole field is extended to a quadrupole field, Fig. 12.14, and if the Boreham experiment would confirm a degree of 30% of polarization dependence of the radial force in a beam (in which case the

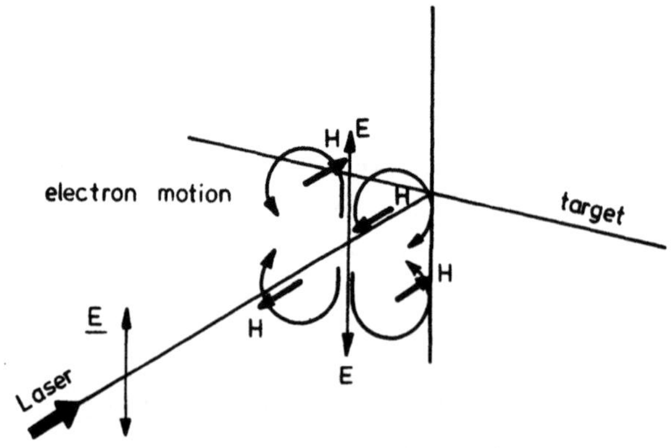

Figure 12.14 Laser radiation incident on a plane target producing a nonlinear force drift motion in the low-density corona along the *E*-field causing a quadrupole magnetic field **H** (double lined arrows.

general theory would have to be extended). The **E**-gradient will cause a quivering drift motion of the electrons according to Eq. (12.21), or according to the relation of Eq. (11.71).

The drift motion in the plasma corona of a density below the cutoff density results in a motion along the **E**-field, which causes a quadrupolelike circular motion of the electrons, as seen in Fig. 12.14. The magnitude of the magnetic field is given by converting the entire energy of oscillation (6.56) into magnetic field energy. This then gives

$$\frac{H^2}{8\pi}=N\frac{I}{c} \tag{12.40}$$

where $N=n_e/n_{ec}$. Using $N=0.1$ and an intensity $I=10^{22}$ cgs $=10^{15}$ W/cm^2. a magnetic field of 1.1 MG is reached in fair agreement with the experiments. It is further remarkable that the increase of the magnetic field follows a square root law on the laser intensity I. Eq. (12.40). The suggestion that a quadrupole field is produced, may not be too far away from the observations [307, 308], and special attention to this fact may lead to more precise experiments. The fact of a quadrupole field may explain why the experiment with the Argus laser [309] did not show a magnetic field, even if linear polarized laser radiation was used. If the direction of the Faraday rotation is very precisely parallel or perpendicular to the **E**-vector, no rotation is observed. Only an inaccuracy, a deviation from this direction results in the observation of the fields. A further confirmation of the theory can be given, if the use of circular polarized laser beams prevents the asymmetric nonlinear force motion described in Fig. 12.14.

We are aware that the model for explaining the spontaneous magnetic field in this subsection is very hypothetical. The fields are a fact, and we mention here for completeness the mechanism of compensating gasdynamic pressures by static magnetic fields on which extensive work of confinement of fusion plasmas (e.g., the tokamak) is based. The force density in a plasma is from Eqs. (8.4), (8.5), and (8.24)

$$\mathbf{f}=-\nabla p+\nabla\cdot\mathbf{T}+\frac{1}{4\pi}(n^2-1)\mathbf{E}\mathbf{E}-\frac{\partial}{\partial t}\frac{\mathbf{E}\times\mathbf{H}}{4\pi c} \tag{12.41}$$

for equilibrium, $\mathbf{f}=0$ and $\partial/\partial t=0$. In a plasma with no HF fields, $\mathbf{E}=0$, and the magnetic field is assumed of the form $\mathbf{H}=(H_x(y), 0, 0)$. This results in Eq. (12.41)

$$\frac{\partial}{\partial y}\left(p-\frac{\mathbf{H}^2}{8\pi}\right)=0 \tag{12.42}$$

or by integration

$$n_e\left(1+\frac{1}{Z}\right)KT=\frac{\mathbf{H}^2}{8\pi} \tag{12.43}$$

where an integration constant has to be added. This compensation of a gasdynamic pressure by a static magnetic field can be important in laser produced plasmas where magnetic fields have been produced spontaneously.

We consider now the generation of the magnetohydrodynamic waves, or Alfvén waves, generated at motion of plasma with a velocity v_0 (y-direction) perpendicular to a static magnetic field H_0 (x-direction). The Lorentz force will then cause a current density j_z (in z-direction) which results in an acceleration of the motion by

$$n_i m_i \frac{\partial v_y}{\partial t} = \frac{1}{c} j_z H_0 \tag{12.44}$$

further differentiation by time gives

$$\frac{\partial j_z}{\partial t} = C \frac{n_i m_i}{H_0} \frac{\partial^2 v_y}{\partial t^2} \tag{12.45}$$

In the diffusion equation (6.8), Ohm's law, we neglect nonlinear terms and the fast oscillations of $j(\partial j/\partial t = 0)$ and collisions ($\nu = 0$)

$$\begin{aligned} \frac{m}{e^2 n_e}\left(\frac{\partial j}{\partial t} + \nu j\right) &= \mathbf{E} + \frac{1}{c}\,\mathbf{v} \times \mathbf{H} + \cdots \\ 0 &= E_z - \frac{1}{c}\, v_y H_0 \end{aligned} \tag{12.46}$$

The electric field E_z is that generated by the current j_z due to the motion of v across $\mathbf{H}_0$. Differentiating (12.46) twice by time results in

$$\frac{\partial^2}{\partial t^2} v_y = \frac{c}{H_0} \frac{\partial^2}{\partial t^2} E_z \tag{12.47}$$

Any fast motion of $\mathbf{E}$ will follow a wave equation (from the Maxwellian equations)

$$\nabla^2 \mathbf{E} = \frac{1}{c^2} \frac{\partial^2}{\partial t^2} \mathbf{E} + \frac{4\pi}{c^2} \frac{\partial j}{\partial t} \tag{12.48}$$

or from (12.45) and (12.47)

$$\nabla^2 \mathbf{E} = \frac{1}{c^2} \frac{\partial^2}{\partial t^2} \mathbf{E} \left(1 + 4\pi \frac{n_i m_i c^2}{H_0^2}\right) \tag{12.49}$$

This is a wave equation with a wave velocity

$$v_A = \frac{c}{(1 + 4\pi n_i m_i c^2 / H_0^2)^{1/2}} \tag{12.50}$$

called the Alfvén velocity. If $4\pi n_i m_i c^2 / H_0^2 \gg 1$, we find

$$v_A = \frac{H_0}{(4\pi n_i m_i)^{1/2}} = \frac{H_0}{\sqrt{4\pi\rho}} \tag{12.51}$$

The Alfvén waves are of importance in plasmas with static magnetic fields. However, there is a formal similarity with the high-frequency fields and the velocity an ion gains by the nonlinear force. Following Eq. (9.21) for high swelling ($1/|n| \gg 1$), the energy gained by the ion after being accelerated along the inhomogeneous plasma surface is

$$\frac{m_i}{2} v_i^2 = \frac{\overline{\mathbf{E}^2}}{8\pi} \frac{Z}{n_{ec}} \tag{12.52}$$

using $n_e = Zn_i$ we arrive at

$$v_i = \frac{|\mathbf{E}|}{\sqrt{4\pi n_i m_i}} \tag{12.53}$$

Here, $\mathbf{E}$ and N_i have to be taken from an area close to the cutoff density. The result (12.53) is similar to the Alfvén velocity, if $\mathbf{E}$ is considered instead of $\mathbf{H}_0$. For densities below cutoff, $|\mathbf{E}| \approx |\mathbf{H}|$ of the laser field. The direction of v_i is that of v_0 in the initial derivation of the Alfvén wave. An interpretation of the connection of the nonlinear-force acceleration by the HF laser field with the Alfvén velocity is possible by considering the stepping through of the plasma along the HF wave maxima. The velocity v_i given by the nonlinear force to the ions is the electric analogy to the Alfvén velocity.

12.5 Conclusions for Medium Laser Intensities

Though the discussion of the self-focusing was described here in a short chapter only, based on the extensively described theory of nonlinear forces in the preceding sections, its influence on the destruction of materials at moderate laser intensities is of eminent importance. This is the reason why discussion of the confusing experiments mentioned in Section 1.4 about the earlier experiments of laser plasma interaction is so complex.

There are no ideal plane wave fronts incident on the targets, but focused laser beams having a certain number of "hot spots" due to insufficiencies of diffraction and birefringent properties in the laser amplifiers. Apart from the fact that in the most cases the power threshold for self-focusing is reached, the hot spots are an additional mechanism for producing more than one self-focusing channel.

Intensities of 10^{15} W/cm^2 are very easily reached in the plasma filaments, and the nonlinear force acceleration produces the keV ions, the nonlinear recoil [69], and the electron emission current densities 1000 times higher than permitted by the space charge limitation laws [71]. The mechanism of overcoming the space charge limitation can be seen very easily in the fact that the electron acceleration by the nonlinear force in the plasma filament works only on the plasma electrons, which are between the space charge neutral ions. This high-frequency acceleration is basically different from the electron acceleration in the surface of a material or in the vacuum above a surface. The nonlinear force acts to accelerate the whole volume of electrons within the space charge neutralizing ions as a volume effect, and all our knowledge of the surface effects of electron emission are not relevant.

The generation of multifilaments by self-focusing in a material is very important, if the destruction of material is considered for laser of moderate intensity. Despite the belief that CO_2 laser beams have a very smooth lateral intensity profile, the fact that moderate laser beams at a very low aperture produce a granulated structure in irradiated solid hydrogen pellets, when used for filling magnetic confinement vessels [310], indicates the complexity of this process.

For laser compression of plasmas, the suppression of self-focusing is a very important condition, if a very homogeneous interaction with a spherical pellet surface is the aim. Therefore we can recommend, for various lower or medium scale experiments, the irradiation of plane surfaces with very large diameter laser beams of low aperture: these will also achieve laser beams with a very smooth and monotonic radial intensity profile. The comparison of these measurements with those of small diameters but the same intensities are then a next step towards clean experimental conditions for any further theoretical investigation. On the other hand, the generation of the filamentary behavior in the best possible way should be the tool for studying the destruction processes of solid targets by laser radiation. One of the aims is the largest possible amount of material destroyed, while other applications are the drilling of holes in materials with the best possible quality. One reason why the laser drilling of holes in ruby crystals for watch jewels did not succeed was because the holes were not smooth, but had very crazy surfaces [311].

12.6 Conclusions for Very High Laser Intensities

As soon as the conditions for relativistic self-focusing are reached, the effects of very high laser intensities in the focused filaments open the door to very interesting high-intensity effects. These effects are indeed not appropriate for laser fusion, and one has to known how to avoid the relativistic self-focusing in the case of laser fusion. For the physics of higher energies rather than for laser fusion, however, the fast shrinking of the laser beams to a diameter of a wavelength is very desirable.

One question is how the oscillation energy of the electrons can be increased in the relativistically self-focused filament, if the fact is taken into account, from Eq. (6.76), that the oscillation energy of the electrons increases by a square root law in the laser intensity only at superrelativistic intensities. This change in the exponent of I is a very important reason to seek to understand the laws of blackbody radiation and to discuss a derivation of the fine structure constant from basic physical laws [312]. This lower power increase of the electron energy, however, results in some disadvantages in reaching the highest possible electron oscillation energies. Another disadvantage is the fact that, if the focusing is performed in plasma densities closer to the cutoff density, the larger the effective wavelength the larger is the effective beam diameter at self-focusing. The relation of short self-focusing length is then influenced negatively by the necessary higher laser power. All these factors together have been evaluated [286] and the result is achieved in Fig. 12.15. The focusing of a neodymium glass laser beam in vacuum is assumed to be down to a diameter of $d_0 = 30$ wavelengths, which seems to be realistic. The resulting maximum oscillation energy of the electrons is then given for the various intensities I_v of the laser beams in such a vacuum focus of 30 wavelengths diameter, where the plasma density has been varied with $N = n_e/n_{ec}$ between 0.1 and 0.99. The resulting oscillation energies and self-focusing lengths are given in the diagram of Fig. 12.15.

It is remarkable that the oscillation energies of 3 MeV, which are necessary for a quantitative production of electron-positron pairs [313], are reached with neodymium glass laser radiation of only 5×10^{17} W/cm^2, corresponding to laser powers of 3×10^{11} W as an absolute minimum. The laser beam then has to be sufficiently smooth that only one filament is being produced. The laser powers, however, are within the state of the art.

A further result [314] is that the ion energy after nonlinear force acceleration from the extremely high intensity of laser beams with a diameter of one wavelength does not follow the relation of Z times the relativistic electron oscillation energy. The fact is that the ions have Z times the oscillation energy of the electrons as if these were following the subrelativistic law as long as the ion energies are subrelativistic. Using the relativistic threshold intensity I_{rel}, Eq. (6.73), the ion energies ε_i^{transl} of translation after being accelerated by the nonlinear forces from the relativistic self-focused filament are [314]

$$\varepsilon_i^{transl} = \frac{Z_i m_e c^2 (I/I_{rel})}{4} \tag{12.54}$$

Depending on the laser power, the ion energy is independent of the wavelength. The result is given in Fig. 12.16. Historically, the measurement of the MeV ion energies [76] by Ehler, in full agreement with the later ones, Fig. 12.16, was very transparent, although several authors could not believe in the MeV ion energies. Hughes et al. [315] measured the MeV ions and

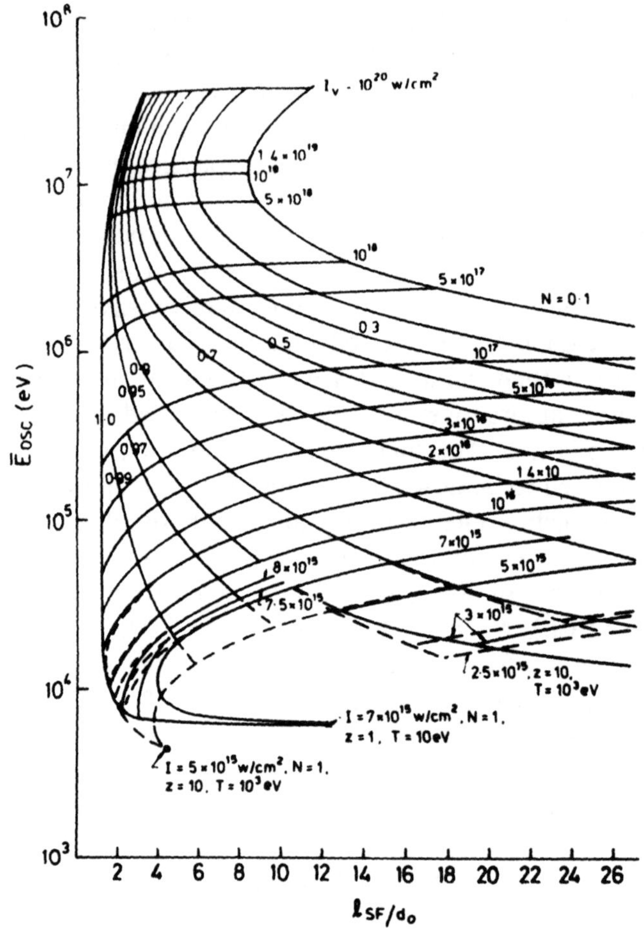

Figure 12.15 Maximum oscillation energy in the relativistically focused neodymium glass laser beam, where the vacuum focusing to 30 wave lengths diameter has reached. The maximum laser intensities I_ℓ are shown, and the plasma densities are given by multiples N from 0.1 to 0.99 of the cut off density [286].

found agreement with the theory of relativistic self-focusing. The discussion of the Z-dependent peaks of the ion probe signals (Fig. 1.9) as obscure stray signals, or of real MeV ion signals, was immediately confirmed by the consideration of a kind of Eq. (12.25) or Fig. 12.16. When Hughes was submitting a paper with the first announcement ever about MeV ions at the Amsterdam Quantum Electronics Conference in 1976, it was rejected with the argument that everyone had seen these MeV ion signals (although nobody had really understood and interpreted them as MeV ions).

It is remarkable that the measured protons from a laser produced plasma of 15 MeV [316] would correspond to 450 MeV ions, which are 30 times

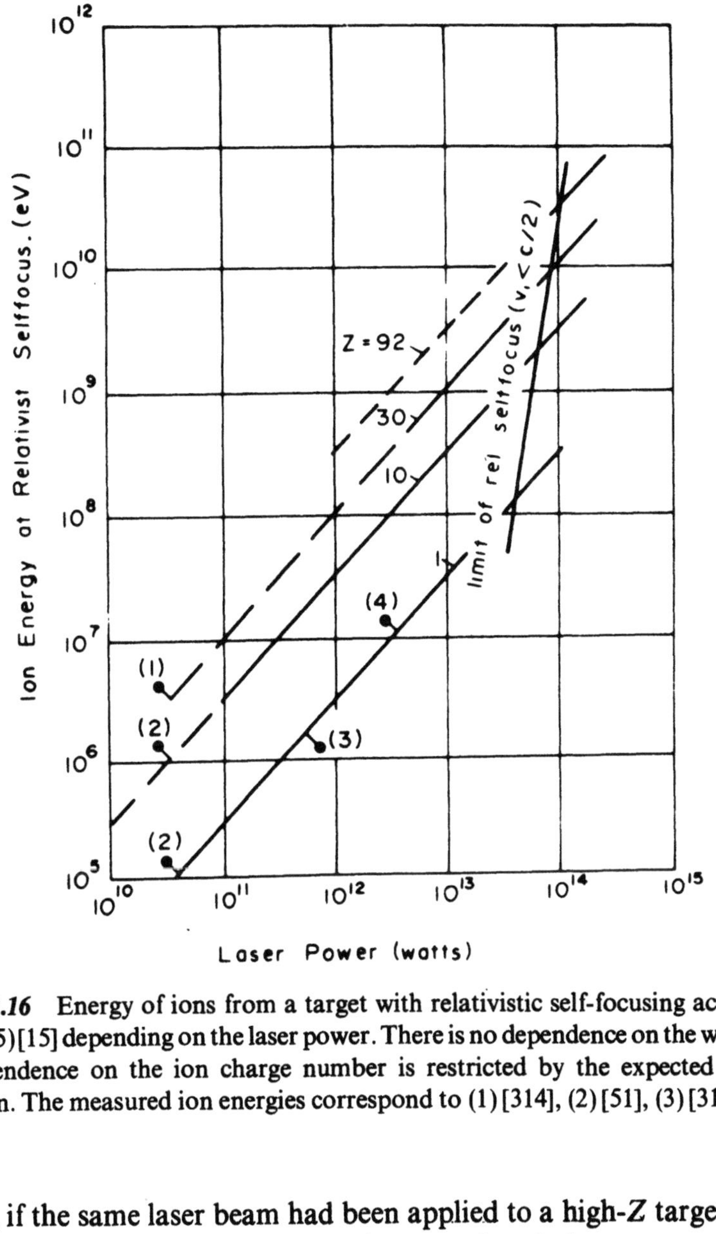

Figure 12.16 Energy of ions from a target with relativistic self-focusing according to Eq. (12.25) [15] depending on the laser power. There is no dependence on the wavelength. The dependence on the ion charge number is restricted by the expected degree of ionization. The measured ion energies correspond to (1) [314], (2) [51], (3) [316], and (4) [317].

ionized, if the same laser beam had been applied to a high-Z target. Ionizations of 40 and more are known from laser produced plasmas.

The production of high Z GeV ions should not be too far beyond the state of the art. Instead of promoting the next middle size accelerators in the range of $100 million of the conventional type for heavy ions, one should first discuss the prospects of producing a laser GeV heavy ion accelerator [318]. The properties of the laser produced multi-MeV ions produced now are quite different from the beams in conventional accelerators, but the disadvantages of a large energy spread and of short pulses for using these ion bursts as sources in accelerators are advantages for other uses, e.g. if the high-Z GeV heavy ions are used to produce not yet known radioactive isotopes within the whole range of nuclei by collisions. If the life time of the new nuclei is then very short, e.g. down to psec, the technique of the laser produced GeV ion bursts would just permit the measurement of the decay time easily. It was pointed out by Sir Ernest Titterton (1985) in his lecture for the 40th anniversary of nuclear explosions at Los Alamos, how important it will be if the present days known chart of nuclei could be drastically extended by this way to get the data on radioactive nuclei mostly in the range of middle atomic weight. It may be expected that new periodicities of decay times or semi-stability could be discovered as very important new input into the rather low developed theory of nuclei.

But nevertheless the use of laser produced MeV ion bursts for accelerators have some superior properties against other methods despite the mentioned problems of monochromaticity. It has been elaborated, how heavy ions as that of lead for the next following experiments at CERN (Haseroth, 1988) may be most preferable. Developments into this direction were known from the Dubna nuclear research center (Ananin et al, 1982; Bykovski et al, 1986), from the accelerator section of the Technological University Munich (Korschinek et al, 1986; Sellmair, 1988), from Arkansas Univ. (Gray et al 1982; Hughes et al 1989) and from Kelly et al (1991) and were reviewed in a very competent and balanced way (Brown, 1989) after Sessler (1983, 1985) underlined the advantages of this type of plasma accelerators (Clark et al, 1985; Joshi, 1989; Cicchitelli et al, 1989).

It was just the later mentioned conference series that the first announcement of achieving 500 MeV ions with about 50 time ionization from the 2 TW pulses of the carbon dioxide laser HELIOS in Los Alamos were disclosed (Ehler, 1982). The numbers fit excellently into the diagram of Fig. 12.16. It was underlined by Joshi (1982) that there is an excellent agreement with the theoretically expected values of the energy and their linear dependence on Z expressed by relativistic self focusing, as we may refer to Fig. 12.16. These important results were, however, a long time not published in order to get a comparison with a model of Gitomer where the disclosure of the number of the degree of ionization of the MeV ions was carefully avoided possibly in order to prevent the comparison and demonstration of the agreement with the relativistic self focusing theory.

In order to provide the theoretical expectations for the then prepared ANTARES carbon dioxide laser (Goldstone et al, 1984) with the then confirmed production of pulses of 80 TW and 70 psec. duration (Ehler, 1982) apart from longer, less powerful pulses, we performed numerical computations following the before mentioned Figures 12.9 and 12.10.

What was evident there, was the result that the laser pulses had to be shorter in order to achieve relativistic self focusing before the radial high nonlinear forces were pushing out the plasma from the laser beam.

The laser pulses had to have a rise time of 1 psec to reach a power of 10 TW for a neodymium glass laser. The beam was focused in vacuum to a diameter of 30 μm (Jones et al, 1982). Fig. 12.17 shows the shrinking of the beam diameter in the plasma at different times down to diffraction limitation of about one wave length diameter. Fig. 11.18 shows as density profile at the time 1.16 psec clearly indicating the shrinking of the beam while strong motion of the plasma in radial direction has been initiated with a wave motion contrary to the drilling of holes with unchanged diameter if the pulses had a rise time of 18 psec (see Fig. 12.10). The generated ion energy for motion in axial direction at the axis has been evaluated in Fig. 12.19 for a plasma of 38 times ionized tin ions arriving at ion energies up to 4 GeV in agreement with Fig. 12.16 (more details are given by Jones et al 1982).

The computation for the cases of carbon dioxide laser arrived at the result that the rise time of the laser pulse has to be less than 25 psec. Just this value was typical for the short pulse mode of ANTARES (Clark et al, 1985). Generation of pulses of up to 10 Amp/cm^2 current density close to the focus for 30 GeV highly charged heavy ion pulses of 20 psec duration could be expected. These would have been ideal for the generation of the large number of new radioactive isotopes as envisaged by Titterton (1985). Instead, the well working ANTARES laser was dismantelled before more shots onto targets were performed. One reason was that laboratory space for a magnetic confinement fusion experiment was needed. Another reason may have been that the rather restricted theoretical understanding of the physics of laser plasma interaction was used as argument against ANTARES obviously ignoring other very positive theoretical aspects which were available at that time.

One very unique experiment for confirmation of the laser generation of the MeV ions in targets by Rode (1984) (Basov et al, 1984, 1986, 1987; Sklizkov et al, 1988) should be highlighted. While in all the earlier measurements of the MeV ions, see subsection 12.2 and this subsection (Hughes et al 1980, 1985), the energy of the ions were determined by the time of flight when emitted from the target, Rode was using the long years developed x-ray spectroscopy (Sklizkov et al, 1988) for the range of 3 to 10 Angstrom to measure the lines emitted from phosphorus targets irradiated by neodymium glass laser pulses of an intensity of few 10^{15} W/cm^2. The spectra of various states of ionization could be identified and the Doppler shift of the lines measured.

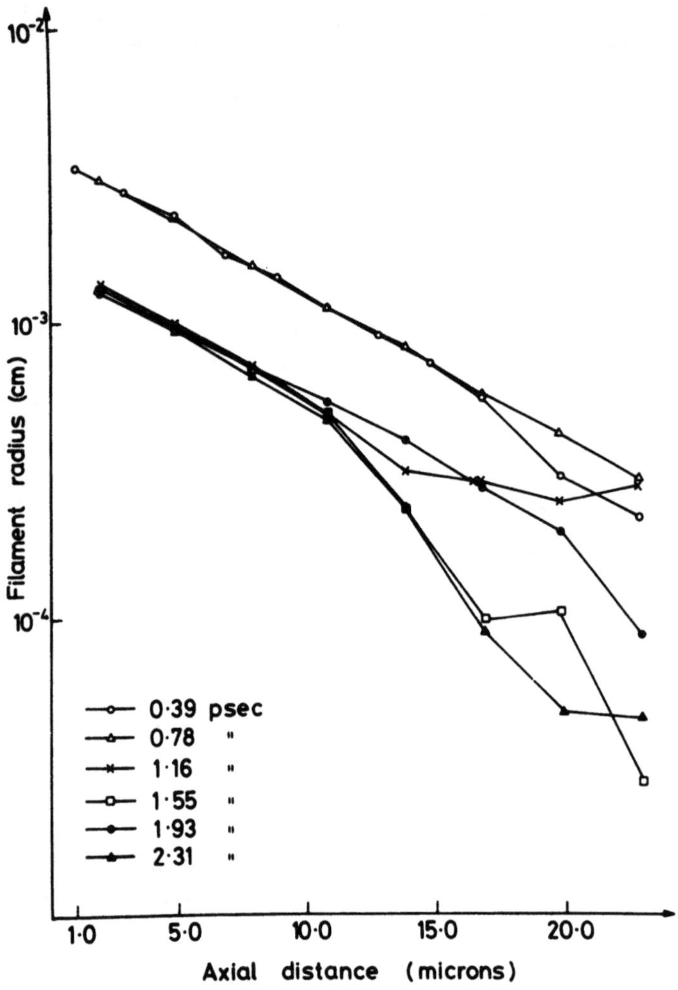

Fig. 12.17: A 10^{13} Watt Nd glass laser pulse from vacuum incient on Sn^{38+} plasma of 10^{21} cm^{-3} density. Filament radius as a function of depth into the plasma for various times.

This was the first time that the velocity of the ions in situ where they are generated in the laser focus could be measured. There was indeed agreement with the result from the time of flight measurement that the ion energy was linear on the charge number Z (Eq. 9.24). 11 times ionized phosphorus had an energy of 1.5 MeV. From the very precise calibration of the experiment it could be concluded, that the focusing of the beam had gone down to a beam diameter of 0.6 times the wavelength (Basov et al

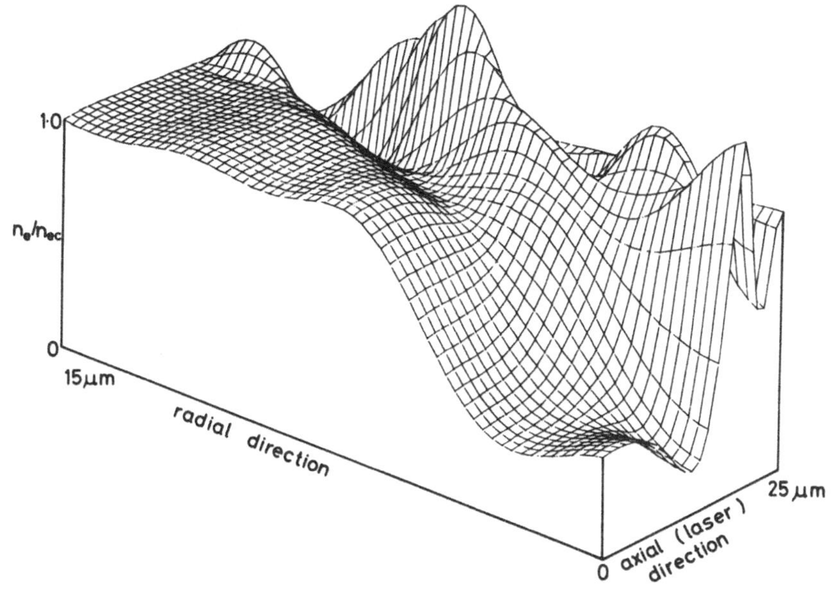

Fig. 12.18. Electron density profile at 1.16 psec after irradiation. Same conditions as for Fig. 12.17.

1987). The exact number is not available from the theory yet, only the rough result of about 1 wave length diameter since the computations are based only on paraxial approximation and not yet on exact diffraction theory. What was new was the result that the higher ionized ions were preferentially emitted in radial and less in axial direction. This could be understood theoretically after a soliton behaviour of relativistic self focusing had been derived (Häuser et al 1988) from an analysis of the preferential direction of ion emission from the focus of interaction (Häuser et al, 1991). Further related work was reported by Bobin (1985); Rubenchik et al (1987); Shvartsburg (1985); König et al (1987); Sack et al (1985); Andreev et al (1989); Singh et al (1987); Kim (1986); Pert (1987); Lin Z.Q. et al (1988); Vertes et al (1989); Phipps et al (1988) and Thompson (1988).

Before finishing the section of relativistic self focusing, one question has to be discussed in retrospect: how is it possible at all that relativistic self focusing is being produced after we have seen from Fig. 12.9 and 12.10 that a laser pulse with a rather long rise time than a few psec only drills holes of rather constant diameter into the plasma and no relativistic shrinking occurs. Only the present days new generation laser with pulse length of few psec for neodymium glass or few ten psec for carbon dioxide with

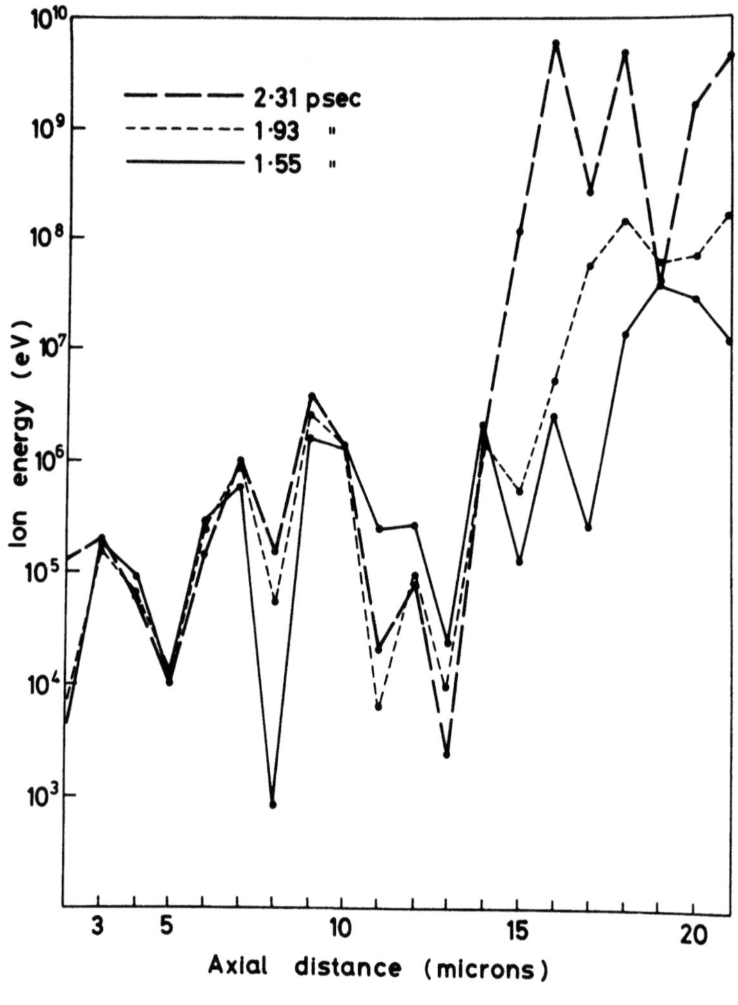

Fig. 12.19. Centre line ion energies as a function of axial depth into the plasma at various times (psec) for the conditions of Fig. 12.17.

maximum intensities above 100 GW would be appropriate to provide the sufficiently fast rise times as seen to be necessary from Figures like 12.17 to 12.19 (Jones et al 1982; Kentwell et al 1987; Limpouch et al 1988; 1990).

Fortunately, the otherwise very disadvantageous pulsation of the laser plasma interaction in the 10 psec range occurred in all experiments in the past where no smoothing was applied (see subsection 10.8). This 10 psec pulsation was just the reason why in so many experiments mentioned before since 1975, relativistic self focussing with ion energies up to 0.5 GeV were measured. The mentioned density rippling and relaxation process within the 10 psec duration was just the necessary condition that the sufficiently fast rise time of the laser interaction was provided such that the relativistic self focusing could occur with laser pulses which nominally were very much longer.

A proof of this situation is possible now when working with a random phase plate RPP, or with ISI or SSD laser irradiation (subsection 10.8). Compared to the initial pulsating laser beam and its MeV ions, no relativistic self-focussing should occur and no MeV ions should be produced only such by nonlinar force (ponderomotive) self focusing (subsection 12.1), however perhaps in larger numbers than under the earlier conditions. When working with ISI and with laser pulses of few psec duration, the relativistic self focusing should then re-appear in a very transparent way.

12.7 Exact Gaussian Beam, Cluster Injection Laser Amplifier, and Laser Acceleration of Particles in Vacuum.

In addition to the results reported in subsection 12.3 there were further developments of interest. One point of basic importance of physics is the consequence we derived that electromagnetic beams in vacuum do have a longitudinal field component. Since the experiment of Young in 1801 it was noted that light (as an example of an electromagnetic wave) has only transversal character. This was very difficult to understand at that time since all other types of waves known then like acoustic waves could well have transversal wave properties but there was necessarily always a longitudinal wave component involved. The fact how waves could have exclusively transversal character was not understood until Maxwell's theory of electromagnetism was discovered where the solutions of his equations resulted in purely transversal waves.

After this triumph of Maxwell's theory it is common knowledge nowadays that light and all other electromagnetic waves are exclusively transversal waves in agreement with the experiments in optics or microwaves physics. One has to remember however that this exact derivation from Maxwell's equations is correct only if one has infinitely spread plane waves or spherical or cylindrical waves excluding the singular points or lines in the latter cases. It was therefore very strange to arrive at the result in subsection 12.3 that light in vacuum does have longitudinal components as a result of exact solutions of Maxwell's equations.

One has to realize that this result was enforced from our studies of the radial emission of electrons in vacuum (or very low density plasma with a Debye length larger than the diameter of the laser beam) [14] where the electrons are accelerated by the nonlinear force to 100 eV or 1 keV energy at neodymium glass laser intensities of 10^{15} or 10^{16} W/cm^2 respectively (experiments by Boreham and Luther-Davies [14][294]). This was very easily understood from the ponderomotive part of the nonlinear force from the radial action of the gradient of the decaying square of the electric field where then according to Eq. (9.24) half of the maximum oscillation energy of the electron when quivering in the laser focus is being converted into translative energy of motion after leaving the laser beam laterally.

This summarizing nonlinear force explanation of the experiment immediately indicates also that the emission of the electrons should not depend on the polarization direction of the light and should be isotropic with respect to the angular cylindrical coordinate of the beam as confirmed experimentally. The dilemma appeared only if one liked to follow up the detailed motion of the electron in the field for deriving the nonlinear force acceleration or when looking to the Maxwellian stress tensor formulation of the force, Eq. (8.21), where the tensor component involved for the lateral acceleration out of the beam has the form of E^2-H^2, i.e. having a zero gradient.

The solution of the dilemma with the single particle motion was elaborated in a transparent way without approximation in subsection 12.3. When calculating the quiverdrift with the transversal components only, one arrives at the measured acceleration in the direction of the E-vector and at no acceleration in the direction of the H-vector, i.e. in a total polarization dependence contrary to the measured polarization independence. We solved this dilemma in subsection 12.3 by realizing that a laser beam of a finite diameter in vacuum does not only have transversal field components. These are some approximation only, and the Maxwellian exact solution - we showed it in elementary formulations for the case of a beam whose amplitude was decaying radially in a linear way - necessarily arrives at additional longitudinal field components (Hora 1981).

The existence of longitudinal components of electromagnetic waves in vacuum was not completely new. Emil Wolf (1951) indicated such behaviour in the focus of a beam and the first general derivation was by Lax et al (1975) at least approximatively using the paraxial approximation. Our exact result (Hora, 1981) showed a very fundamental property of nonlinear physics: when adding the rather very small longitudinal field to the transversal components in the calculation of quiver drift of the electrons when emitted radially from the laser beam in the Boreham experiment [14], the nonlinear force acceleration in the H-direction was no longer zero but arrived at the same value as in the E-direction.

The very little addition of the longitudinal component of the laser light changed a result from "no into yes" or from "wrong into right". This is insofar of fundamental importance as it shows a principle in nonlinear physics: the need for more precise knowledge of linear physics for predictions in nonlinear physics than in linear Physics. Usually one expected the contrary. When going to treat nonlinear problems based on higher order extension one thought it was justified to permit even more neglections and approximation than in linear physics because one was only performing "a higher order approximation". We know now that this can be fatally wrong. The neglection of the very little longitudinal field gave a very wrong result.

If theoretical physics continues in the task to analyse existing physics laws and to predict new phenomena to be verified experimentally later, and if linear physics is going to be more or less saturated, the step into nonlinear physics is then most challenging: it requires a re-confirmation of all linear physics laws and their continuous establishment by higher accuracy, while any theoretical work for such predictions or analyses needs the unapproximated mathematical or numerical treatment. This is indeed a very difficult task in the future but it explains that physics is going into a new and by orders of magnitudes richer area with enormously more applications than known before.

The problem of Maxwellian exact solutions for optical beams is well known from the theory of the Goos-Hänchen effect (Lotsch, 1970, Beauregard 1977) where always approximations for optical wave bundles had been used before. When establishing the theory of this effect for electron waves, the second order difference to the solution of the Schrödinger equation permitted a mathematically exact formulation of a wave bundle (Hora, 1960; Renard, 1964) where the exact evaluation for the case of grazing incidence arrived at the possibility how one may be able to detect a phase property of electron waves (Carter et al, 1971).

Being aware of these difficulties for finding an exact solution for an optical wave bundle, an extensive generalization of the first exact derivation of the longitudinal field from a triangular optical beam (Hora, 1981) towards the case of a Gaussian beam was performed (Cicchitelli, 1988). In this case one starts as before for the triangular case with the transverse components of the field known from the infinite plane wave but truncated in a Gaussian way radially. Using this solution in the Maxwellian equation arrives at a first approximation of the longitudinal components with which the transversal components have to be modified causing in retrospect a further modification of the longitudinal field etc. It was shown by Castillo (1986) that this iteration converges only if the beam diameter is larger than about one wave length. This is an interesting result for the diffraction limitation of this theory.

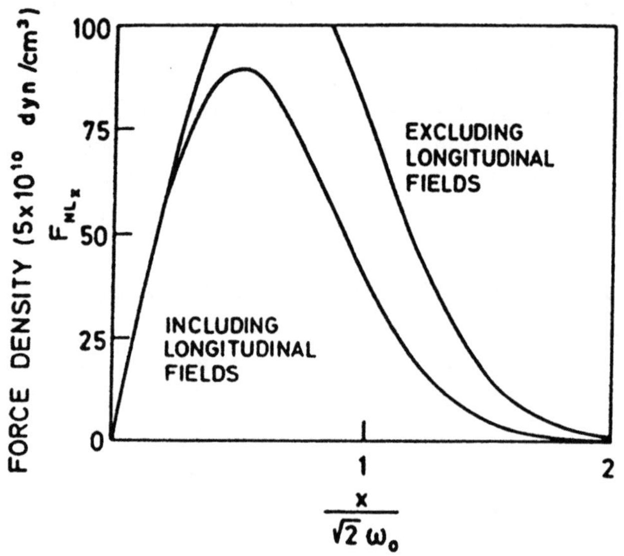

Fig. 12.20. Radial force density (dyn/cm^3) in a linearly polarized Gaussian neodymium glass laser beam of 13 μm diameter (half maximum intensity width) of 1 x 10^{15} W/cm^2 intensity in the direction of the E vector for the two cases including and excluding longitudinal laser field components, plotted against the distance, relative to the spot-size parameter from the center of the beam. [Note that (0,0,z) is the axis of symmetry of the beam].

After Cicchitelli (1988) was able to derive the Maxwellian exact transversal and longitudinal field components of a laser beam in the focus in agreement with the approximative result of Lax et al (1975), this exact electromagnetic field was applied to calculate the electron motion in radial direction using the forces given by the Maxwellian stress tensor. The addition of the very little longitudinal component arrived then at non-zero tensor components and the final force was polarization independent only (as measured) if the longitudinal field was included. Working with the transversal components only again arrived at a polarization dependence, Fig. 11.20 and Fig. 11.21 (Cicchitelli et al, 1990).

Having shown how the Boreham experiment [14] led to the exact derivation of the longitudinal field component of laser beams and to the principle of high accuracy for nonlinear physics, there were important applications of this experiment. One was the first possibility to measure the tunnel ionization of helium atoms (Baldwin et al 1981) confirming the theory of Keldysh [196] contrary to the multiphoton ionization (Kruit et al, 1983; McIlath et al 1987; Chyba et al 1988). In the Boreham experiment (Baldwin et al, 1981), it was possible to detect the time delay of the emission of the first and of the secondly ionized keV electron of the helium atom emitted by the nonlinear (ponderomotive) force from the focus of the laser beam. When using much lower laser intensities which were permitting electron energies of several eV only for the nonlinear force acceleration, a basically different energy spectrum of the emitted electrons was detected (Kruit et al, 1983) which showed a series of peaks

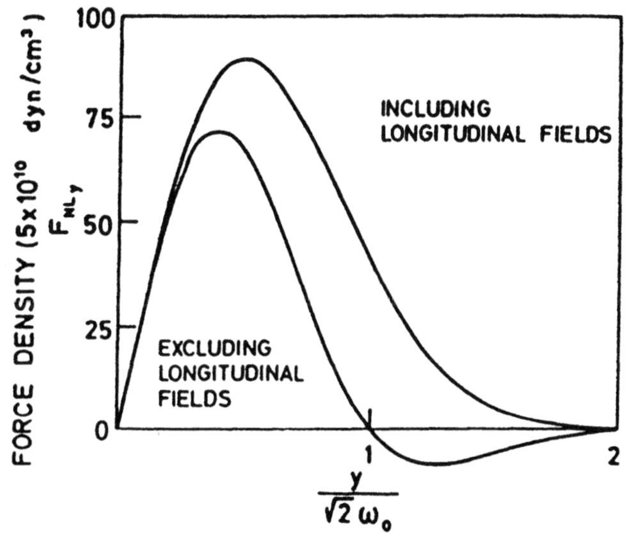

Fig. 12.21. Same as Fig. 12.20, for the direction of the **H** vector.

immediately expressing the multiphoton ionization process. These peaks are washed out when the laser intensity is being increased (Mainfray, 1990).

However nowhere yet has there been a systematic experiment performed with a very wide range change of the laser intensity in order to see the transition from the multiphoton into the Keldysh ionization. Based on other observations of quantum modulation of electron beams by lasers, we derived a criterion of a threshold intensity (Hora et al, 1987a; Oganesian et al, 1989) below which quantum mechanisms and above which classical processes are expected. This threshold intensity (Hora et al 1987; Meyers 1987) is

$$I^* = \frac{h}{4\pi}\frac{m\omega^3 c}{8\pi e^2} = 1.66/\lambda^3 \; [\mathrm{W/cm^2}] \tag{12.55}$$

where the wave length of the radiation (λ) is given in cm. This threshold (which is different but rather close to the one defined by Keldysh including specific ionization energies) has been interpreted as the **correspondence principle of electromagnetic interaction** (Hora et al 1987). The interpretation of the Boreham experiments for tunnel ionization consists very simply in the fact that this kind of classical tunneling is within the classical high intensity range of this criterion while the multiphoton case of Kruit et al (1983) is on the quantum mechanical side of lower laser intensity than in Eq. (12.55).

The threshold of Eq. (12.55) immediately explains why long wave length (e.g. radio wave) interaction with electrons is classical: only an extremely low intensity would be able to show quantum interaction as long as the long wave limit due to the Handel effect has not been passed (Handel 1975). It also explains why the molecular level electromagnetic interaction in the 10μm to 10 mn wave length range in biology for communicationn, e.g. in telepathy or measured with SQIDS as signals from rats a few meters away from others when being killed (Tankovich et al 1990), are those of quantum effects for very low intensities. These are specific molecular line quantum effects and have to occur with respect to generation and detection at the intensity level below the threshold of Eq. (12.55). The present days far infrared physics usually operates with intensities far beyond this threshold and therefore covers the classical effects only. Thanks to SQIDS it may be possible now under intentional attention of criterion (12.55) to learn about the quantum molecular exchange between individuals in biology.

The other application of the Boreham experiment [14] was mentioned in subsection 12.3: a free electron laser without wiggler fields. The Boreham experiment showed how the emission of the electrons from a laser beam was a collisionless nonlinear optical absorption process where the kinetic part of the quiver energy of the electrons was converted by the quiver drift due to the ponderomotion by nonlinear forces into translative energy of electron motion. In this case optical energy had to be taken out of the laser beam and was converted into kinetic energy of electrons. The idea to invert this process was obvious: fire electrons into a laser beam and convert their kinetic energy into optical energy. It was clarified soon that this process will not work with stationary laser beams as in the mechanism described by Kibble (1966): the electrons will be reflected by the beam or transmitted without energy loss (if one neglects Thomson scattering).

If one however used laser pulses, the switching on and off process does permit the amplification of laser beams by the kinetic energy of injected electrons [303]. This scheme of a free electron laser amplifier has been developed in more details (Kentwell et al 1986; Wang et al 1986) where the advantage consists in the fact that an appropriately precise synchronisation and adjusting of the required parameters may lead to laser amplification with 80% efficiency and more, whatever wave length is used. The amount of amplification, however, depends strongly on the density of the injected particles and decays similar to all free electron lasers with the square of the wave length (Eq. 12.39). While amplification with electrons beams is possible in the range of 0.4 mm wave length with high power electron guns (Hefei, 1983) the limitation of electron densities and space charge effects are too disadvantageous for lower wave lengths. One way out is the use of clusters of condensed pellets injected into the laser pulse (Wang et al 1986, Kentwell et al 1986) where the amplification is by a factor 2 in the visible or near UV range, and even for 10 Angstrom wave length interesting amplification as well as manipulation of the x-ray beams with respect to correcting phase anomalies of changing the beam direction can be combined with the amplification.

One fundamental problem has been solved with respect to the momentum transfer. It has been shown that the emission or injection of the electrons in the beam has not to be exactly perpendicular but oblique. This is due to the fact that the optical energy added or subtracted from the laser beam has a momentum and just this is to be compensated by the mv_e of the electrons in the axial direction (see reference Viera in citation Kentwell et al 1986). This result also confirmed the otherwise derived forward velocity according to Klima and Petrzilka [184] derived from their first transient theory of the nonlinear force. The result is that the electrons emitted from a neodymium glass laser beam of 10^{17} W/cm^2 intensity will have a forward direction of about 6 degrees.

While this was the clarification of the axial momentum transfer it still has to be clarified how the radial momentum is being exchanged. From the results of Cicchitelli (1988) with the exact transversal and longitudinal components of the laser field, it can be seen how the phase shift betwen the components are producing a quiver drift of the electrons which causes a spiralling motion due to the beam. The electrons emitted by this spiralling in the Boreham experiment [14], are exchanging angular momentum to the electromagnetic beam when taking optical energy out of the beam. By this way of explaining the radial momentum exchange, one has a first direct experimennt available how to measure the angular momentum and the spin of photons (Jackson 1975, Sokolov 1986) which preferably would use a Boreham type experiment in the microwave range in order to be able to see the spiralling and eccentric emission of the electrons from the beam.

The acceleration of electrons by the nonlinear force of laser beams in vacuum raises the question how this at all is possible in view of the knowledge (Sessler, 1988) that an electron can never exchange energy when being passed by a laser pulse of internal symmetric properties. This is e.g. the result of an exact solution of the relativistic motion of electrons passed by a plane wave optical pulse (Scheid et al 1989). In the case of the Boreham experiment the situation is different in such a way that the electron can never exchange its energy enters the laser pulse in the bound state within the helium atom and only after ionization within the focus, the quiver motion and acceleration process begins. Furthermore, the plane wave condition is no longer valid and the radial beam properties are involved for the energy exchange raising even the problem with the longitudinal components. This is indeed different to the conditions mentioned before with the symmetric plane wave laser pulses.

An acceleration of electrons is possible by the mechanism of the Boreham experiment by sidewise kicking from a laterally sweeping laser beam, Fig. 12.22. In the same way as the electron is reflected from the laser beam in the way of the Kibble (1966) mechanisms (Sakai, 1989) if its kinetic energy is less than the maximum quiver energy in the laser beams (left hand side of Fig. 12.22), a resting electron (right hand side of Fig. 12.22) when being hit by a sweeping beam having the same relative velocity, will receive twice the sweeping velocity. In view of the spiralling motion with respect to the angular momentum of the optical beam and to the photon spin, the sidewise kicking becomes a rather sophisticated mechanism.

Acceleration of electrons hopefully to such high energies as TeV with high luminosity is of interest for high energy physics. One scheme is the beat wave accelerator (Tajima, 1985) where a superposition of two laser waves with little different frequencies irradiate a homogeneous plasma whose plasma frequency is equal to the difference of the laser frequencies. The then created longitudinal Langmuir electron wave acclerates injected

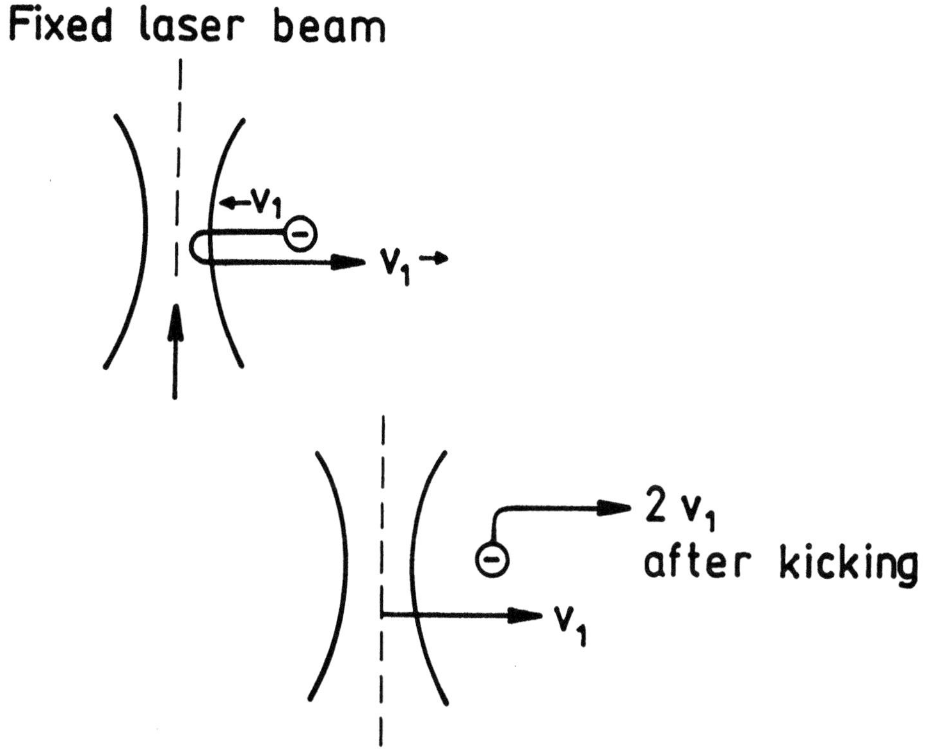

Fig. 12.22. Acceleration of a resting electron by a sidewise sweeping of a laser beam and kicking on the electron due to the nonlinear force mechanisms [14](Sakai, 1989).

electrons during a half wave length interaction. This is the same half wave mechanism as the relativistic solution of the electron motion by a half wave (after "recitifcation" of the wave) mechanism can well accelerate electrons to energies of TeV (Scheid et al 1989; Cicchitelli et al 1989). Since the needed driving energy of the half wave mechanism is inversely proportional to the wave length, the rather long wave mechanism of the beat wave accelerator is disadvantageous. The more disadvantages arise from the plasma interaction process: the beams destroy the necessary condition of a homogeneous plasma, cause relativistic self focusing and instabilities and further complications which are avoided if one uses acceleration of electrons by lasers in vacuum without any plasma (Evans, 1988).

In order to achieve laser acceleration in vacuum, several mechanism are being discussed (Hora, 1988; Sakai 1989; Kawata, 1989; Cicchitelli et al,

1990). One concept is to trap electrons in the intensity minima of laser fields, e.g. in standing wave fields or in the interference pattern e.g. of a fresnel mirror, and to move and accelerate then the location of these minima. The proposal for moving of the standing wave minima by varying of the frequency of one of the two superimposed laser fields was elaborated (Hora, 1988) after Gover et al (1987) performed such an experiment with microwaves and confirmed that injected keV electrons gained energy in the range of few eV. To understand this, one has to include the lateral beam mechanism in order to avoid a confusion with the result that infinitely spread symmetric plane wave pulses cannot transfer energy to the electrons.

The further question remained how electrons can be trapped in the intensity minima of laser fields and whether they would slip through the maxima of the wave field if this is being accelerated (question by Lawson, 1988). While the simplified ponderomotive force picture (Weibel [161]) would immediately result in a positive answser about trapping and accelerating of the electrons, a careful reconsideration of the single particle motion was indicated. When evaluating the simple equation of motion one realized that the trapping is the result of higher order effects. The sufficiently precise numerical treatment was then not a trivial task and it turned out (Cicchitelli et al 1990) that the generation of a density maximum of electrons (or plasma) in the nodes of the standing waves due to the nonlinear force is due to the pendling of the electrons between points of equal laser intensity such that the density maximum is the result only of an averaging process over a large number of electrons. Furthermore, the trapping at acceleration has been confirmed and the net acceleration of the electrons with the moving "standing" wave fields was immediately reproduced (Cicchitelli et al 1990).

The half wave acceleration mechanism using the longitudinal component of the laser beam was discussed by Caspars et al (1990) and was included in the total half wave acceleration for oblique injection of electrons into a powerful laser beam such that the electrons will have crossed the beam after the first half wave at relativistic interaction (Cicchitelli et al 1990). Using a neodymium glass laser of PW (10^{15} Watts) power now available, an upshift of the electron energy by nearly GeV is expected.

THIRTEEN

Laser Compression of Plasma for Nuclear Fusion

After the laser was discovered [57] and the possibility of extremely high spatial and temporal concentration of energy was evident (Collins et al 1960), the question arose whether or not the laser could be used to irradiate frozen mixtures of the hydrogen isotopes deuterium and tritium to ignite a nuclear fusion reaction with a very high gain of energy. It is remarkable that Andrei Sakharov (1983) in his comments to his collected papers, mentioned that he had thoughts in this direction immediately after the discovery of the laser in 1960. Several estimations to this question appeared after the disclosure by Basov and Krokhin (1964) Febr. 1963 at a Quantum Electronics Conference in Paris (Kastler, 1964; Dawson, 1964; Hora, 1964) while extensive work by Nuckolls (1974) at this time was classified (Teller, 1972).

Thirty years later it has been clarified what difficult problems have to be solved. However, the conclusion can be drawn that the concept is now mature, the physics problems are solved and an economic, clean power station can be built within ten years if a very intense engineering development program is initiated. Knowledge of the required physics is even going beyond this goal: it is possible that within 30 years, the costs of this clean and inexhaustive energy could be a factor three or five times below the today's lowest cost energy, as given by the use of light water fission reactors. There is the possibility then within 80 years, production of energy from light hydrogen and boron may be a reality. This will mean the production of **nuclear energy without radioactivity**, or at least with less radioactivity per unit of produced energy than is released by burning coal and oil (Weaver et al 1973).

About the difficulties, the long years and very expensive work with most advanced technologies ("at a level of the next century") matured now. The physicists involved may, by then, have become nearly complete experts following a definition by Edward Teller (1972a) he referred to Niels Bohr (however not confirmed by Werner Heisenberg at this state): "An expert is a person who knows all mistakes one can do in a special field". Teller said then in 1972 to a leading plasma theoretician that this person is not an expert, obviously expressing doubts whether the knowledge of plasma physics was complete.

The difficulties are in two directions. One is that one needs laser pulses of very high energy, values which may be available with the present days technology if very large laser systems are built (Manes et al 1986; Campbell, 1991). The other difficulty is the interaction of the laser light with the plasma. It seems as if the plasma "does everything" to prevent the input of energy from lasers.

All preceding Chapters of this book described the long story how complicated the interaction of the laser radiation with plasmas and the subsequent plasma dynamics is and where new relativistic effects (Chapts. 6.5, 12.2 and 12.6), quantum effects (Chapts. 2.6 and 6.6), and highly nonlinear effects appeared in the force densities and the plasma dynamics (Chapts. 8 to 10), in the optical constants (Chapt. 6), in self focussing (Chapt. 12), instabilities (Chapt. 9.5), particle acceleration (Chapt. 12.7), and resonances (Chapts. 11.2 and 11.3). Furthermore properties of the laser beams had to be clarified with respect to longitudinal wave structure in vacuum (Chapt. 12.3) and angular momentum (Chapt. 12.7) apart from new technology of lasers (Chapt. 1.5) which in no way could be imagined before.

It is a result of the long years research to get these phenomena under control and it seems that the very last years' discoveries about the pulsation of the laser-plasma interaction and the empirical methods to suppress these difficulties by smoothing are now available. This pulsation is now understood theoretically (Chapt. 10.8) providing now sufficient knowledge about the physics tools to solve the problem of the greenhouse catastrophy: and to provide an increase of energy production while reducing the emission of carbon dioxide into the atmosphere. There is now extensive and highly refined research work available which scientists are presenting to the people of business and law (Suzuki, 1987) who determine the political decisions for appropriate action.

13.1 Nuclear Fusion Reactions

The reactions of interest are deuterium with deuterium D-D and deuterium with tritium (D-T), as it was discovered by Oliphant, Harteck, and Rutherford 1934.

$$D+D \rightarrow 50\% \begin{array}{l} ^3He+n+3.27MeV \\ ^3T+p+4.03MeV \end{array} \tag{13.1}$$

$$D+T \rightarrow {}^4He+n+17.6MeV \tag{13.2}$$

The necessary tritium can be produced (bred) from ^7Li by the emitted neutrons from this reaction or by reaction (13.1).

It should be noted that the understanding of the results of these reactions in the first experiment (Oliphant et al 1934) was not easy. It was indeed a success that the high current 100 keV deuterium discharge based on the long years development of Oliphant succeeded in a reaction at all after an earlier attempt in Berkeley did not show any reaction (Oliphant 1987). But the understanding was even a difficulty for Rutherford who in history was the most successful physicist this century to decipher radioactivity and nuclear reactions. The difficulty was that the reactions (13.1) were branched and that the superheavy hydrogen isotope tritium, T, appeared which was not at all known before. And this new isotope performed hefty reactions with the heavy hydrogen isotope (deuterium). This all to analyze was the merit of Rutherford. It was his last active involvement in an experiment (Oliphant, 1987).

The DT reaction (13.2) is the only one which can be used exothermally by the magnetic confinement method of the tokamak, as the other interesting reactions produce too high a loss of energy by cyclotron radiation. The inertial confinement fusion is not restricted to reaction (13.2), and there is even reasonable hope that clean nuclear reactions could be realized where no neutrons are produced, which damage the reactor or produce radioactive waste in the reactor material. Preferable are further those reactions where the reaction products are charged particles, whose kinetic energy can then be converted directly into electrical power. Heat pollution through thermomechanical conversion is avoided. Examples for the clean fuel fusion reaction are (Miley, 1976; 1981; Best 1977; Weaver et al, 1973)

$$D+{}^3He \rightarrow {}^4He+p+18.3 \text{ MeV} \tag{13.3}$$

where the necessary ^3He can be bred by the fast protons of the reaction from ^6Li. Another example is the reaction of light hydrogen H with ^{11}B (Oliphant et al, 1933; Okamoto, 1974)

$$H+^{11}B \rightarrow 3\,^4He+8.9\ \mathrm{MeV} \tag{13.4}$$

or the reaction of D with ^6Li, for which several branches exist. The valuation of the cross sections of the 2 ^4He branch D ^6Li reaction and the comparison with D-D, D-T, and D-3He for use in fusion reactions has been done by Clark et al (1978), Fig. 13.1. The average of the cross section σ for a thermalized plama of temperature T for the spectrum of the thermal velcoity v is for an averaged mass m_i of the nuclei

$$<\sigma v> = \frac{\sqrt{m_i}}{2\sqrt{\pi}(KT)^{3/2}} \int_0^{\infty} \frac{m_i}{2} v^2 \sigma(v) \exp\left(\frac{m_i v^2}{KT}\right) dv^2 \tag{13.5}$$

is given in Fig. 13.1.

The reaction of the nuclei is not possible in a simple way for achieving exothermal nuclear energy if the nuclei of one species is fired with the interesting energy of several 10 keV into cold targets with the other nuclei because most of the kinetic energy of the irradiated nuclei is lost by interaction with electrons instead of the fusion reactions. The conclusion was then that one has not to use cold targets but has to heat up the whole reacting system into a plasma of temperatures of the few keV energy - corresponding to few hundred million degrees Kelvin. These plasmas have to be confined for a sufficiently long time at a sufficiently high temperature such that the expected number of nuclear fusion reactions will occur.

If gravitational forces are used for the confinement of plasmas one arrives at stationary fusion reaction conditions for the diameters of stars. Kippenhahn (1980) in 1958 used one of the first electronic computers for calculating the equilibrium conditions for this gravitational confinement and arrived at the diameters of the stars for different conditions of brightness and age such that the Hertzsprung-Russel diagram could be explained. For producing fusion energy on earth- a method for confinement of plasma - other than gravitational confinement has to be used. Confinement by magnetic fields (toroidal geometries, magnetic bottles, pinches, rotamaks (Tuczek, 1979; Hugrass et al 1980) etc. are used but do not yet arrive at the physics solution for a sufficiently long confinement. Even if confinement would be reached, the very high costs of a reactor following from these systems (Pfirsch and Schmitter, 1989) are of a prohibitive level.

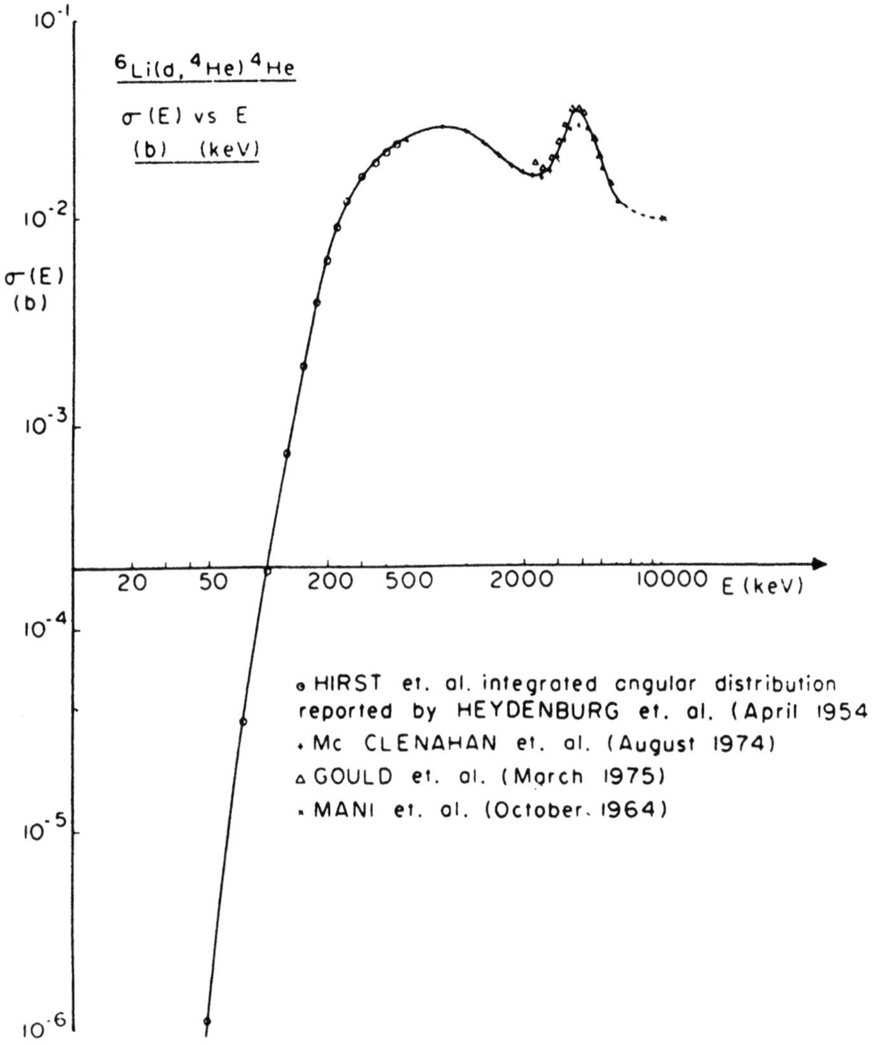

Fig. 13.1. Measured and best fit cross sections for the 6Li(d,α)α reaction using the results of Heydenburg et al (1954), McClehan et al (1975), Gould et al (1975), and Mani et al (1965) following Clark et al (1977).

Confinement by inertia has been verified by the thermonuclear explosions (Teller, 1987). Using an explosive nuclear fission reaction of a type of an A-bomb, the released energy is so high that it heats and compresses solid material with light nuclei so fast that the subsequent expansion and cooling is slow enough and much more energy is being released from the fusion reaction than one had to put in by the fission reaction for ignition. After this well demonstrated success of the

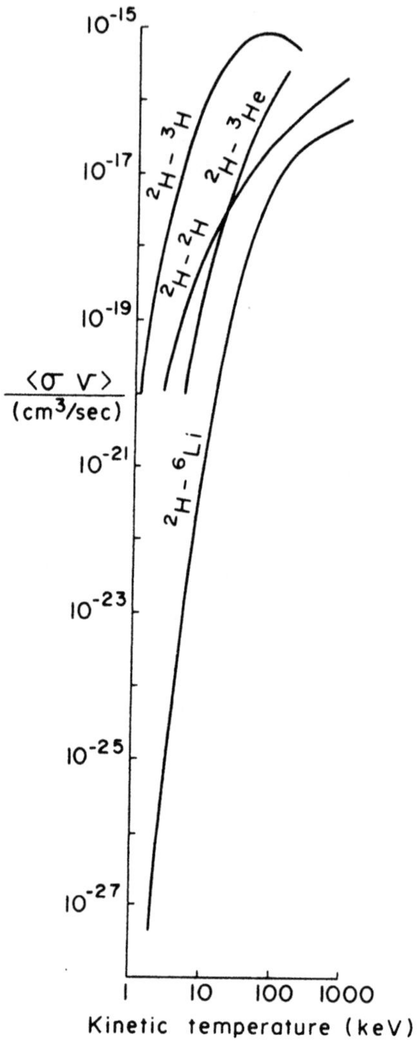

Fig. 13.2 Velocity averaged cross section for plasma of various temperatures T for the reasction as in Fig. 13.1 compared with reactions of hydrogen isotopes and ^{3}He (Feldbacher et al 1988; Clark et al 1978).

uncontrolled (explosive) inertial confinement fusion, the current aim is to simulate this process in a controlled way by using pulses of laser energy or of intense particle beams for ignition of the fusion reaction. This type of inertial confinement fusion - after long years of research - has proved that the physics solution can be reached with today's technology of lasers and possibly also with particle beams.

13.2 Adiabatic Volume Compression and Volume Ignition

In order to get information about the necessary conditions of inertial confinement fusion (ICF) the following numerical evaluations were performed. We consider a spherical plasma of a volume V_o or radius R_o and an initial density n_o which, by not specified mechanisms, has been heated up by an input laser (or particle beam) energy E_o to an initial temperature T_o. We calculate then the fusion reaction energy during the subsequent expansion and adiabatic cooling (sometimes called the "adiabatic losses"). The gain G is then given by this energy divided by E_o. We shall discuss this problem up to the last year's results though some earlier conclusions went into another direction of computation where instead of the adiabatic volume type self similarity hydrodynamics (Schmalz, 1986) (see Chapter 5) the fusion reaction by a spark ignition of the center of a laser irradiated spherical fusion pellet (or capsule) is being considered. The spark ignition will be discussed separately in the following subsection.

The fusion reaction gain is defined by the ratio

$$G = \frac{\text{Reaction energy}}{\text{Input energy } E_O} = \frac{\varepsilon_R}{E_O} \int dt \int d^3r \frac{n_i^2}{A} <\sigma v> \qquad (13.6)$$

where ε_R is the energy per fusion reaction (see Eqs. (13.1) or (13.2), n_i is the ion density and $<\sigma v>$ (Eq. 13.5) is the velocity averaged fusion cross section with a constant A=4 for binary reactions, otherwise A=2. Gains depending on the initial volume, density, and input energy can be calculated, Fig. 13.3. The highest gains are those where the initial temperature for DT is 17.3 keV and for $H^{11}B$ is 150 keV. The optimum gains (Hora, 1964) follow from the tangential line to the parabolic curves of Fig. 13.3. and result in the relation (Hora and Pfirsch, 1970; 1972)

$$G_O = \left(\frac{E_O}{E_{BE}}\right)^{1/3} \left(\frac{n_O}{n_{ec}}\right)^{2/3} \qquad (13.7)$$

where n_s is the solid-state density of the fusion fuel, and E_{BE} is the break-even energy, which initially was few MJ (Hora, 1964; Hora et al, 1972) and arrived at 7.6MJ for adiabatic compression and expansion after using more precise cross sections and the usual definitions for DT (Kasotakis et al, 1989). This formula has the ablity to show immediately how an increase of the initial plasma density n_O decreases the necessary input energy E_O by the quadratic power, if the same gain G has to be produced, indicating the need for compression of the plasma. If one expresses the input energy $E_0 = 4R_0^3 n_{i0}(1+Z)kT_0/3\pi$ by the initial radius R_0 of the

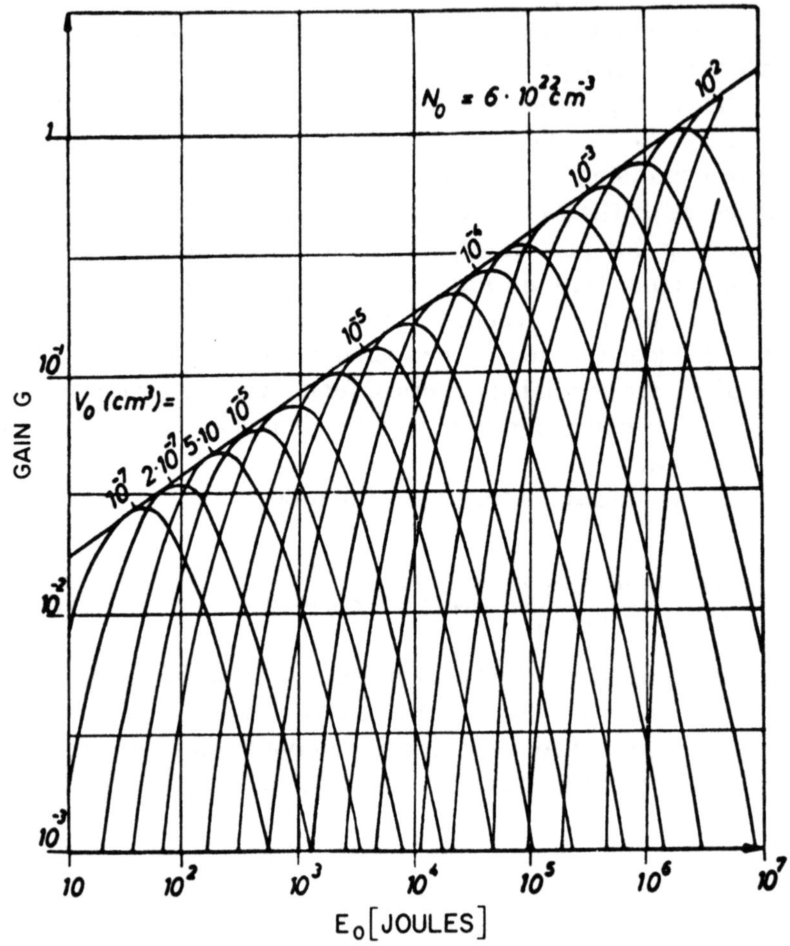

Fig. 13.3 Fusion gains G from Eq. (13.6) for DT depending on the input energy E_0 into the spherical pellet of solid state density of varous initial volume V_0 (Hora, 1964).

plasma and the initial temperature T_0 derived from the optimum calculation, formula (13.7) is then (ρ_0 is the initial density)

$$G = \text{const}\ n_{i0}\ R_0; \ \text{const} = 8.78 \times 10^{-23}\ \text{cm}^2 \quad (13.8)$$

$$G = D\rho_0 R_0\ ; \ D = 21 g^{-1}\ \text{cm}^2 \quad (13.8a)$$

as first given by Kidder (1974) with a constant for DT differing by a minor amount only from our earlier derivation from Eq. (13.7).

The next step for a generalization of the gain calculations is to take into account the depletion of the fusion fuel and the losses by bremsstrahlung as long as the absorption length is larger than the plasma size. Furthermore, the reheat or selfheat of the plasma by the generated alpha particles must be included. Because the alpha production is of a large intensity at a fusion reaction of the following ranges, the Fokker-Planck approximation for the binary stopping power cannot be used, since the approximation works with only small perturbation and with the first Fokker-Planck coefficients. Based on a concept that was successful for high-intensity electrons, and based on polarization effects, a collective model was used for calculating the stopping power (Ray et al, 1976; 1977).

It must be emphasized that Ray was first discussing the Fokker-Planck range, reproducing the Winterberg (1969) approximation of the range $R = T^{3/2}$ (T= plasma temperature) but more generally showing a kink at temperatures near 1 keV, Fig. 13.4. There was no change due to a quantum

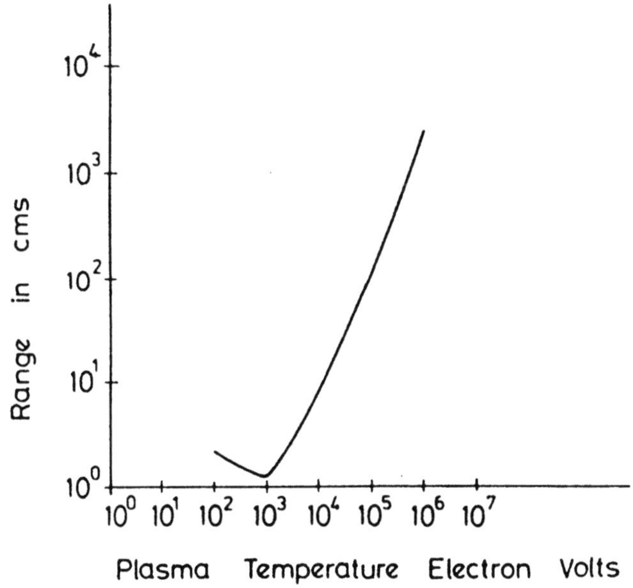

Fig. 13.4. Fokker-Planck approximation of the range R of 14.7 MeV protons in a DT plasma of solid-state density depending on the plasma temperature T (Ray et al 1977).

electrodynamic generalization (Ray et al, 1977). The low-temperature part merges into the ranges calculated from the collective model (Ray et al 1976) where the range R is

$$R = \frac{e^2}{2KT}\frac{m_H}{m} \cdot E_i[\ln(\lambda E_H)^2]; \quad Ei(X) = \int_{-\infty} \frac{\exp(t)}{t} dt \tag{13.9}$$

where

$$\lambda = \left(\frac{m}{m_H}\right)^2 \frac{KT}{\pi n_e Z^2_H e^6} \tag{13.10}$$

for an initial energy E_H and mass m_H of the high-energy particle of charge Z_H and the plasma temperature T. An example for the range of the alphas of the $H^{11}B$ reaction is given in Fig. 13.5, which always corresponds to the range left of the kink of the Fokker-Planck result. To the right hand side of the kink (for temperatures above 10^3 to 10^4 eV) the discrepancy between the collective model and the Fokker-Planck approximation could be large, (Ray, 1977). It is evident that one has to use the shortest possible R if there were competitive models for R for calculating the reheat. With this stopping power (which was always that of the collective model), we arrived at fusion gains given in Fig. 13.6 for DT (Hora et al 1978; Kasotakis et al 1989).

When computing the fusion gains shown in Fig. 13.6, an unexpected behaviour was observed. The results were similar to that of the early computations (Fig. 13.3) only if the gain was not much larger than 7 and the fully drawn lines were linear. These were the asymptodes to the parabolas (as in Fig. 13.3) as noted with dashed plots in Fig. 13.6 for low gains. The "parabolas" for higher gains, were strongly deformed (see dashed curves in Fig. 13.6) having nearly a jump or a discontinuity, changing the gain by more than an order of magnitude when the initial temperature changed a few percents. We found out - as explained in the following - that a volume ignition occurred (Hora et al 1978).

What was remarkable was that the superlinear deviation from the linear plots in Fig. 13.6 occurred always at a gain of about 8 independent of the plasma density. The slope of the gain curves were similar for the various densities and one could derive the following numerical-empirical generalized formulas for the gain G using the result of G_o, Eq. (13.7) (Hora, 1987).

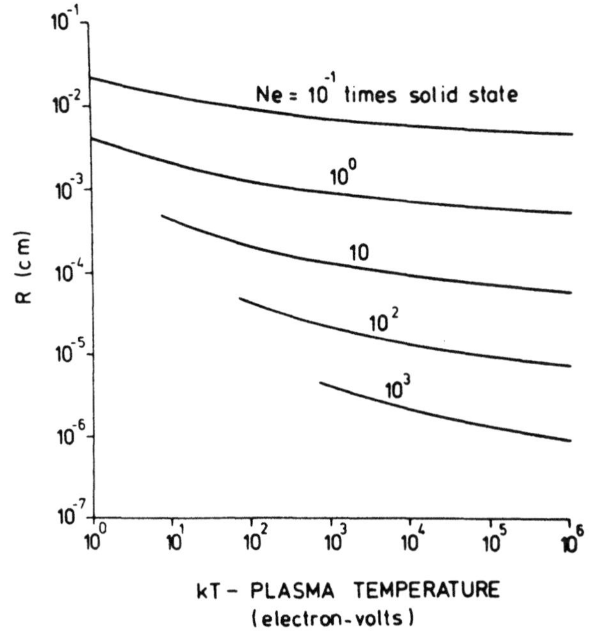

Fig. 13.5. Range of 2.89 MeV alphas from the $H^{11}B$ reaction on the temperature of plasmas of various electron densities n_e (Ray et al 1976).

$$G = G_o \frac{7.2x10^{-4}G_o{}^3}{[1+1.44x10^{-3}G_o{}^3]^{1/2}-1} \tag{13.11}$$

This formula is correct only up to gains of 200. Above that value, fuel depletion lowers the increase in G with the input energy, approaching saturation, see Fig. 13.6. The optimum temperature is given by

$$T_{opt} = 17.2\frac{3.52\{[1+0.142G_o]^{1/2}-1\}}{G_o} \tag{13.12}$$

An analytical derivation of these formulas was given by Tan Weihan et al (1990) in agreement with the results of Eqs. (13.11) and (13.12). A simplified discussion of the volume ignition with several approximations was presented by Basko (1990,1990a).

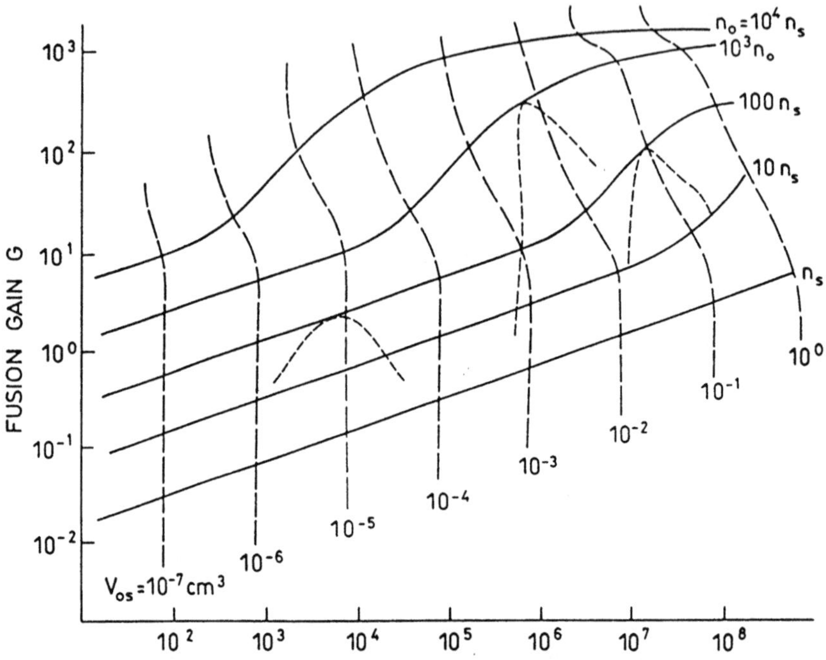

Fig. 13.6. Optimum fusion gains G calculated for an adiabatic compression and expansion of a DT plasma of an initial volume at solid state density with a maximum compression to a density n_o in multiples of the solid state density. When reaching the density n_o, the compression velocity is zero and the input laser energy E_o is fully converted into thermal energy representing the optimum temperature as given by Eq. (7). The dashed lines are the gains for a fixed volume for various input energy touching the optimum gains at the straight lines. The dashed lines are parabolas for low gains and show a jump-like increase in the case of volume ignition (Hora et al 1978; Kasotakis et al 1989).

To understand the results of the computations of volume ignition (Hora et al, 1978; Cicchitelli et al, 1988; Kasotakis et al 1989; 1980a) with regard to simplified earlier known facts, Figure 7 shows the time dependence of the plasma temperature T for the case of an initial density of 100 times the solid state with an initial volume of one cubic millimeter. Three cases are given in Figure 13.7 where the input laser energy was 11, 12, 13 MJ respectively. The fusion gains G were 2.01, 22.38 and 103.3. The ignition process can be seen clearly, not only from the strong increase of the gain G accompanying little increase of the input energy defining the initial plasma temperature, but from the increase of the temperature with time. In the first case (11 MJ), the temperature is nearly constant for a long time, since the temperature loss by bremsstrahlung and by the adiabatic expansion is nearly compensated by the alpha self-heat until a strong drop of the temperature occurs with the fast expansion. We call such a case of monotonically decreasing temperature a 'simple burn', similar to the pure burn cases of the early computations (Fig. 13.3), Eq. (13.7), where the contribution of self-heat is not so influencial.

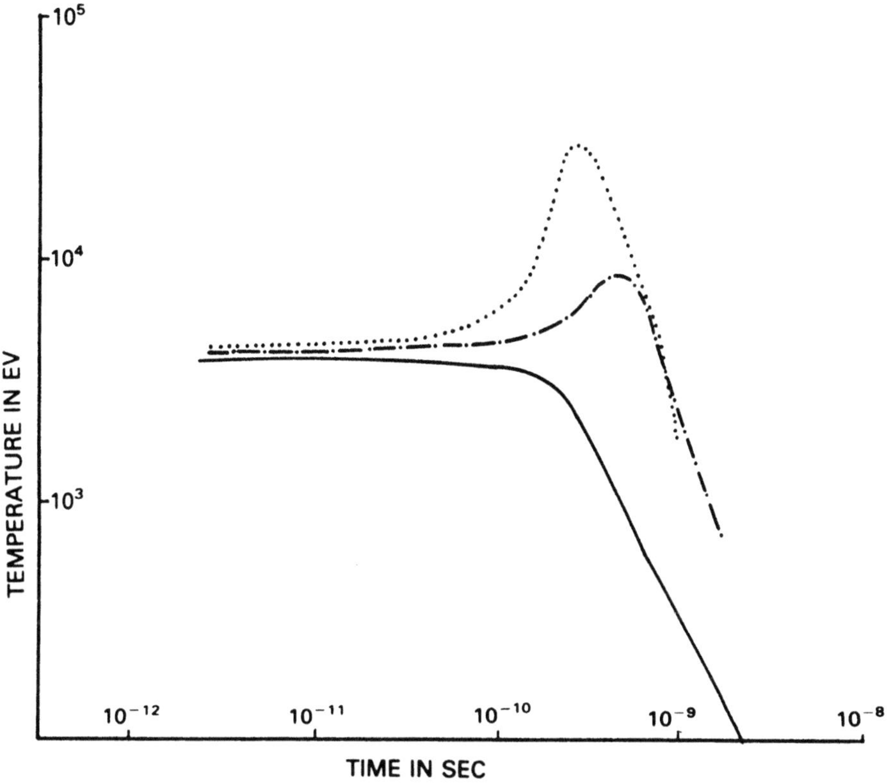

Fig. 13.7. Dependence of the temperature T of a pellet on the time t at initial compression to 100 times the solid state and initial volume of 10^{-3} cm^{-3}. For the cases of the (fully drawn; dashed; dotted) curves respectively, the input energies were 11, 12, 13 MJ, the fusion gains G were 2.01, 22.38, 103.3, the initial temperatures were 3.95, 4.31, 4.67 keV and the deuterium tritium fuel depletion was 0.28, 3.33, 16.65% (Kasotakis et al, 1989).

A ten percent higher energy input results in the appearance of ignition instead of the simple burn. The alphas inject more heat into the plasma than the losses and the temperature of the pellet increases after the input of the driver energy of 12 MJ. A maximum temperature of nearly 10 keV is reached before the fast expansion occurs and later adiabatic cooling drops the temperature. Increasing the input again by about 10 percent results in strong ignition with a maximum temperature of 31 keV and a high gain above 100. The initial temperature in the three cases of figure 13.7 are 3.95, 4.31 and 4.67 keV. This is just a little bit less than the simple case (Glasstone et al, 1960) where the generated fusion energy is equal to the generated bremsstrahlung (T = 4.5 keV). This is an easy global confirmation that the computations are perfect.

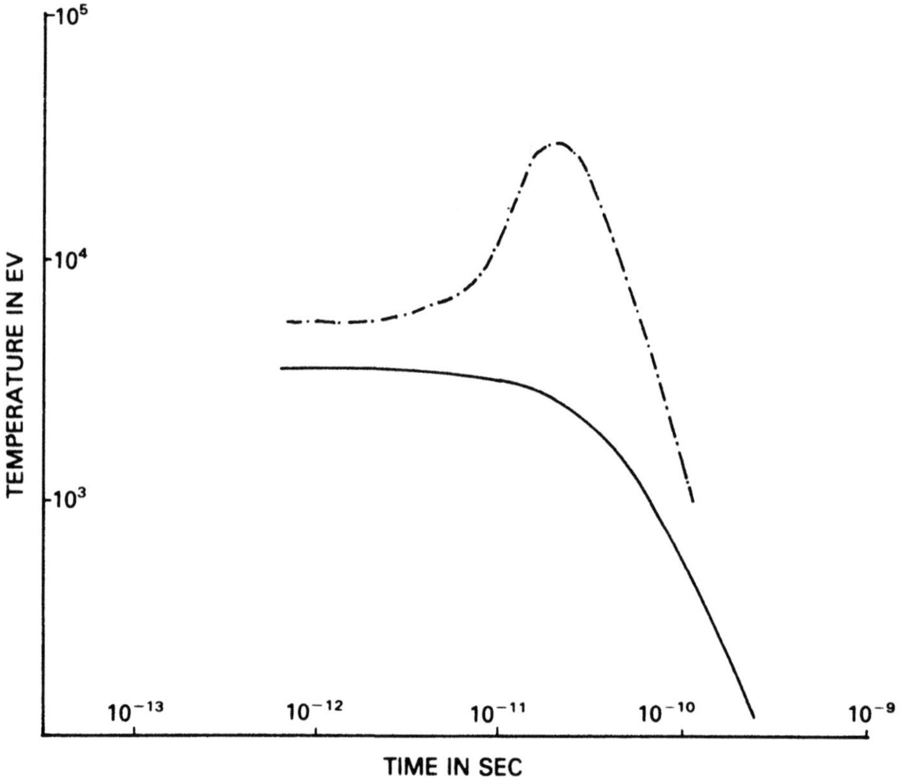

Fig. 13.8. Same as in Figure 13.7 for a compression to 1000 times solid state density and initial volume of 10^{-6} cm^3. The (fully drawn; dashed) curves respectively correspond to initial laser energies of 100; 150 KJ, the gains G of 1.69, 99.55, the initial temperatures were 3.59, 5.39 keV and the fuel depletion was 0.21, 18.51% (Kasotakis et al 1989).

Obviously a small part of the bremsstrahlung is re-absorbed in the pellet in Fig. 13.7. This fact can be expected since the maximum wavelength of bremsstrahlung is at 4.5 keV so the absorption length in the pellet of 100 times solid-state density is large compared to the typical diameter.

Figure 13.8 reports the results of a case as in Figure 13.7 but for a compression to 1000 times the solid state density using an initial volume of 10^{-6} cm^3. There are only two cases shown, one with an input laser energy of 100 KJ and another one with 150 KJ. The respective gains G are 1.69 for simple burn and 99.5 for ignition. The initial temperatures were 3.59 and 5.39 keV. Once again no strong re-absorption of the bremsstrahlung can be observed with an initial temperature of about 5 keV even for ignition. This can be confirmed by examining the initial radius and density and taking the maximum x-ray wave length for the 5 keV temperature: the absorption length is not shorter than the initial

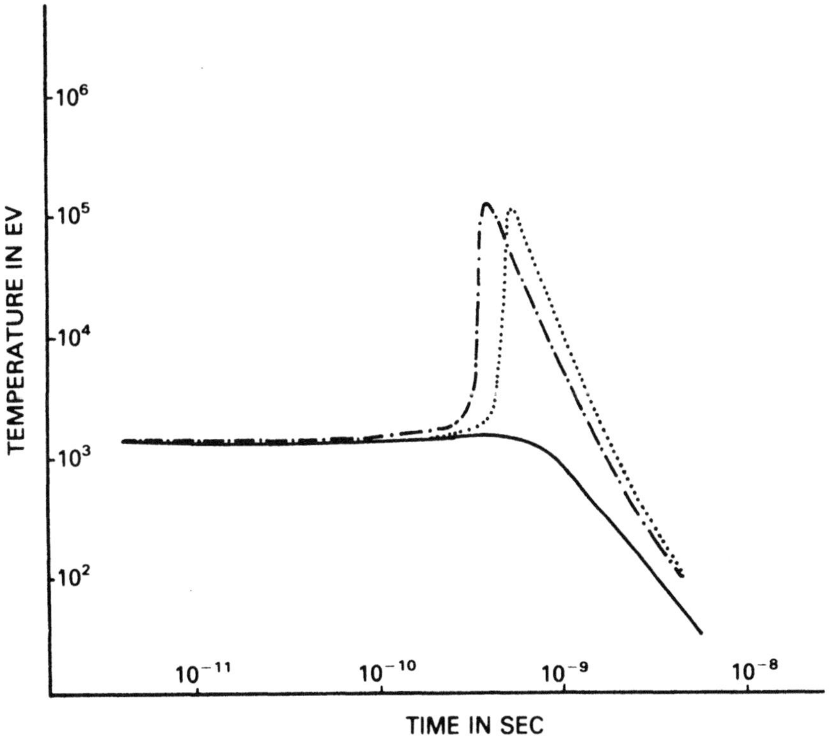

Fig. 13.9. Same as in Fig. 13.7 for a compression to 1000 times solid state density and initial volume of 10^{-3} cm^{-3}. The fully drawn; dashed; dotted curves correspond respectively to initial laser energies of 40, 41, 42 MJ; the gains G of 2.41, 1170, 1199, the initial temperatures were 1.44, 1.47, 1.51 keV and the fuel depletion was 0.12, 59.52 and 62.39%.

radius. The fuel depletion in the two cases of this figure were 0.21% and 18.51% respectively (Kasotakis et al 1989).

We are now showing a case with strong re-absorption of bremsstrahlung, Figure 13.9, with similar time dependence of the pellet temperature T(t) as in Figure 13.7, after driver energy of 40, 41 and 42 MJ has been deposited, with an initial volume of one cubic millimeter and initial density of one thousand times the solid state. The fusion gains G in these cases are 2.41, 1170 and 1199 respectively. The first case is practically a case of simple burn while the second and third cases are volume ignition as seen from the strong rise of the gain G by a factor of nearly 500 and with the strong increase of the pellet temperature to a maximum above 100 keV.

It is important to note that the initial temperatures are only 1.44, 1.47 and 1.51 keV respectively. The re-absorption of the bremsstrahlung is then so strong that the former threshold of 4.5 keV - when the bremsstrahlung is just compensated by the fusion energy generation - is strongly undercut. The evaluation of the fuel depletion results in the percentages of depletion by the reactions of 0.12, 59.52 and 62.39% respectively.

Another case with strong re-absorption is seen in figure 13.10. This is indeed an extreme case of 10 000 times the solid state density with an initial volume of 10^{-5} cm^3. The input laser energies are 3, 3.5 and 4 MJ with gains G of 0.5 (simple burn), 1731 and 1585 (volume ignition). The initial temperatures had the remarkably low values of 1.08, 1.26 and 1.44 keV. The fuel depletions were 0.02, 75.08 and 78.59%.

The calculations permit an evaluation of the fuel depletion in order to see how very high completeness of fuson burns could be expected. To show the cases with a fuel depletion above 80%, we plotted the calculations for a set of volumes for the cases of 1000 times and of 10 000 times solid state density in figure 13.10a

The computations of volume compression and volume ignition were for a long time academic only as long as it could not be shown experimentally what has to be done to get a fusion pellet into the mode of an ideally adiabatic compression. This ideal compression was even avoided when all laser-fusion experiments in getting were intended a shock compression of the pellet center. In order to empirically find the highest fusion gains, the experiments on laser compression of pellets at ILE Osaka (T. Yamanaka, 1985; Yamanaka et al 1986; Yamanaka et al 1986a) have proved how the ideal conditions of volume compression can be achieved by avoiding shocks and stagnation. In this case they measured the highest fusion gains. These conditions of volume compression required for the best fusion gains have been calculated by other authors

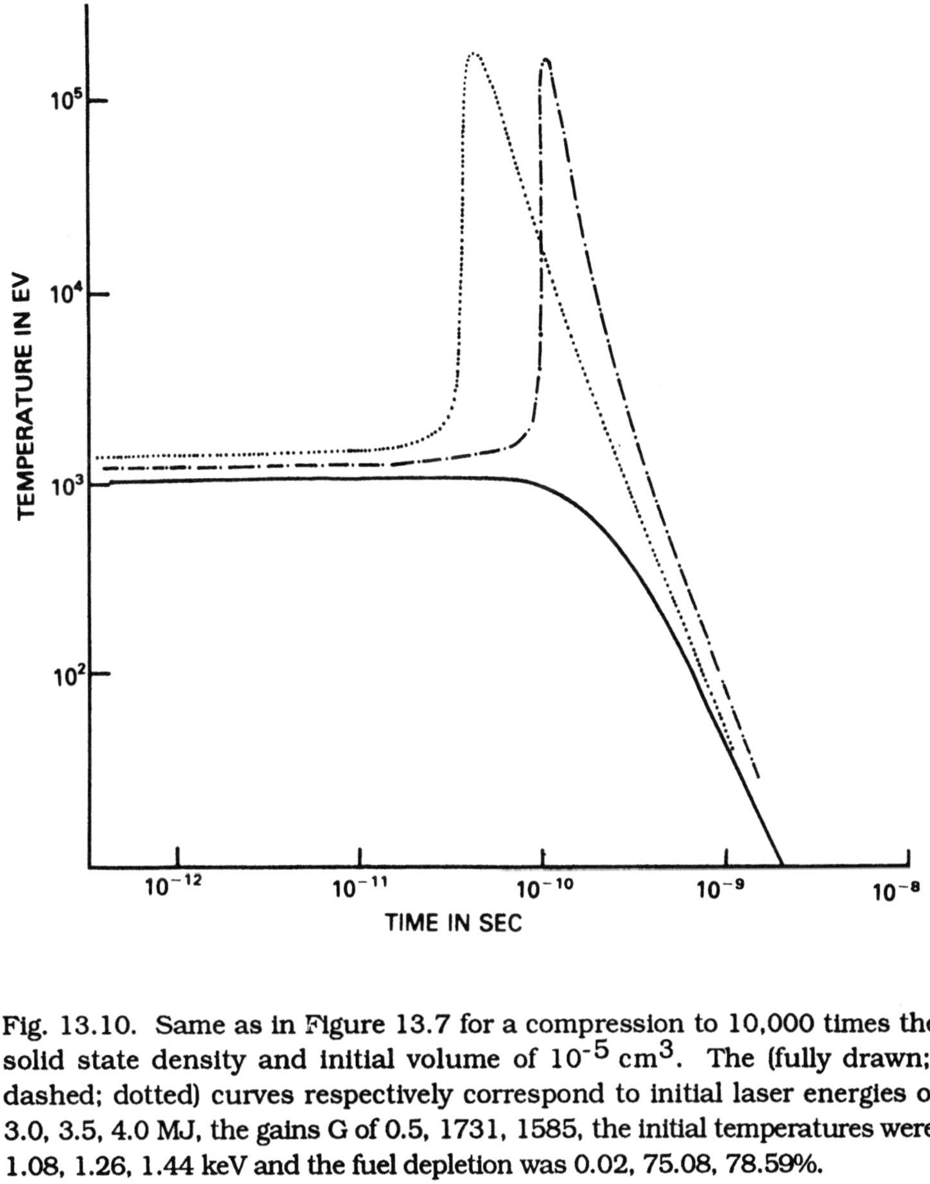

Fig. 13.10. Same as in Figure 13.7 for a compression to 10,000 times the solid state density and initial volume of 10^{-5} cm^3. The (fully drawn; dashed; dotted) curves respectively correspond to initial laser energies of 3.0, 3.5, 4.0 MJ, the gains G of 0.5, 1731, 1585, the initial temperatures were 1.08, 1.26, 1.44 keV and the fuel depletion was 0.02, 75.08, 78.59%.

(Velarde et al 1986; Andre et al 1988) and a special comparison of our earlier results (Hora et al 1978; 1987) with specific computations by Mima et al (1989) is of interest. Their computations are much more detailed than ours (Kasotakis et al 1989) since the laser interaction with the pellet plasma and the ablation is included at least with mechanisms known at present. Recent understanding about the pulsation and smoothing of the

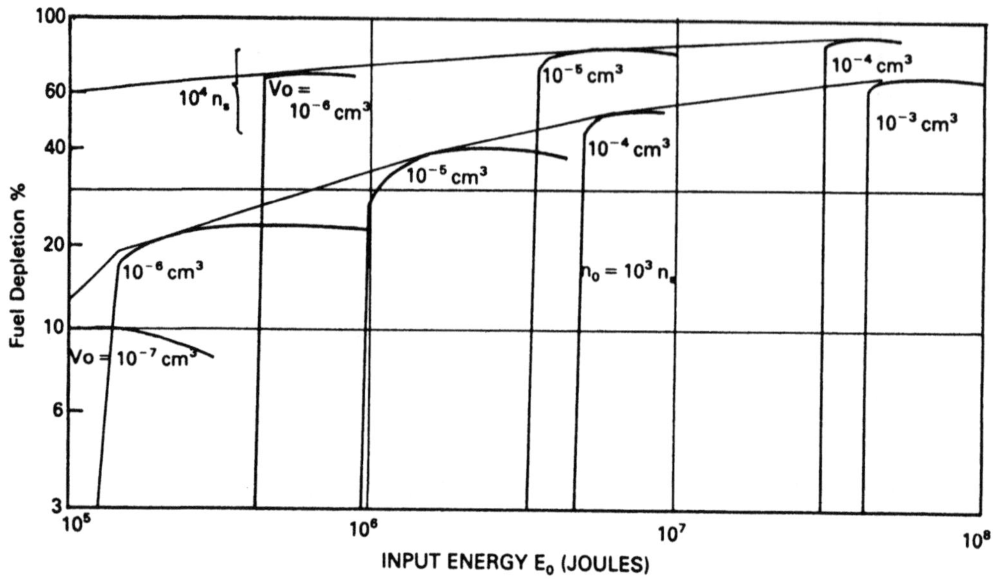

Fig. 13.10a. Dependence of the ionic depletion of the fuel on the input laser energy for 1,000 and 10,000 times the solid state density and initial pellet volumes V_0 from 10^{-7} cm^3 to 10^{-3} cm^3.

interaction may cause modifications to the basic importance of mechanisms. However, the usual results of the energy transfer may not change. In the experiments (Yamanaka et al 1986), smoothing by random phase plates and suppression of pulsation was included (Yamanaka et al, 1986).

The calculations of Mima et al (1989) show the amount of plasma involved in the ablation and how much energy enters the compressed pellet (determined by the hydrodynamic efficiency). From studies of the shock mechanisms in the compressed plasma, a very detailed treatment of the dynamics of each partial shell is included, taking into account thermal conduction and radiation mechanisms. Our calculations start from the energy input into the pellet. The laser energy is first reduced by the hydrodynamic efficiency; further, adiabatic compression and expansion dynamics are assumed such that no details about the shock effect in each shell are necessary. The experimental conditions (Yamanaka et al 1986a; Nakai, 1989) may be close to these cases so that the shorter computation time for each case (Hora et al 1978; Kasotakis et al 1989) permits a more detailed investigation of the dependence on the various parameters as derived from the large number of treated cases.

One important point in the analysis of Mima et al (1989) of the general behaviour is that for fusion gains larger than 8 (corresponding to a $\rho_o R_o$ of 0.4 [corresponding to a gain G = 5.3 in Eq. (13.8a)], above which the volume ignition occurs), our result for the exponent of the fusion gain dependence on the input energy

$$G = E_0^a \tag{13.13}$$

results in a value a = 0.9 - derived from our numerical plots - while Mima et al (1987) derived a value a = 1. The subsequent number (N) of neutrons generated per shot is then

$$N = E_0^{a+1} \tag{13.14}$$

where the exponent in our case is 1.9 and the value obtained by Mima et al (1989) is 2.

For the case evaluated by Mima et al (1989) for a total gain of 3 (related to the total input laser energy of 100 kJ) and 100 times compression of the pellet after an ablation given by 7% hydrodynamic efficiency, a $\rho_o R_o$ value of equation (13.8) of 0.9 was found by Mima et al. (1987) while our result is 0.8, based on achieving the same total gain of 3 with a compression of the core of 100 n_s. This example indicates what gains may be expected from the next generation of experiments with 100 kJ pulses of laser energy if volume compression is applied and just the volume ignition is being reached.

If the above mentioned conditions can be verified experimentally, the present parametric framework for pellet fusion, based on volume ignition provides (Cicchitelli et al. 1988) fusion reactor pellet operation with 1 MJ driver pulses with a moderate compression to only 100 times the solid state density. This will relax any problem with Rayleigh-Taylor instabilities (see subsection 8.9) because of the larger geometry of volume ignition than for spark ignition.

The shock free and stagnation free ideal adiabatic volume compression and volume ignition provides a very transparent and reliable model with respect to reheat and re-absorption of bremsstrahlung. It corresponds to the thermodynamically ideal adiabatic compression dynamics and the ignition is similar to that of a Diesel engine which has the desirable feature of working within the whole volume simultaneously in contrast to the combustion front mechanisms of the Otto engine. The initially expected complicated mechanism of laser interaction required to achieve a shockfree compression, has been solved by C. Yamanaka et al experimentally (Yamanaka et al 1986; 1986a) using a simple burn only without ignition and low applied laser pulse energies in the range of 12 KJ.

For the next generation design for laser fusion according to the ATHENA (Storm et al, 1988) scheme, volume ignition may offer a transparently understandable, very efficient and experimentally more easily realized scheme. The reduction of the Rayleigh-Taylor instability (Cicchitelli et al 1988) especially causes a more efficient direct drive where smaller capsules and low laser pulse energies down to 1 MJ may be applied simplifying the basic reactor technology.

13.3 Solution of Laser Fusion by Spark Ignition and Indirect Drive

The concept of volume compression, ingeniously verified by the team at Osaka University (T. Yamanaka 1985; Yamanaka et al 1986; 1986a; Nakai 1989) with the highest nuclear fusion gains and the prospect of volume ignition as described in the preceding subsection, do not represent the main stream of laser fusion research of the last 20 years. Aspects of volume ignition were developed rather marginally only since 1985 (Yamanaka et al, 1986a; Nakai, 1989) though the theoretical concept was known since 1978 (Hora et al 1978).

The main stream of laser fusion research followed the basically different **concept of spark ignition** (Ahlstrom, 1983, Meldner, 1981) which was invented and treated by Nuckolls (1974) at least since 1969 and promoted also by Brueckner (1974). The first experiments to detect fusion neutrons from laser irradiated targets were reported (Basov et al 1968; Lubin 1969) but a convincingly large emission of fusion neutrons from laser irradiated frozen hydrogen using the trick of a million times suppression of laser prepulses (Lubin 1969a) was reported by Francis Floux (1969; 1970).

The initial model of volume compression of plasma and its ideal adiabatic hydrodynamic expansion (Basov et al 1974, Dawson, 1974) with the evaluation of the optimized conditions for the highest fusion gains with the formula (13.7) (Hora 1964, see Fig. 13.3), had a lot of disadvantages though it was fitting experiments very well (Hora, 1970 (see Fig. 5.1); Engelhardt et al [72]) while the following experiments were then much more sophisticated (Basov et al 1973; Manes et al [139]; Haas et al [316]). The disadvantages resulted in a very disappointing conclusion from Eq. (13.3): Taking these results with its optimum initial temperature of 17.2 keV and assuming a rather good output of 25% fuel being burned, the achievable gain G_O is only 41. All the advantages of an increase of the gain by compression of plasma [known to designers of nuclear explosives (Neddermeyer, 1942)] and the desired strong lowering of the input energy for interesting high gains by increasing the compression (Eq. 13.3) could not change the disappointing maximum gain of 41.

In view of the very low efficiency of lasers and other losses in laser fusion reactors, the use of nuclear fusion burn by volume compression had to be given up (the later known ignition and gains above 1000, were not known then). Instead, Nuckolls (1974) invented the spark ignition: irradiating the fusion pellet by a tailored laser pulse, slowly growing within one nsec to a peak within orders of magnitudes of laser power and fast switching off, should produce ablation of the plasma corona and compression of the plasma interior. The mechanical recoil (see Figures 10.1 and 10.2 by Mulser [88]) causes a compression of the plasma interior in a programmed way of a Guderley [220] mechanisms. Instead of having a compression to four times the solid state density for the one-dimensional plane interaction (Fig. 10.1) or 60 times the solid state for the spherical case, Nuckolls (1974) (see also Brueckner et al 1974) could compute densities of 10,000 times the solid state density. The only requirement was that the laser pulse had to be tailored appropriately.

Then, only a very small amount of the DT pellet material had to be compressed to this density and to be heated to temperatures of about 10 keV (perhaps better: 17 keV) such that a fusion detonation was initiated from this central spark in the pellet and the surrounding low density and low temperature DT fuel would then burn by the self sustained fusion combustion wave. In this way, much larger gains than 41 could be expected. Values of 1000 were computed such that even very low efficiency lasers should produce the right gains for a commercial reactor.

On the basis of this spark ignition in contrast to the ideal adiabatic volume compression and ignition process, the large amount of research in the field (Ahlstrom, 1983; Basov et al, 1986; Yamanaka et al 1986a) was performed. One problem to be brought under control was the fusion wave itself (Teller, 1987). Similar processes were known from thermonuclear explositions but even under these conditions, no final physics solution was at hand. Only most refined engineering type solutions comparing computer simulations with a large number of experiments in underground nuclear explosion tests were achieved about the radiation wave and about the interpenetration (mix) of the hot plasma of the combustion front with the cold fuel to be burned. The physics of the mix could not be followed up by the shock theory of hydrodynamics where the hydrodynamic fronts are not interpenetrable, neither could kinetic theory be used for a complete solution. Nor could the problems mentioned for the complex question (Chape. 3.5) be solved by the simulations of 1 million plasma particles for the processes of interpenetration and the energy deposition from the hot plasma to the cold plasma since the correct inclusions of the collisions were numerically too complicated (Meyers, 1987) - apart from the question of the quantum deviation of the collision frequency (subsection 2.6).

The most difficult question of the fusion wave is to determine the correct theory for the stopping power was of the nuclear reaction products in high density hot plasmas especially in the case of very high current densities of reaction products including nonlinear deviations etc.. It is known (Nuckolls 1978) that the binary collision theory of MeV ions with keV plasma (Winterberg, 1969) does not reproduce the experimental facts of the underground tests and stronger stopping powers derived empirically in a very indirect way have to be used, see e.g. Ray et al (1976; 1977). This is one of the reasons also, why the intensive experiments with the spark ignition in laser fusion is so interesting for exploring the physics of large scale nuclear reactions.

Several models for the fusion wave were developed (Ahlborn, 1971; Basov et al, 1972; Bobin, 1974) and the later collective model of the stopping power was applied (Hora, 1983a) but these were very rough first steps into the understanding of the very complex problem. The result is that the energy per area required for such a fusion wave must be approximately

$$E_F = 10^{10}\, (n/n_s)^2\ J/cm^2 \qquad (13.15)$$

following models previously published (Basov et al 1972; Bobin 1974; Ahlborn 1971). The process of the interpenetration of the hot fusion plasma has only been treated approximately in these cases. The strong thermal sink of the "heat wave" of precursory electron clouds can no longer be considered as a valid assumption since the repulsion of the electrons at the interface between the hot and the cold plasma is dominated by a strong electric double layer (see subsection 8.9).

Another question is what value should be taken for the penetration depth of the thermal ions into the cold plasma. If the strong stopping power according to the collective model is used (Ray et al 1978), the value of the numerical factor for E_F in equation (13.15) is about 100 times less (Hora 1983). This is an optimistic and not the final answer to the problem. This analysis also results in a necessary time scale of the build-up of the fusion wave of 10 to 100 psec (Hora, 1983). When these considerations are taken all together - still using the conservative values of equation (13.15) - an energy transfer at the surface of the igniting core in the pellet of about 10^{19} W/cm^2 is required as mentioned by Emmett et al (1974). In contrast, the laser intensities for a purely gas dynamic interaction without non-linearities must be 100,000 times less with neodymium glass laser radiation or 10,000 times less with excimer laser frequencies.

Using the simplified density for the compressed pellet core given in Figure 13.11 as the point where the fusion wave should be ignited for penetration into the less dense outer part of plasma, we obtain the following relation from equation (13.15).

$$GE_1 = E_F\, 4\pi R_1{}^2 = 4\pi 10^{10} \left(\frac{n_s}{n_e}\right)^2 R_1{}^2 \tag{13.16}$$

If the optimum gain G is assumed according to equation (13.7) or (13.11), the temperature in the core is assumed to be at an optimal value (it is difficult to obtain the exact parameters; a small deviation will cause a large reduction in gain) and using the relation for the optimum conditions of (all energies in Joules and all lengths in cm):

$$R_1 = 1.075 \times 10^{-3} E_1{}^{1/3} \left(\frac{n_s}{n_0}\right)^{1/3}, \tag{13.17}$$

we arrive at the relation

$$E_1 \left(\frac{n_1}{n_s}\right)^2 \left(\frac{n_2}{n_s}\right)^3 = 6.9 \times 10^{10}\,\text{joules.} \tag{13.18}$$

With further substituting of Eq. (13.7) we obtain the additional relation

$$G\left(\frac{n_1}{n_s}\right) = 35.05 \tag{13.19}$$

Postulating that the gain should not be less than 10, we see that the plasma outside the core has to be compressed at least to 3.5 times the solid state density. With a reasonable energy input E_1 of 10 kJ (based on a laser input energy of 1 MJ and a realistically assumed transfer of 1% of the energy into the compressed core), Eq. (13.18) results in a minimum compression of the core

$$n_1/n_s = 401 \tag{13.20}$$

which is about half of a thousand times compression. These data are selected in order to arrive at a relatively low compression. In a real case for volume ignition, the outer density may be lower than assumed, changing the other parameters to obtain a higher core density and higher gains.

If however, the difference between the densities is less, as e.g. in numerical cases (Kidder 1979), the properties of the central ignition are reduced to a volume ignition or even a mixture of volume compression and volume ignition processes. It should be noted that the use of the realistic equations-of-state (Storm 1986; Eliezer et al 1986a) causes further difficulties for an inhomogenous plasma system as it does for the central

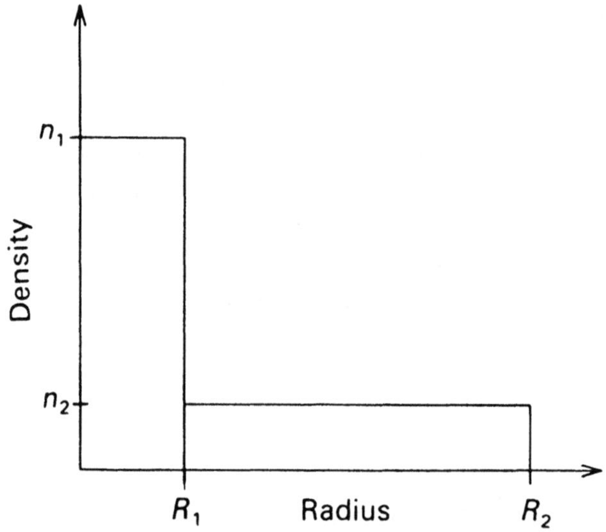

Figure 13.11. Scheme for radial density distribution in the pellet after laser irradiation where a central high density core of radius R_1 with a density n_1 and an energy input of E_1 is produced. The resulting fusion reaction (gain G) produces an energy which corresponds to the ignition of the fusion combustion wave according to Eq. (13.16). The density of the plasma into which the wave penetrates is n_2 , assumed to be homogeneous up to the pellet boundary R_2.

ignition. There is always the possibility that central ignition may be verified experimentally, but it is no surprise that this scheme produced the well-known difficulties outlined here that prevented faster progress in laser fusion. In the usual cases of computations of spark ignition, densities up to 10,000 times the solid state are used which conditions relax the mentioned accuracy of the parameters.

Another problem than that of the high gains - solved by the spark ignition since volume ignition was not available at the early stage - was the problem of the very complicate interaction of the laser light with the plasma. When the large scale experiments were started in beginning of the seventies, one realized that the interaction was in not smooth. While it had been realized that the nonlinear (ponderomotive) force effects (Chapt. 8) had to be included into the plasma dynamics, it was assumed that the parametric instabilities (Subsection 9.5) and the Försterling-Denisov resonance (Subsection 11.2) were the main evils of the interaction processes. It was evident that there were nonlinear groups of plasma apart from thermally behaving plasma (Fig. 1.4) and that there were the suprathermal energetic electrons apart from the thermal electrons to be detected in the x-ray spectra. The possibility that the energetic electrons may have been simply the result of quiver motion or quiver drift in several types of resonance structures (p.180) was less popular and the other kind of complex interaction as pulsation and "stuttering" interaction (subsection 10.8) was not known.

Since there was not any understanding for the complicated interaction nor any method known how to achieve a smooth interaction, an ingenious way out was proposed by Nuckolls (1982) with the indirect drive. Instead of driving the plasma ablation and compression by the direct interaction of the laser radiation with the pellet surface, the laser energy was converted in a high-Z covering layer of the pellet into x-rays. This radiation subsequently produced a very smooth and uniform compression of the fusion fuel of the pellet when being absorbed in the entire fusion fuel pellet. The conversion of the laser radiation into x-rays was very efficient (up to 70%) and half of this isotropic radiation was lost by going outside such that finally, up to one third of the laser radiation was available for the x-ray compression of the DT fuel. The x-radiation had nearly a Planck spectrum of black body (or hohlraum) radiation. The extensive knowledge of how such radiation of about 300 eV temperature is able to compress and ignite DT fusion reactions was well available. For the whole dynamics, the spark ignition scheme was aimed for.

One modification of this indirect drive became known as the cannon ball scheme which was invented by Yabe (1975) and studied extensively by the ILE in Osaka (Nakai, 1989). The fusion pellet is imbedded in a high-Z outer mantle with an empty space in between. The laser radiation is brought through thin holes through the outer mantle into the empty space where the laser radiation converts nearly instantly into hohlraum radiation and produces then a very smooth and highly uniform compression of the inner pellets. The study of the conversion of the laser radiation within an empty mantel arrived at the confirmation that black body radiation of 200 eV temperature has been produced (Sigel et al, 1990).

The fusion gains achieved with the indirect drive were not so high only up to 10^{11} neutrons per interaction (Lindl, 1988) were reported while the shockfree direct drive arrived at 10^{13} neutrons (Yamanaka et al 1986; Storm 1986). The indirect drive, however arrived at a higher compression. Compressions as high as 600 (or 1000) times the solid state for carbon hydrates have been achieved (Nakai, et al 1990) producing then materials with a specific weight of 1000 kg per litre.

The compression of DT fusion pellets by very intense x-rays not from laser pulses but from underground test experiments, was studied in the Centurion Halite experiment (Broad, 1988). These results had to be declassified for the discussion with the magnetic confinement fusion research in the USA in order to demonstrate that laser fusion has the physics solution for developing a commercial power station, which could not be reached by magnetic confinement fusion. It was demonstrated that x-ray energy pulses of 40 to 100 MJ energy could very successfully compress and ignite DT fusion capsules.

For the more controlled and variable irradiation by lasers, the conditions for such production of fusion energy may be performed with lower laser pulse energies applying all the knowledge gained from the extensive and very detailed and most advanced experimental and numerical research of the last 20 years. Putting these results all together (Storm et al 1988), it has been elaborated how a glass laser pulse of 10 MJ energy can produce 1000 MJ fusion energy by irradiating a capsule using the techniques of indirect drive and spark ignition. The costs for the laser driver based on the present days technology of neodymium glass lasers (ATHENA laser) are approximately $700 Mill. The step from this laser system to such one with a repetition rate of few shots per sec and with an operation routine for years is indeed another question of large scale engineering development; however, the physics principles have been shown to be settled.

The indirect drive for compression and ignition of DT fuel by very intense x-rays is the scheme how a large scale thermonuclear reaction is being ignited (Teller 1987). This is the reason why the laser indirect drive studies are of interest for the large scale explosions too. Teller (1987) mentions hohlraum radiation corresponding to a black body temperature of 1 keV needed for this purpose. It is expected that for the controlled ICF energy production a temperature of 300 eV is sufficient. After 200 eV have been reached (Sigel et al 1990) the expectations in the next steps of experiments are in the focus of interest.

It should be mentioned that indirect (x-ray hohlraum) drive is of interest now as an alternative to lasers for ICF, if driving is done by light or heavy ion beams. The direct driving is complicated (Goodlove et al 1978; Herrmannsfeldt et al, 1990; Bangerter, 1991) where the ions are irradiating the pellet or capsule at the whole surface and where the stopping of the particles produces a plasma for ablation and compression of the pellet. Since the use of heavy ion drivers may profit from advanced technology including the concept of non-Liouvillian

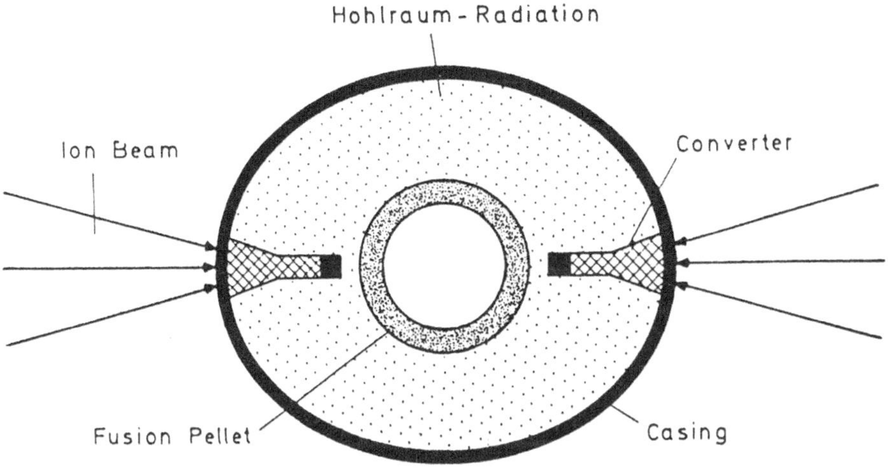

Fig. 13.12. Indirect drive heavy ion beam fusion (Bock, 1990). A fusion pellet or capsule is incorporated into the ellipsoidial high -Z casing (mantle) with a hohlraum in between. Heavy ion beams are absorbed in the connecting wires which convert the ion energy into black body radiation in the hohlraum for compression of the pellet (Meyer-ter-Vehn, 1989).

switching (Rubbia; Hofmann, 1990) the scheme has been elaborated by Meyer-ter-Vehn (Bock, 1990), Fig. 13.12, where a kind of a canon ball is used with an internal fusion pellet and an outer high-Z mantle (casing) connected by two diametrical high-Z wires between the empty space. The GeV heavy ion beam is absorbed in the wires producing hohlraum radiation in the empty space such that if 300 eV black body temperature is reached, the pellet should be compressed and ignited for generation of fusion energy. The light ion beam fusion (Van Devender et al, 1985; Kessler et al, 1989; Leeper, 1989; Maron et al, 1989; Yatsui, 1989; Moses et al, 1989; Westermann et al, 1989) arrived at a remarkable level of the driver systems. For fusion, a scheme of indirect drive is offered by getting such a discharge through a tube (liner) in which the interior is producing hohlraum radiation of 200 eV (Smirnov, 1990).

The costs for the ICF reactor are in the same order of magnitude as the driver of $1 Bill. For example, if one uses the concept of the CASCADE reactor (General Atomics, 1985). Since any reactor has to be most simple in its concept, this is ideally fulfilled by CASCADE. In order to avoid all the wall problems and the damage by radiation and shocks, the blanket for absorbing the fusion neutrons consists of a 50 cm thick layer of lithium ceramic pebbles which are kept on the walls by centrifugal forces of an approximately 10 meter large rotating vessel. At one end, the cold pebbles are put in and at the points of the highest speed the 1500^0 C hot pebbles are released into heat exchangers for driving a power engine. The pebbles have to be either returned or after some time of use to be regenerated where the generated tritium has to be removed by chemical processing and re-melting such that any problems of radiation damage in the pebbles are always solved implicitly within this process and in a straight forward way.

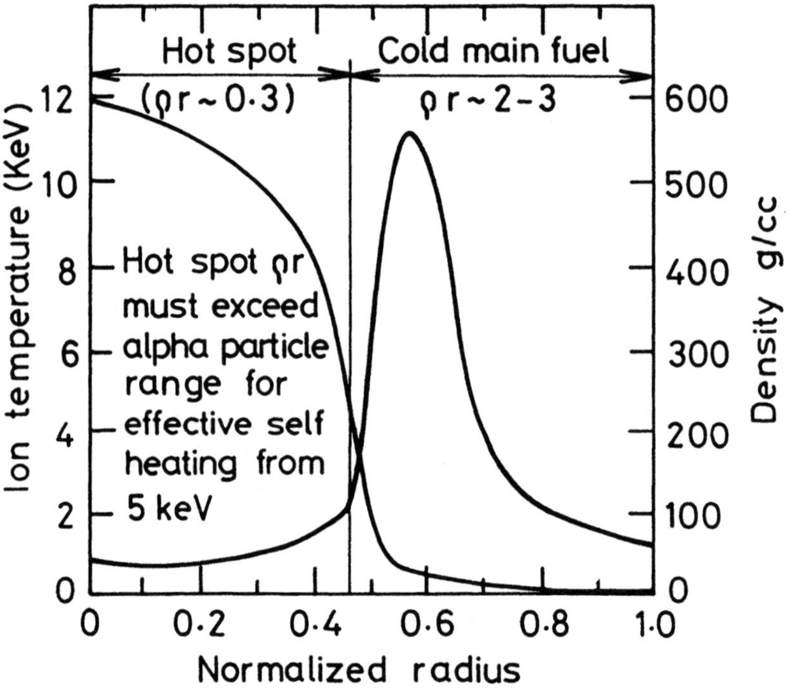

Fig. 13.13. Radial profile of density and ion temperature in a compressed pellet at "central spark" ignition (Storm et al (1988)) with incident laser energy of 10 MJ and 1 GJ fusion energy generation.

The hot spot spark ignition process has been evaluated numerically and is shown in Fig. 13.13. Obviously, the hot spot is not longer a very small spark but a quite extended area within 40% of the radius from where the ignited fusion plasma expands into a specially high density chosen cold plasma. The density there, is then nearly 2500 times the solid state.

Since this is a kind of a large spot ignition one may compare this result with our results of volume ignition of the preceding subsection.

Starting from the results of Fig. 13.6, for a density of 1000 times the solid state and an energy E_0 of 1.5 MJ (corresponding to a hydroefficiency of 15% relating the 10 MJ laser energy converted into the reacting plasma after subtraction of 85% of the laser energy which went into the ablated plasma) the resulting gain G in the definition of Fig. 13.6 is 600, corresponding to an output of total fusion energy of 600 times 1.5 MJ = 900 MJ which is quite close to the result of Fig. 13.13 with 1 GJ output (Storm et al, 1988).

In summary it has been clarified that on the basis of experimentally confirmed underground x-ray compression experiments with high gain DT fusion capsules, extended to the long years of highly sophisticated experimental and numerical results on laser fusion, large scale economic clean fusion energy can be generated on the basis of indirect drive and spark ignition.

An appropriate technology program is now needed to develop laser and reactor systems with the necessary reliability required for long years of effective operation using today's technology.

13.4 Improvement by Volume Ignition and Direct Drive

The successful elaboration of the physics concept for a laser fusion reactor on the basis of spark ignition and indirect drive and today's available neodymium glass laser technology of the ATHENA type (Storm et al 1988) including a lithium pebble reactor of the CASCADE type (Maya, 1985), may be used as the foundation for developing an economic fusion energy source. If a political decision could be taken along the lines e.g. of the Tsongas MacCormack Bill in the USA in coordination with a worldwide effort, to proceed with the development and construction of fusion power plants. Only then, will the necessary applied knowledge be gained to bring engineering technology upon to par with available physics.

The effect of such a conscious political decision could be likened to the decision taken in 1960 to launch a manned space flight to land on the moon. The basic physics for such a venture was available, however the engineering details had to be elaborated and further refinements made to physics resulted as the project matured.

This is comparable to the situation of 1960 with the decision for the manned flight to the moon that the basic physics was available but the engineering details had to be elaborated and even further improvements in more refined physics elaboration were to be included in due course.

While there must be no disturbance of this procedure of decision findings, one should be aware from the examples shown in this subsection as well as from further yet unknown but well expected further improvements, that physics permits a further increase of efficiencies in due course which will effect further developments. It should be possible to expect **generation of energy to the cost three or five times below that of the light water reactors**. The examples to be given are with respect to volume ignition and direct drive.

One important question for laser fusion physics, is whether or not the ignition can be reached under the condition elaborated from the spark ignition (Storm et al, 1988) and to what degree of accuracy one can conclude this from the underground tests (Broad, 1988). There is some gap of variability and the next steps of laser fusion using glass laser pulses of 0.5 to 1 MJ energy are scheduled to confirm the ignition process. As we have shown in the preceding subsection, there are similar results with the quasi-spark ignition elaboration (Fig. 13.13) and the close conditions of volume ignition. Since the volume ignition is based on rather transparent and simple properties of physics, this should be another argument in support of the otherwise very well founded physics in the elaboration of Storm et al (1988).

The use of the volume ignition has, however, not only the advantage of arriving at more reliable mechanisms and less sensible selection of parameters. There is also the physics of Rayleigh-Taylor instabilities at the compression (Takabe et al, 1983; Kull, 1986). Any deviation from the spherical symmetry of the laser driven compression front causes bumps in the sphere which, as an instability, are amplified during the subsequent compression. It is assumed that the uniformity of the laser intensity for the compression has to be within 3% just to tolerate deviations. A mechanism counteracting these instabilities apart from collisional damping (Mulser, 1988) by straightening back deviations and by suppressing of surface waves, is the surface tension (subsection 8.9), a process which not at all has been included sufficiently in the computations of the compression. Some first steps were done by Nishihara (1988). However, it is evident is the fact that the sensibility against such instablities is much larger for the spark ignition with the smaller radius of the final compression than for the volume ignition where a much larger radius of compression is involved.

A further advantage for the volume compression and ignition with very safe predictions of the gains (see subsection 13.2) is the fact that it is not connected to the problems of the self sustained fusion combustion waves such that research in this direction cannot longer be denounced by the magnetic fusion community as being a military topic which has to be pushed into the corner of classification and be eliminated from financial support.

After the details of volume compression and volume ignition have been elaborated in subsection 13.2, we can discuss some developments from the experiments of T. Yamanaka (1985), Yamanaka et al (1986; 1986a) and Nakai (1989). These results were gained with the Gekko-12 glass laser with a little different laser pulses for the different regimes of indirect drive (Cannonball) and direct drive using neodymium glass laser light or its second harmonics (wave length 530 nm) (Fig. 13-14). The data includes the achievement of nearly 1000 times solid state density with direct drive single shell pellets.

The group of experiments of direct drive single shells (using glass spheres filled with DT gas) with neutron gains from 10^{11} to 10^{13} per shot were illuminated with second harmonics laser pulses of roughly constant incident total laser energy of E_{tot} = 10 kJ. Considering these then as a group of experiments (T. Yamanaka, 1985) of some correlation, the drawn straight line in Fig. 13.14 may be considered. This corresponds to the fact that if the same energy E_o is transferred into the compressed plasma which is about 10% of E_{tot} since 90% is lost into the ablated plasma of the corona, it would be understood that the neutron yield is increasing on the square of the compression (see Eq. 13.6). This is shown with the line rather agreeing with the experiments. This simplified assumption would indeed include a rather similar temperature of the maximum compression during the volume compression. The possibility of achieving higher neutron gains with the same rather low energy input of 10 kJ would require an improvement of uniformity and synchronization of the laser radiation and a systematic modification of the tailoring of the time dependence of the laser pulse what seems to be possible experimentally (T. Yamanaka, 1990). The expected gain of 2 x 10^{14} neutrons would then correspond to break even with respect to the E_o if 6% of hydrodynamic efficiency is involved.

All these evaluations agree with the volume compression model (subsection 13.2) and with the more sophisticated computations (Mima et al 1988) and correspond to the results of measured and prospected fusion gains as given in Fig. 13.15 using the results of the various Gekko lasers (Yamanaka et al 1986a; Nakai 1989) and of the experiment with the NOVA laser (Storm, 1986). The last experiment produced the highest neutron yield ever reported (2x10^{13}) using blue (third harmonics) laser light with an energy of 15 kJ. It has to be taken into account that this result was not due to an optimized selection of parameters in order to produce the best

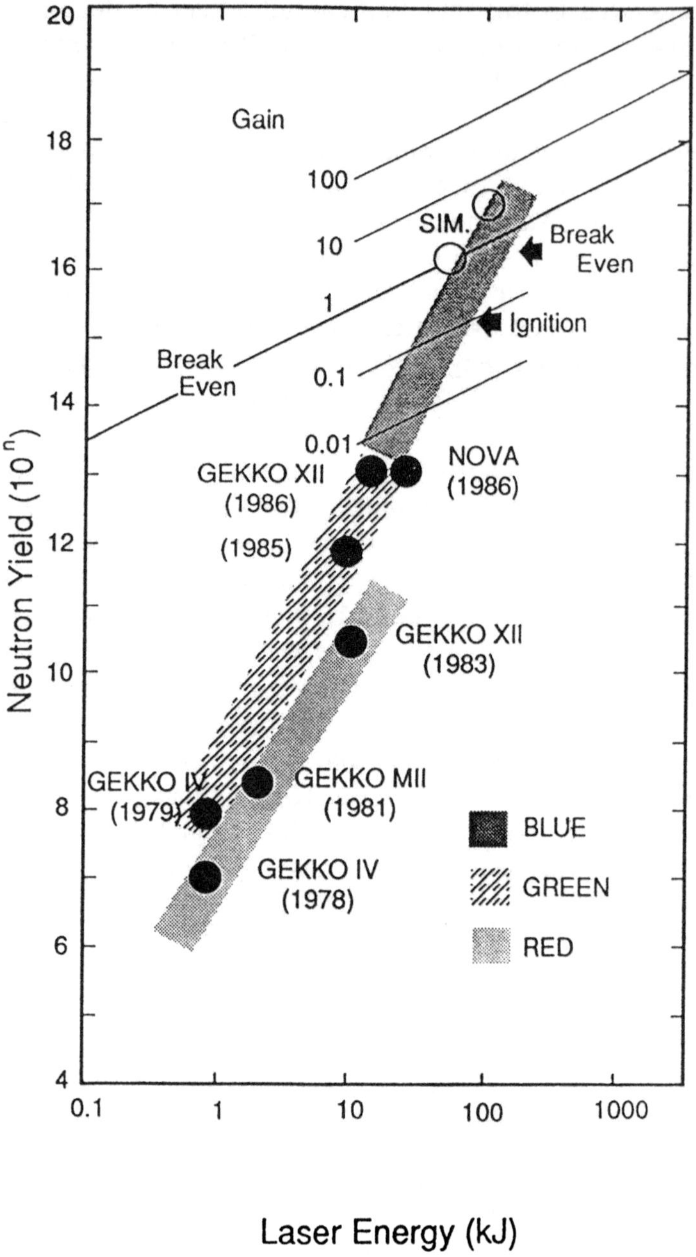

Fig. 13.14. Summary of density of DT fuel in laser compressed pellets and of the neutron yields, according to Nakai (1989) see (Meyers, 1991).

volume compression and to receive a smooth laser irradiation. This fact implies that the parameters could have well been varied in order to arrive at much higher neutron gains with the same laser pulse energy. Remarkably high fusion gains were achieved with the forth harmonics of glass laser beams (Fabre, 1988) though the laser pulses had only about 100 J energy.

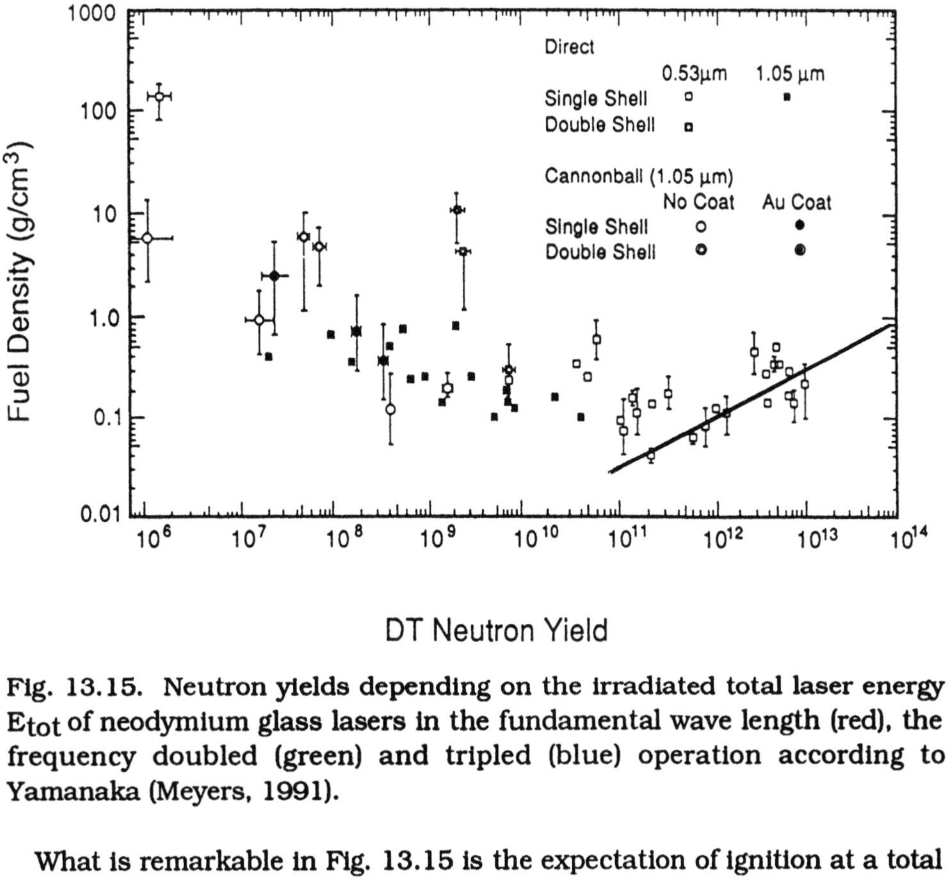

Fig. 13.15. Neutron yields depending on the irradiated total laser energy E_{tot} of neodymium glass lasers in the fundamental wave length (red), the frequency doubled (green) and tripled (blue) operation according to Yamanaka (Meyers, 1991).

What is remarkable in Fig. 13.15 is the expectation of ignition at a total gain

$$G_{tot} = \text{Fusion Energy/total energy of Laser Pulse} \qquad (13.21)$$

(identical with the gain plotts of Fig. 13.15) of 0.3. Taking into account a hydrodynamical efficiency of about 5% (experimental values are usually 5% with glass micro balloon surfaces while theoretically a G of 20% is possible for pellets with hydrogen (or DT) atoms at the surface for direct drive). The gain as defined by Eq. (13.6) is then 6. This result was calculated by detailed numerical codes (Mima, et al 1989) and this is just the value from Fig. 13.6 where volume ignition begins (where the lower straight lines are to be bent upwards as explained in subsection 13.2).

The further improvement of laser fusion is that of direct drive if smoothing of the laser pulses by the random phase plate (RPP), or by induced spatial incoherence (ISI), or by broad band irradiation, or if smoothing by spatial dispersion (SSD) is being used as explained in subsection 10.8. This smoothing can be understood by suppression of the pulsation and stuttering interaction of the plasma at laser irradiation where the coherent generation of standing wave ripples, the peripheric Bragg reflection and the relaxation is eliminated.

It is most important to note that the high fusion gains with green laser light observed at the Osaka University (Yamanaka et al 1986a, Nakai, 1989) were achieved only when smoothing with the RPP was used.

The advantage of direct drive can be seen from the following estimation (excluding here the case of the cannon ball geometry). If the laser radiation has to be converted in a high-Z material on the outside of the pellet by up to 70% efficiency and if half of this radiation has to be used for the pellet compression, only 30% of the laser radiation, will at best, be available for laser fusion. In the case of direct drive with very low reflectivity after smoothing (working then always in the mode where the light penetrated to the critical density and where Bragg reflection is avoided, see subsection 10.8) nearly 100% of the laser radiation will be available for compression. This improvement in efficiency reduces the costs of a laser fusion power station by much more than a factor of 3 since all components can then be made smaller. Further, this type of operation has no relation to the x-ray hohlraum compression and ignition of fusion fuel in the case of nuclear explosions as described by the illustration in the article by Teller (1987). This is one reason more to encourage an unrestricted handling of research for laser fusion and for the open development of the laser fusion reactors.

13.5 Estimations of Future Clean Fuel Fusion

Once it has been clarified that laser fusion offers a physical solution such that with today's technology a commercial and low cost power station could be developed within 10 years and after it is evident that the physics involved permits a further decrease of the cost of electricity during the further development, we may estimate how this development could produce the ideal clean nuclear energy production practically without any radioactivity. This question is indeed a point of the future and can be based only on the assumption that the normal laser fusion of deuterium tritium would be developed on a large scale for a number of years.

The question is whether the ideal fuels could be treated at all by ICF after a 50 or 80 years intense worldwide use of laser fusion reactors. The following estimations seem to show a possibility for this ideal option in the future.

Let us consider the following estimations. The problem with the DT reactor is still that radioactivity is being produced inside the reactor and inside the whole cycle of operation. Fusion energy is much more "clean", compared with the uranium, plutonium or thorium fission reactor with respect to the ash. The product of the fission reactor, the exhaust, is always highly radioactive material of middle weight atoms. Huge amounts of waste material are produced which have to be deposited carefully such that the radioactivity does not damage other materials or the biosphere. This radioactive waste problem is not a simple one and is well expensive but it has been solved such that it is economical for the low cost energy production of light water reactors. The solution involves a controlled confinement of the waste material while the reaction product from burning coal or oil, the carbon dioxide emission, cannot be confined and will lead to the "greenhouse catastrophy" of the earth if it is not being strongly reduced (Schneider, 1989).

For fusion energy, the ash is not at all radioactive and simply can be released to the air. It is helium an otherwise expensive gas for other uses. Its amount of release is many million times less in weight per produced energy than the release of carbon dioxide gas because nuclear energy produces several million times more energy per unit of fuel than chemical energy obtained from burning fossil fuel.

The remaining problem with DT fusion is the radioactive tritium (half life of the beta decay: 12.26 years) which is produced by the fusion neutrons in the lithium containing blanket of the reactor in order to breed the tritium fuel from the heavy hydrogen of water and from lithium. Using the lithium pebble blanket of a CASCADE (Maya, 1985) reactor the action of the neutrons is nearly completely confined to the blanket such that the neutron damage and neutron reactions are well under control by the refining of the pebbles when the tritium is extracted and the pebbles regenerated. This also is one of the important arguments against magnetic confinement reactors, in which the neutron reaction in lithium is in a blanket which has to be confined by the first wall of the reactor and has to the thermally isolated against magnetic coils (even at superconducting low temperatures). The first wall is often heavily damaged and eroded, e.g. 1 cm per day of operation (Vielder et al, 1989; Harrison et al 1989).

The neutron interaction in the pebble system, results in strongly reduced neutron interaction in the wall material of the rotating vessel. The neutrons are absorbed in the pebbles where they produce heat and tritium. Therefore the problem of radioactivity is eliminated from the rigid materials in contrast to the magnetic fusion reactor. Nevertheless there are a lot of precautions necessary to handle the tritium and the rest of radioactivity remains in the rigid reactor materials. However, these problems are much less than in magnetic fusion and very much less than in fission reactors (apart from its enormous waste problems) and would be ideal in future, if fusion reactions without neutrons could be used.

Ideal fusion reactions are available: the reaction of hydrogen and boron, Eq. (13.4) is free of any neutrons and the reaction product are nearly monoenergetic alphas of 2.9 MeV energy. Their energy can be converted directly into electricity, almost without heat pollution using a Mosely type reactor: the alphas from the pellet are moving in a spherical electric field against 2.5 MeV voltage and convert the nuclear energy directly into electricity. Losses by switching and providing the contacts between the elements can be limited.

The only disadvantage, is that the reaction (13.4) has a rather small fusion cross section and normally needs an ignition temperature between 100 and 160 keV. This would be nearly impossible for a fusion reactor to achieve.

We shall see how volume compression and ignition can reduce the optimum initial temperature to 25 keV and under what conditions one may consider this fuel in the later future as the ideal nonradioactive nuclear energy production. The rest of radioactivity is produced by the reaction of the produced alphas with the boron-11 isotopes resulting in some radioactive nitrogen and other nuclei. It has been evaluated by Weaver et al (1973) that the amount of radioactivity produced per energy is less than the radioactivity released from burning fossil fuels since coal, e.g., contains 2 parts per million uranium as a natural mix like all other materials in the Earth's crust.

When the volume ignition code (Kasotakis et al 1989) was extended to the case of p-^{11}B, we found that the initial adiabatic burning during compression and expansion, required an optimum temperature of about 150 keV. This can be reduced by self heating of an exceedingly high density plasma (1×10^5 times solid state) such that the optimum initial temperature for volume ignition due to reheat and self-absorption of bremsstrahlung is only about 20 keV.

We performed an extensive series of computations similar to those for the case of the DT reactions. Parameters have been examined where a drastic decrease of the initial temperature for good gains with reasonable energy input of up to GJ was possible. In the following reported cases, the re-absorption of bremsstrahlung was not strong. Indeed we found in such cases much higher gains with very high fuel depletion and much lower initial optimum temperatures for ignition. However, input energies were a little beyond our limit of input energy.

The results of Fig. 13.16 immediately show the conditions required for volume ignition. The time dependence of the plasma temperature during the reaction is shown. As for DT, we have also illustrated the conditions at a slightly lower initial temperature which did not permit ignition, so that the temperature dropped monotonously with time. With a little higher input energy, the temperature rose to a sharp maximum due to

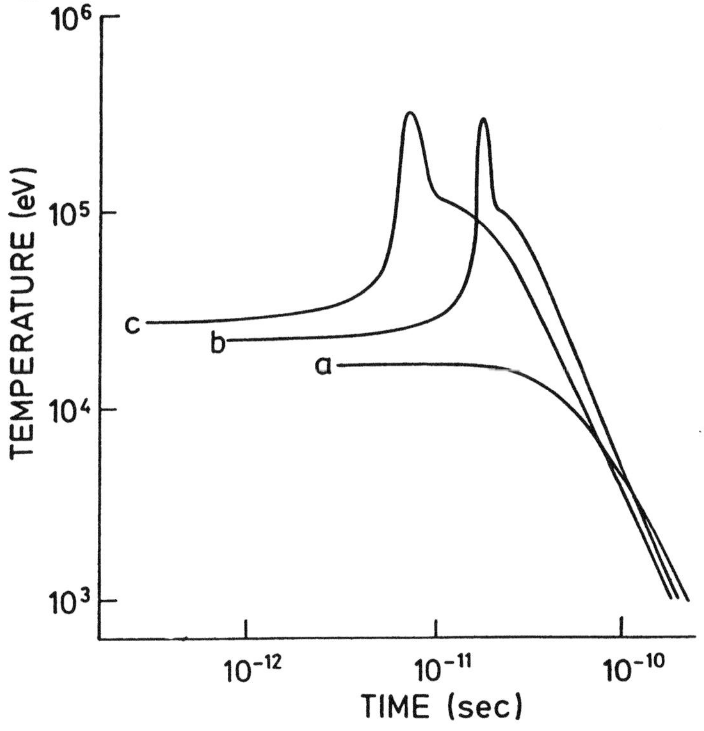

Fig. 13.16. Time dependence of the temperature p-^{11}B pellets of solid state density and 10^{-5} cm^3 initial volume with intake of the following energies, a: 1.5 GJ, b: 2.0 GJ, c: 2.5 GJ, resulting in gains of 0.09, 25.3, 21.0 respectively with a fuel burn of 0.21%, 77.7%, 80.7% and initial temperatures of 16.6, 22.2, 27.7 keV.

volume ignition. The gain achieved was 25.3 with respect to the input energy. This is indeed a low value and could only be considered in the far future if much more improved drivers with high efficiency are developed. At the same time, the methods used to obtain uniform irradiation must be drastically improved if there is to be any hope of achieving the enormous compressed density implied here, in the range of 100,000 times solid state density. The third case of Fig. 13.16 shows that a gain of 21.0 is achieved. The conditions for the future would then be that a laser of at least 30% efficiency is needed and that ways must be found to achieve an ultra high compression of 10^5 times solid state.

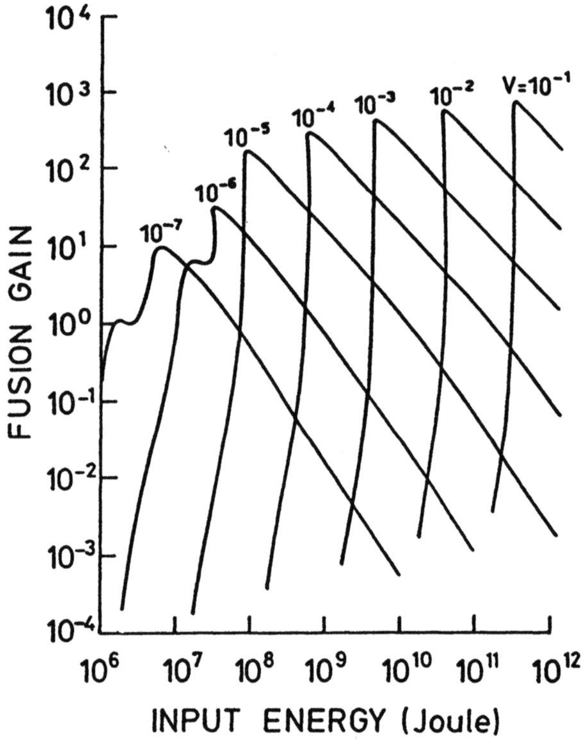

Fig. 13.17. The energy dependence of the fusion gain of the D-^{3}He fusion reaction for initial pellet volumes in the range of 10^{-7} - 10^{-1} cm^3. The initial plasma density was chosen as 3.0 x 10^4 times solid state density. Reheat phenomena were taken into account for both protons and alpha particles. A temperature - time analysis of the V=10^{-5} cm^3 case for selected initial energies is plotted in Fig. 13.18.

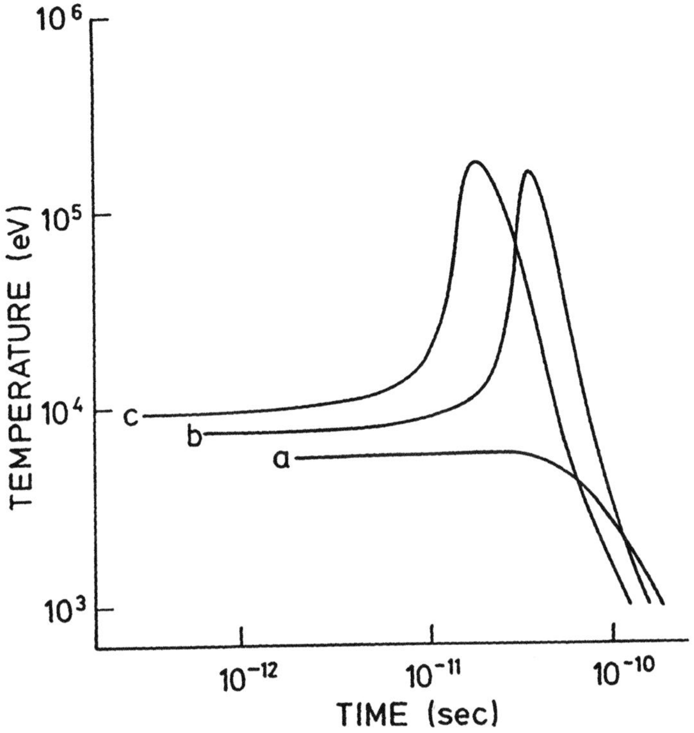

Fig. 13.18. Time dependence of the temperature of D-^{3}He pellets of 3.0 x 10^4 times solid state density and 10^{-5} cm^3 initial volume with intake of the following energies, a: 60 MJ, b: 80 MJ, c: 100 MJ, resulting in gains of 0.27, 190.2, 164.1 respectively with a fuel burn of 0.06%, 59.7%, 64.4% and initial temperatures of 5.7, 7.7, 9.6 keV.

The predicted compression of the p-^{11}B laser fusion requires 10 times the maximum estimate for compression of DT (to 10,000 solid state density (Ahlstrom, 1983)). 1000 times solid state density has been achieved experimentally (Nakai, 1990). Thus the p-^{11}B requirement may not be too difficult to reach within the next 80 years if intensive technological activity on laser fusion with the present or next step DT reactions is initiated. Compared with the predicted scales of DT reactors, the size of the p-^{11}B reactors indeed would be about one order of magnitude larger with respect to the energy production compared with the present scales of DT reactors. This one order of magnitude should not be insurmountable. The Moseley energy conversion can be done in a very clean and static way. The reactor per shot has to absorb about 25 GJ energy, corresponding to 2.8 tons TNT explosive energy. The momentum transferred at the nuclear fusion reactor however is 3000 times less than by the chemical explosion given by the square root of the energy released chemically per shot. Therefore the mechanical shock corresponds to 1 kg of chemical explosive which may simply be absorbed by elastically fixed electrodes in the spherical converter.

In principle it is known for laser amplifiers that an efficiency of 80% or more can be expected, while providing high gains in the visible or far UV optical spectrum from cluster injection FEL (Wang et al, 1986). The kinetic energy of condensed speckles with Mach 300 velocity is converted into optical energy.

The "lean" neutronic reaction of deuterium with the helium-3 isotope Eq. (13.3), has been discussed before by numerous authors for use as clean fuel (Miley, 1976; 1981). It could well be that after a first laser fusion phase with deuterium tritium, a phase with deuterium helium-3 may provide an intermediary stage before the ideal proton boron-11 reaction is used. The necessary amount of helium-3 for the fuel is available in the surface layers of the moon from where it can be economically mined in the future (Wittenberg, 1986; Miley 1988; 1988a). Reaction (13.1) implies D-D reactions with neutron and tritium which can be limited to 10% of the whole D-He(-3) reaction.

Generalizing our computations we have established the calculation of the D-^{3}He case where, contrary to numerous published and unpublished claims with more analytical estimations in this direction, we are following the consequent and detailed computational model. The case of the two different charged reaction products and the then separate stopping power mechanisms have been elaborated now in our codes. A rough result is that the proton component of the reaction changed the ignition conditions by about 10% to 20%. This may confirm that the estimations of other authors, ignoring the correct two component self heat mechanisms, are not too far from the reality.

We now have the unrestricted result without any approximation of the two branch stopping power mechanisms for the protons and for the alphas as reaction products. From a plot of the fusion gains for various input energies for constant uncompressed initial volume, we see that the parabolic plots are similar to the case of DT for low gains where only adiabatic expansion and simple fusion burn occurs without ignition.

For the density of helium-3 mixed with deuterium 1:1 the reasonable approximation to use the density of DT of 5.8×10^{22} cm^{-3} is used for the condensed matter state of the pellets before compression. From the plots similar to Fig. 13.6 for DT of the parabolas, we derived a break even energy of 1.25×10^{13} J and an optimum temperature for (ignitionless) burn of 51 keV, including self-heat and re-absorption of bremsstrahlung. One should take into account that at very large volumes, self heat is rather effective compared to optimum temperatures without self heat.

What is a little surprising is the result that the deviation of the gain law from the cubic root dependence on the input energy for simple burn, Eq.(13.7), shows a superlinear deviation even at gains around one. This is shows the beginning of volume ignition at low gains of one, while the volume ignition in the case of DT appeared at gains above 8 to 10. That means 500 times higher input energy in relation to the two systems.

The resulting gain for compression to 30,000 times solid state density for D-^{3}He is shown in Fig. 13.17 for initial volumes of 10^{-7} to 10^{-1} cm^3. The rather strong jump of the gain indicates volume ignition. However there is a minimum (a "nose") in the plot for smaller volumina. That might be understood from the two component stopping powers, though a more detailed analysis of the D-^{3}He fusion is still to be done. This "nose" vanishes at higher and at smaller volumina. It is remarkable that gains of a few hundred with ignition temperatures of 7 keV, can be reached with an input energy below one to two GJ. The typical ignition process, as in Fig. 13.16 for p-^{11}B, is shown in Fig. 13.18 from plots of the time dependence of the temperature.

Since compression by lasers for volume ignition to 1000 times the solid state was achieved (Nakai et al, 1990), the possibility of producing 100 times higher compression after many years of technological experience of laser fusion seems not to be unrealistic. Under these conditions we have shown that volume ignition does occur in hydrogen boron and in deuterium-helium-3 such that fusion gains of 25 or 300 respectively are possible with a laser energy input of few MJ or 30 MJ respectively. Therefore, if the DT laser fusion is the accepted solution for large scale energy production, the subsequent development of technology may well provide the use of the ideal clean fuel for **nonradioactive nuclear fusion energy production.**

13.6. Responsible Politics

It may be against all ethics, for a physicist to comment on things other than his discipline. Perhaps it is a matter of freedom, or lack of it, which prevents a humble physicist from commenting on factors affecting his discipline which remain separate from that discipline. Rosseau defined freedom as having the right not to do what one would not like to do. One must ask whether or not it is freedom if a physicist talks outside physics. Was it, e.g., freedom when Blokhinzev (1964) in an otherwise very high level book about quantum mechanics wrote a chapter about Leninistic theories and Marxism for explaining the so called philosophical position of quantum mechanics.

The author of this book feels a lack of freedom if remarks about politics with regard to laser fusion have to be mentioned. Indeed it is a loss of freedom if it is considered that it does not help enough to underline how important it is to the society. Regrettably this is the case about the future of environment and energy and the survival of mankind. It seems to be against the principle of freedom if the following statements are suppressed when attempting to warn our contemporaries that it is time to find crucial solutions in politics with respect to world energy production and to do this in time to avoid mistakes which may have cataclysmic consequences.

a) Need for Energy and Need for Safe Environment

There is undoubtedly a need for low cost energy while there is simultaneously a need to stop wasting our planet and the environment if mankind is to survive in future. Burning coal, fossil oil and petrol for the purpose of energy production has to be stopped or at least strongly reduced if the expected "greenhouse" catastrophy of the atmosphere of the earth is to be prevented. While we must be grateful that we have had these fuels as the basis for the development of the industrial age, based on Newton's mechanics and based on Steven Watt's steam engine. While all this, including the electrical and the internal combustion engines up to the computer is the basis of the extraordinary achievment of the industrialized civilization, we must now accept that a change of the large scale energy sources is unavoidable. The phenomena of global warming (Schneider, 1989) is well confirmed after the most sophisticated doubts about the action of clouds for compensation of the radiation effects were excluded (Ravel et al 1989; Levi 1990).

The need for low cost energy is partly illustrated by recent statistics (IAEA, 1988). There is a remarkable correlation between the nearly zero growth of the population of the industrialized countries, such as North America or Western Europe and the nearly zero growth of energy production. North America is doubling its energy consumption within 231 years and Western Europe in 77 years. What the most radical environmentalists could not expect when demanding a reduction of energy consumption has been realized: no growth. In contrast to this, there is Asia where the energy production doubles every 8.4 years or Africa and Middle East where the doubling occurs after 9.9 years. Flying over India 30 years ago, it was dark; now each little village has its street illumination. The same is going on in China. Should this development of using energy not be welcomed against all the protests of environmentalists?

Responsible environmentalists who are known as propagators of green arguments and used for the propaganda of naive green movements, now argued that "we cannot renounce the use of nuclear energy in Europe in the near future" (Stichel et al, 1989; Michaelis 1989). There is no doubt that despite the regrettable disasters produced by irresponsible human interference to the otherwise safe technological control systems at Three Mile Island or in Chernobyl, nuclear energy production is strongly growing. Even in North America where the cleverest lawyers are executing the most extreme arguments of some antinuclear manipulators, nuclear energy production is currently doubling every 7.6 years (IAEA, 1988) despite nearly zero growth of energy consumption. In Western Europe with a similar situation of nearly zero growth, the nuclear power capacity is doubling every 5.9 years. France produces about 80% of its electricity from nuclear sources and its economy is nearly independent of Middle East Oil and its associated problems. In the developing areas of the world, energy production from nuclear sources is increasing at an even much faster rate. Asia currently doubles its nuclear energy production every 2.6 years (IAEA, 1988) while Africa and the Middle East double their nuclear energy production each 3.8 years.

It is no secret that one of the antagonists of nuclear power are in the "green" movement arguments and claims made by elements of the "green" movement are sometimes rather ridiculous, but cannot be overturned. E.g. there is a strong need for electricity in the Los Angeles area especially for cooling in summer. They have the rare advantage in that energy can be received from an environmentally most ideal souce: from water power harnessed in the rocky mountains. The distance of many hundred miles is no problem since high voltage direct current (dc) transmission provides the solution for transferring this energy. The highly needed increase of the capacity, however, is blocked by the "greeny" dominated media with the argument that a mountaineer may get a heart attack when seeing the power transmission lines. Therefore a second line could not be built and the existing lines are now overloaded.

The main, very hidden and quiet enemy of nuclear power has a nearly unexhaustable funding: the oil industry. Indeed, a responsible policy has to be developed if energy production may have to change from the fossil energy to another one like nuclear energy, since one has to respect the huge number of people involved in the oil industry with their jobs and investment commitments. It is therefore necessary to open another direction for this industry to gradually develop without economic and social friction: It should be considered that instead of using timber and cutting forests for producing paper, furniture and building material, that this all should be produced from coal, natural gas and petrol. Taking into account that the industrialized countries need $100 per year and capita for paper, producing this from mining and the petrochemical industry may have a volume of $300 Bill. to $1 Trill.. On top this all will prevent the cutting of the most important and valuable trees. This would counteract the greenhouse effect for compensating the still always necessarily remaining little sector of society, burning fossil fuels.

Much is being touted about by the dominating media that "renewable" energy is the future, ie., solar, wind, wave and geothermic energy. There is no doubt that a lot has to be done in this direction which automatically will be paid by investors on the normal scale if it is profitable. Remarkable progress can be noted about the efficiency of rather not too expensive solar cells developed (Green et al: Hannah 1990). This kind of development will definitely become an enormous business in the many ten billion dollars range when this type of solar energy production will substitute up to 10% of all energy production (Böer, 1986). What had to be admitted however even by one of the most engaged proponants of solar energy (Winter, 1989) is the fact that energy from light water nuclear fission reactors is at least 10 times less expensive.

There is also the argument that due to the law of entropy (Niu, 1989), solar energy can never be competitive against energy production of light water reactors.

Reference can be given also to a study of the otherwise environmentally most sensitive German Ministry of Research and Technology where laws have been established which threaten the ability of competition of the German industry against other countries because of too heavy (and sometimes unnecessary) restrictions. It was elaborated (BMFT-ABB, 1988) what the cost for transporting the energy would be, if solar cells in the North African Sahara produce the energy. In this case the most advanced technology of high voltage dc transmission, or the best developed hydrogen gas pipe energy transport and other systems were used.

The notoriously repeated argument that technology has been neglected due to granting too much for nuclear projects is not valid in this case. The mentioned energy transport systems have the top level of technology and can practically not be furthermore improved since the limits of natural phenomena are reached. It turned out that alone the transport of the energy from North Africa into the industrial centers of the Ruhr river would cost more than the present day relatively high costs of local energy production at the Ruhr.

b. Difficulty of Political Decisions

Since energy from uranium fission reactors is seriously expected as a substitute for the large scale energy production at least for the near future if the emission of carbon dioxide has to be reduced (by international agreement until the year 2005), the question arises whether fission energy in the future could not be substituted by the environmentally cleaner fusion energy. While fission reactors are cleaner than that of fossil fuel furnaces, because of the absence of carbon dioxide emission, the handling of the nuclear waste products (the ash) of fission is a well known difficulty. In view of this problem with fission energy, fusion may provide again a much cleaner energy production at least with respect to the end product (ash) which in the case of fusion is helium, a harmless inert gas.

These conclusions seem to be so very transparent but the fact that any political decision is far away from following this view, requires a discussion how difficult it is to find a political consensus and a will. It is a regrettable fact that in most cases despite the very best will of the persons involved, numerous political decisions are rather short sighted, insufficient or even wrong and may cause a lot of damage instead of the needed improvements.

Let us first mention examples of good and successful decisions. After Japan pioneered the design of the fast railways solving the difficult mechanical oscillation problems by a small team of engineers, it was France where a broad based systematic development arrived at the train system with speeds of 300 km/h. In some sense, however, this is not much improvement if one remembers that an electric Siemens locomotive reached the speed of 210 km/h at an experimental track near Königswusterhausen in 1911.

It was the political will of some decisionmakers in France who, to their merit, have witnessed a wholesale change to the economic system by the fast trains. A thousand years rivalry between Paris and Lyon was broken by a train connection of less than two hours. The tunnel under the English Channel will provide a kind of a double city Paris-London with some link to the European administration in Brussels. These trains will determine where the weight is in Europe. The people in Geneva - always more French than the French - use the low cost three hours fast train to go shopping in Paris disconnecting the German Zürich; and the Swiss customs officers are powerless against this development.

France has also developed a new kind of telephone system, not by basic new discoveries, but only by putting together what all is known in the field of electronics. When big countries in the second and third world are now establishing the necessary infrastructure by introducing new telephone systems, they will buy the French solution ignoring the antiquated systems from Sweden or Germany where all disadvantages of a nationalized telecommunication system has frustrated progress. A typical case is illustrated. When the Bavarian government brought back Nobel Laureate Rudolf Mössbauer in the sixties with many million dollars, he received the standard form from the post office that his telephone will be installed after four years. This was printed without comment in the Spiegel magazine which issues were spread by railway freight from Hamburg. Before the first copies came to Munich, it was read half way in Bonn by the minister of telecomunication who ordered that his men installed a microwave connection to Mössbauer's house: a second Mössbauer effect.

Positive decisions resulted in a system of safe and fast (200 to 250 km/h) car traffic on freeways in Germany for which further development was a little reduced during recent years by the notorious influences of left wing and green minorities. The low speed and highly taxed French "free" ways are a contrasting example.

As a good example, Dr. Kohl's decision should be mentioned to give high tax incentives to developers in the new federal states in central Germany after liberation from communism. This will save unemployment payments there and will get the country moving. This ingenious decision is rather rare in politics.

There are more political decisions to be praised even where technocrates and administrators have been overruled. Until 1949, the powerful Oppenheimer committee refused to recommend the development of the H-bomb based on the most naive argument that the Soviets would not go ahead with this development if the USA would not do so. (Teller 1987). President Truman nevertheless followed the minority arguments of Edward Teller and finally began to develop the H-bomb at a rather late stage. The Soviets were far ahead by the beginning of 1953. Stalin was in a strong position with his H-bomb developments to attack the West. Some unique indications for this intentions are known though the history surrounding Stalin's death remains hidden. It may have been very close that the free world would have been lost.

A splendid decision was that of President Kennedy, who in 1960, ordered that humans should be sent to the moon and returned safely before 1970. This decision stimulated enormous research - apart from the defence research stimulated by the Presidents Truman and Eisenhower - which resulted in golden economic years in North America and in the free world. Kennedy took the decision though not all scientists were fully convinced about the likelihood of success. It was Nobel Laureate Max Born who wrote at this time in the Physikalische Blätter that he could not see how it would be possible to send a man to the moon safely before the year 2000. The mechanics and the physics problems were solved in 1960 for the Apollo project, only the expensive technological constructions and developments had to be performed. Nobody knew precisely in 1960 what the 100 meter high Saturn-5 rocket would look like, to fire a space craft to the moon.

The situation changed later dramatically. There was Vietnam and Ho Chi Minh was able with his manoeuvres to shutter the giant USA and to damage the whole world's economy and wealth. There were so many lost chances. The US troops knew in all details the preparation of the Tet offensive and the commanding US-general tried to do what one has to do in such a case: to attack beforehand but the US Defence Minister McNamara in presence of Senator E.J. Gurney who visited the front lines, strictly forbid it. But even after the Tet case was lost, the success of the US conventional bombing almost brought Hanoi (including secret negotiations of Henry Kissinger) to the point of capitulation. The final result in Vietnam was then organized by some US media mobilizing the street. The financing of the Vietnam operations was controlled by Senator Mansfield who succeeded by simply cutting expenses especially for research and development. The money saved was insignificant but the damage for the future was devastating.

Nobody could stop Mansfield not even President Nixon. The last two flights to the moon were cancelled because each one would have cost $140 Mill. (Gell-Mann, 1987, 1988). There were the wisest of the wise scientists, the US Nobel Laureates invited at the White House who entreated Mr. Nixon, not to stop the flights, but nothing helped. Most of the knowledge of this technology including the blueprints are lost now. In 1990 we can rather say that it will not be possible (again) to send persons to the moon and get them back safely before the year 2000.

Another example of poor decision making, concerns the central slaughterhouse in Paris. Acknowledging the excellence of the French cuisine it was an important question for the government to do something special. Under General de Gaulle as president everything had to be as great as possible. The idea was then obvious instead of having all the numerous slaughterhouses spread over Paris, to build a single central one with the most advanced technologies. This was done in the North-East of Paris and more than one Billion dollars was invested into this project, the land was purchased, the huge concrete buildings were established after designs of the most prominent architects and expensive changes to the railway system were finished for the mass transport of the animals to the slaughterhouse. The project was nearly finished until the clear arguments of the opposition were accepted: with the todays air conditioned trucks it is much easier, less expensive and much more flexible for the distribution, if the slaughterhouses outside of Paris at the countryside are used and the products are transported more individually to the consumers. The billion dollar buildings in the North East of Paris were now changed into a science museum.

A lot of mistakes are made in the day to day management of governments and in industrial institutions e.g. in the continuous struggle of the US and European computer companies to keep up with the Japanese position in microelectronics and computer design. It is sometimes the component of human personality and ingenuity against all the mechanical transfer of rules that is essential. When after 1960, the Volkswagen car producers tried to take over the design of cars from the ingenious Porsche designer team for an inhouse development, a typical mechanical management decision was made. One knew that the number of employees was 50,000 and the textbooks of economics and management prescribed that a designer bureau for such a car company had to have 2000 men. These persons were hired in a very hectic way and the result was as expected: they were not assimilated as in a continuously growing organization and team work, psychology, management and leadership could not simply be reached by producing numbers of recruitment.

When the first in-house designed cars appeared on the market (the Canadians celebrated them as the car of the year) the results of insufficient cooperation and management became evident: e.g. the motion of the hand brake bar was blocked by the gear shift when positioned for reverse drive. The designer of the gear shift part did not know what the designer of the hand brake was doing.

Similar problems arise if an interchange of experiences has to be solved in industry, e.g. if an optics industry follows the need to use electronics, care should be taken to spend sufficient means when hiring the right electronics manager and experts. A wrong decision results in enormous losses, if poor staff from another electronics company are hired and if those persons determine who else is to be hired. The same situation occurs if a classical electrical company has to move into microelectronics and has to use most advanced optics. Wisdom and management consulting are the keys in these steps, though losses are unavoidably similar to situations in government organizations.

These examples of good and bad political decisions were given to illustrate how difficult it is to find the truth and the best decisions and that one should not have illusions if priorities are not found in the best way for the future of energy. There is not only the problem of straight forward thinking, the problems of nonlinear phenomena (subsection 12.7, p.321) and logics result in even more difficult problems. If on top of this all human interests and influences by groups, investors, or social factors are involved, one may imagine that it is sometimes a miracle that a good decision can be reached.

c) Decision about Magnetic Confinement Fusion

In view of the need for low cost and large scale energy generation, for the need to avoid environmental damage, and to reach a long term solution, fusion energy will always be of interest though it has been produced exothermally on earth only in the H-bomb in uncontrollable way and no other scheme of controlled energy production has succeeded as yet. This is the reason that it was possible to gain huge amounts of research support worldwide and hopes were stimulated again and again to find a solution. It sounds like a story of non-success if one remembers all steps of history.

After discovery of the deuterium-tritium fusion reaction of Oliphant, Harteck, and Lord Rutherford (1934) and other similar reactions, where light nuclei burn to heavier ones releasing more than a million times energy than by chemical burning, it was realized that this was the energy source of the stars (Bethe, 1937; von Weitzsäcker 1935). When trying to repeat the experiment by firing 100 keV deuterium nuclei in a gas discharge on deuterated targets (Oliphant et al 1934) with the aim to get an exothermal reaction, i.e. to produce more energy than one had to put in for the ignition of the nuclear reaction, Rutherford strictly forbade this research (Oliphant, 1972, 1987). This particle beam technology of fusion reactions has now been developed as a commercial energy source (Häfeli, 1985). Regrettably, nothing has been achieved yet to get more energy out than is put in apart from the explosive reactions as a merit of the Ulam-Teller and the Sakharov mechanisms.

When in 1951, the discussion started again to produce fusion energy in a controlled way, the use of the particle beam irradiation of a target was rejected by the following argument of Lyman Spitzer Jr. (Hora, 1984): The cross section for the deuterons in the gas discharge with an electron in the cold target is even under the best conditions 300 times higher than the cross section of a nuclear fusion reaction with a tritium nucleus. With a deuterium nucleus, the ratio is 3000 times different. While one achieves a large number of fusion reactions as demonstrated by the excellent neutron output, Spitzer indicated that this ratio 300 completely forbids an exothermal fusion reaction by impact of deuterium ions on a cold target.

There were important physicists at this time, e.g. Nobel Laureate E.O. Lawerence at Berkeley and many others including Sir Mark Oliphant to argue that one should use higher deuterium beam currents to get an exothermal reaction, but this all was rejected by Spitzer. The solution was considered to heat and confine a plasma to equilibrium temperatures of several ten million degrees such that the collisions of the deuterium and tritium ions with electrons are elastical and do not absorb energy as in the cold target and that one can then wait until the much less probable thermonuclear fusion collisions happen and produce more energy out than one had to put in. The only problem was that the confinement e.g. by magnetic fields is not long enough. Lawson (1957) calculated the time for this confinement (Lawson criterion) arriving at a minimm time (τ) and a minimum ion density n for the optimum temperatures near 100 Mill. degrees of

$$n\tau = 10^{14}\ \mathrm{cm}^{-3}\ \mathrm{sec} \tag{13.22}$$

Since then, toroidal and other magnetic field confinements have been tried to produce plasmas of sufficient density and temperatures for a sufficient length of time. Spitzer used the toroidal twisted field with zero plasma current (stellarator) while Yawlenski and Artsimovich used the strong transformer type current in the same configurations (tokamak).

The argument of Spitzer not to use beam currents for irradiation of cold targets for gaining fusion energy was **correct with respect to physics, mathematics and logics.** Nevertheless **it is wrong: it is linear and not nonlinear** and correct only in linear physics. This is a typical case as we experienced in Chapter 12.7 with respect to the longitudinal part of electromagnetic waves how a prediction can change from no into yes, from incorrect into correct if only small additional nonlinear parts are added. The same nonlinearity exposes Spitzer's linear arguments. It should indeed (in future) be possible with very high beam current densities when irradiating cold targets to produce exothermal fusion energy in the so called beam fusion (Yonas, 1979). Nonlinear mechanisms are an integral part of the complicated plasma dynamics of the interaction with heating and expanding and compressing of the target.

With magnetic confinement, the first partial success with a linear Z-pinch discharge "Zeta" in 1957, resulted in predictions that in 20 years from then, fusion energy would be commercial. The hope was postponed again and again, until it was clarified that the toroidal magnetic fields were the main interest. Today, about 2 Billion dollars per year are spent worldwide on magnetic confinement fusion, mostly for tokamak research. The official prediction of the most optimistic protagonists (Pinkau, 1989;

Law 1990; Bruhns 1990; Spiegel 1990) is that by continuing the $2Bill. real value expenses it may be possible in 50 to 60 years to demonstrate that the tokamak power station will work sufficiently well, not clarifying whether it will be financially competitive. It may have to be considered that the energy costs have risen to very high levels such that the estimation by Pfirsch and Schmitter (1989) may be acceptable that produced energy will be ten times more expensive than energy produced from light water fission reactors.

It should be mentioned that the paper of Pfirsch and Schmitter uses very optimistic assumptions with respect to physics which have not been solved yet for the tokamaks. It does not include the more recent downcasting result (Vieider et al 1989) that the tokamak plasma does destroy at least 1 cm wall per day of operation. The procedures for the paper of Pfirsch and Schmitter were reported in the following way: while both authors as Directors of the Max-Planck-Institute of Plasmaphysics in Garching, Germany, have the permission to publish what they consider is correct and do not need to obey any censorhip, it happened that when the preprints for the mentioned publication were finished, the Scientific Director ordered that the preprints had to go through the shredder. A few copies, however, came to the European Parliament in Strassburg from where some uncomfortable questions were asked at Offices of the European Community in Brussels (Euratom).

What followed was an expensive tribunal against the authors where many very prominent plasma physicists were flown in for one week. The result was that every argument and result of Pfirsch and Schmitter (1989) was correct and the publication should not be prevented.

There is the difficulty to find a decision in favour of providing further funding for magnetic confinement fusion research. This does not only depend on the scientific facts but also on the influence of the powers to lobby the governments. How easy this can be was demonstrated by the Society to Advance Fusion Energy, New York, (La Mer Slaner, 1979), which had access to influential Congressmen, and President Carter. It was finally possible that with the help of the administration of the Departent of Energy (Ed. Kintner) a Bill was approved by the congress and signed by the president, the Tsongas-McCormack Bill, for spending $20Bill. within ten years for fusion energy research. While the initial steps for this bill were for fusion energy generally, some partners involved moved this exclusively to the study of magnetic confinement fusion.

The Bill, though being law in the USA, is however ignored. The following congress with President Reagan's majority of republicans froze the law simply by not approving the necessary money. In view of the later confirmed problems with magnetic confinement fusion, it may have been a correct decision not to waste money for a concept which may be successful after the year 2040 and whose economic competitiveness even will be questionable.

d) What Can Inertial Confinement Fusion (ICF) Offer?

Fusion energy seems to remain attractive and despite the enormous disappointments in the past with magnetic confinement fusion, still there may be support. We ask whether it is possible that laser fusion or more generally inertial confinement fusion (ICF) could offer more than the magnetic confinement fusion? Could one - following the report of results of the preceding subsections claim that

- laser fusion offers the physics solution for producing energy,

- a gigawatt fusion power station with DT fuel could be built in a crash program within 10 years for the present days costs of $3 Bill. (apart from development costs),

- that the cost of electricity will be equal or less than that of light water fission reactors,

- that the physics is flexibily enough to possibly permit a price of energy 3 to 5 times less than that of light water fission reactors within 20 to 30 years of development,

- that use of clean fuel may be possible in the distant future such that nuclear fusion energy can be produced with less radioactivity than burning coal?

It seems so that all these arguments are at least developed to such a point as it was with the flight to the moon in 1960. There may be highly regarded and precise thinking experts who nevertheless may have doubts, but in view of the underground nuclear explosion results and of the other arguments explained before, one could share the positive view about the mentioned five points. This was summarized also in the carefully formulated and checked sentence of the director for ICF in the US-Department of Energy, Sheldon Kahalas (1988), when he said

FUSION ENERGY BY LASERS IS NO LONGER A QUESTON OF 'IF' BUT OF 'WHEN'

Despite the argument and the mentioned problems concerning facts about magnetic confinement fusion (MCF), there is still a lot of infighting and intrigue. There have been many years of argument between MCF and ICF and the fact was overemphasised by MCF that ICF has a large scale military application especially with respect to the physics studied with indirect (hohlraum) driving of pellets by lasers using x-rays and the physics of the nuclear fusion detonation waves for spark ignition, which are similar processes as in bombs. The preceding chapters underline that these methods are not necessary for laser fusion. Even better results may be reached by using the direct driving because of our recent understanding of pulsation-free and smooth direct drive of pellets by lasers with high efficiencies. We have further developed the volume ignition which is much more transparent and better understood than the fusion detonation processes.

Nevertheless, the easiest way to exclude laser fusion from public discussions by MCF protagonists was always the neighbourhood of ICF with classified military applications. This is the reason that the European Community's budget for fusion energy contains 99% for magnetic fusion putting ICF always into the corner of undesired classified work. In view of the confirmation of the physics solution of ICF from the underground tests of the Centurion Halite project (Broad, 1988) confirmed in the Harper report of the US Academy of Sciences, there have been steps within the Department of Energy to get the correct position of ICF. It was R.O. Hunter the Director of Energy Research (Goodwin, 1990) who was trying to get ICF and MCF into a more equable position. The result was that the Representative from New Jersey protected the MCF project in Princeton and succeeded together with other lobyists to discredit Hunter so much at his Department Head, Admiral Watson, and in the presence of the President, that Hunter had to resign.

Though the goal of Hunter did not come through that ICF and MCF are to be considered on equal positions and that competition for funds must be introduced with a first shift of funds from MCF to ICF, a motion into this direction was the establishment of a committee by Admiral Watson to discuss the position of ICF and MCF simultaneously. In view of the mentioned success of ICF it was unavoidable that ICF received a favourable recommendation at least within the modest lines of the present days budget situation.

Again the MCF experts succeeded to move ICF into the corner of the military projects but at least the first steps for an objective consideration have been taken. The next steps are planned to evaluate the costs of energy produced by the different fusion schemes and there is hope that the truth will determine the result. The pressure to the governments with respect to long term decisions about the greenhouse effect, and accepted restriction of carbon dioxide emission may open the re-activation of the Tsongas-McCormack Bill. However under the much more realistic aspects given from the success of ICF due to the underground explosions: to develop fusion energy from ICF, based on the result that laser drivers with few MJ pulse energy for a GW power station can be produced for $1 Bill and a reactor of the most simplified function as the Cascade type may cost the same, the technological steps to a power station is then only a question of funding.

Nevertheless the aforementioned examples of poor decision making are well known and it is possible that the enormous potential for developing a low cost, clean energy source from laser fusion or with the potential of heavy ion beam fusion is being missed or unnecessarily postponed causing environmental damage in the trillions of dollars. Thiequestion of damage is to be seen also with the loss of the opening of the golden age of negligibly low cost of energy based on the economic dilemma (La Mer Slaner, 1979) which then initiated the Tsongas-McCormack Bill as expressed by Arthur Grey Jr., George Miley and George Brumlik (1979):

"The history of man is a record of progress directly tied to a continuing reduction in the cost of energy. From the discovery of fire and wheel, the harnessing of coal, oil and natural gas for the generation of electricity, man's material lot has improved in indirect proportion to the cost and availability of energy. Now that fossil fuels have reversed their curve of cost efficiency, and other sources of energy (winds, tides, solar energy, fission, geothermal and hydroelectric power) have become economically attractive only in comparison of the ever-increasing costs of fossil fuels, we must look to what is the cheapest source of future energy in our universe -- fusion."

APPENDIX A

The Effective Mass

In Section 2 we used the effective mass m^* of electrons. Though this is a property of condensed matter, a marginal discussion for high-density plasmas is useful under the aspect of the quantum properties of high-density plasmas. As it is the endeavor of this book to derive the physics *ab initio*, as, for example, the hydrodynamics or electrodynamics, a similar derivation of the quantum mechanics is given.

Quantum physics was due to the discovery of the fact that all quanties with the dimension of an action can appear only in multiples of Planck's number $h=6.67\times10^{-27}$ erg sec, or $\hbar=h/2\pi$. This observation was not as easy as the observation of the atomistic structure of the electric charge (given by the electron charge), which everyone could see immediately in the Millikan experiment. The history of the discovery of the atomistic structure of action was pronounced where energies E [of electrons at photoemission or in gas discharges (Franck–Hertz effect)] were related to (optical) frequencies ν, where

$$\frac{E}{\nu}=h \tag{A.1}$$

Since the product of a momentum p and a length x is of the dimension of an action, there were difficulties on how to keep the beautiful knowledge of mechanics, which was formulated so successfuly by Newton, d'Alembert, Lagrange, and Hamilton. If we write the Hamilton function of a simple system as the sum of kinetic energy $p^2/2m$ and potential energy $V(x)$

$$H=\frac{p^2}{2m}+V(x)=\mathrm{E} \tag{A.2}$$

as the total energy E, then we have to be aware of the background of the Lagrangian and aware that this is not simply the difference of kinetic and potential energy for nonconservative forces [115].

One way to quantize Eq. (A.2) is to not use the quantities of p and x directly, but to describe them by a distribution function as shown in Section 3. Instead of getting the average value of a set of quantities q_n one could use a distribution function f_n to arrive at the average value of (see Eq. 3.2)

$$q = \frac{\sum f_n q_n}{\sum f_n} \tag{A.3}$$

Knowing this, Eq. (A.2) can be written by differential operators, where, however, only a distribution function will be defined from which the physical quantities have to be derived similarly to Eq. (A.3). If the operators

$$p = -i\hbar \frac{\partial}{\partial x}; \qquad E = -\frac{\hbar}{i}\frac{\partial}{\partial t} \tag{A.4}$$

are used we are in agreement with quantization

$$p\,\partial x = \text{“}\hbar\text{”}; \qquad E\,\partial t = \text{“}\hbar\text{”} \tag{A.5}$$

where the quotes are a symbolism which nobody would have accepted if the following steps were not performed historically by wave equations (de Broglie, Schrödinger). Only in retrospect, the motivation for (A.4) from a quantization as in (A.1) or (A.5) should be understood. Using (A.4) in Eq. (A.2), the Hamilton function becomes a Hamiltonian operator for a differential equation for a distribution function Ψ (Schrödinger equation)

$$\left\{-\frac{\hbar^2}{2m}\frac{\partial^2}{\partial x^2} + V(x)\right\}\Psi(x, t) = -\frac{\hbar}{i}\frac{\partial}{\partial t}\Psi(x, t) \tag{A.6}$$

Stationary (time-independent) solutions of this wave equation can be expressed by

$$\Psi = \psi(r)\exp\left(-\frac{i}{\hbar}Et\right) \tag{A.7}$$

where E is an eigenvalue representing an energy in the time-independent Schrödinger equation expressing the spatial dependence now by the coordinates of

$$\left(-\frac{\hbar^2}{2m}\nabla^2 + V(r) - E\right)\psi(r) = 0 \tag{A.8}$$

If the potential $V=0$, electrons in vacuum can be described from (A.8) by plane waves

$$\psi(r) = A\exp(i\mathbf{k}\cdot\mathbf{r}); \qquad \Psi = A\exp\left(\mathbf{k}\cdot\mathbf{r} - \frac{i}{h}Et\right) \tag{A.9}$$

where $E/\hbar=\omega$ is a radian frequency, and the wave vector $\mathbf{k}$ is from Eq. (A.8)

$$|\mathbf{k}|=\frac{1}{\hbar}\sqrt{2mE} \tag{A.10}$$

In order to arrive at a physical quantity (expectation value) from the distribution functions, one has to proceed as in Eq. (A.3); however, as Ψ can be complex we have then to include the conjugate complex value Ψ^*

$$Q=\frac{\int\Psi^* q\Psi\, d^3\tau}{\int\Psi^*\Psi\, d^3\tau} \tag{A.10a}$$

integrating over the whole space. The normalization of the amplitude A in Eq. (A.9) is to fulfill $\int\Psi^*\Psi\, d^3\tau=1$. For example, to arrive at the momentum of the electron, the quantity q in (A.10a) is the operator p of Eq. (A.4), and we find the momentum by spatial differentiation of (A.9)

$$\mathbf{p}=\int\psi^*\frac{\hbar}{i}\nabla\psi\, d^3\tau=\hbar\mathbf{k} \tag{A.11}$$

using the result (A.10), we find the point mechanical relation between momentum and energy of a free electron

$$p=\sqrt{2mE} \tag{A.12}$$

It is a classical example in this method of quantum mechanics to use the Coulomb potential of a proton for $V(r)$ in Eq. (A.8) to arrive at the stationary (bound) states of the electron. The solutions of the equation for the distribution functions ψ arrive by mathematical reasons at eigenvalues $E_n(n=1, \ldots, \infty)$ which are the energy levels of the electrons in the atom. The spatial distribution of $\Psi^*\Psi$ corresponds to the electron density in the atom arriving at the diameter of twice the Bohr radius for $n=1$, or eight Bohr radii at $n=2$, and so on (see Section 2.3).

For the theory of condensed matter, the case of a periodic potential

$$V(\mathbf{r}+\mathbf{d})=V(\mathbf{r}) \tag{A.13}$$

with a periodicity vector

$$\mathbf{d}=\mathbf{i}_1 d_1 a_1+\mathbf{i}_2 d_2 a_2+\mathbf{i}_3 d_3 a_3 \tag{A.14}$$

is of interest, where d_i are distances of atoms in the three crystal directions $i=1, 2, 3$, and the a_i are integers. Bloch discovered that the solutions of Eq. (A.8) for the periodic potential (A.13) are of the form

$$\psi(\mathbf{r})=u(\mathbf{k}, \mathbf{r})\exp(i\mathbf{k}\cdot\mathbf{r}) \tag{A.15}$$

The Schrödinger equation has then the form

$$\left\{-\frac{\hbar^2}{2m}\nabla^2+V(\mathbf{r})-E(\mathbf{k})\right\}u(\mathbf{k},\mathbf{r})\exp(i\mathbf{k}\cdot\mathbf{r})=0 \tag{A.16}$$

Mathematically, the mechanical problem is determined, if $V(\mathbf{r})$ is given, resulting in a uniquely defined $\psi(\mathbf{r})$. This is uniquely related to the function $E(\mathbf{k})$. Instead of describing electrons in a crystal by $V(\mathbf{r})$, or $\psi(\mathbf{r})$, one can uniquely describe them by considering $E(\mathbf{k})$. $\mathbf{k}$ is defining the momentum of the electrons; therefore the energy momentum relation is unique for the description. It was the discovery of Bloch that instead of the parabolic relation $p^2/2m=E$ of Eq. (A.12) for free electrons, there are forbidden gaps for E. The $E(\mathbf{k})$ functions are periodic also, and for $k=0$, that parabolic relation for free electrons can be approximated by the $E(\mathbf{k})$ functions. The parabolas only are more or less curved. The dimensionless factor between the curvatures is simply given by

$$m^*=\frac{1}{\hbar^2}\frac{[\partial E/\partial(p^2)]_{\text{vacuum}}}{\partial E/\partial(\mathbf{k}^2)} \tag{A.17}$$

which is the effective mass.

The structure of energy bands for electrons can occur in very high-density low-temperature (degenerate) plasmas. The physics of laser compressed plasmas is now going into these conditions which will need more detailed study in the future.

To complete the conceptual framework of quantum mechanics drawn in this Appendix, it should be mentioned that the discussion of the quantum mechanical problems with distribution functions and expectation values to conserve the Newton–Hamiltonian mechanics by Schrödinger's differential equation has a mathematical equivalence with an integral equation problem (transformation theory by Weil). The eigenvalues of the differential equation correspond then to the elements of the infinite matrices of the integral equation problem. The matrices alone can be used in a Hamiltonian as it was described by Heisenberg (matrix mechanics).

APPENDIX B

The Maxwell–Boltzmann Distribution

The distribution function of the energy to the particles of a plasma or a gas at equilibrium that was used in Section 3 is derived now. There is a correlation between the entropy S_{12} and the probability W_{12} of the microscopic structure of the states of two thermodynamic systems with the respective values S_1, S_2, W_1, and W_2 [394, 395]

$$S_{12} = S_1 + S_2 \tag{B.1}$$

$$W_{12} = W_1 W_2 \tag{B.2}$$

from their definition. The function that reproduced the correlation [396]

$$f(x_1 x_2) = f(x_1) + f(x_2) \tag{B.3}$$

is given by Boltzmann's relation

$$S = K \ln W \tag{B.4}$$

using the Boltzmann constant K as the gas constant per particle.

The philosophy for describing a plasma by the probabilities of the distribution of energy to its individual particles is an extremistic picture and may not cover all facts of reality. It implies, for example, that the forces between the particles are small or negligible in first-order or only during negligible times, while the interactions are necessary on the other hand to achieve equilibrium. The other extreme with its insufficiencies is the description of phenomena by differentiable or by analytic (holomorphic) functions which may run into a superdeterminism (Laplace, Cauchy). This can even be a consequence of quantum mechanics (not only in the Schrödinger picture) if the correlation between object and measuring apparatus is considered [397].

In the Boltzmann statistics—in contrast to the quantum statistics—the possibility to distinguish between the particles of an ensemble is assumed. Using six-dimensional volume elements $\Delta\tau_i = \Delta x\ \Delta y\ \Delta z\ \Delta v_x\ \Delta v_y\ \Delta v_z$, the number N_i of particles in this element is given by a distribution function $f(i)$

$$N_i = f(i)\Delta\tau_i \tag{B.5}$$

The total number N of particles should be constant

$$N = \sum f(i)\Delta\tau_i; \qquad \delta N = \sum \delta f(i)\Delta\tau_i = 0 \tag{B.6}$$

where the constancy of $\Delta\tau_i$ at any variation (due to the Liouville theorem) has been used. The energy $U(\delta)$ of the particles in the ith cell results is an energy $N_i u(i)$ in the cell. The total energy U should be constant

$$U = \sum u(i)\, f(i)\Delta\tau_i; \qquad \delta U = \sum \delta f(i) u(i)\Delta\tau_i = 0 \tag{B.7}$$

The probability of the system whose states of each cell are weighted by

$$G_i = \Delta\tau_i \tag{B.8}$$

is given by the number of the combination of all cases $N!\Pi G_i$ that are possible by permutation without repetition if we use distinguishable particles:

$$W = \frac{N!\Pi G_i^{N_i}}{\Pi N_i!} \tag{B.9}$$

Using Eqs. (B.5), (B.8), and the Stirling formula (on approximation for large N)

$$N! = \frac{N^N}{e^N} \tag{B.10}$$

we find from (B.9)

$$W = \frac{N^N \Pi \Delta\tau_i^{\,f(i)\Delta\tau_i}}{\Pi (f(i)\Delta\tau_i)^{f(i)\Delta\tau_i}} \tag{B.11}$$

and the entropy from Eq. (B.4)

$$S = KN \ln N - K \sum f(i)\Delta\tau_i \ln f(i) \tag{B.12}$$

Equilibrium corresponds to the value of the highest probability W, or $\delta S = 0$ at the secondary conditions of constant total particle number N, Eq. (B.6), and total energy U, Eq. (B.7), from Eq. (B.12)

$$0 = \sum \delta_i f(i)\Delta\tau_i \ln f(i) + \sum \delta f(i)\Delta\tau_i \tag{B.13}$$

To include the secondary conditions, the method of multiplicators is used by adding conditions (B.6) after multiplying with α to Eq. (B.13) and to proceed with (B.7) after multiplying with β in the same way. The result is

$$\ln f(i) + 1 + \alpha + \beta u(i) = 0 \tag{B.14}$$

This leads immediately to the desired energy distribution function

$$f(i)=A\exp(-\beta u(i)) \tag{B.15}$$

where

$$A=\exp[-(1+\alpha)] \tag{B.16}$$

is given by the constant number N of all particles from Eqs. (B.6) and (B.15)

$$N=A\sum\exp(-\beta u(i))\Delta\tau_i \tag{B.17}$$

or

$$f(i)=\frac{N\exp(-\beta u(i))}{\sum\exp[-\beta u(i)]\Delta\tau_i} \tag{B.18}$$

The dominator is the equipartition function (sum of states)

$$\sigma=\sum\Delta\tau_i\exp(-\beta u(i)) \tag{B.19}$$

The physical interpretation of the multiplicator β is given from the definition of the entropy. Using Eqs. (B.18) and (B.19) in (B.12)

$$S=K\ln N-K\sum\frac{N}{\sigma}\Delta\tau_i\exp(-\beta u(i))[\ln N-\beta u(i)-\ln\sigma] \tag{B.20}$$

and Eq. (B.7)

$$\frac{N}{\sigma}\sum\Delta\tau_i\exp(-\beta u(i))=U \tag{B.21}$$

Eq. (B.20) reduces to

$$S=K\beta U+KN\ln\sigma \tag{B.22}$$

Thermodynamics defines the relation between S, U, and the temperature T for conditions of constant volume V

$$\frac{1}{T}=\left(\frac{\partial S}{\partial U}\right)_v \tag{B.23}$$

U is a function of β by Eq. (B.21), therefore

$$\left(\frac{\partial S}{\partial U}\right)_v=\frac{dS}{d\beta}\left(\frac{\partial\beta}{\partial U}\right)_v=\frac{dS}{d\beta}\frac{1}{(\partial U/\partial\beta)_v} \tag{B.24}$$

After differentiating Eq. (B.22) we find

$$\frac{dS}{d\beta}=KU+K\beta\frac{\partial U}{\partial\beta}+\frac{KN}{\sigma}\frac{\partial\sigma}{\partial\beta} \tag{B.25}$$

by substitution of the differential quantities of Eqs. (B.19) and (B.21)

$$\frac{\partial\sigma}{\partial\beta}=-\sum u(i)\Delta\tau_i \exp[-\beta u(i)]=-\frac{U\sigma}{N} \tag{B.26}$$

and from Eq. (B.25)

$$\frac{dS}{d\beta}=K\beta\frac{\partial U}{\partial\beta} \tag{B.27}$$

Taking the differential form and Eq. (B.24) we find

$$\left(\frac{\partial S}{\partial U}\right)_v=\beta K=\frac{1}{T} \tag{B.28}$$

and finally

$$\beta=\frac{1}{KT} \tag{B.29}$$

The distribution function (B.18) arrives then at the Maxwell–Boltzmann distribution

$$f(i)=\frac{N\exp(-u(i)/KT)}{\sum\Delta\tau_i \exp(-u(i)/KT)} \tag{B.30}$$

as used in Eq. (3.20).

It should be noted that the discussion of a correction of the dimension of the quantity σ in Eq. (B.22) which has to be of no dimension, while σ is of the dimension of an action to the cube, led Planck [398] to the addition of an arbitrary constant h in the entropy of Eq. (B.22)

$$S=\frac{U}{T}+KN\ln\frac{\sigma}{h^3} \tag{B.31}$$

arriving at a free energy

$$F=U-TS=-KNT\ln\frac{\sigma}{h^3} \tag{B.32}$$

This conclusion of the atomistic structure of action (quantization) from the classical statistics required that oscillators of a frequency ν can gain multiples $h\nu$ of energy only. This caused a reformulation of the sum of states σ (for the oscillator only)

$$\sigma=\frac{h}{1-\exp(-h\nu/KT)} \tag{B.33}$$

and in combination with the Rayleigh–Jeans density of low frequency modes of blackbody radiation led to Planck's radiation law from which the comparison with the fully fitting experiments resulted in the number of h. Remembering that Einstein discovered the stimulated emission (the laser) from his derivation of Planck's radiation law [399], these far ranging facts may be considered [400] as a consequence of Boltzmann's statistics.

APPENDIX C

Derivation of the General Two-Fluid Equations

The direct derivation of the two-fluid equation of motion, (8.3) or (8.6), from the Euler equation of electrons and that for ions is presented in this appendix even though it contains several trivial steps that are usually omitted in textbooks. The derivation is similar to Schlüter's original work [136]. We use the Euler equation with a general viscosity term (determined by the collision frequency ν) for ions [as Eq. (6.1)]

$$m_i n_i \left[\frac{\partial}{\partial t} \mathbf{v}_i + \mathbf{v}_i \cdot \nabla \mathbf{v}_i \right] = -\nabla n_i K T_i + Z n_i e \mathbf{E} + \frac{Z n_i e}{c} \mathbf{v}_i \times \mathbf{H}$$

$$- \frac{m_i n_i m n_e}{m_i n_i + m n_e} \nu (\mathbf{v}_i - \mathbf{v}_e) + \mathbf{K}_i \tag{C.1}$$

and a Euler equation for electrons [as Eq. (6.2)]

$$m n_e \left[\frac{\partial}{\partial t} \mathbf{v}_e + \mathbf{v}_e \cdot \nabla \mathbf{v}_e \right] = -\nabla n_e K T_e - n_e e \mathbf{E} - \frac{n_e e}{c} \mathbf{v}_e \times \mathbf{H}$$

$$+ \frac{m_i n_i m n_e}{m_i n_i + m n_e} \nu (\mathbf{v}_i - \mathbf{v}_e) + \mathbf{K}_e \tag{C.2}$$

which both may be considered to be based on hydrodynamics, as derived from the kinetic theory in Section 3. The viscosity terms will be canceled when we add both equations later.

In this section we shall use the net pressure of the plasma

$$p = n_i K T_i + n_e K T_e \approx n_i (1 + Z) K T \tag{C.3}$$

with the (optional) simplification

$$T_i \approx T_e \approx T \tag{C.4}$$

of thermal equilibrium.

The discussion that follows basically involves the question of quasi-neutrality

$$n_e \approx Z n_i \tag{C.5}$$

which is used to such an extent that it usually is not perfect [sign of equality in Eq. (C.5)], but it is so well realized within all known properties of fluctuations that the net plasma velocity

$$\mathbf{v} = \frac{m_i n_i \mathbf{v}_i + m n_e \mathbf{v}_e}{m_i n_i + m n_e} \tag{C.6}$$

can be approximated by [see Eq. (6.3)]

$$\mathbf{v} \approx \frac{m_i \mathbf{v}_i + Z m \mathbf{v}_e}{m_i + Z m} \tag{C.7}$$

The velocity difference permits a similar sufficient approximation

$$\mathbf{v}_e - \mathbf{v}_i \approx \frac{n_e \mathbf{v}_e - Z n_i \mathbf{v}_i}{n_e} \frac{e}{e} = - \frac{\mathbf{j}}{e n_e} \tag{C.8}$$

where the definition of the electric current density [see Eq. (6.4)]

$$\mathbf{j} = e(Z n_i \mathbf{v}_i - n_e \mathbf{v}_e) \tag{C.9}$$

is included. A derivation without all these assumptions was performed by Lüst and a result is given in Ref. 401. Agreement with this result will not only justify the assumptions used here, but will also emphasize the essential properties of the relations.

In order to achieve the equation of motion of the plasma by adding Eqs. (C.1) and (C.2), we get for the right-hand side

$$-\nabla p + (Z n_i - n_e) e \mathbf{E} + (Z n_i e \mathbf{v}_i - n_e e \mathbf{v}_e) \times \frac{\mathbf{H}}{c} + \mathbf{K}_i + \mathbf{K}_e \tag{C.10}$$

or using Eq. (C.9),

$$-\nabla p + \mathbf{E} e (Z n_i - n_e) + \mathbf{j} \times \frac{\mathbf{H}}{c} + \mathbf{K}_i + \mathbf{K}_e \tag{C.10a}$$

where the viscosity terms have canceled.

By adding (C.1) and (C.2) and by using an identity for the left-hand side of Eq. (C.2), we get for the left-hand side

$$m_i n_e \left(\frac{\partial}{\partial t} \frac{m}{m_i} \mathbf{v}_e + \frac{m}{m_i} \mathbf{v}_e \cdot \nabla \mathbf{v}_e \right) \tag{C.10b}$$

We arrive at the following expression after adding several more terms and subtracting them again

$$
\begin{aligned}
& m_i n_i \frac{\partial}{\partial t} \mathbf{v}_i + m_i n_i \frac{\partial}{\partial t} \frac{m}{m_i} Z\mathbf{v}_e + m_i n_i \mathbf{v}_i \cdot \nabla \mathbf{v}_i + m_i n_i \frac{Zm}{m_i} \mathbf{v}_e \cdot \nabla \mathbf{v}_e \\
& + m_i N_i \frac{Zm}{m_i} \mathbf{v}_e \cdot \nabla \mathbf{v}_i - m_i n_i \frac{Zm}{m_i} \mathbf{v}_e \cdot \nabla \mathbf{v}_i + m_i n_i \frac{Zm}{m_i} \mathbf{v}_i \cdot \nabla \mathbf{v}_e \\
& - m_i n_i \frac{Zm}{m_i} \mathbf{v}_i \cdot \nabla \mathbf{v}_e - m_i n_i \left(\frac{Zm}{m_i}\right)^2 \mathbf{v}_e \cdot \nabla \mathbf{v}_e \\
& - m_i n_i \frac{Zm}{m_i} \mathbf{v}_i \cdot \nabla (\mathbf{v}_e - \mathbf{v}_i) + m_i n_i \frac{Zm}{m_i} \mathbf{v}_i \cdot \nabla (\mathbf{v}_e - \mathbf{v}_i)
\end{aligned} \tag{C.11}
$$

Here the term with the quadratic mass ratio m/m_i can be neglected as well as the very last term with $\mathbf{v}_i \cdot \nabla \mathbf{v}_i$ in the third term because of the mass ratio. All other terms of Eq. (C.11) can be combined to result in

$$
m_i n_i \left[\frac{\partial}{\partial t} \mathbf{v} + \mathbf{v} \cdot \nabla \mathbf{v}\right] + m_i n_i \frac{Zm}{m_i} (\mathbf{v}_e - \mathbf{v}_i) \cdot \nabla (\mathbf{v}_e - \mathbf{v}_i) \tag{C.12}
$$

The last term in Eq. (C.12) can be rewritten by using Eqs. (C.8) and (C.9)

$$
mZn_i(\mathbf{v}_e - \mathbf{v}_i) \cdot \nabla (\mathbf{v}_e - \mathbf{v}_i) = \frac{m\mathbf{j}}{e} \cdot \nabla \frac{\mathbf{j}}{en_e} = \frac{4\pi}{\omega_p^2} \mathbf{j} \cdot \nabla \mathbf{j} - \frac{m}{e^2 n_e^2} \mathbf{jj} \cdot \nabla n_e \tag{C.13}
$$

From Eqs. (C.10), (C.12), and (C.13), the result of adding Eqs. (C.1) and (C.2) arrives at the net force density $\mathbf{f}$ in the plasma

$$
\begin{aligned}
\mathbf{f} = m_i n_i \left[\frac{\partial}{\partial t} \mathbf{v} + \mathbf{v} \cdot \nabla \mathbf{v}\right] = & -\nabla p + \mathbf{E}e(Zn_i - n_e) + \mathbf{j} \times \frac{\mathbf{H}}{c} + \mathbf{K}_i + \mathbf{K}_e \\
& - \frac{4\pi}{\omega_p^2} \mathbf{j} \cdot \nabla \mathbf{j} + \frac{4\pi}{\omega_p^4} \mathbf{jj} \cdot \nabla \omega_p^2
\end{aligned} \tag{C.14}
$$

This is the complete result of Schlüter at an appropriate interpretation of his velocities [136] where we have permitted ions of a general charge Z, as one point of generalization. The same result was reported as the outcome of an exact treatment by Lüst [401] without our assumptions of eqs. (C.5) to (C.9). Additional terms were negligible because of the space charge neutrality.

Although Schlüter [136] regarded forces by high-frequency electromagnetic fields such as radiation pressure to be included in the unspecified forces $\mathbf{K}_i + \mathbf{K}_e$, the treatment of the high-frequency fields is possible directly by Eq. (C.14). We neglect $\mathbf{K}_i$ and $\mathbf{K}_e$ and use then monochromatic time dependence given by a frequency ω for all quantities $\mathbf{E}$, $\mathbf{j}$, and $\mathbf{H}$ as in Eq.

(6.19). We further use Eq. (6.8)

$$\frac{\partial}{\partial t}\mathbf{j}+\nu\mathbf{j}=\frac{\omega_p^2}{4\pi}\mathbf{E} \tag{C.15}$$

which is the result of subtracting Eq. (C.1) from (C.2). Nonlinear terms have been neglected in Eq. (C.15) as is permissible for subrelativistic high-frequency fields.

The steps of subtracting Eq. (C.2) from (C.1) which led to Eqs. (C.15) and (6.7) should be shown. Equation (C.1) is multiplied by Zm and (C.2) by m_i. After subtraction, the left-hand side is

$$m_i m\left[Zm_i\left(\frac{d}{dt}\right)_i \mathbf{v}_i-n_e\left(\frac{d}{dt}\right)_e \mathbf{v}_e\right]\approx m_i m Z n_i \frac{d}{dt}(\mathbf{v}_i-\mathbf{v}_e) \tag{C.16}$$

where use was made of Eq. (C.5). This implies also approximate equality of $(d/dt)_{e,i}=\partial/\partial t+\mathbf{v}_{e,i}\cdot\nabla$ and $d/dt=\partial/\partial t+\mathbf{v}\cdot\nabla$ using $\mathbf{v}_e\approx\mathbf{v}_i\approx\mathbf{v}$. The necessary neglect of coherent quiver motion in $\mathbf{v}_e$ does not affect the following conclusions. Another limiting case is where the spatial derivations are less compared with the $\partial/\partial t$ terms and thus no restriction is given to the amplitudes [if not Eq. (C.5) is violated]. Using Eq. (C.8), the left-hand side of the result of subtraction is then

$$\frac{m_i m}{e}\left[\frac{d}{dt}\mathbf{j}-\frac{\mathbf{j}}{n_e}\frac{d}{dt}n_e\right] \tag{C.17}$$

The second term in brackets is usually neglected in order to achieve the well-known result that follows [Eq. (C.20)].

The right-hand side after adding and subtracting one more term is

$$\begin{aligned}&-m_e\nabla Zn_iKT_i+m_i\nabla n_eKT_e+\mathbf{E}e(mn_iZ^2+n_em_i)\\&+(Z^2mn_ie\mathbf{v}_i+m_in_ee\mathbf{v}_e)\times\frac{\mathbf{H}}{c}\\&+[Zm^2n_e\nu+m_imn_e\nu](\mathbf{v}_e-\mathbf{v}_i)\\&+m_iZn_ie\mathbf{v}_i\times\frac{\mathbf{H}}{c}-m_iZn_ie\mathbf{v}_i\times\frac{\mathbf{H}}{c}\end{aligned} \tag{C.18}$$

Dividing both sides by m_i and neglecting mn_iZ^2 as compared to n_em_i results in

$$\frac{m}{e}\left[\frac{d}{dt}\mathbf{j}+\nu\mathbf{j}\right]=\nabla n_eKT_e+\mathbf{E}en_e-\mathbf{j}\times\frac{\mathbf{H}}{c}+Zn_ie\mathbf{v}_i\times\frac{\mathbf{H}}{c}$$

Dividing by en_e, using $\mathbf{v}_i\approx\mathbf{v}$, and using $p_e=p/(1+1/Z)$ at thermal equilibrium, we arrive at the generalized Ohm's law (diffusion equation)

$$\frac{4\pi}{\omega_p^2}\left[\frac{d}{dt}\mathbf{j}+\nu\mathbf{j}\right]=\mathbf{E}-\frac{1}{en_ec}\mathbf{j}\times\mathbf{H}+\mathbf{v}\times\frac{\mathbf{H}}{c}$$
$$+\frac{1}{en_e}\nabla\frac{p}{1+1/Z} \tag{C.20}$$

which is identical with Eq. (6.7) and the derivations of Schlüter [136] and Lüst [401].

If we had not neglected the second term in brackets of expression (C.17), the generalized Ohm's law would be

$$\frac{4\pi}{\omega^2}\left[\frac{d}{dt}\mathbf{j}+\mathbf{j}\left(\nu-\frac{1}{n_e}\frac{d}{dt}n_e\right)\right]=\mathbf{E}-\frac{1}{en_ec}\mathbf{j}\times\mathbf{H}+\mathbf{v}\times\frac{\mathbf{H}}{c}+\frac{1}{en_e}\nabla p_e \tag{C.21}$$

This shows that an additional damping mechanism appeared, given by an effective collision frequency

$$\nu_{\text{eff}}=-\frac{1}{n_e}\frac{d}{dt}n_e=-\frac{1}{n_e}\left[\frac{\partial}{\partial t}n_e+\mathbf{v}_e\cdot\nabla n_e\right] \tag{C.22}$$

This damping—which obviously has not been recognized before—will find a very special interpretation for the special case where one has a one-dimensional variability of n_e only (using the x-direction) and especially if

$$\frac{\partial}{\partial t}n_e+\mathbf{v}_e\frac{\partial}{\partial x}n_e=-\mu\frac{\partial^3}{\partial x^3}n_e \tag{C.23}$$

The damping is due to Langmuir solitons where Eq. (C.23) is the Korteweg–de Vries equation for the electron density n_e (Langmuir waves). The dispersion relation μ is not necessarily the usual dispersion but can be much more complex as the example in Section 10.6 has shown, where for the numerically derived dynamic solitons [$\mathbf{v}$ instead of n_e in Eq. (C.23)] a dispersion appeared that was not of the usual value but of the same formulation as Denisov derived for resonance absorption.

Ohm's law [Eq. (C.21)] determines the high-frequency electromagnetic waves in a plasma, and the new damping process, given by an effective collision frequency [Eq. (C.22)] or—in the special case of Eq. (C.23)

$$\nu_{\text{eff}}=-\frac{1}{n_e}\frac{d}{dt}n_e=\mu\frac{\partial^3}{\partial x^3}n_e \tag{C.24}$$

indicates a dissipation of transversal waves by longitudinal (Langmuir) waves. Remembering that these electrostatic oscillations are always damped by Landau damping (Section 3.4), a direct relation for the damping of transversal (electromagnetic) waves in a plasma by Landau damping has been achieved.

Let us return to our discussion of the equation of motion. Using the monochromatic time dependence [Eq. (6.20)], one arrives from Eq. (C.15) at

$$\mathbf{j}\left(1-i\frac{\nu}{\omega}\right)=\frac{\omega_p^2}{4\pi\omega^2}\frac{\partial E}{\partial t} \tag{C.25}$$

and

$$\frac{\partial \mathbf{E}}{\partial t}\frac{\partial \mathbf{E}}{\partial t}=-\omega^2\mathbf{EE} \tag{C.26}$$

We assume $\nu \ll \omega$ which is always possible at very high amplitude fields, therefore we use

$$1-\frac{i\nu}{\omega}\approx 1 \tag{C.27}$$

The two last terms in Eq. (C.14) can then be written with Eqs. (C.25) and (C.26)

$$-\frac{4\pi}{\omega_p^2}\frac{\omega_p^2}{4\pi\omega^2}\frac{\partial \mathbf{E}}{\partial t}\cdot\nabla\frac{\omega_p^2}{4\pi\omega^2}\frac{\partial \mathbf{E}}{\partial t}+\frac{4\pi}{\omega_p^4}\frac{\omega_p^4}{(4\pi)^2\omega^4}\frac{\partial \mathbf{E}}{\partial t}\frac{\partial \mathbf{E}}{\partial t}\cdot\nabla\omega_p^2 \tag{C.28}$$

Using Eq. (C.26) and the refractive index [Eq. (6.28)]

$$\mathbf{n}^2=1-\frac{\omega_p^2}{\omega^2(1-i\nu/\omega)}$$

including Eq. (C.27), we find for the expression (C.28)

$$-\frac{1}{4\pi}\mathbf{E}\cdot\nabla\mathbf{E}(1-\mathbf{n}^2)+\frac{1}{4\pi}\mathbf{EE}\cdot\nabla(1-\mathbf{n}^2)=-\frac{1}{4\pi}(1-\mathbf{n}^2)\mathbf{E}\cdot\nabla\mathbf{E} \tag{C.29}$$

This is the Schlüter term, Eq. (6.7), as the high-frequency result of the two last terms in the equation of motion (C.14).

Up to this stage, the discussion of the two-fluid theory of plasma followed the convenient derivations, with some modifications only with respect to Z-time ionized ions, or with respect to fields oscillating with a high radian frequency ω. The only operation we had to carry out for arriving at the most general equation of motion (8.3) or (8.6) was with respect to the second term on the right-hand side of Eq. (C.14), which describes the action of electric fields **E** due to space charges. In the cases preceding Ref. 138, these space charges in plasmas had been neglected because of space charge neutrality. We had to adhere strictly to this convention with regard to static or stationary distributions of space charges. A difference exists, however, with respect to (time-averaged vanishing) oscillating high-frequency space charges. This is due to the knowledge of the exact description of plane electromagnetic waves in stratified plasmas at oblique incidence and p-polarization, where longitudinal optical fields appear that drive oscillations of high-frequency space charges [138].

The important question was then [138], how should the second term on the right-hand side of the equation of motion (C.14) be interpreted? To avoid any difficulty with the convention of dielectric or diamagnetic properties in the description of plasma, a consequent Lorentz picture with $\varepsilon=\mu=1$ was used. Nobody, however, will doubt that the high-frequency oscillation of space charges are due to polarization currents, determined by a complex refractive index $n=\varepsilon^{1/2}$, as it determined the refraction of the propagation of the optical waves and their change of direction in inhomogeneous plasmas at oblique incidence. As it was the confirmation of the correctness of the subsequent treatment [138], the space charge term in Eq. (C.14) had to be formulated with inclusion of the dielectric displacement,

$$\frac{1}{4\pi}\,\mathbf{E}e(Zn_i-n_e)=\frac{1}{4\pi}\,\mathbf{E}\nabla\cdot n^2\mathbf{E}=+\frac{1}{4\pi}\,\mathbf{E}\nabla\left(1-\frac{\omega_p^2}{\omega^2}\right)\mathbf{E} \tag{C.30}$$

$$=\frac{1}{4\pi}\,\mathbf{E}\nabla\cdot\mathbf{E}-\frac{1}{4\pi}\,\mathbf{E}\nabla\cdot\mathbf{E}\,\frac{\omega_p^2}{\omega^2}$$

$$\frac{1}{4\pi}\,\mathbf{E}e(Zn_i-n_e)=\frac{1}{4\pi}\,\mathbf{E}\nabla\cdot\mathbf{E}-\frac{1}{4\pi}(1-n^2)\mathbf{E}\nabla\cdot\mathbf{E}-\frac{1}{4\pi}\,\mathbf{E}\mathbf{E}\cdot\nabla(1-n^2) \tag{C.31}$$

Using this result in the equation of motion (C.14) and the Schlüter term [Eq. (C.29)] instead of the last two terms in Eq. (C.14), we arrive at

$$f=m_in_i\left[\frac{\partial}{\partial t}\mathbf{v}+\mathbf{v}\cdot\nabla v\right]=-\nabla p+\mathbf{j}\times\frac{\mathbf{H}}{c}+\frac{1}{4\pi}\,\mathbf{E}\nabla\cdot\mathbf{E}$$

$$-\frac{1}{4\pi}(1-n^2)\mathbf{E}\nabla\cdot\mathbf{E}-\frac{1}{4\pi}\,\mathbf{E}\mathbf{E}\cdot\nabla(1-n^2)-\frac{1}{4\pi}(1-n^2)\mathbf{E}\cdot\nabla\mathbf{E} \tag{C.32}$$

which is identical with Eq. (8.3) or (8.6) [138].

The final proof of the correctness of Eq. (C.32) was given (as shown in Section 8) by the correct result of the nonlinear forces at oblique incidence. Using $n\equiv 1$ in Eq. (C.26) gave wrong results. The unsolved question was, how the electrostatic field (or very low frequency fields) have to follow, which usually are described by

$$\mathbf{E}\nabla\cdot\mathbf{E}=\frac{1}{4\pi}\,\mathbf{E}e(Zn_i-n_e) \tag{C.33}$$

where there is no refractive index n.

The solution of this question may be related to Novak's answer to experiments where the low-frequency description of the field is given by the

Abraham tensor, while the high-frequency case is given by the Minkowski tensor [179] as the then valid approximation of the Abraham tensor. The description of plasmas using quantities **E** and space charges is strange to the earlier concepts of plasma theory. There seems, however, a much more general aspect given by astrophysics than by our question of laser plasma dynamics in favor of this new development as Alfvén has derived [402].

Notes Added in Proof

As a consequence of a discussion with the responsible series editor after completion of the manuscript, it seems desirable to add the following notes:

To page 170

Our presentation of the electric double layers and the dynamic internal electric fields inside of inhomogeneous plasmas is restricted to the numerical and semi-analytical procedures in the same manner as these results were elaborated for high-density plasmas for the first time. This was just the minimum for preparation of the subsequent derivation of the surface tension in plasmas and the application of this plasma theory to metals and nuclei. For a more extensive presentation of double layers in low-density plasmas in astrophysics (Alfvén 1988) or in tri-plasma devices (Hershkovitz, 1985) the reader is referred to specific reviews of this field (Eliezer and Hora, 1989; 1989a).

To page 193:

For a better understanding of the analogy of the pendulum (Fig. 9.8) to a plasma for the parametric instabilities, it should be mentioned that the eigenfrequency of the pendulum corresponds to intrinsic frequencies of the plasma, mostly to the plasma frequency, while the frequency of the interaction from the outside, the frequency of disturbing the pendulum at its origin, corresponds to the frequency of the light irradiating the plasma.

To page 231:

With respect to solitons this subsection only presents the numerical-empirical result describing how the plasma dynamics at very intense laser irradiation adopts the behavior given by the Korteweg–de-Vries equation or by the Benjamin–Ono equation. This transition is due to plasma friction and optical dissipation (absorption) and is one of the first examples – if not the only one according to Kruskal (private comm. 1990) – where the transition from non-soliton behavior to soliton behavior has been followed up. This result has not yet been elaborated with respect to the Schrödinger soliton. The latter appeared in laser–plasma interactions as a direct result of the properties of the nonlinear forces as proposed in the explanation of relativistic self-focusing (Häuser et al. 1988).

Additional Remarks:

To the interpretation of the "question mark experiment" (pages 253 and 254) it was suggested that the picture of a question mark reflected from the laser-irradiated plasma should be upside down if the density-ripple-produced Bragg reflection occurs, as observed (Eidmann et al. 1974), but that for short intermediate times the mirror reflection may happen when reflection at the critical density occurs. Eidmann confirmed in a private communication (March 1991) that this indeed has been observed [R. Sigel, K. Eidmann, C.H. Pant, and P. Sachsenmeier,

Phys. Rev. Lett. 36, 1369 (1976)]: some sign of the question mark in the other direction has been detected but since the time resolution of the experiment was only 30 psec it can only be vaguely confirmed in retrospect that the alternating of the up and down of the question mark did really happen corresponding to the pulsation with a period of 10 to 30 psec.

Since this was an important confirmation of the 20 psec pulsation and since this is of crucial importance, it should be especially highlighted. Apart from the suggestion after the numerical result of Fig. 10.10a and b, its experimental appearance was crucial. It was first observed by Lubin (1974) [see also S. Jackel, B. Perry, and M. Lubin, Phys. Rev. Lett. 37, 95 (1976)], later, generally and in more detail, by Maddever and Luther-Davies (1990), and further by the pulsation of the double layer in the plasma corona [A. Ludmirsky et al., Laser and Particle Beams, 2, 245 (1983) see also Eliezer et al. (1989)] or by the same pulsation of the third half harmonics (Giulietti et al. 1989) or of the K-alpha emission (A.V. Rode et al., AINSEE Plasma Physics Conf. Lucas Heights, Australia, Feb. 1991). This phenomenon should be called the "Lubin effect" or the LESLEGMLR effect (for the names Lubin, Eidmann, Sigel, Ludmirsky, Eliezer, Giulietti, Maddever, Luther-Davies, and Rode).

LIST OF SYMBOLS

$\mathbf{a}$	acceleration of a particle (4.1)
a'	constant for absorption (6.48)
A	second order term (8.48)
c	vacuum velocity of light, see, for example, Eq. (6.16)
c_p	specific heat at constant pressure
c_s	velocity of sound (ion acoustic waves) (4.36)
c_v	specific heat at constant volume
c_ψ	phase velocity (6.26)
E	energy
$\mathbf{E}$	electrical field strength
E_F	Fermi energy (2.47)
E_{kin}	kinetic energy of plasma (10.17)
E_r	amplitude of laser field (6.20) depending on $\mathbf{r}$ only
E^r	relativistic oscillation energy of electron (threshold, 6.71)
E_V	Vacuum amplitude of E
E^*	threshold for predominance of nonlinear force (9.4)
f	distribution function (3.3)
$\hat{f}_M$	Maxwell–Boltzmann distribution (3.20)
F	exponent in WKB approximation (7.9)
F_0	real part of F (8.31)
$\mathbf{F}$	force (4.1)
G_0	length (5.23)
G	for WKB solution (7.29)
G'	Eq. (8.64)
$\mathbf{H}$	magnetic field strength
I	laser intensity (6.55)

I_r	relativistic threshold laser intensity (6.73)
k	wave number
$\mathbf{k}$	wave vector (7.19)
$\mathbf{k}_s$	wave vector for electrostatic waves (9.69)
$\bar{k}$	average absorption constant (7.13)
K	Boltzmann constant
$\bar{K}$	optical absorption constant
$\tilde{\bar{K}}$	effective wave number (11.51)
$\mathbf{K}_e$	force density to electrons by gravitation
$\mathbf{K}_i$	force density to ions by gravitation
$\bar{K}_{NL}$	nonlinear absorption constant
l	mean radius of a Gaussian density profile (5.21)
$\bar{l}$	mean free path (3.69)
L	Lagrangian for parametric resonance (9.56)
$\mathscr{L}$	Lagrangian of Proca equation (9.50)
L'	Denisov length (11.62)
m	mass of electron
$\bar{m}$	mass of particle (4.1)
m_i	mass of ion
m_0	rest mass of electron
M	total mass of plasma from pellet
$\bar{M}$	radially averaged mass (5.14)
n	particle density
n_e	electron density
n_i	ion density
n	(optical) refrective index (6.27)
n'	real part of the refractive index
N_i	total number of ions in sphere (5.26)
p	pressure
$\bar{p}$	classical Coulomb cross section (2.32)
p_A	Abraham momentum of photon (9.31)
p_M	Minkowski momentum of photon (9.32)
p_0	initial pressure
p_ϕ	momentum per photon (9.29)
$p_{\phi,pl}$	p_ϕ in plasma (9.30)
P	momentum of electromagnetic energy

P_{inh}	P in inhomogeneous plasma
P_{int}	internal radiation pressure
P_0	P in vacuum (9.10)
P'	laser power (12.22)
r	radius
r_F	focus radius (5.41)
r_0	Coulomb impact parameter (2.29)
R	radius of spherical pellet (5.1)
R_0	initial radius of pellet
$\dot{R}$	expansion velocity of pellet radius
$\dot{R}_0$	initial expansion velocity of pellet radius
Re_{kr}	Reynolds number (11.42)
$\tilde{R}$	Fresnel reflection coefficient (9.34)
$\mathbf{S}$	Poynting vector (8.7)
t	time
t_0	initial time
t_{TP}	time at which laser irradiated pellet becomes transparent (5.31)
T	temperature
$\tilde{T}$	transmission coefficient (9.34)
T_{th}	thermokinetic temperature (6.57)
$\mathbf{T}$	Maxwellian stress tensor (8.26)
$T^{\mu\nu}$	canonical energy-momentum tensor (9.50)
u	angle of incidence
u_x, u_y, u_z	angles for the direction of plane waves (7.19)
U	blackbody radiation density (9.40)
U_p	Planck's value of U for vacuum (9.41)
$\mathbf{v}$	macroscopic (drift) velocity (3.34)
v_r	radial velocity
v_{r0}	initial radial velocity (5.13)
v_ϕ	phase velocity (9.70)
V	volume
V_0	initial volume
W	input power density (4.39)
W_1	special input power density (5.24)

x	Cartesian coordinate
y	Cartesian coordinate
z	Cartesian coordinate
Z	number charges of ions
α	parameter for Rayleigh profile (7.29)
α_0	angle of incidence in vacuum (8.43)
$\alpha(x)$	angle of incidence in plasma
$\bar{\alpha}_{(n)}$	Fokker–Planck coefficients (3.18)
β	angle of polarization (8.42)
γ	ratio of specific heats (4.23)
γ_e	Spitzer's correction for collisions (2.35)
ε	dielectric constant
$\bar{\varepsilon}$	complex dielectric constant (6.27)
ε_D	Debye energy (2.15)
ε_{osc}	oscillation energy of electrons (6.54)
ε_{osc}^{kin}	average kinetic energy of quivering electron (6.53)
ζ	spatial variable (7.83)
η	viscosity (4.4; 11.38)
κ	imaginary part of the refractive index
$\bar{\kappa}$	compressibility (4.21)
κ_T	thermal conductivity (4.39)
λ_D	Debye length (2.14)
μ	magnetic permeability
μ'	dispersion function (10.21)
ν	collision frequency
ν_e	collision frequency of electrons
ν_{ei}	electron ion collision frequency
ρ	density of plasma (4.2)
ρ_0	initial density
σ	electric conductivity (2.42)
σ_o	Stefan–Boltzmann constant (9,46)
σ_{op}	radiation constant in plasma (9.45)
τ_{col}	electron collision time (10.1)
τ_0	duration of laser pulse (half-width) (10.4)
τ_R	rise time of laser pulse (8.7)
τ^*	minimum rise time of laser pulse for thermalization (10.5)

ϕ	potential (4.1)
ω	laser radian frequency
ω_e	Bohm–Gross frequency (9.69)
ω_p	plasma frequency (2.6)
Ω_0	vacuum resistivity (6.55)

REFERENCES, 1st Part (by Numbers)

1 H. Hora, *Laser Plasmas and Nuclear Energy* (Plenum, New York, 1975), 424 pages.

2 T. P. Hughes, *Plasmas and Laser Light* (Adam Hilger, Bristol, 1975).

3 H. Motz, *The Physics of Laser Fusion* (Academic Press, London, 1979), 290 pages.

4 H. Hora, "Nonlinear Plasma Dynamics at Laser Irradiation," Lecture Notes in Physics (University of Berne, Springer, Heidelberg, 1979).

5 *Laser Interaction and Related Plasma Phenomena*, H. J. Schwarz *et al.*, Eds. (Plenum, New York) Vols. 1 to 5 (1971–1980).

6 R. Castillo, H. Hora, E. L. Kane, G. W. Kentwell, P. Lalousis, V. F. Lawrence, R. Mavaddat, M. M. Novak, P. S. Ray, and A. Schwartz, see Ref. 5, Vol. 5 (1980) p. 399

7 H. Alfvén, *Phys. Today* **24** (February 1971), 29.

8 R. T. Young, C. W. White, G. J. Clark, J. Narayan, W. H. Christie, M. Murakami, P. W. King, and S. D. Kramer, *Appl. Phys. Lett.* **32** (1978), 139; H. Hora, *Naturwissenschaften* **48** (1961), 641; Z. Angew Phys. **14** (1962), 9; S. Hinckley, H. Hora, and J. C. Kelly, *Phys. Status Solidi* **51A** (1979), 523.

9 P. D. Maker, R. W. Terhune, and C. M. Savage, *Proc. 3rd Int. Quant. Electron. Conf., Paris*, February 1963, N. Bloembergen and M. Grivet, Eds. (Dunod, Paris, 1964), Vol. 2, p. 1559.

10 R. G. Meyerand and A. F. Haught, *Phys. Rev. Lett.* **11** (1963), 401; C. DeMichelis, *IEEE J. Quantum Electron.* **5** (1969), 181, **6** (1970), 630; E. Panarella and P. Savic, *Can. J. Phys.* **46** (1968), 183; S. A. Ramsden, *Physics of Hot Plasmas*, B. J. Rye et al., Eds. (Oliver & Boyd, Edinburgh, p. 346.

11 R. Papoular, see Ref. 5, (1972) Vol. 2, p. 79.

12 G. V. Ostrovskaya and A. N. Zaidel, *Usp. Fiz. Nauk* **111** (1973), 579; *Sov. Phys. Uspekhi* **16** (1974), 834.

13 P. Kolodner and E. Yablontovich, *Phys. Rev. Lett.* **37** (1976), 1754.

14 B. W. Boreham and H. Hora, *Phys. Rev. Lett.* **42** (1979), 776.

15 H. Hora, E. L. Kane, and J. L. Hughes, *J. Appl. Phys.* **49** (1978), 923.

16 H. Hora, *J. Opt. Soc. Am.* **65** (1975), 882.

17 D. J. Bradley, *Phys. Bull.* **29** (1978), 418; J. C. Diels, *Laser Weekly* **12**, no. 5 (1978), p. 1.

18 J. Trenholme, E. Bliss, J. Emmett, J. Glaze, T. Gilmartin, R. Godwin, W. Hagen, J. Holzrichter, G. Linford, W. Simmons, and R. Speck, see Ref. 5, 1977, Vol. 4A, 1.

19 J. L. Emmett, *Proc. IAEA Conf. Nuclear Fusion, Innsbruck*, 1978, paper B-1.

20 N. G. Basov, O. N. Krokhin, Yu. A. Mikhailov, G. V. Sklizkov, and S. I. Fedotov, see Ref. 5, 1977, Vol. 4A, 15.

21 V. V. Korobkin, V. M. Ovchinnikov, P. P. Pashinin, Yu. A. Pirogov, A. M. Prokhorov, R. V. Serov, *Digest 8th Nat. Conf. Laser and Nonlinear Optics, Tibilisi*, May 1976, p. 248.

22 W. Kroy, "CW pumped 1000 Hz Nd Glass Laser," (Messerschmitt-Bölkow-Blohm GmbH, Ottobrunn).

23 P. P. Pashinin, see Ref. 21, postdeadline paper.

24 R. B. Allen and S. J. Scalise, *Appl. Phys. Lett.* **14** (1969), 188; W. Koechner, *Solid-State Laser Engineering* (Springer, Heidelberg, 1976), p. 277.

25 G. V. Sklizkov, paper P1, presented at *13th Europ. Conf.* Laser Interaction with Matter, Leipzig, December 1979.

26 R. R. Jacobs and W. F. Krupke, *IEEE J. Quantum Electron.* **13** (1977), 103.

27 R. B. Perkins, *Plasma Physics and Controlled Nuclear Fusion* (IAEA, Vienna, Vol. 3 (1978), p. 41.

28 S. Singer, *Development in High Power Lasers*, C. Pellegrini ed. (E. Fermi Sch. Vol. 74, North Holland, Amsterdam, 1980).

29 H. S. Kwok and E. Yablontovitch, *Appl. Phys. Lett.* **30** (1978).

30 N. G. Basov, V. A. Boiko, V. A. Danylichev, V. D. Zvorykin, A. N. Lobanov, A. F. Suchkov, T. V. Holin, and A. Y. Chugunov, *Kvant. Elekt.* **4** (1977), 1761.

31 C. Yamanaka, "0.5 nsec Pulses from 50 Atm. Compact CO_2 Lasers," (Osaka 1978).

32 W. Kroy, J. Langhole, T. Halderson, *Appl. Opt.* **19** (1980), 6. H. K. Koebner, *Laser in Medicine*, (Wiley, New York, 1980).

33 G. H. Miley, see Ref. 5 (1977), Vol. 4A, p. 181.

34 K. Hohla, G. Brederlow, E. Fill, R. Volk, and K. J. Witte, see Ref. 5 (1977), Vol. 4A, p. 97.

35 K. Hohla and K. L. Kompa, *Chem. Phys. Lett.* **14** (1972), 445.

36 K. Witte, G. Brederlow, E. Fill, K. Hohla, and R. Volk, see Ref. 5 (1977) Vol. 4A, p. 155.

37 S. Witkowski, *Laser & Elektro-Optik*, 10, no. 3 (1978), 47.

38 R. Lüty and K. Witte, Dept. Laser Physics, Univ. Berne, Report (1978).

39 N. G. Basov and V. S. Zuev, *Nuovo Cim.* **31B** (1976), 129.

40 R. J. Jensen, *Laser Focus* **12** (May 1976), 51.

41 F. G. Houtermans, *Helv. Phys. Acta* **33** (1960), 933.

42 M. H. Hutchinson, C. C. Ling, and D. J. Bradley, *Opt. Com.* **26** (1978), 273.

43 D. Jacobi, G. J. Pert, S. A. Ramsden, L. D. Shorrock, and G. J. Tallents, *Phys. Rev. Lett* **45** (1980) 1826; P. Jaegle, G. Jamelot, A. Carillon, and A. Sureau, see Ref. 5, (1977), Vol. 4A, p. 229.

44 M. H. Key, M. J. Lamb, C. L. S. Lewis, J. G. Lunney and A. K. Roy, *Opt. Comm.* **18** (1976), 156.

45 H. Hora, G. V. H. Wilson, E. P. George, *Aust. J. Phys.* **31** (1978), 55.

46 K. Okamoto, see Ref. 5, (1977), Vol. 4A, p. 283; *J. Nucl. Sci. Tech.* **14** (1977), 762.

47 K. Okamoto, see Ref. 5 (1980), Vol. 5, p. 299.

48 Y. Izawa, H. Otari, and C. Yamanaka, see Ref. 5 (1980), Vol. 5, p. 289

49 D. A. G. Deacon, L. R. Elias, J. M. J. Madley, G. J. Ramian, H. A. Schwettman, and T. J. Smith, *Phys. Rev. Lett.* **18** (1977), 892.

50 H. Motz, *J. Appl. Phys.* **22** (1951), 527; **24** (1953), 826.

51 S. Pellgrini, presented at 4th General Conf. European Physical Soc., York, England, September 1978, invited paper.

52 S. B. Segall, *Laser and Elektro-Optik* **10**, no. 3 (1978), 27.

53 H. Hora, presented at 2nd Int. Conf. Energy Storage, Electron and Laser Beams, Venice, Dec. 1978, invited paper (in print); H. Hora, B. W. Boreham, and J. L. Hughes, *Sov. J. Quant. Elect.* **9** (1979), 464.

54 H. Hora, see Ref. 5 (1971), Vol. 1, p. 383.

55 H. J. Schwarz, *Phys. Rev. Lett.* **42** (1979), 1141.

56 H. Schwarz and H. Hora, *Appl. Phys. Lett.* **15** (1969), 349; H. Hora, *Nuovo Cim.* **26B** (1975), 295; E. T. Jaynes, in *Novel Sources of Coherent Radiation, Physics of Quantum Electronics* Vol. 5, S. F. Jacobs, M. Sargent and M. O. Scully, Eds. (Addison–Wesley, Reading, 1978), p. 1.

57 T. H. Maiman, *Nature* **187** (1960), 493.

58 K. H. Steigerwald, *Chem. Ing. Tech.* **33** (1961), 191; H. Hora, *Chem. Rundschau* **14** (1960), 395.

59 J. F. Ready, *Effects of High-Power Laser Radiation* (Academic Press, New York, 1971); R. E. Honig, *Appl. Phys. Lett.* **3** (1963), 8.

60 H. Zahn and H. J. Dietz, *Exp. Techn. Phys.* **20** (1972), 401.

61 M. von Allmen, W. Lüthy, and K. Affolter, *Appl. Phys. Lett.* **33** (1978), 824.

62 H. Schwarz and H. A. Tourtellotte, *J. Vac. Sci. Techn.* **6** (1969), 373.

63 F. J. McClung and R. W. Hellwarth, *Proc. IEEE* **51** (1963), 46.

64 W. I. Linlor, *Appl. Phys. Lett.* **3** (1963), 210.

65 N. R. Isenor, *Appl. Phys. Lett.* **4** (1964), 152.

66 H. Schwarz, *Laser Interact. Rel. Plasma Phenomena*, H. Schwarz and H. Hora, Eds. (Plenum, New York, 1971), Vol. 1, p. 207.

67 S. Namba, H. Schwarz and P. H. Kim, *Proc. IEEE Sympos. Electrons, Ion and Laser Beam Technol.*, Berkeley, May 1967, p. 861.

68 D. W. Gregg and S. J. Thomas, *J. Appl. Phys.* **37** (1966), 4313.

69 S. A. Metz, *Appl. Phys. Lett.* **22** (1973), 211; H. Hora, *Appl. Phys. Lett.* **23** (1973), 39; J. E. Lowder and L. C. Pettingill, *Appl. Phys. Lett.* **24** (1974), 204; P. T. Rumsby, M. M. Michaelis and M. Burgess, *Opt. Comm.* **15** (1975), 422; B. Steverding and A. H. Werkheiser, *J. Phys.* **D4** (1971), 545; J. E. Lowder, *Appl. Phys. Lett.* **24** (1974), 204; S. Zweigenbaum, Y. Gazit, and Y. Paiss, *J. Phys.* **E11** (1978), 830.–The driving of solid foils by laser irradiation up to velocities of 2×10^7 cm/sec resulted in the fastest moving objects produced by man corresponding to a speed of 600 Mach: B. H. Ripin, R. Decoste, S. P. Oberschain, S. E. Bodner, E. A. McLean, F. C. Young, R. R. Whitlock, C. M. Armstrong, J. Grun, J. A. Stamper, S. H. Gold, D. J. Nagel, R. H. Lehmberg and J. M. McMahon, *Phys. Fluids* **23** (1980) 1012. This can be used for impact fusion (F. Winterberg, *Z. Naturforsch.* **19A** (1964) 231) where the concept of laser driven foils was proposed by W. Kaiser, H. Opower, and B. H. Puell, German Patent No. 1 279 859 (1966) and the collapsing and adiabatic heating at compression will follow in the same way as achieved by the fast acceleration of thick blocks of plasma by nonlinear fores (see section 13.4). The use of the nonlinear force for propulsion by lasers was described by H. Hora, German Patent 1933 409 (1971).

70 S. Zweigenbaum, Y. Gazit, and Y. Komet, *Plasma Physics* **19** (1977), 1035; D. Salzmann, Y. Gazit, Y. Komet, A. D. Krumbein, H. M. Loebenstein, M. Oron, Y. Paiss, M. Rosen-

blum, H. Szichman, A. Zigler, H. Zmora, and S. Zweigenbaum, see Ref. 5 (1977), Vol. 4A, p. 407.

71 G. Siller, K. Büchl, and H. Hora, see Ref. 5 (1972), Vol. 2, p. 252.

72 A. G. Engelhardt, T. V. George, H. Hora, and J. L. Pack, *Phys. Fluids* **13** (1970), 212.

73 K. Eidman and R. Sigel, see Ref. 5 (1974), Vol. 3B, p. 667.

74 H. Hora, see Ref. 1, p. 4.

75 K. B. Büchl, K. Eidmann, P. Mulser, H. Salzmann, and R. Sigel, see Ref. 5, (1972), Vol. 2, p. 503.

76 A. W. Ehler, *J. Appl. Phys.* **46** (1975), 2464.

77 J. W. Shearer, J. Garrison, J. Wong, and J. E. Swain, *Phys. Rev.* **A8** (1973), 1582.

78 C. Yamanaka, T. Yamanaka, T. Sasaki, K. Yoshida, and M. Waki, *Phys. Rev.* **6A** (1972), 2342.

79 F. J. Mayer, R. K. Osborn, D. W. Daniels, and J. F. McGrath, *Phys. Rev. Lett.* **40** (1978), 30.

80 B. Luther–Davies and J. L. Hughes, *Opt. Com.* **18** (1976), 351; M. Siegrist, B. Luther–Davies, and J. L. Hughes, *Opt. Comm.* **18** (1976), 605.

81 H. Hora, E. L. Kane, and J. L. Hughes, *Nucl. Inst. Meth.* **150** (1978), 589.

82 P. Wägli and T. P. Donaldson, *Phys. Rev. Lett.* **40** (1978), 875.

83 T. P. Donaldson and I. J. Spalding, *Phys. Rev. Lett.* **36** (1976), 467.

84 R. A. Haas, W. C. Mead, W. L. Kruer, D. W. Phillion, H. N. Kornblum, J. D. Lindl, D. MacQuigg, V. C. Rupert, and K. G. Tirsell, *Phys. Fluids* **20** (1977), 322; E. B. Goldman, L. M. Goldman, J. Delettrez, J. Hoose, S. Jackel, G. W. Leppelmeier, M. J. Lubin, A. Nel, I. Pelak, E. Thorsos, D. Woodall, and B. Yaa kobi, see Ref. 5 (1977), vol. 4B, p. 535.

85 R. G. Evans, M. H. Key, D. J. Nicholas, F. O'Neil, A. Raven, P. T. Rumsby, I. N. Ross, W. T. Tower, P. R. Williams, M. S. White, C. L. S. Lewis, J. G. Lunney, A. Moore, J. M. Ward, T. A. Hall, J. Murdoch, D. J. Kilhenny, T. Goldsack, J. D. Hares, *Plasma Phys. Contr. Nucl. Fusion Res.*, 1978 (IAEA, Vienna, 1979) Vol. 3, p. 87.

86 C. Yamanaka, M. Yokoyama, S. Makar, T. Yamanaka, Y. Izawa, Y. Kato, T. Sasaki, T. Mochizuki, Y. Kitagawa, M. Matoba, and K. Yoshida, see Ref. 5 (1977), Vol. 4B, p. 577.

87 A. Bekiarian, E. Buresi, A. Coudeville, R. Dautray, F. Delobeau, P. Guillaneux, C. Patou, J. M. Reisse, B. Sitt, J. M. Vedel, and J. P. Watteau, *Plasma Physics Contr. Nucl. Fusion Res.*, 1978 (IAEA, Vienna, 1979), Vol. 3, p. 65.

88 P. Mulser, *Z. Naturforsch.* **25A** (1970), 282.

89 E. J. Valeo, W. L. Kruer, *Phys. Rev. Lett.* **33** (1974), 750.

90 D. Biskamp and H. Welter, *Plasma Phys. Contr. Nucl. Fusion Res.*, Tokyo, 1974 (IAEA, Vienna, 1975) Vol. 2, p. 507.

91 For the change from Gaussian cgs-units preferred in Plasma Physics, to MKQS units, see, for example, D. L. Book, *Formulas for Plasma* (Naval Res. Lab., Washington, 1975).

92 P. Debye and E. Hückel, *Phys. Zeitschr.* **24** (1923), 185.

93 S. R. Milner, *Phil. Mag.* **23** (1912), 551; **25** (1913), 743.

94 T. P. Donaldson, *J. Phys.* **D10** (1977), 1589; *Plasma Phys.* **20** (1978), 1279.

95 G. Ruthemann, *Naturwissensch.* **29** (1941), 648.

96 G. Möllenstedt, *Optik* **5** (1949), 499.

97 D. Bohm and D. Pines, *Phys. Rev.* **92** (1953), 609.

98 H. Ringler and R. A. Nodwell, *Phys. Lett.* **29A** (1969), 151; D. Ludwig and C. Mahn, *Phys. Lett.* **35A** (1971), 191.

99 L. A. Godfrey, R. A. Nodwell, and F. L. Curzon, *Phys. Rev.* **A20** (1979), 567.

100 H. Thomas, *Z. Phys.* **147** (1957), 395.

101 P. Görlich and H. Hora, *Optik* **15** (1958), 116.

102 H. Fröhlich, *Ann Phys.* **7** (1930), 103.

103 B. Yaakobi and S. Goldsmith, *Phys. Lett.* **37A** (1970), 408; M. Neiger and H. Griem, *Phys. Rev.* **A14** (1976), 289; D. R. Inglis and E. Teller, *Astrophys. J.* **90** (1939), 439; F. L. Mohler, *Astrophys. J.* **90**, (1939), 429; H. Margenau and M. Lewis, *Rev. Mod. Phys.* **31** (1959), 569; S. Volonté, *J. Phys. D* **11** (1978), 1615.

104 G. Ecker and W. Kröll, *Phys. Fluids* **6** (1963), 62; H. R. Griem, *Plasma Spectroscopy* (McGraw-Hill, 1964), p. 139.

105 D. Salzmann and A. Krumbein, *J. Appl. Phys.* **49** (1978), 3229.

106 B. I. Henry, "Polarization Shift," Honours Thesis, University of New South Wales (Sydney, 1980); B. I. Henry, and H. Hora (to be published).

107 L. Spitzer, Jr., *Physics of Fully Ionized Gases*, 2nd ed. (Wiley Interscience, New York, 1962).

108 L. Spitzer and R. Härm, *Phys. Rev.* **89** (1953), 977.

109 P. S. Ray and H. Hora, see Ref. 5 (1977), Vol. 4B, p. 1081.

110 J. M. Blatt and A. H. Opie, *J. Phys.* **A7** (1974), 1895.

111 J. M. Blatt, *Prog. Theor. Phys.* **22** (1959), 745; *About Theories of Superconductivity* (Academic Press, New York, 1959).

112 J. M. Dawson, *Phys. Fluids* **7** (1964), 981.

113 E. Hinnov and J. Hirschberg, *Phys. Rev.* **125** (1962), 795.

114 A. F. Haught and D. H. Polk, *Phys. Fluids* **13** (1970), 2825.

115 H. Hora, *Higher Mechanics*, Dept. Theor. Phys. Rept. No. 24 (Univ. New South Wales, Sydney, 1980).

116 L. D. Landau, *J. Phys. USSR* **10** (1946), 25.

117 J. Meixner, *Ann. Physik* **39** (1941), 39; about fluctuations in the electric field, see M. S. Sodha, and L. A. Patel, *J. Appl. Phys.* **51** (1980) 2381.

118 D. Enskog, *Svenska Akademia* (1928), 21.

119 R. Castillo, M.Sc. Thesis, University New South Wales, 1979; R. Castillo, H. Hora, E. L. Kane, V. F. Lawrence, M. B. Nicholson-Florence, M. M. Novak, P. S. Ray, J. R. Shepanski, R. Sutherland, A. I. Tsivinsky, and H. A. Ward, *Nucl. Inst. Meth.* **144** (1977), 27.

120 W. J. Fader, *Phys. Fluids* **11** (1968), 2200.

121 E. A. Milne, *Z. Astrophys.* **6** (1933), 1.

122 O. Heckmann, *Theorie der Kosmologie* (Springer, Heidelberg, 1965).

123 L. L. Lengyel and M. Salvat, *Z. Naturforsch.* **30A** (1975), 1577.

124 Ya. B. Zeldovich and Yu. P. Raizer, *Physics of Schock Waves and High Temperature Hydrodynamic Phenomena* (Academic Press, New York, 1966).

125 N. G. Basov and O. N. Krokhin, *3rd Int. Quantum Elect. Conf. Paris, 1963*, P. Grivet and N. Bloembergen, Eds. (Dunod, Paris 1964), Vol. 2, p. 1373; S. Kaliski, *Bull. de l'Academis Polonaise des Sciences-Serie des Sciences Technique* **20** (1972), 297; **23** (1975), 881; V. V. Demchenkov and N. M. El-Siragy, *Physica* **67** (1973), 336; T. P. Donaldson, J. E. Palmer, J. A. Zimmermann, *J. Phys.* **D13** (1980) 1221; K. E. Lonngren, *Plasma Phys.*

22 (1980) 511. Yu. V. Afanasyev, N. G. Basov, O. N. Krokhin, V. V. Pustovalov, V. P. Silin, G. V. Sklizkov, V. T. Tikhonchuk, and A. S. Shikanov, *Interaction of Strong Laser Light with Plasmas*, (Radiotekhnika Vol. 17, Moscow, 1978).

126 H. Hora, *Inst. Plasmaphys. Garching, Rept. 6/23*, 1964.

127 H. Hora, see Ref. 5 (1971), Vol. 1, p. 365; H. T. Suji, K. Sato, and T. Sekiguchi, *Jap. J. Appl. Phys.* **18** (1979), 1807; S. O. Dean, *Rept. NRL PRO* (1971); Y. Ohwadano and T. Sekiguchi, *Jap. J. Appl. Phys.* **16** (1977), 1025; G. J. Pert, *J. Phys.* **A5** (1972), 506; L. L. Lengyel, *Nucl. Fusion* **17** (1977), 805.

128 A. F. Haught and D. H. Polk, *Phys. Fluids* **9** (1966), 2047; R. G. Tuckfield and F. Schwrizke, *Plasma Phys.* **11** (1969), 11; H. J. Kunze, *Z. Naturforsch*, **A20** (1965), 801; A. Cavalieri, P. Guipponi, and R. Gratton, *Phys. Lett.* **A25** (1967), 636.

129 J. Jacquinot, C. Leloup, and F. Waelbrock, *Rapp. CEA No. 12.2617* (1964).

130 M. Mattioli, *Euratom-CEA-Fontenay-Report*, EUR-CEA-FC-523C (1969).

131 E. Fabre and P. Vasseur, *J. Physique* **29** (1968), 123.

132 T. V. George, A. G. Engelhardt, J. L. Pack, H. Hora, and G. Cox, *Bull. APS* **13** (1968), 1553; H. Hora, see Ref. 5, Vol. 1 (1971) p. 273.

133 R. Sigel, K. Büchl, P. Mulser, and S. Witkowski, *Phys. Lett.* **26A** (1968), 498; H. Puell, *Z. Naturforsch*, **A25** (1970), 1807.

134 P. Mulser and S. Witkowski, *Phys. Lett.* **28A** (1969), 703.

135 K. Hohla, *unpublished measurements* (1969); J. Tulip, K. Manes, and H. J. Seguin, *Appl. Phys. Lett.* **19** (1971), 433.

136 A. Schlüter, *Z. Naturforsch.* **5A** (1950), 72.

137 See for example, H. Alfvén and C. G. Fälthammer, *Cosmical Electrodynamics*, 2nd ed. (Oxford University Press, London, 1973).

138 H. Hora, *Phys. Fluids* **12** (1969), 182.

139 P. A. M. Dirac, *Directions of Physics*, H. Hora and J. R. Shepanski, Eds. (Wiley Interscience, New York, 1978).

140 H. Hora and H. Wilhelm, *Nucl. Fusion* **10** (1970), 111; J. L. Bobin, *Phys. Fluids* **14** (1971), 2341; N. H. Burnett, *Can. J. Phys.* **50** (1972), 3184.

141 C. W. Allen, *Astrophysical Quantities* (Athlon Press, London, 1955).

142 J. A. Gaunt, *Proc. Roy. Soc. (London)* **A126** (1930), 654.

143 S. F. Smard and K. C. Westfold, *Phil. Mag.* **40** (1949), 831.

144 J. M. Dawson and C. Oberman, *Phys. Fluid* **5** (1962), 517.

145 G. W. Spitzer and H. Y. Fan, *Phys. Rev.* **108** (1957), 268.

146 H. Hora, *Opto Electronics* **2** (1970), 202.

147 S. Rand, *Phys. Rev.* **B136** (1964), 231; T. P. Hughes and M. B. Nicholson-Florence, *J. Phys.* **A2** (1968), 588.

148 C. Max and F. Perkins, *Phys. Rev. Lett.* **27** (1971), 1342.

149 H. Schwarz and R. Tabenski, see Ref. 5 (1977), Vol. 4B, p. 961.

150 V. L. Ginzburg, *The Propagation of Electromagnetic Waves in Plasma* (Pergamon, Oxford 1964), p. 205.

151 H. Hora, *Jenaer Jahrbuch*, P. Görlich Ed. (Fischer, Jena, 1957), p. 131; *Inst. Plasmaphysik*, Garching, Rept. 6/5 (1963).

152 H. Osterberg, *J. Opt. Soc. Amer.* **48** (1950), 513.

153 H. Hora and V. F. Lawrence, see Ref. 5 (1977), Vol. 4B, p. 877; V. F. Lawrence and H. Hora, *Optik* **55** (1980), 291.

154 J. Lindl and P. Kaw, *Phys. Fluids* **14** (1971), 371.

155 N. G. Watson, *A Treatise on the Theory of Bessel Functions* (Cambridge University Press, 1922).

156 M. E. Marhic, *Phys. Fluids* **18** (1975), 837; F. F. Chen, *Plasma Physics* (Plenum, New York, 1975); R. Dragila, *J. Phys.* **D11** (1978), 683.

157 J. A. Stamper and D. A. Tidman, *Phys. Fluids* **16** (1975), 2024.

158 H. Hora, D. Pfirsch, and A. Schlüter, *Z. Naturforsch.* **22a** (1957), 278.

159 H. A. H. Boot, S. A. Self, and R. B. Shersby–Harvie, *J. Elect. Contr.* **22** (1959), 434.

160 V. A. Gapunov and M. A. Miller, *Sov. Phys. JETP* **7** (1958), 168.

161 S. Weibel, "TRW Report May 1957," *J. Electr. Contr.* **5** (1958), 435.

162 H. Hora, Second term in Eq. (25b) in Ref. 138. By this way the use of the nonlinear force has arrived in a correct formulation of the equation of motion even with absorption in agreement with the following described experiments. The very complex open problems with the ponderomotive force are described by V. I. Pavlov, *Sov. Phys. Uspekhi* **21** (1978), 171. See also Ref. 179: F. Panarella, *Can. J. Phys.* **46** (1969), 183; G. P. Banfi and P. G. Gobbi, *Plasma Phys.* **21** (1979), 845.

163 J. A. Stamper, see Ref. 5 (1977), Vol. **4B**, p. 721.

164 R. D. C. Miller and H. Hora, *Plasma Phys.* **21** (1979), 183.

165 G. W. Kentwell and H. Hora, *Plasma Phys.* **22** (1980), 1051.

166 L. D. Landau and E. M. Lifshitz, *Electrodynamics of Continuous Media* (Pergamon, Oxford, 1966), p. 242.

167 L. P. Pitaevskii, *Sov. Phys. JETP* **12** (1961), 1008.

168 H. L. Berk, D. L. Book, and D. Pfirsch, *J. Math. Phys.* **8** (1967), 1611.

169 G. W. Kentwell, "Resonance Absorption and Striated Motion at Laser Plasma Interaction," *Honours Thesis*, University New South Wales (1979).

170 S. Hinckley, H. Hora, E. L. Kane, G. W. Kentwell, J. C. Kelly, P. Lalousis, V. F. Lawrence, R. Mavaddat, M. M. Novak, P. S. Ray, A. Schwartz, H. A. Ward. *Experim. Tech. Phys.* **28** (1980), 417.

171 J. W. Shearer, R. E. Kidder, and J. W. Zink, *Bull. Amer. Phys. Soc.* **15** (1970), 1483; J. W. Shearer, *LLL-Report*, UCID-15745 (December 1970).

172 R. B. White and F. F. Chen, *Plasma Phys.* **16** (1974), 565.

173 E. Valeo, *Phys. Fluids* **17** (1974), 1391; J. J. McClure, *Bull. Amer. Phys. Soc.* **19** (1974), 869.

174 J. A. Stamper and S. E. Bodner, *Phys. Rev. Lett.* **37** (1976), 435. R. G. Tuckfield, and F. Schwirtzke, *Plasma Phys.* **11** (1969) 11; E. Schmutzer, and B. Wilhelmi, *Plasma Phys.* **19** (1971) 799; P. Mulser, and C. van Kessel, *Phys. Rev. Lett.* **38** (1977) 902; *J. Phys.* **D11** (1978) 1085; P. Mulser, and H. Tasso, *Z. Naturforsch.* **33A** (1978) 85; A. Ng, L. Pitt, D. Salzmann, and A. A. Offenberger, *Phys. Rev. Lett.* **42** (1979) 703; P. Chandra, *J. Appl. Phys.* **47** (1976) 3447; J. R. Saraf, *Z. Naturforsch.* **31A** (1976) 1038; A. T. Lin, and J. M. Dawson, *Phys. Fluids* **18** (1975) 201; F. Winterberg, Z. Naturforsch. **30A** (1975) 976; N. E. Andreev, V. P. Silin, and G. L. Stenchik, *Zh. Ex. Teo. Fiz.* **78** (1980) 1396; J. L. Hughes, *Izvest. Akad. Nauk SSSR, Ser. Fiz.* **42** (1978) 2593; **43** (1979) 1523.

175 R. E. Kidder, *Nucl. Fusion* **14** (1974), 797.

176 K. A. Brueckner and S. Jorna, *Rev. Mod. Phys.* **46** (1974), 325.

177 C. E. Max, *Phys. Fluids* **19** (1976), 74.

178 R. S. Craxton and M. G. Haines, *Plasma Phys.* **20** (1978), 487.

179 M. M. Novak, "Interaction of Photons with Electrons in Dielectric Media," Ph.D. Thesis, University New South Wales (February 1979), *Forschr. Phys.* **28** (1980), 339.

180 L. C. Steinhauer and H. G. Ahlstrom, *Phys. Fluids* **13** (1970), 1103.

181 S. L. Shapiro, M. A. Duguay, and L. B. Kreuzer, *Appl. Phys. Lett.* **12** (1968), 36.

182 H. P. Weber, *Phys. Lett.* **27A** (1968), 321.

183 See Eq. (36) in Ref. 156, (Marhic).

184 R. Klima, *Plasma Phys.* **12** (1970), 123; R. Klima and V. A. Petrzilka, *Cz. J. Phys.* **B22** (1972), 896; *J. Phys.* **A11** (1978), 1687; V. A. Petrzilka, *Cz. J. Phys.* **B26** (1976), 115.

185 H. Hora, *Phys. Fluids* **17** (1974), 1042.

186 H. Bebié, Seminar Lecture, Deptartment of Laser Physics, University of Bern, November 1978.

187 G. Bekefi, *Radiation Processes in Plasma* (John Wiley, New York, 1966)

188 J. M. Dawson, *Adv. Plasma Phys.* **1** (1968), 1.

189 A. Einstein, *Phys. Z.* **18** (1917), 121.

190 G. Badertscher (private communication).

191 F. J. Belinfante, *Physica* **6** (1939), 887.

192 H. Hora, *Lett. Nuovo Cim.* **22** (1978), 55; *Atomkernenergie* **34** (1979), 297.

193 F. F. Chen, see Ref. 5, Vol. 3A, p. 291; J. A. Stamper, *Phys. Fluids* **18** (1975), 735; Y. Sakamoto, *Jap. J. Appl. Phys.* **16** (1977), 1015; M. S. Sodha, R. P. Sharma, and S. C. Kaushitz, *Plasma Phys.* **18** (1976), 879.

194 F. F. Chen, *Comments on Mod. Phys.*, Part E **1**, no. 3 (1972), 81.

195 V. N. Oraevski and R. Z. Sagdeev, *Sov. Phys.-Tech. Phys.* **7** (1963), 955.

196 V. P. Silin, *Sov. Phys. JETP* **21** (1965), 1127.

197 D. F. DuBois and M. V. Goldman, *Phys. Rev.* **164** (1967), 201.

198 K. Nishikawa, *J. Phys. Soc. Japan* **24** (1968), 1152.

199 F. Cap, *Plasma Instabilities* (Academic Press, New York, 1980).

200 L. D. Landau and E. M. Lifschitz, *Mechanics* (Pergamon, New York, 1969) p. 80.

201 W. Paul and M. Raether, *Z. Phys.* **140** (1955), 262; H. Hora, and H. J. Schwarz, *Jap. J. Appl Phys.* **12**, Suppl. 2 (1974) 69.

202 E. Kamke, *Differentialgleichungen, Lösungsmethoden und Lösungen* (Akademie Verlag Ges. Leipzig, 1943), Vol. 1, p. 397.

203 D. Biskamp and H. Welter, Inst. f. Plasmaphysik, Garching, Report (1972); see Ref. 90.

204 D. Bohm and E. P. Gross, *Phys. Rev.* **75** (1949), 1851.

205 G. A. Askaryan, *Sov. Phys. JETP* **15** (1962), 1088.

206 H. Hora, *Z. Phys.* **226** (1969), 156.

207 D. F. Dubois, see Ref. 5, Vol. 3A (1974), p. 267.

208 R. L. Dewar, *Phys. Fluids* **16** (1973), 431.

209 J. P. Watteau, see Ref. [87]. (IAEA, Vienna, 1979) *Vol. III, p. 65*.

210 A. Y. Wong, see Ref. 5, Vol. 4B (1977), p. 783.

211 J. L. Bobin, W. Woo, and J. S. DeGroot, *J. Physique* **38** (1977), 769; J. Weiland and H. Wilhelmsson, *Coherent Non-linear Interaction of Waves in Plasma* (Pergamon, New York, 1977).

212 R. Balescu, *Developments in High Power Lasers*, C. Pellegrini, ed. (E. Fermi Sch. Vol. 74, North Holland, 1980).

213 H. H. Chen and C. S. Liu, *Phys. Rev. Lett.* **37** (1976), 693; **39** (1977), 881; C. S. Liu and M. N. Rosenbluth, *Phys. Fluids* **17** (1974), 778.

214 H. H. Chen, C. Grebogi, C. S. Liu, V. K. Tripathi, *Plasma Physics and Controlled Nuclear Fusion Research 1978* (IAEA, Vienna, 1979), Vol. 3, p. 181.

215 R. G. Rehm, *Phys. Fluids* **13** (1970), 282; F. Winterberg, *Z. Naturforsch*, **30A** (1975), 976; C. Yamanaka, T. Yamanaka, J. Mizui, and N. Yamaguchi, *Phys. Rev.* **A11** (1975), 2138; M. S. Sodha, S. Prasad, and V. K. Tripathi, *J. Appl. Phys.* **46** (1975), 637; V. A. Volkov, F. V. Grigorev, V. V. Kalinovski, S. B. Korner, L. M. Lavrov, Y. V. Maslov, V. D. Urkin, V. P. Chudinov, *Sov. Phys. JETP* **42** (1975) 58; E. B. Goldman, *J. Appl. Phys.* **45** (1974), 5211; V. D. Leuthauser, *Atomkernenergie* **24** (1974), 193; J. R. Saraf, *Naturforsch.* **31A** (1976), 1038; D. Baboneau, G. diBona, P. Chelle, M. Decroissette, and J. Martineau, *Phys. Lett.* **A57** (1976), 247; R. Dragila, *J. Phys.* **D11** (1978), 1683; V. H. Kulkarni, *Ind. J. Phys.* **A51** (1977), 356.

216 Yu. A. Afanasyev, O. N. Krokhin and G. V. Sklizkov, *IEEE J. Quant. Electron.* **2** (1966), 483.

217 A. Caruso and R. Gratton, *Plasma Phys.* **10** (1968), 867.

218 G. J. Pert, *J. Phys.* **A5** (1972), 506; **B12** (1979) 2067; G. J. Tallents, *J. Phys.* **B13** (1980) 3057; **B10** (1977), 1763.

219 J. Nuckolls, see Ref. 5 (1975), Vol. 38, p. 399.

220 G. Guderley, *Z. Luftfahrtforschung* **19** (1942), 302.

221 L. L. Lengyel, *AIAA J.* **11** (1973), 1347; *Nucl. Fusion* **17** (1977), 805

222 H. Hora, *Aust. J. Phys.* **29** (1976), 375.

223 H. Hora, see Ref. 5, Vol. 2 (1972), 341.

224 J. W. Shearer, see Figs. 6 to 8 in Ref. 171.

225 E. Goldman, H. Hora, and M. Lubin, presented at 7th Europ. Conf. Laser Interaction with Matter, Garching, April 1974.

226 H. Hora, *Atomkernenergie* **24** (1974), 187.

227 A. Y. Wong and R. L. Stenzel, *Phys. Rev. Lett.* **34** (1975), 727; R. L. Stenzel, *Phys. Fluids* **19** (1976), 865.

228 H. C. Kim, R. L. Stenzel, and A. Y. Wong, *Phys. Rev. Lett.* **33** (1974), 886; G. Farkas, *Opt. Comm.* **21** (1977), 408.

229 M. E. Marhic, see Fig. 10, Ref. 156. Indications of the Nonlinear Force were reported before by G. Beaudry and J. Martineau, *Phys. Lett.* **43A** (1973), 331.

230 Yu. A. Zakharenkov, N. N. Zorev, O. N. Krokhin, Yu. A. Mikhailov, A. A. Rupasov, G. V. Sklizkov, and A. S. Shikanov, *Sov. Phys. JETP* **43** (1976), 283.

231 R. Fedosejevs, I. V. Tomov, N. H. Burnett, G. F. Enright, and M. C. Richardson, *Phys. Rev. Lett.* **39** (1977), 932.

232 H. Azechi, S. Oda, K. Tanaka, T. Norimatsu, T. Sasaki, T. Yamanaka, and C. Yamanaka, *Phys. Rev. Lett* **39** (1977), 1144.

233 J. F. Lam, B. Lippman, and F. Tappert, *Phys. Fluids* **20** (1977), 1176.

234 W. Gekelman and R. L. Stenzel, *Phys. Fluids* **20** (1977), 1316.

235 B. Luther-Davies, *Opt. Comm.* **23** (1977) 98.

236 T. P. Donaldson, J. E. Balmer, P. Wägli, and P. Lädrach, *Plasma Physics and Controlled Nuclear Fusion Research 1978* (IAEA, Vienna 1979) Vol. 3, p. 157; see also Ref. 82.

237 D. C. Slater, *Appl. Phys. Lett.* **31** (1977), 196; see also Ref. 79.

238 B. Luther–Davies, *Appl. Phys. Lett.* **32** (1978), 209; J. C. Samson, and A. J. Alcock, *Phys. Rev Lett.* **A51** (1975), 315.

239 K. R. Manes, H. G. Ahlstrom, R. A. Haas, and J. F. Holzrichter, *J. Opt. Soc. Amer.* **67** (1977), 717; see also Ref. 84.

240 E. B. Goldman, W. Leising, A. Brauer, and M. Lubin, *J. Appl. Phys.* **45** (1975), 1158.

241 D. J. Bradley, presented at 4th General Conf. Europ. Phys. Soc. York, 1978, invited paper.

242 R. Castillo, H. Hora, E. L. Kane, V. F. Lawrence, M. B. Nicholson–Florence, M. M. Novak, P. S. Ray, J. R. Shepanski, R. Sutherland, A. I. Tsivinsky, and H. A. Ward, *Nucl. Inst. Meth.* **144** (1977), 27.

243 V. F. Lawrence, Ph.D. Thesis, Univ. New South Wales, March 1978.

244 H. Hora, R. Castillo, R. G. Clark, E. L. Kane, V. F. Lawrence, R. D. C. Miller, M. F. Nicholson–Florence, M. M. Novak, P. S. Ray, J. R. Shepanski, and A. I. Tsivinsky, *Plasma Physics and Controlled Nuclear Fusion Research 1978* (IAEA, Vienna 1979) Vol. 3, p. 237.

245 G. B. Lubkin, *Phys. Today* **30**, No. 9 (1977) p. 19.

246 H. Izeki, invited lecture, *Proc. 13th Int. Conf. Phen. Ionized Gases* (Berlin, 1977), p. 139

247 R. Godfrey, Honours Thesis, Theoret. Phys., University New South Wales, Sydney 1978.

248 N. G. Denisov, *Sov. Phys. JETP* **4** (1957), 544.

249 See Fig. 7.8b of Ref. 1.

250 C. Yamanaka, *Progress in Inert. Conf. Fusion*, IAEA, San Francisco, Tech. Comm., February 1978 (Science Application, McLean Va, 1978), p. 29; A. L. Peratt and R. L. Watterson, *Phys. Fluids* **20** (1977), 1911; J. L. Bocher, J. P. Flie, J. Martineau, M. Rabeau, and C. Patou, see Ref. 5 (1977), Vol. 4B, p. 657; V. V. Blazhenkov, *Zh. E. TF.* **78** (1980), 1386.

251 K. Sauer, N. E. Andreev, and K. Baumgärtel, *Plasma Physics and Controlled Nuclear Fusion Research 1978* (IAEA, Vienna, 1979) Vol. 3, p. 187.

252 H. Hora, in *Advances in Inertial Confinement Systems*, Chiyoe Yamanaka, Ed. (Institute of Laser Engineering, Osaka University, 1980) p. 263.

253 P. Lalousis, see Section 5) of Ref. 6.

254 See H. H. Chen, and Y. C. Lee, *Phys. Rev. Lett.* **43** (1979), 264; R. Klima and V. A. Petrzilka, *Cz. J. Phys.* **29** (1979), 863.

255 H. Hora, *Sov. J. Quant. Elect.* **6** (1976), 154.

256 H. Hora, *Phys. Fluids* **17** (1974) 939.

257 J. L. Carter, and H. Hora, *J. Opt. Soc. Am.* **61** (1971) 1640.

258 V. G. Borodin, A. V. Cjarukchev, V. K. Chevokin, A. A. Girokhov, M. F. Danilov, V. D. Dyatlov, V. M. Komarov, A. A. Mak, V. A. Malinov, R. N. Medvedev, G. V. Obraztsov, P. P. Pashinin, A. M. Prokhorov, M. Ya. Schelev, and A. D. Starikov, *Summaries of the 8th Natl. Conf. Laser and Nonlin. Optics* (Tbilisi, 1976), p. 245.

259 R. B. White and F. F. Chen, Eq. (31) of Ref. 172.

260 See p. 119, Ref. 202.

261 P. Lädrach (private communication 1979).

262 H. Kogelnik and H. P. Weber, *J. Opt. Soc. Amer.* **64** (1974), 174.

263 I. P. Kaminov, W. M. Mammel, and H. P. Weber, *Appl. Opt.* **13** (1976), 396; R. Renard,

J. Opt. Soc. Amer. **54** (1964), 1190; see also Ref. 257.

264 S. Eliezer and Z. Schuss, *Phys. Lett.* **A70** (1979), 307.

265 P. Lädrach and J. E. Balmer, *Opt. Comm.* **31** (1979), 350.

266 See Ref. 4, p. 182.

267 H. Maki and K. Niu, *J. Phys. Soc. Japan* **45** (1978), 269.

268 J. P. Freidberg, R. W. Mitchell, R. L. Morse, and L. J. Rudsinksi, *Phys. Rev. Lett.* **28** (1972), 795; D. W. Forslund, J. M. Kindel, K. Lee, E. L. Lindman, and R. L. Morse, *Phys. Rev.* **A11** (1975), 679; K. G. Estabrook, E. J. Valeo, and W. L. Kruer, *Phys. Fluids* **18** (1975), 1151.

269 H. Maki, *J. Phys. Soc. Japan* **46** (1979), 653; Y. Sakamoto, *Jap. J. Appl. Phys.* **16** (1977), 1015

270 P. Koch and J. Albritton, *Phys. Rev. Lett.* **32** (1976), 1420.

271 J. L. Bobin, see Ref. 5 (1980), Vol. 5, p. 000.

272 H. Hora, *Ann. Physik* **22** (1969), 402.

273 H. Hora, see Ref. 5 (1977), Vol. 4B, p. 841.

274 M. V. Goldman, and D. R. Nicholson, *Phys. Rev. Lett.* **41** (1978) 406.

275 P. Debye, *Ann. Physik* **30** (1909), 755.

276 H. Hora, *Optik* **17** (1960), 409.

277 The difference between optical waves and matter waves in the second order is seen in the long beating wavelength of modulated electron waves: H. Hora, *Light Scattering in Solids*, M. Balkanski, Ed. (Flamarion, Paris, 1972), p. 128.

278 R. Y. Chiao, E. Garmire, and C. H. Townes, *Phys. Rev. Lett.* **13** (1964), 479; A. G. Litvak, *Sov. Phys. JETP* **30** (1970), 364; S. A. Akhmanov, D. P. Krindakh, A. P. Sukhorokov, and R. V. Khokhlov, *JETP Lett.* **6** (1977) 38.

279 H. A. Ward, MSc Thesis, Univ. New South Wales, 1980.

280 A. Schlüter, *Plasma Phys.* **10** (1968), 471.

281 M. S. Sodha, R. S. Mittal, *Opto-Electron.* **6** (1974), 167; M. S. Sodha, A. K. Ghatak, and V. K. Tripathi, *Progress in Optics*, E. Wolf, Ed. (Academic Press, New York, 1976), Vol. 13, p. 171; W. Engelhardt, *Appl. Phys. Lett.* **15** (1969), 216; D. P. Tewari and A. Kamar, *Plasma Phys.* **17** (1975), 133; M. Y. Yu, K. H. Spatschek, and P. K. Shukla, *Z. Natf.* **A29** (1974), 1736; M. S. Sodha, A. K. Chakravarti, U. P. Phadke, G. D. Gautama, and I. Rattan, *Appl. Phys. Lett.* **22** (1973), 121; M. S. Sodha, L. A. Patel, and R. P. Sharma, *J. Appl. Phys.* **49** (1978), 3707.

282 A. J. Palmer, *Phys. Fluids* **14** (1971), 2714; J. W. Shearer and J. L. Eddleman, *Phys. Fluids* **16** (1974), 1753; E. Valeo, *Phys. Fluids* **17** (1974), 1391; G. J. Tallents, *J. Phys.* **B10** (1977), 796; B. Bhat and V. K. Tripathi, *J. Appl. Phys.* **46** (1975), 1141; W. M. Mannheimer, *Phys. Fluids* **17** (1974), 1413; V. del Pizzo, B. Luther–Davies, and M. R. Siegrist, *Appl. Phys.* **14** (1977), 381; Yu. My, and K. H. Spatschek, *Z. Naturforsch*, **39A** (1974), 1736; R. Dragila and J. Krepelka, *J. Physique* **39** (1978), 617; *J. Phys.* **D11** (1975), 217; A. J. Alcock, C. De Michelis, and M. C. Richardson, *IEEE J. Quant. Electron.* **6** (1970), 622; N. A. Amherd and G. C. Vlases, *Appl. Phys. Lett.* **24** (1974), 93; M. Hugenschmidt, K. Vollrath, and A. Hirth, *Appl. Opt.* **11** (1972), 339; L. C. Steinhauer and H. G. Ahlstrom, *Phys. Fluids* **14** (1971), 1109; V. S. Soni and V. P. Nayyar, *J. Phys.* **D13** (1980), 361; S. K. Sinha and M. S. Sodha, *Phys. Rev.* **A21** (1980), 633; W. M. Mannheimer and E. Ott, *Phys. Fluids* **17** (1974), 1413.

283 V. V. Korobkin and A. J. Alcock, *Phys. Rev. Lett.* **21** (1968), 1433; M. C. Richardson and A. J. Alcock, *Appl. Phys. Lett.* **18** (1971), 357.

284 N. Ahmad, B. C. Gale, and M. H. Key, *J. Phys.* **B2** (1969), 403; R. G. Tomlinson, *IEEE J. Quant. Electron.* **5** (1969), 591; F. V. Grigoev, V. V. Kalinovski, S. B. Kormer, L. M. Lavrov, Yu. V. Maslov, V. D. Orlin, and V. P. Chudinov, *Sov. Phys. JETP* **42** (1976), 58.

285 N. L. Tsintsatse and E. G. *Tsikarishvili, Astrophys, Space Sci.* **39** (1976), 191.

286 H. Hora and E. L. Kane, *Appl. Phys.* **13** (1977), 165.

287 K. H. Spatschek, *J. Plasma Phys.* **18** (1978) 293.

288 H. Hora, E. L. Kane, and J. L. Hughes, see Ref. 81, Fig. 1.

289 M. R. Siegrist, *J. Appl. Phys.* **48** (1977), 1378.

290 E. L. Kane, Ph.D. Thesis, University of New South Wales, April 1979.

291 D. A. Jones, E. L. Kane, P. Lalousis, P. R. Wiles, and H. Hora (unpublished); see Ref. 170

292 M. J. Hollis, *Opt. Comm.* **25** (1978), 395.

293 See Ref. 14 and B. W. Boreham, R. Mavaddat, J. L. Hughes, and H. Hora, *Proc. 11th Symp. Rarefied Gas Dyn.*, R. Campargue ed. (CEA, Paris 1979) Vol. 1, p. 505.

294 B. W. Boreham and B. Luther-Davies, *J. Appl. Phys.* **50** (1979), 2533; B. Luther-Davies, B. W. Boreham, and V. E. del Pizzo, see Ref. 5, Vol. 5 (1980), p. 000; M. S. Sodha, and D. Subbarao, *Appl. Phys. Lett.* **35** (1979), 851; R. Mavaddat, and K. Ghatak, *J. Appl. Phys.* **31** (1980) 3501.

295 B. W. Boreham (in print); B. W. Boreham, *J. Opt. Soc. Amer.* **68** (1978), 698.

296 V. L. Keldysh, *Sov. Phys. JETP* **20** (1965), 1307; G. V. Ostrovski, *Usp. Fi Nauk* **111** (1973), 579; S. A. Ramsden, *Physics of Hot Plasma*, B. J. Tye, et al., Eds. (Oliver & Boyd, Edinburgh 1970).

297 N. K. Bereshotshaya, G. S. Voronov, G. A. Delone, and G. K. Piskova, *Sov. Phys. JETP* **31** (1970), 403; A. V. Phelps in *Physics of Quantum Electronics* (McGraw-Hill, 1966) p. 538.

298 P. A. M. Dirac, *The Principles of Quantum Mechanics*, 3rd ed. (Oxford University Press, 1949) p. 284.

299 R. Castillo (unpublished, 1977).

300 See pp. 386–390 of Ref. 54.

301 H. Hora and J. L. Hughes, German Pat. 2 832 100; about a forward momentum of the electrons to take account for the increase of the momentum of the laser pulse, see G. Viera, and H. Hora (to be published).

302 D. A. G. Deacon, L. R. Elias, J. M. J. Madey, G. J. Ramian, H. A. Schwettman, and T. I. Smith, see Ref. 49; J. M. J. Madey and D. A. G. Deacon, Stanford Univ. Rept. HEPL-797 (1976).

303 H. Hora, J. L. Hughes, and B. W. Boreham, *Sov. J. Quant. Elect.* 9 (1979), 464.

304 J. A. Stamper, K. Papadopoulos, R. N. Sudan, S. O. Dean, E. A. McLean, and J. M. Dawson, *Phys. Rev. Lett.* **26** (1972), 1012; F. Schwirzke, see Ref. 5 (1974), Vol. 3A, p. 234.

305 J. A. Stamper, NRL-Report 7411 (May 1972); see Ref. 158; D. A. Tidman, *Phys. Rev. Lett.* **32** (1974), 1179

306 J. A. Stamper, E. A. McLean, and B. H. Ripin, *Phys. Rev. Lett.* **40** (1978), 1177; J. J.

Thomson, C. E. Max, and K. Estabrook, *Phys. Rev. Lett.* **35** (1975), 663; S. P. Obenschain and N. C. Luhmann, *J. Phys. Rev. Lett.* **42** (1979), 311; A. Hasegawa, M. Y. Yu, P. K. Shukla, K. H. Spatschek, *Phys. Rev. Lett.* **24** (1978), 1656.

307 M. H. Key, presented at 12th Europ. Conf. Laser Interact. Matter, Moscow, December 1978

308 T. Sakagenu, *Advances Inertial Conf. Fusion*, C. Yamanaka Ed. (Institute of Laser Engineering, Osaka 1980) p. 122.

309 H. A. Ahlstrom, "Inertial Confiement Fusion," *Opt. Soc. Amer.*, San Diego (February 1978), paper ThA2-1.

310 A. C. Walker, S. Kogoski, T. Samatak, I. Spalding *Opt. Comm.* **27** (1978) 247.

311 H. P. Weber (private communication 1979).

312 H. Hora and M. M. Novak, see Ref. 5, 1977, Vol. 4B, p. 999.

313 J. W. Sheaerer, J. Garrison, J. Wong and J. E. Swain, see Ref. 5 (1974), Vol. 3B, p. 803; H. Hora, *Opto-Electron* **5** (1973), 491; A. I. Tsivinsky, M.Sc. Thesis, University of New South Wales, 1977; F. V. Bunkin and A. M. Prokhorov, in *Polarization, Matiere et Rayonnement* (Soc. France de Physique, Paris, 1969).

314 H. Hora, E. L. Kane, and J. L. Hughes, Fig. 1 in Ref. 15.

315 See Ref. 80; B. Luther–Davies, *Opt. Comm.* **23** (1977), 98.

316 R. A. Haas, J. F. Holzrichter, H. G. Ahlstrom, E. Storm, and K. R. Manes, *Opt. Comm.* **18** (1976), 105.

317 Los Alamos, press release, 4th March 1977; R. P. Godwin and F. Engelmann, *Europhys. News* **8**, No. 5 (1977), 11.

318 W. Willis, see Ref. 5, 1977, Vol. 4B, p. 991.

394 G. Joos, *Theoretical Physics* (Blackie & Son, London, 1949), p. 554.

395 W. Weizel, *Lehrbuch der Theoretischen Physik* (Springer, Heidelberg, 1950) Vol. 2, p. 1172.

396 W. Weizel, *Z. Phys.* **135** (1953) 270.

397 R. Sutherland, Ph.D. Thesis, University of New South Wales, 1980.

398 M. Planck, *Ann. Physik* (*4*) **1** (1900), 719.

399 A. Einstein, *Mitt. Phys. Ges. Zürich*, No. 18, 1916; *Physikalische Zeitschr.* **18** (1917), 121.

400 H. Hora, *General Relativity and Gravitation One Hundred Years After the Birth of Albert Einstein*, A. Held, Ed. (Plenum, New York, 1980) Vol. 1, p. 17.

401 R Lüst, *Fortschr. Phys.* **7** (1959) 503.

402 H. Alfvén, *Cosmic Plasma* (Reidel, Doordrecht, 1981).

REFERENCES, 2nd Part (Alphabetical)

Ahlborn, B. 1971, Phys. Lett. 37A, 227.

Ahlstrom, H.G., 1983, Physics of Laser Fusion (Nat. Tech. Info. Service, Springfield, Va.)

Aleksandrova, I.V., W. Brunner et al, 1985, Laser and Particle Beams 3, 197.

Alexeff, I., and M. Rader, 1990, Int. J. Electronics 68, 385.

Alfvén, 1942, Nature 150, 425.

Alfvén, Hannes, 1981, Cosmic Plasmas (Van Reidel, Dordrecht).

Alfvén, Hannes, 1988, Laser and Particle Beams 6, 389.

Ananin, O.B., Yu.B.Bykovski et al, 1982, Sov. Phys. Tech. Phys. 27, 903.

Andre, M., D. Babonneau, E. Berthier, J.L. Bocher, J.C. Bourgade, M. Busquet, P. Combis, A. Sondeville, J. Coutaset, R. Dautray, M. Decroissette et al, 1988, Laser Interaction and Related Plasma Phenomena, H. Hora, and G.H. Miley eds. (Plenum, New York) Vol. 8, p.503.

Andreev, N.E., V.P. Silia, and G.L.. Stenchikov, 1980, Zh.E.T.Fiz. 78, 1396.

Aydin, M., and H. Hora, 1990, Bull. Am. Phys. Soc. 35, 2021.

Baldwin, K.G.H. and B.W. Boreham, 1981, J. Appl. Phys. 52, 2627.

Bangerter, R.O., 1991, Phys. Fluids, 33, (issue with review papers of APS-Meeting Nov. 1990).

Baranov, V. Yu. 1990, Lectures in Australia.

Basko, M.M., 1990, Nucl. Fusion, 33, Dec.

Basko, M.M., 1990a, Laser and Part. Beams, 8, 409.

Basov, N.G., and O.N. Krokhin, 1964, Sov. Phys. JETP 19, 123.

Basov, N.G., P.G. Kriukov, B.D. Zakharov, Yu. V. Senatski, and S.V. Tchekalin, 1968, IEEE, J. Quant. Electr. 4, 864.

Basov, N.G., O.N. Krokhin, and G.v. Sklizkov, 1972, Laser Interaction and Related Plasma Phenomena, H. Schwarz and H. Hora eds. (Plenum, New York), Vol. 2, p.389.

Basov, N.G., A.E. Danilov, M.P. Kalashinkov, B.V. Kruglov, Yu.A. Mikhailov, A.V. Rode, G.V. Sklizkov, and S.I. Fedotov, 1984, Laser and Part. Beams, 2, 103.

Basov, N.G., Yu A. Zadharenkov, N.N. Zorev, A.A. Rupasov, G.V. Sklizkov and A.S. Shikamov, 1986, Heating and Compression of Thermonuclear Targets by Laser Beams (Cambridge Univ. Press).

Batha, S.H., D.S. Montgomery, K.S. Bradley,

R.P. Drake, K.E. Stabrook, D.W. Phillon, and B.A. Remington, 1990, Bull. Am. Phys. Soc. 35, 1943.
Beauregard, O. Costade, 1977, Phys. Rev. D15, 3553.
Bell, M., 1979, Nucl. Fusion 13, 33.
Best, R.W.B., 1977, Nucl. Instr. Meth. 144, 1.
Bethe, H.A., 1934 Handbuch der Physik H. Geiger and L. Scheel Eds. (Springer, Berlin) Vol. 24, part 1, p.497.
Bethe, H.A., and C.L. Critchfield, 1938, Phys. Rev. 54, 248.
Blokhintsev, D.I., 1964, Principles of Quantum Mechanics (Allyn & Bacon, London), paragraph 129.
BMFT-ABB, Bundesminist. Forschung & Techn., Bonn, 1988, Report on Costs of Transport of Solar Energy from the Sahara to the Ruhr River.
Bobin, J.L., 1974, Laser Interaction and Related Plasma Phenomena, H. Schwarz and H. Hora eds. (Plenum, New York) Vol. 3B, p.456.
Bobin, J.L., 1985, Phys. Rept. 122, 173.
Bock, R., 1990, Laser and Part. Beams 8, 563.
Böer, K.W., 1990, Survey of Semiconductor Physics (Van Nostrand, New York).
Briand, J.V. Adrian, M. El Tamer, A. Gomes, Y. Quemener, J.P. Disguard, and C. Kieffer, 1985, Phys. Rev. Letters 54, 38.
Broad, W.J., 1988, New York Times, 137, No. 47, 451, March 21.
Brown, J.G., 1989, The Physics and Technology of Ion Sources (John Wiley, New York)
Brueckner, K.A. and S. Jorna, 1974, Rev. Mod. Phys. 46, 325.
Brueckner, K.A., 1975, private communication.
Bruhns, Hardo, 1990, Physics World , 3, Oct,p.17.
Bykovski, et al, 1986, Joint Inst. Nucl. Res. Dubna, Rept. P9-86-3.
Campbell, E. M. 1991, Laser and Particle Beams 9, No. 2.
Campbell, E.M. (Editor) 1991a Laser and Particle Beams, 9, No. 1.
Carter, J.L., et al, 1971, J. Opt. Soc. Am. 61, 1640.
Cartwright, D.C. 1990, Laser Interaction and Related Plasma Phenomena, H. Hora and G.H. Miley eds. (Plenum, New York) Vol. 9, p.
Caspars, F., and E. Jensen, 1990, Laser Interaction and Related Plasma Phenomena, H. Hora and G.H. Miley eds (Plenum, New York) Vol. 9, p.
Castillo, R., 1986, Ph.D. Thesis, Univ. New South Wales.
Chernikov, V. et al 1989, Phys. Rev. A40, 4070.
Chyba, T.H., and L. Mandel, 1988, J. Opt. Soc. Am. B5, 1305.
Cicchitelli, L. et al 1984, Laser and Particle Beams 2, 467.
Cicchitelli, L., 1988, Ph.D. Thesis, Univ. New South Wales.
Cicchitelli, L. et al, 1988, Laser and Part. Beams, 6, 163.

Cicchitelli, L., H. Hora, and W. Scheid, 1989, Laser Acceleration of Particles, AIP Proceedings No. 193, C. Joshi ed. (Am. Inst. Phys., New York) p.17.
Cicchitelli, L. and H. Hora, and R. Postle, 1990, Phys. Rev. A41, 3727.
Cicchitelli, L., and H. Hora, 1990a, IEEE J. Quantum Electr. 26, 1833.
Clark, R.G., H. Hora, P.S. Ray, and E.W. Titterton, 1978, Phys. Rev. C18, 1127.
Clark, P.J., S. Eliezer, F.J.M. Farley, M.P. Goldsworthy, F. Green, H. Hora, J.C. Kelly, P.Lalousis, B. Luther-Davies, R.J. Stening and J.-C. Wang, 1985, Laser Acceleration of Particles, AIP Proceed. No. 130, T. Katsouleas ed. (Am. Inst. Phys. New York) p.380.
Corkum, P., and C. Rolland, 1988, J. Opt. Soc. Amer. 5B, 642.
Collins, R.J., D.F. Nelson, A.L. Schawlow, W. Bond, C.G.S. Garret and W. Kaiser, 1960, Phys. Rev. Lett. 5, 303.
Coutant, J., 1990, Contribution ECLIM Conference Schliersee 1990.
Dawson, J.M. 1964, Phys. Fluids, 7, 981.
Deacon, D.A.G., L.R. Elias, J.M.J. Madey, G.J. Ramian, H.A. Schwettman, and T.L. Smith, 1977, Phys. Rev. Letters 38, 892.
Deng Ximing, and Yu Wenyan, 1984, Advances in Inertial Confinement Fusion, IAEA 1983, Ctte. Proceed., C. Yamanaka ed. (ILE, Osaka) p.66.
Drake, R.P. 1988, Laser and Particle Beams 6, 235.
Drska, L., 1987, ECLIM Conf. Prague.
Ehler, A.W., 1982, Lecture at 1st Workshop on Laser Acceleration of Ions, Los Alamos, Febr.
Eidmann, K., and R. Sigel, 1974, Laser Interaction and Related Plasma Phenomena, H. Schwarz and H. Hora eds. (Plenum, New York) Vol. 3B, p.667.
Eliezer, S., and A. Ludmirski, 1983, Laser and Part. Beams 1, 251.
Eliezer, S., and A. Loeb, 1986, Phys. Rev. Letters 56, 2252.
Eliezer, S., A.K. Ghatak, H. Hora, and E. Teller, 1986a, Equations of State, (Cambridge University Press).
Eliezer, S., and Hora, H. 1989 Physics Reports 172, 339.
Eliezer, S. and H. Hora , 1989a, Fusion Technology, 16, 419.
Emmett, J.L., J.H. Nuckolls, and L. Wood, 1974, Scient. Amer. 230, 24.
Evans, R., 1988, Nature 333, 296.
Fabre, E., P. Fews, M. Koenig, A.Michard, and J. Virmont, 1989, Laser Interaction with Matter, G. Velarde, E. Minguez, and J.M. Perlado eds. (World Scientific, Singapore) p.433.
Fälthammer, 1988, Laser and Part. Beams 6, 437.
Feldbacher, R., 1988, Nucl. Instr. Meth. A271, 22.
Floux, F., 1969, Laser Conference, Belfast, Sept.

Floux, F., 1970, Laser Interaction and Related Plasma Phenomena, H. Schwarz, and H. Hora eds. (Plenum, New York) Vol. 1, p.447.
Förster,E. 1991, Laser and Particle Beams 9, No. 1.
Försterling, K., 1950, Arch. El. Übertr. 5, 209.
Freeman, P.R. et al, 1986, Phys. Rev. Lett. 57, 3156.
Gazit, Y., J. Deletttrez, T.C. Bristow, A. Entenberg, and J. Soures, 1979, Phys. Rev. Lett. 43, 1943.
Gell-Mann, M., 1987, Lecture at The Princeton Institute of Advanced Studies, Sept.
Gell-Mann, M., 1988, Nucl. Instr. Meths., A271, 178.
Giulietti, A., T.Afshar-Rad, S. Coe, O. Willi, Z.Q. Lin, and W. Yu, 1989, Laser Intraction with Plasmas, G. Velarde, E. Minguez, and J.M. Perlado eds., (World Scientific, Singapore)., p.208.
Glasstone, S., and R.H. Loveberg, 1960, Controlled Thermonuclear Reactions (NRC, New York).
Goldsworthy, M.P., F. Green, P. Lalousis, R.J. Stening, S. Eliezer, and H. Hora, 1986, IEEE Trans. Plasma Sci. 14, 823.
Goldsworthy, M.P., 1988, Ph.D. Thesis, University of New South Wales.
Goldsworthy, M.P., H. Hora, and R.J. Stening, 1988, Plasma Physics and Controlled Fusion Research (Nice Conf. IAEA, Vienna) Vol. III, p.181.
Goldsworthy, M.P., H. Hora, and R.J. Stening, 1990, Laser and Part. Beams, 8, 33.
Goodlove, T.F., and D.F. Sutter, 1978, Plasma Physics and Controlled Nuclear Fusion Research, (IAEA, Vienna) Vol. III, p.211.
Goodwin, I, 1990, Phys. Today, 43, Jan., 49.
Gould, C.R. and J.M. Joyce, 1975, NBS Rept. 425, Vol. II, p.697.
Gover, A., R.Z. Olshan, S. Ruschin, and Al. Kleinman, 1987, Phys. Rev. Lett. 58, 443.
Grey, A. jr., G. Miley and G. Brumlik, 1979, National Rev. (150E, 95th St., New York) Feb. 2.
Grieger, G. and Wendelstein VII-Team, 1981 Plasma Physics and Controlled Thermonuclear Fusion Research 1980 (IAEA, Vienna) Vol. I, p.173 and p.185.
Gu Min, and H. Hora 1989, Journal of Chin. Laser, 16, 656.
Häfeli, 1985, Prospectus, Häfeli Corp., Basel.
Hannah, Rell, 1990, Alumni Papers Univ. New South Wales, Spring, p.19.
Handel, P.H., 1975, Phys. Rev. Lett. 34, 1492.
Harrison, M.F.A., and E.S. Hotston, 1989, "Modelling of the Edge Plasma and Divertors of ITER", (Net-Report Garching, Febr.)
Haseroth, H., 1988, CERN, Private Communication.
Häuser, T., W. Scheid and H. Hora, 1988, J. Opt. Soc. Amer. B5, 2029.
Häuser, T. and W. Scheid, 1991, Laser and Particle Beams 9, No.3.

Hebst, M.Y. 1981, Laser Focus, 17, (No. 12) 104.
Hefei Inst. Fine Mechanics and Optics, Academia Sinica, 1983, Cheng, Gan-Zhen, Yang, De-Yu, Shoa Han-Feng, and Gu Ze-Yu.
Herrmansfeldt, W.B., and D. Keefe, 1990, Laser and Part. Beams 8, 81.
Hershkovitz, N., 1985, Space Sci. Rev. 41, 351.
Heydenburg, N.P., C.M. Hudson, D.K. Inglis, and W.D. Whitehead, 1954, Phys. Rev. 74, 405.
Hofmann, I., 1990, Laser and Part. Beams 8, 527.
Hofstadter, R. 1987, Aust. Physicist, 24, 236.
Hora, H., 1960, Optik 17, 409.
Hora, H., 1964, Max-Planck-Inst. Plasma Phys., Garching, Rept. 6/23.
Hora, H. 1970, Laser Interaction and Related Plasma Phenomena, (Plenum, New York) Vol. 1, p.273.
Hora, H. and D. Pfirsch, 1970, 6th Int. Quantum Electr. Conf., Kyoto, Sept., Conf. Digest, p.10.
Hora, H. 1971, Laser Interaction and Related Plasma Phenomena, H. Schwarz et al eds. (Plenum, New York) Vol. 1, p.386.
Hora, H., and D. Pfirsch, 1972, Laser Interaction and Related Plasma Phenomena, H. Schwarz et al eds. (Plenum, New York) Vol. 2, p.515.
Hora, H., and P.S. Ray, 1978, Z. Naturforsch. 33A, 890.
Hora, H. 1981, Nuovo Cimento 64B, 1.
Hora, H., 1981, Physics of Laser Driven Plasmas (Wiley, New York).
Hora, H. 1982, Laser and Particle Beams 1, 151.
Hora, H. 1982a, Opt. Comm. 41, 268.
Hora, H., P. Lalousis, and D.A. Jones 1983, Phys. Lett. A99, 89.
Hora, H., 1983, Laser and Particle Beams 1, 151.
Hora, H., 1983a, Atomkernenergie 42, 7.
Hora, H., P. Lalousis and S. Eliezer, 1984, Phys. Rev. Lett. 53, 1650.
Hora, H., 1984, Laser Interaction and Related Plasma Phenomena, H. Hora and G.H. Miley eds. (Plenum, New York) Vol. 6, p.927.
Hora, H. and A.K. Ghatak, 1985, Phys. Rev. A31, 3473.
Hora, H. 1985, Phys. Fluids 28, 3706.
Hora, H., 1987, Z. Naturforsch 42A, 1239.
Hora, H., and P.H. Handel, 1987, Advances in Electronics and Electron Physics, P. Hawke ed. (Acad. Press, New York) Vol. 69, p.55.
Hora, H., 1988, Nature 333, 337.
Hora, H., 1988a, Short Wavelength Lasers and their Application, C. Yamanaka ed., (Springer, Heidelberg) p.87
Hora, H., Gu Min, S. Eliezer, P. Lalousis, R.S. Pease and H. Szichman, 1989, IEEE Trans. Plasma Sc. PS-17, 284.
Hora, H. 1989, Naturwissenschaften, 76, 241.

Hora, H., L. Cicchitelli, Gu Min, G.H. Miley, G. Kasotakis, and R.J. Stening, 1990, Laser Interaction and Related Plasma Phenomena, H. Hora and G.H. Miley eds., (Plenum New York) Vol. 9, p.95.
Hora, H. G.H. Miley, L. Cicchitelli, G. Kasotakis, P.W. Pieruschka, R.J. Stening, and Tan Weihan, 1991, Plasma Physics and Controlled Nuclear Fusion, Washington 1990 (IAEA, Vienna) Vol. p.
Hughes, R.H. et al, 1980, J. Appl. Phys. 51, 4088.
Hughes, R.H. et al, 1985, Appl. Phys. Lett. 47, 1282.
Hughes, R.H., and R.J. Anderson, 1989, The Physics and Technology of Ion Sources, I.G. Brown ed. (John Wiley, New York) p. 294.
Hugill, J. 1983, Nuclear Fusion, 23, 331.
Hugrass, W.N., I.R. Jones, K.F. McKenna, M.G.R. Phillips, R.G. Storer, and H. Tuczek, 1980, Phys. Rev. Lett. 44, 1676.
IAEA, 1988, Bulletin, No. 4, p.52.
Jackel, S., B. Perry, and M. Lubin, 1976, Phys. Rev. Lett. 37, 95.
Jackson, J.D., 1975, Classical and Electrodynamics (John Wiley, New York) 2nd edition, p.604-6.
Jones, D.A. et al, 1982, Phys. Fluids 25, 2295.
Joshi, C., T. Tajima, J.M. Dawson, H.A. Baldis, and M.A. Ebrahim, 1981, Phys. Rev. Letters 47, 1285.
Joshi, C., 1982, Discussion at 1st Workshop on Laser Acceleration of Ions, Los Alamos, Febr.
Joshi, C., ed., 1989, Laser Acceleratzion of Particles, AIP Proceed. No. 193 (Am. Inst. Phys., New York).
Kahalas, S., 1988, ECLIM Conference, Madrid, Oct.
Kaiser, N., J. Meyer-ter-Vehn and R. Sigel, 1989, Phys. Fluids B1, 1747.
Karpman, V.I., and A.G. Shagalov, 1982, Plasma Phys. 27, 215.
Kasotakis, G. et al., 1989, Laser and Part. Beams 8, 531.
Kasotakis, G., et al, 1989a, Nucl. Instr. Meth. A278, 110.
Kastler, A., 1964, Compt. Rend. Ac. Sc. Paris 258, 489.
Kato, Y., K. Mima et al, 1984, Phys. Rev. Lett. 53, 1057.
Kawata, S., M. Matsumoto, and Y. Masubudi, 1989, Laser and Part. Beams 7, 267.
Kelly, J.C. et al, 1991, AINSE Plasma Physics Conference, Lucas Heights Australia, Feb. 4-6.
Kentwell, G.W. et al, 1986, Laser Interaction and Related Plasma Phenomena, H. Hora and G.H. Miley eds. (Plenum, New York) Vol. 7, p.109.
Kentwell, G., and D.A. Jones, 1987, Phys. Rept. 145, 319.
Kessler, G. and H.U. Karov, 1989, Laser and Part. Beams 7, 647.
Key, M.H. et al, 1990, Laser and Particle Beams 8, 19.
Kibble, R.W.B., 1966, Phys. Rev. Lett. 16, 1054.
Kidder, R.E., 1974, Nucl. Fusion 14, 797.

Kidder, R.E., 1979, Nucl. Fusion 19, 223.
Kim, S.H., 1986, J. Plasma Phys. 36, 195.
Kirillov, G.A. et al, 1990, Laser and Particle Beams 8, No. 4.
König, R., K.-H. Kolk, and H.J. Kunze, 1987, Phys. Fluids 30, 3579.
Kono, M., M.M. Skoric, and D. ter Haar, 1981, J. Plasma Phys. 26, 123.
Kono, M. (1989) private communication.
Korschinek, G., and J. Sellmair, 1986, Rev. Sci. Instr. 57, 745.
Krall, N.A. and A.W. Trivelpiece, 1973, Principles of Plasma Physics (McGraw Hill, New York).
Kruer, W.C., and K. Estabrook, 1985, Phys. Fluids, 28, 243.
Kruer, W., 1988,\Physics of Laser Plasma Interaction, 182p., Addison Wessley.
Kruit, P.K. et al, 1983, Phys. Rev. A28, 248.
Kull, H.J. 1986, Laser and Part. Beams, 4, 473.
Kulsrud, 1983, Physics Today, 34, No. 4, 56 (3rd Col. 7th line).
Laboune, C., E. Fabre, A. Michard and F. Briand, 1985, Phys. Rev. A32, 577.
Lalousis, P. 1983, Ph.D. thesis, Univ. New South Wales.
Lalousis, P. and H. Hora, 1983, Laser and Particle Beams 1, 283.
Lalousis, P. and H. Hora 1984, Computational Techniques and Applications, J. Hoye, and C. Fletcher eds (North Holland, Amsterdam) p.699.
La Mer Slaner, L., 1979, Prospectus (Society to Advance Fusion Energy, 10 Nomandy Lane, Scarsdale, New York, 10583).
Landau, L.D., and E.M. Lifshitz, 1959, Fluid Mechanics (Pergamon Press, London) p.230.
Landau, L.D., and E.M. Lifshitz, 1966, Electrodynamics of Continuous Media (Pergamon Press, London).
Laud, N.G. and W.I. Kohn, 1970, Phys. Rev. B1, 4555.
Law, S., 1990, Physics World, Oct., p.13.
Lawson, J., 1957, Proc. Phys. Soc. B70,6.
Lawson, J., 1988, private communication.
Lax, M., W.H. Louisell and W.B. Knight, 1978, Phys. Rev. A11, 1365.
Lee, N.C. and G.K. Parks, 1983, Phys. Fluids 26, 724.
Leeper, R.J., 1989, Laser and Part. Beams 7, 649.
Lehmberg, R.H., and S. P. Obenschain, 1983, Opt. Comm. 46, 27.
Lerche, R.A., D.R. Kania, and S.M. Lane, 1988, Laser Interaction and Related Plasma Phenomena, H. Hora and G.H. Miley eds. (Plenum, New York), Vol. 8, p.541.
Lerche, R.A., D. Ress, R.J. Ellis, S.M. Lane, and K.A. Nugent, 1991, Laser and Part. Beams 9, No.1.
Levi, B.G., 1990, Phys. Today 43, Febr. 17,

Limpouch, J., G. Loncar, I.G. Lebo, and V.B. Rozanov, 1988, Laser and Part. Beams, 6, 295.
Limpouch, J., G. Loncar, and R. Dragila, 1990, Laser and Part. Beams 8, 143.
Lin, Z.Q. et al, 1988, Opt. Comm. 65, 445.
Lindl, J.D., 1987, Bull. Am. Phys. Soc. 34, No, 9.
Liu Renhang, and Tan Weihan 1990, Chinese Journal of Lasers 17, 658.
Ljamov, P.P. 1967, Zh.Tekhn.Fiz, 37, 629.
Lotsch, H., 1970, Optik 32, 116.
Lowdermilk, W.H. 1991, Laser and Particle Beams 9 No. 2.
Lubin, M.J., 1969, Gordon Conference on Laser Interaction with Matter, August.
Lubin, M.J., 1969a, Discussion with G.V. Sklizkov and the author, 16 June.
Lubin, M.J. et al, 1974 ECLIM '74, Garching Conference Abstracts, p.34.
Luther-Davies, B.A. Perry, and K.A. Nugent, 1987, Phys. Rev. A35, 4306.
Maddever, R.A.M., B. Luther-Davies and R. Dragila, 1990, Phys. Rev. A41, 2154.
Maddever, R.A.M., 1988, Ph.D. thesis, Austral. Nat. University, Canberra.
Maddever, R.A.M., B. Luther-Davies, and R. Dragila 1990, Phys. Rev., A41, 2154.
Mainfray, G., 1990, ECLIM Conference, Schliersee.
Manes, K., 1986, Laser Interaction and Related Plasma Phenomena, H. Hora and G.H. George eds. (Plenum New York) Vol. 7, p.21.
Mani, G.S. et al, 1965, Proc. Phys. Soc. 85, 281.
Maron, Y. et al, 1989, Laser and Part. Beams, 7, 665.
Marshak, R.E., 1941, Ann. New York Acad. Sci. 41, 49.
Maya, I, K.R. Schulz et al, 1985, General Atomics, Technol. Dept. San Diego Report (GA-A 17842, Oct.).
McClenahan, C.R., and R.E. Segal, 1975, Phys. Rev. C11, 370.
McCrory, R.L. et al, 1990, Laser and Part. Beams 8, 27.
McIlrath, T.J., P.H. Bucksbaum, R.R. Freeman, and M. Bashkansky, 1987, Phys. Rev. A35, 4611.
Meldner, H.W., 1981, Encyclopedia of Physics, R.G.. Lerner and G.L. Trigg eds. (Addison-Wesley, Reading, Mass) p.497.
Mendel, C.W. and J.N. Olsen, 1975, Phys. Rev. Lett. 34, 859.
Meyers, R.A. ed., 1987, Encyclopedia of Physical Sciences and Technology (Acad. Press, New York) Vol. 7, p.91.
Meyers, R.A. ed. 1991, Encyclopedai of Physical Sciences and Tecnology, 2nd Ed. (Acad. Press, New York) Vol. 7, p.125.
Meyer-ter-Vehn, J., 1989, Plasma Phys. Contr. Fusion 31, 1613.
Meyer-ter-Vehn, J., 1990, Laser and Particle Beams, 8, 523.
Michaelis, Hans, 1989, Frankf. Allg. Ztg. No. 66, 18 March, p.15.

Miley, G.H., 1976, Fusion Energy Conversion, (Am. Nucl. Socl, La Grange Park, Ill.) Chapter 2.
Miley, G.H., 1981, Laser Interaction and Related Plasma Phenomena, H. Schwarz et al eds. (Plenum, New York) Vol. 5, p.313.
Miley, G.H., 1988, Rev. Sci. Instr. A271, 197.
Miley, G.H. 1988a, Fusion Technology, 15, 197.
Mima, K. et al, 1988, "Short-Wavelength Lasers", C. Yamanaka ed (Springer Heidelberg) p.128.
Mima, K., H. Takabe, and S. Nakai, 1989, Laser and Particle Beams 7, 249.
Mima, K., H. Takabe, and S. Nakai, 1989, Laser and Part. Beams, 7, 487.
Montes, A., and O. Willi, 1982, Plasma Physics 24, 671.
Moses, G.A. et al, 1989, Laser and Part. Beams, 7, 721.
Mulser, P. and H. Schwabe, 1983, Laser and Part. Beams 1, 379.
Mulser, P., 1988, Laser and Part. Beams, 6, 119.
Nakai, S., 1980, private communication.
Nakai, S., Y. Kato, T. Sasaki, T. Mochizoki, and C. Yamanaka, 1980, Advances in Inertial Confinenent Systems, C. Yamanaka ed. (ILE, Osaka Univ.).
Nakai, S., 1989, Laser and Part. Beams, 7, 467.
Nakai, S. et al, 1990, Laser Interaction and Related Plasma Phenomena, H. Hora and G.H. Miley eds. (Plenum, New York) Vol. 9.
Neddermeyer, J., 1942, Los Alamos Nat. Lab.
Nicholson, D.R. 1983, Plasma Theory (John Wiley, New York).
Nishihara, K. 1988 (private communication).
Nishihara, K. 1988, US-Japan Conf. Hawaii, August 8-12.
Niu, K., 1989, Fusion Energy, Cambridge Univ. Press, BMFT-ABB, 1988, Report on Cost of Transport of Solar Energy from the Sahara to the Ruhr River.
Nuckolls, J.H., 1974, Laser Interaction and Related Plasma Phenomena, H. Schwarz, and H. Hora ed. (Plenum, New York), Vol. 3B, p.399.
Nuckolls, J.H. 1978, IAEA Inertial Confinement Fusion Meeting, Livermore, Febr. 6.
Nuckolls, J.H., 1982, Phys. Today 35 (No. 9) 24.
Nugent, 1991, Laser and Particle Beams 9, No. 1.
Obenschain, S.P. et al, 1986, Phys. Rev. Letters 56, 2807.
Obenschain, S.P., C.J. Pawley, A.M. Moistovych, J.A.
Sigmar, D.I., J.F. Clarke, R.V. Neidigh, and K.I. Vander-Slinz, 1974, Phys. Rev. Letters 33, 1376.
Okamoto, K., 1977, J. Nucl. Sci. Tech. 14, 762.
Oliphant, M.L.E., and Lord Rutherford, 1933, Proc. Roy. Soc. A141, 692.
Oliphant, M.L.E., P. Harteck, and Lord Rutherford, 1934, Proc. Roy. Socl. A144, 692.

Oliphant, M., 1972, Rutherford: Recollections of his Cambridge Days (Elsevier, Amsterdam) p.144.
Oliphant, Sir M., 1987, Endeavour, 11, 133.
Oganesian, S.G., and S.V. Abadin, 1989, Opt. Comm. 73, 380.
Peratt, A., 1988, Laser and Particle Beams 6, 385.
Peratt, A. 1989, IEEE Trans. Plasmas Sc. PS ,
Perry, A.J., 1986, Ph.D. thesis, Austral. Nat. Univ., Canberra.
Pert, G.J., 1987, Laser and Part. Beams 5, 643.
Pfirsch, D., and K.H. Schmitter, 1989, Fusion Technology, 15, 1471.
Philberth, B. and Philberth K. 1982, Das All, (Christina, Stein/Rh, Switzerland).
Phipps, C.R. et al, 1988, J. Appl. Phys. 64, 1083.
Phipps, C.R., Jr., 1989, Laser and Particle Beams, 7, 835.
Pinkau, K, 1989, 4th Europ. Phys. Soc. Conf. Internat. Res. Facil., Zagreb 17-19 March, Ivo Slaus, ed. (Tech. Univ. Press, Zagreb) p.223.
Pismenny, V.D. 1990, Lectures in Australia.
Prossnitz, 1984, Laser Interaction and Related Plasma Phenomena, H. Hora, and G.H. Miley eds. (Plenum, New York) Vol. 6, p.193.
Ravel, A., and V. Ramstian, 1989, Nature 342, 758.
Ray, P.S., et al, 1976, Nucl. Fusion 16, 535.
Ray, P.S. et al, 1977, Z. Naturforsch. 32A, 538.
Ray, P.S., 1977, Ph.D. Thesis, Univ. New South Wales.
Razumova, K.A., 1983, Plasma Physics, 26, 37.
Renard, R., 1964, J. Opt. Soc. Am. 54, 1190.
Rode, A.V., 1984, Ph.D. Thesis, Moscow.
Rode, A.V. et al, 1990, private communication.
Rowlands, T. 1990, Plasma Physics and Controlled Fusion 32, 297.
Rowlands, T. 1991, Book of Abstracts, AINSE Plasma Phys. Conf., Lucas Heights, Australia, Febr. 4-6.
Rubbia, C., 1989, Nucl. Instr. Methods A278, 253.
Rubenchik, A.M. and S.K. Turitsyu, 1987, Laser and Part. Beams 5, 3.
Sack, C., and H. Schamel, 1985, Plasma Phys. 27, 717.
Sakai, K., 1989, J. Phys. Soc. Japan 68, 2325.
Sakharov, A.D., 1983, Collected Scientific Works (Marcel Decker, New York).
Schäfer, F.P. 1986, Physik. Blätter 42, 283.
Schäfer, F.P. 1990, ECLIM Conference Schliersee 1990.
Scheid, W., and H. Hora, 1989, Laser and Part. Beams 7, 315.
Schmalz, R.F. 1986, Phys. Fluids 29, 1389.
Schneider, S.H., 1989, Global Warming, (Sierra Club Books, San Francisco).

Scully, M. and H. Fearn, 1990, Laser Interaction and Related Plasma Phenomena. H. Hora and G.H. Miley eds. (Plenum, 1990) Vol. 9, p.23.
Scully, M. 1990, private communication.
Sellmair, J., 1988, Dr. rer. nat. Thesis, Technol. Univ. Munich.
Sessler, A., 1983, Laser Acceleration of Particles, AIP Proceed. No. 91, P. Chanell ed., (Am. Inst. Phys. New York), p.10.
Sessler, A., 1985, Laser Acceleration of Particles, AIP Proceed. No. 130, T. Katsouleas ed. (Am. Inst. Phys. New York) p.350.
Sessler, A.M., 1988, Phys. Today 41 (No.1) 26.
Shapiro, R. 1979, Rev. Geophys. & Space Res. 8, 359.
Shaw, M.J., 1991, Laser and Particle Beams 9, No. 2
Shvartsburg, A.B., 1985, Phys. Rept. 125, 187.
Sigel, R., et al, 1990, Phys. Rev. Lett. 65, 587.
Singh, T., et al, 1987, Nuovo Cim. D9, 987.
Sklizkov, G.V. ed. 1988, The Delfin Laser Installations (Nova Sc., Cormack, NY).
Smirnov, V.P. et al, 1990, Beam '90, Conf. Novosibirsk, July
Solokov, A.A., 1986, Radiation from Relativistic Electrons (Am. Inst. Phys., New York).
Spiegel, Dec, 1990, NullWatt, No. 47, p.82.
Stamper, and S.E. Bodner, 1989, Phys. Rev. Letters 62, 768.
Stichel, P. et al, 1989, The Manifest of Basil, International Convention of Safeguard the Survival of Mankind, European Scientists at the Ecumenic Assembly "Peace with Justice", May 15-21, Phys. Blätter, 45, 340.
Storm, E., 1986, Press Conference, Livermore, 16 Jan.
Storm, E., J. Lindl, E.M. Campbell, T.P. Bernat, L.W. Coleman, J.L. Emmett, W.J. Hogan, Y.T. Horst, W.F. Krupke, and W.H. Lowdermilk, 1988, Progress in Laboratory High Gain ICF: Prospects for the Future: (LLNL Livermore) Rept. No. 47312, August.
Storm, E., 1991, see Fig. 3 of article Laser and Particle Beams in Encyclopedia of Physical Sciences and Technology (Academic Press, New York, 2nd Edition) Vol. 7.
Stratham G., and D. ter Haar, 1983, Plasma Phys. 25, 681.
Suzuki, D., 1987, Lectures in Australia (Australian Broadcasting Commission, Sydney) Tapes, ISBN0642128464.
Szichman, H. 1988, Phys. Fluids 31, 1702.
Tajima, T., 1985, Laser and Part. Beams 3, 351.
Takabe, H., L. Monterth, and R.L. Morse, 1983, Phys. Fl. 26, 2299.
Tan Weihan, Liu Renhong, 1990, Chinese J. of Lasers 17, 658.
Tankovich, N.I. et al, 1990, Lectures in Australia, Aug/Sept.
Teller, E. 1972, IEEE, J. Quantum Electronics 8, 1972.

Teller, E. 1972a, Colloquium, Max-Planck-Inst. Plasmaphys., Garching, Oct.
Teller, E., 1987, Encyclopedia of Physical Science and Technology, R.A. Meyers ed., (Academic Press, New York) Vol. 5, p.723.
Thompson, W.B., 1988, Laser and Part. Beams 8, 307.
Titterton, E, 1985, Report Res. School Phys. Sc. Austral. Nat. University.
Troup, G., 1990, private communication.
Tsakiris, G.D., et al, 1990, Phys. Rev. 42, No. 10.
Tsakiris, G.D. et al 1991, Laser and Particle Beams 9, No. 1.
Tskhakaya, D.D., 1981, Plasma Phys. 25, 233.
Tuczek, H., 1979, "The New Scheme of Rotamak", Colloq. Univ. New South Wales, Febr.
Vandevender, J.P., J.A. Sweegle, D.J. Johnson, K.W. Bieg, E.J.T. Burns, J.W. Poukey, P.A. Miller, J.N. Olsen and G. Yonas, 1985, Laser and Part. Beams, 3, 93.
Velarde, G., J.M. Aragones, J.A. Gago, L. Guanes, et al, 1986, Laser and Part. Beams, 4, 349.
Vertes, A. et al, 1989, Analyt. Chemistry 61, 1028.
Vieider, G. et al, 1989, "ITER Divertor Engineering Design", Rept. ITEK-TN-PC-3-9-1, Oct. 20.
Wang, J.C., H. Hora, P.J. Clark, and R.J. Stening, 1986, Laser and Part. Beams 4, 83.
Weaver, T., G. Zimmermann, and L. Wood, 1973, Exotic CTR Fuels; Nonthermal Effects and Laser Fusion Applications", Lawrence Livremore Lab. Rept. UCRL - 74938.
Weitzsäcker, C.F. von, 1935, Z. Phys. 96, 431.
Westermann, Th., and W. Schimassek, 1989, Laser and Part. Beams, 7, 675.
Willis, W.J., 1977, Laser Interaction and Related Plasma Phenomena, H. Schwarz and H. Hora eds, (Plenum, New York) Vol. 4B, p.991.
Winter, C.-J., 1989, Physik Blätter, 45, 264.
Winterberg, F., 1969, Desert Res. Inst. Rept. 64, March.
Wittenberg, L.J. ewt al, 1986, Fusion Technology 10, 167.
Wolf, E., 1951, Proc. Roy. Soc. A204, 533.
Yabe, T., K. Nishihara and J. Mizui, 1975, Rept. J. - 235 (Inst. Plasma Phys., Univ. Nagoya) p.6.
Yabe, T. 1984, Jap. J. Appl. Phys. 22, L.51.
Yabe, T., 1988 (private communication).
Yabe, T., and K.A. Tanaka, 1989, Laser & Particle Beams 7, 259.
Yamanaka, C., and S. Nakai, 1986, Nature 319, 757.
Yamanaka, C., S. Nakai, T. Yamanaka, Y. Izawa, K. Mima, K. Nishihara, Y. Kato, T. Mochinzuki, M. Yamanaka, M. Nakatsuka, and T. Yabe, 1986a,

Laser Interaction and Related Plasma Phenomena, H. Hora, and G.H. Miley eds. (Plenum, New York) Vol. 7, p.395.

Yamanaka, T., 1985, Personal presentation at ILE, Osaka, Febr.

Yamanaka, T., 1990, Elaboration during a JSPS collaboration.

Yatsui, K., 1989, Laser and Part. Beams 7, 733.

Yonas, G., 1979, Scientific Am., 239, No. 5, 40.

Zehrfeld, H.P. and B.J. Green, 1972, Nucl. Fusion 12, 569.

Zeidler, A., H. Schwab, and P. Mulser, 1985, Phys. Fluids 28, 372.

Subject Index

GPSR Compliance

The European Union's (EU) General Product Safety Regulation (GPSR) is a set of rules that requires consumer products to be safe and our obligations to ensure this.

If you have any concerns about our products, you can contact us on ProductSafety@springernature.com

In case Publisher is established outside the EU, the EU authorized representative is:

Springer Nature Customer Service Center GmbH
Europaplatz 3
69115 Heidelberg, Germany

Batch number: 10838062

Printed by Printforce, the Netherlands